A Guide to Psychosocial and Spiritual Care at the End of Life

Henry S. Perkins

A Guide to Psychosocial and Spiritual Care at the End of Life

Springer

Henry S. Perkins
Ecumenical Center for Religion and Health
The University of Texas Health Science
Center
San Antonio, TX
USA

ISBN 978-1-4939-6802-2 ISBN 978-1-4939-6804-6 (eBook)
DOI 10.1007/978-1-4939-6804-6

Library of Congress Control Number: 2016956836

Printed on acid-free paper

This Springer imprint is published by Springer Nature
The registered company is Springer Science+Business Media LLC
The registered company address is: 233 Spring Street, New York, NY 10013, U.S.A.

To the three most important women in my life: my wife, Helen; my mother, Jean; and my sister, Suz

Preface

Albert R. Jonsen—a gentle, compassionate, superbly educated former priest and perhaps the first clinical ethics consultant ever—had just helped two young parents and the neonatology staff at Moffitt Hospital reach a painful decision. He had guided them to the sad, inescapable conclusion that life support could not keep the parents' premature, multiply handicapped newborn alive. The baby was inexorably dying. The parents and the staff reluctantly decided to stop life support and allow the baby to die.

I was a fully trained internist and brand-new bioethics fellow of Professor Jonsen's at the time. I had heard about the baby's predicament and observed from afar Professor Jonsen's ethics consultation about it. I naturally believed (as any clinically focused doctor would) that his consultative responsibilities in the case had ended with the decision to stop all but comfort care.

I happened to notice Professor Jonsen in his office the day after the formal consultation. He appeared weary and subdued. I asked whether he felt ill. He said, No, he had just returned from visiting this baby's parents again. Naively surprised, I asked why. He replied that, although he had served as the ethics consultant, not the hospital chaplain in the case, he had sensed something of a therapeutic relationship between himself and the parents. He, therefore, had felt a pastoral duty to see them through the ordeal of the baby's death.

His reply challenged me to step out of the narrow, strictly rational perspectives of both scientific medicine and academic ethics. Over the subsequent months of my bioethics fellowship, I began to see the importance of health professionals' recognizing and somehow addressing the spiritual dimension of illness. Later experiences as a missionary doctor in Africa and as a clinical ethics consultant in my own right back in the USA confirmed that impression and extended it to the psychological, social, and cultural dimensions of illness, too. I came to believe that *every* illness has multiple important dimensions, not just the bioscientific, and they all deserve attention.

Healthcare professionals who care for dying patients have an especially important responsibility to see the nonbioscientific dimensions of terminal illness addressed. That idea, of course, is not new. The modern hospice movement has successfully drawn attention to them, and that attention now reaches into mainstream medical settings. But I wonder whether the progress has stalled. Other authors do, too, suggesting that the bioscientific has already co-opted the psychological, social, and spiritual [1] as palliative care practice conforms increasingly to the standards of an established medical specialty [2] and as scientific evidence begins to drive all clinical practice even at the end of life.

I wrote this book out of the conviction that, as important as the bioscientific dimension is, end-of-life care must redouble its efforts to attend to those other dimensions, too. I envisioned this book as an overview of current knowledge and practices in psychological, social, and spiritual care at the end of life. I wrote primarily with hospital chaplains and social workers in mind. But all the health professionals involved in terminal care—doctors, nurses, hospital chaplains, parish-based clergy, social workers, clinical psychologists, and others—should find new information here they can use in attending dying patients and their survivors. I certainly did in researching the book. Even patients and survivors may find information they can use. Aware of this potentially broad readership, I tried to avoid technical medical terms and concepts as much as possible. However, whenever I needed to use them, I defined them explicitly.

This book's chapters after the introductory Chap. 1 fall into three groups. Chaps. 2 through 5, present basic information that the nonmedically trained might need to understand the process of medical care. I wrote these chapters because a clergy friend once told me that medical care bewilders the patients and families he pastors in the hospital, and they ask *him* to explain the clinical process. He says he always feels unprepared to do so. These four chapters cover end-of-life pathologic processes and symptoms, doctors' clinical thought processes, patients' risk/benefit assessments of treatment options and their preferences concerning those options, and doctors' prognoses. Even the medically trained including doctors may find these chapters helpful reviews of aspects of care that may have become unthinkingly routine in practice.

Chapters 6 through 14, address specifically the psychological, social, and spiritual dimensions of end-of-life care. These chapters cover advance care planning, proxy decision-making, care of family survivors in general and of family caregivers in particular, cultural aspects of end-of-life care, the "right" time and way to die, beliefs about the death transition and the afterlife, bereavement and grief, and spiritual care by health professionals. This book's last chapter, 15, returns to my point that the bioscientific dimensions of end-of-life care may again be overwhelming the nonbioscientific dimensions. The social and spiritual dimensions are at most risk of neglect. I, therefore, believe that health professionals of all kinds must take steps to reestablish a balance between the bioscientific and the nonbioscientific in end-of-life care. Healthcare researchers should also take up the challenge of addressing the many, still unexplored aspects of the nonbioscientific dimensions of such care.

Each chapter here presents a few key concepts and an illustrative case from real life. This format emphasizes the dynamic interplay between concepts and experiences. Concepts shape our perceptions of reality, and real-life experiences hone our concepts. I occasionally offer my opinions, explicitly labeled, on controversial points. I have also changed a few details of the cases to promote clarity and concision of the text and to protect the anonymity of my sources. Some readers will notice that all the cases depict adult patients. Because I am an internist, my clinical experience involves only adults. Nonetheless, I believe most of the concepts illustrated by these dying adults also apply to dying children. Each chapter concludes with lists of major points, of a few especially worthwhile references for readers who want to read more, and of all references cited in the chapter.

I sometimes use tables to summarize findings from one article or to compare findings of multiple articles on a topic. The format of these tables will be familiar to frequent readers of the clinical literatures but possibly not to other readers. I have, therefore, explained the main points of each table in the text. Readers who wish may skim the tables and concentrate on the explanations in the text; other readers may wish to examine the tables directly, to check my conclusions, and to draw their own conclusions.

I, of course, did not create this book completely on my own. I needed help from others. My wife, Helen, read and critiqued every chapter. A trained actress and portraitist, she reminded me when the numbers began to overload the text or tables. (I tried to heed her advice to revise out as many numbers as possible, but I did not succeed in satisfying her completely.) She and other immediate family members of mine persevered through the long writing process and encouraged me along the way. Of course, I depended on patients and family members who graciously offered their stories of end-of-life experiences as illustrative cases. In addition, years ago Professor Jonsen took a chance by accepting me—though a traditionally trained, bioscientific-style doctor—as one of his bioethics fellows. He then taught me the rich, multifaceted nature of the human condition during illness. I owe my bioethics career to him. My research colleagues in San Antonio, sociologist Helen P. Hazuda at the University of Texas Health Science Center and anthropologist Josie D. Cortez at the Intercultural Development Research Association, helped nurture many of the ideas in this book over our 30-year interdisciplinary collaboration. And, of course, my new colleagues at Springer assisted in many important ways. Kathy Cacace initiated the idea for this book; Margaret Burns shepherded it patiently through development; and Margaret Moore helped with editing, printing, publicity, and distribution. All these people have my heartfelt thanks.

San Antonio, USA Henry S. Perkins

References

1. Clark D. Between hope and acceptance: the medicalization of dying. BMJ. 2002; 324(13 April):905–7.
2. Von Gunten CF, Lupu D. Development of a medical subspecialty in palliative medicine: a progress report. J Palliat Med. 2004;7(2):209–19.

Foreword

Dr. Henry Perkins opens the Preface of this book by describing a clinical bioethics consultation I apparently conducted many years ago. I was Dr. Perkins' bioethics mentor at the time, and he was observing the consultation. It involved a newborn with severe respiratory distress syndrome. The doctors had concluded the baby would never breathe on its own and was, in fact, dying. They and I conveyed this dreadful news to the parents. The doctors, the parents, and I together sadly agreed on the only course of action left: to stop respiratory support and to let the baby die as humanely as possible. We did stop the respirator, and the baby died the next day. I met with the bereaved parents afterward. That visit, coming as it did after the baby had already died, surprised Dr. Perkins because it involved no active medical decisions. He quotes me as saying I made the visit because I felt an important pastoral duty to see the parents through to the end of their ordeal.

I do not specifically recall that consultation. Still, it reminds me about a feature, now often forgotten, of the early days of bioethics. When bioethics emerged as a field in the late 1960s, there was much uncertainty about who should develop it, what form it should take, and whether it could ever be applied usefully to clinical problems. Many of the first bioethicists were, in fact, trained as philosophers or theologians. The philosophers such as Daniel Callahan, H. Tristram Englehardt, and K. Danner Clouser formulated clear concepts and a persuasive logic for the new field. Some theologians such as Joseph Fletcher, Paul Ramsey, and Richard McCormick contributed important perspectives grounded in particular religious traditions. Those theologians discussed such concepts as the personhood of patients and the sanctity of life. Still other theologians including James Childress, LeRoy Walters, Stanley Hauerwas, James Drain, and me contributed additional content by drawing simultaneously from multiple particular religions and philosophies. We explored such ideas as patient autonomy and just allocation of medical resources. All of us, whether philosophers or theologians, tried to understand the human experiences of illness and healing through the new concepts of bioethics.

Like me, many of the theologians in early bioethics had served as pastors; some had even taught pastors. While pastors know that complete health care addresses both body and soul, they focus their efforts on "care for souls." Pastors devote themselves to listening sensitively to humans in distress; to giving prudent guidance in life crises; and sometimes simply to "being with" those who are alone, frightened, or depressed. Pastors often journey alongside people experiencing life's problems until some resolution is found. Thus, the theologians of bioethics who had trained as pastors brought to the field a firsthand acquaintance with the precarious and fragile human condition.

The consultation Dr. Perkins describes also reminds me that the powerful life-saving technologies introduced in the 1960s presented difficult life-or-death decisions people had never faced before. More than any other, those decisions prompted bioethics' move from the study and the classroom into the hospital and the clinic usually by way of teaching medical students. Clinical bioethics consultations arose in time. But they had a style different from medicine's typical clinical consultations. Doctors had long sought informal clinical advice from other, more experienced doctors in difficult cases, but formal clinical consultations became common only when medical specialties emerged. A general practitioner might then call a cardiologist, say, to advise about managing some complex heart disease. The cardiologist would visit the patient, perhaps several times, and give his opinion. The cardiologist would then step back, leaving further management to the requesting practitioner. As a rule, the consultant would focus on the disease, not the patient; he would advise but not directly treat. The clinical consultant, therefore, would remain at a distance from the patient's emotional and spiritual struggles. In contrast, bioethics consultants, like pastors, often found themselves drawn into those struggles by the very nature of bioethics issues: truth-telling, keeping confidentiality, judging decision-making capacity, respecting autonomous treatment refusals, allocating resources, caring for the dying, and knowing when to allow death to happen. Influenced by the theologians' commitment to pastoral "care for souls," bioethics consultants soon realized the frequent need to engage patients, family members, and health professionals on a deeply personal level as they faced bioethics issues together.

Dr. Perkins recognizes, of course, that every patient encounters many kinds of health professionals, not only doctors. Some health professionals are acutely aware of the pastoral dimensions of their work. Nurses, for example, provide considerable pastoral care. They visit patients frequently, give intimate bodily care, and often get to know patients well. In the process they address many patient needs beyond the disease. All forms of patient discomfort and distress, physical and spiritual, concern nurses. Other therapists and technicians may also develop close relationships with patients and attend to their pastoral needs while performing normal care duties. Nurses and these other health professionals may even serve as an important source of clinical information for patients and their families or friends.

I, therefore, believe that doctors as well as other health professionals around the patient's bed have two significant responsibilities related to bioethics. One is to understand the patient as a whole person, to learn the patient's health-related wishes, and to try to implement those wishes whenever possible. The other is to ensure faithful, "pastor-like" care for the patient and his or her closest supporters.

That care must attend to the psychological (or emotional), the social (or interpersonal), and the spiritual (or existential) dimensions of illness. Dr. Perkins' book takes up those critical dimensions with a focus on dying patients. But his aim is, in the long run, to encourage *all* health professionals to contribute to a "pastoral care" atmosphere that should prevail in every situation in the hospital or the clinic.

Albert R. Jonsen, Ph.D.
Professor Emeritus of Ethics in Medicine, School of Medicine,
University of Washington; Senior Scholar in Bioethics
Program in Medicine and Human Values
California Pacific Medical Center San Francisco CA USA

Contents

Chapter 1
Psychosocial and Spiritual Care with the End of Life in Sight

> *O Lord, support us all the day long of this troublous life until the shad(ow)s lengthen, the evening comes, the busy world is hushed, the fever of life is over, and our work is done. Then, Lord, in your mercy grant us safe lodging, a holy rest, and peace at the last. Amen.*
>
> John Henry Newman [1].

This famous prayer, often called the "peace at the last" prayer, concisely and elegantly expresses basic wishes for safety, rest, and peace at the end of life [1]. In an attempt to realize those wishes, end-of-life care has changed dramatically in the last 25–40 years. Widespread dissatisfaction with the end-of-life care before that and the emergence of hospice in response prompted much of the change. A broad public dialogue about balanced curative and palliative care, new advance care planning techniques (including advance directives), ever more humane tests for diagnosing life-threatening diseases, individualized protocols for delivering and monitoring palliative treatments, concerted efforts to relieve pain, and the new palliative care medical specialty all contributed to the improvements. Yet, because most end-of-life research has focused on bodily symptoms, functional decline, life-support treatments, and patient wishes concerning those treatments, much of the change has been confined to the physical domain of dying.

Considerable wisdom about the art of dying well (the *ars moriendi*)—which medieval Europeans thought an essential part of living well—appears forgotten. Medieval people saw dying as both a physical and a spiritual event [2]. Living to them had only one main purpose: attaining eternal salvation. European Christianity used stories of the saints to inspire the living to cultivate such spiritual virtues as faith, hope, charity, patience, and humility. Exhibiting those virtues in life earned believers entry into heaven after death. The dying process, therefore, was important as people's last chance to renounce sin, embrace virtue, and lay claim to God's salvation. So they cultivated practices of dying well, and those practices provided psychological and social as well as spiritual benefit. For example, they offered the dying a guide for assessing one's life and for communicating memories and lessons to those who would remain behind on earth.

H.S. Perkins, *A Guide to Psychosocial and Spiritual Care at the End of Life*,
DOI 10.1007/978-1-4939-6804-6_1

The rise of science from the sixteenth century on, however, changed people's attitudes about dying. It became ever less a spiritual event and ever more strictly a physical one. Some modern authors now rue scientific medicine's ascendancy and the resulting dominance of the physical domain of dying over its psychological, social, and spiritual domains [2–4]. One of these authors calls this ascendancy "the spell of science" and attributes its power to "our intoxication with the rational." He asks, Can we ever again acknowledge the nonrational parts of life and death? [3]. I believe we can: A renewed appreciation for the mysteries of death *can* rebalance today's overemphasis on the physical domain and underemphasis on the non-physical domains in end-of-life care. Unfortunately, even the modern hospice movement [5] has not yet been able to accomplish that rebalancing fully.

One Stimulus for This Book: The "Peace at the Last" Prayer

I wrote this book as my contribution to that rebalancing effort. The "peace at the last" prayer provided one important stimulus.

The famous Christian theologian and author John Henry Newman (1801–90) wrote the prayer. Newman led an influential but controversial life in the Church. He began as an Anglican priest but converted to Roman Catholicism at midlife and then rose to become a Catholic cardinal. The prayer reflects Newman's deep inner spirituality and attraction to ancient Christian wisdom [1].

I like the prayer because it captures in just two sentences some basic themes about the human predicament of life and death. The prayer begins by addressing a supportive, beneficent God, actively engaged in the lives and deaths of individuals. (Nearly all Westerners, Christians or not, embrace some form of this belief.) The prayer also acknowledges the difficulties of earthly life; its brevity (by using the image of a day); and the constant, hard work involved. The prayer imagines a time when the living have completed their earthly labors and ends by expressing those basic, universal wishes for a safe, holy, and peaceful rest after bodily death. I believe this simple prayer can speak to any human death regardless of its details.

A Second Stimulus for This Book: Data About Doctors, Patients, and the Nonphysical Domains of End-of-Life Care

But this book is not primarily religious. It is clinical. I have seen many patients die in my 40 years of practicing and teaching general internal medicine. In that time I have become steadily more convinced about the importance of the nonphysical domains in end-of-life care. Many other people, of course, share that conviction [5–8]. Data even support it. One study tabulated doctors' reactions to the deaths of their patients on general internal medicine wards and in intensive care units similar to where I trained and taught for many years [9]. Most of these doctors did not know the patients before hospitalization and attended them only a short time before their deaths. Still, the doctors gained great professional satisfaction from the care they gave. But they split nearly evenly over whether they felt upset by the deaths: 55 % did not and 47 % did. (The sum here curiously exceeds 100 %.) Twenty-four percent felt emotionally "numb." Among the doctors who were upset, 46 % sought support from colleagues, 36 % tried to recast the deaths in some positive light, and 36 % tried to distract themselves by reimmersing in their work. Patient deaths obviously disturb many doctors, and their reactions must affect the patients and survivors they care for. Some authors believe (and I agree) the distressed reactions of doctors often mirror the reactions of dying patients and their survivors: Deaths *are* distressing [10]. Another study suggested that whatever the personal reactions of health professionals (HPs), patients and survivors expect and need HPs' respect, sensitivity, and compassion in end-of-life situations [11]. These and other similar data provided a second stimulus for this book.

A Third Stimulus for This Book: My Own End-of-Life Cases and My Developing Interest in Death and Dying

Typical biomedical education trains HPs to believe their professional tasks end just after the patient dies. Once the body has succumbed to its pathology, death has been pronounced, the family has been notified, and the body has been prepared for final transport, HPs quickly redirect their attention elsewhere to living patients. The acute care workflow simply discourages further immediate pursuit of questions about dying and death. (Many HPs eventually develop an interest in palliative or hospice care anyway from their accumulated end-of-life care experiences.) My own interest, however, developed "backwards," beginning with a few patients' deaths and then reminding me about the process of dying for others I knew. That progression taught me the importance of the psychological, social, and spiritual domains of dying. Three cases illustrate my developing interest in death and dying. Such cases provided a third stimulus for this book.

Case 1

The first case awakened my interest in the spiritual or existential nature of death itself. The case involved my first patient to die during internship. I attended this elderly man hospitalized for chronic alcoholism. The other doctors and I treated him aggressively until his late-stage cirrhosis finally pushed him into a permanent coma. The family and I agreed to comfort measures only.

The patient died several nights later. The nurses called me to pronounce him dead. The door to his hospital room was closed when I arrived. Feeling nervous and alone, I entered and closed the door behind me. The eerie silence there made me feel like a trespasser on holy ground. This man, who had moved and talked only a few days before, lay completely still. The life force that had animated his body had now left it. I asked myself what really had just happened to him. What had he experienced, if anything, in his last moments of life and his first moments of death? Had he suffered? Did he have a spirit after death? If so, was that spirit there in the room? Was it watching me? I soon set aside those questions for the moment and began the routine pronouncement tasks. I listened to the man's heart and lungs as I had done many times before when he was living. They both were silent. Then (as I recall) my pager went off, calling me to some task among the living. The spell broken, I finished quickly, returned to the nurses' station, wrote a death note in the man's record, and telephoned the family. My spiritual and existential questions remained unanswered.

This first encounter with a dead patient initiated my curiosity about the spiritual and existential mysteries of death. The encounter taught me the importance of analyzing my professional experiences that involved patients who die [8, 12]. I realized any analysis had to include how those experiences affect me. The encounter also gave me a heightened awareness of life's transience. As a patient quoted in one article says pithily, "Tomorrow is not promised by today" [8]. I decided that, while I planned for the future, I had also to appreciate the present [13].

Case 2

This case taught me primarily about the importance of the psychological and social domains of dying. The case involved a cancer patient later in my internship. This middle-aged man, a recovering alcoholic, had achieved sobriety for many years through Alcoholics Anonymous. He had meanwhile maintained his marriage, raised four children with his wife, achieved some success as a businessman, and attained a local leadership role in the Boy Scouts.

I admitted him to the cancer ward for a rare, especially aggressive acute leukemia. Working under a renowned hematologist, I had the task of administering the chemotherapy ("pushing the poisons") to this man. Chemotherapy at the time produced especially horrible toxicities. As I gave this man round after round of it, I knew each one would make him vomit, lose his hair, ulcerate the inner lining of his mouth, and possibly contract life-threatening infections. When each regimen failed, an ever more toxic and less promising one replaced it. Still, the man never complained.

I attended this man off and on for the next 18 months. I saw his physical condition deteriorate and sensed his remaining life span growing short. One day I came to the bedside to give him yet more chemotherapy. I noticed he was reading his Bible. Discouraged about inflicting the toxicities on him yet again, I interrupted his reading and asked, "Why don't you just quit these horrible treatments and go home to your family?" His reply astonished me. "The time I have gained through these treatments has been the greatest blessing to me," he said. He then explained how fighting the leukemia had brought him closer to God and to his family than ever before and had helped him value the brief time he had left to live. He also mentioned his pride at having participated in chemotherapy research that might help future patients. He finished by saying, "I would not have traded this time for any other in my life."

I was dumbstruck: While the technical details of preparing and administering the chemotherapies had consumed my attention over those months, this man had been experiencing profound psychological, social, and spiritual benefits I had not perceived. Perhaps my bioscientific training had blinded me to them.

The man died several months after I rotated off that cancer ward for the last time, but his wife sent me a card the next Christmas. She thanked me for my care and confirmed what he had told me. She repeated how important those months during chemotherapy had been to him and to the family. They all had felt blessed. Still I could not quite grasp what she told me. I probably did not yet have enough life experience to do so.

Nonetheless, I had seen this man's remarkable strength and resilience of spirit during great suffering [8]. I had learned that not all clinical problems can be solved [7, 12], and that modern medicine eventually fails everyone. Still, being emotionally present for patients and listening carefully to them are always important. Even when medicine exhausts all its technical treatments, such personal attention can benefit them and me. This man also taught me what I later learned Dame Cicely Saunders had described as seeing dying "not as defeat but as fulfillment" [5]. In the end, his personal growth through terminal illness had enriched him, his family, and eventually me.

Case 3

The third case was personal, not professional, and involved my dad as he died. I had completed my internal medicine internship, internal medicine residency, and bioethics fellowship years before. Since then I had tried to practice what I had learned about the psychological, social, and spiritual domains of end-of-life care [14, 15]. But I often could not assess whether my efforts to address those nonphysical domains mattered to patients and their families. But Dad's case made me experience end-of-life care as patients and families do. That personal experience proved to me more than had all my prior professional experience how important dying patients and their families consider those domains.

Dad presented with bladder cancer in his late 80s. Because initial biopsies showed minimal invasion of the bladder wall, he chose treatment with periodic superficial scrapings of the wall lining. Biopsies over subsequent months eventually showed more extensive invasion, requiring palliative irradiation. The oncologist who assumed Dad's treatment at that point estimated he would live only six more months. My wife and I moved close to my parents to help oversee Dad's care. His condition stabilized at first. My wife and I ate dinners with my parents, and I assisted Dad in walking for exercise. He gained a few pounds. The family celebrated his ninetieth birthday. All the while I hoped Dad would be the exception to the oncologist's discouraging survival estimate.

But Dad deteriorated right on schedule. My mom, my sister, my wife, and I grieved as he progressively lost everyday functions. He began having trouble feeding himself; his appetite diminished. He stopped walking and became wheelchair-bound literally overnight. And he required ever more assistance with daily tasks such as toileting, bathing, and dressing. The four family members pitched in to help as his care began to overwhelm the hired aides. We found the caregiving physically and mentally exhausting. Unable to accept Dad's deteriorating condition, Mom wanted to believe he was feigning helplessness only to get attention. To Mom's distress Dad also began seeing visions of his long dead parents. And, although he had fulfilled the role of family patriarch for many years, he unexpectedly asked me one day, "Who's the head of the family now?" I replied, "You are, of course." because I still hoped for a miracle survival.

Dad's last outpatient doctor appointment several months before he died was telling. I wheeled him into the examination room and stayed as he talked with the primary care geriatrician. The doctor immediately noticed Dad's sickly condition and asked, "How do you think you're doing?" Dad did not give his usual automatic reply of "Fine." Instead, he admitted, "I'm not doing well. In fact, I'm declining quickly." The geriatrician agreed. Dad was not kidding himself, and I could no longer realistically kid myself either. I finally had to release my hope for an unusually long survival.

As Dad's illness progressed from there, the nurses and aides continued to support him and the family, but the doctors began to keep their distance. The care he needed eventually became more than the hired aides and the family together could give in my parents' apartment. My sister, my wife, and I decided to move Dad to a skilled nursing facility over Mom's objections. He lived there in a hospital-like room for about a month.

I was traveling out of town on business when Dad suddenly turned imminently terminal. He stopped eating and complained of abdominal pain. My sister called in hospice. Dad vomited up blood. It horrified her. Seeing his suffering and realizing his end had come, my sister undertook perhaps the most difficult conversation of all. In it she gave him permission to abandon his fight to live. "It's OK," she said, holding his hand. "Just let yourself go." Dad may have needed that final permission from the family to release himself. He died only minutes later.

I returned home within hours and viewed Dad's body at the funeral home. Some of the same thoughts I had when I pronounced my first patient dead came back. Dad lay still and expressionless on a table in clinic-like room. The forces that had animated him most in life no longer did so: his deep religious faith, his pride in work, his sense of order in everyday activities, and his dry sense of humor. Some of the same questions as before occurred to me again. What had he experienced, if anything, in his last moments of life and his first moments of death? Had he suffered? Did he still have a spirit? If so, was it there? Was it watching me?

Dad's death gave me a new, greater appreciation for what the family endures as a patient dies. During Dad's steady deterioration toward death, I experienced many of the griefs loving families must feel as a dying patient loses ever more physiologic "normals," functional capacities, social roles, and personal relationships [8]. Those losses created great disappointments. Nonetheless, the other family members and I learned to accept Dad's each "new self" that emerged as his illness changed him.

Dad's dying also taught me that both negative and positive changes can occur during the dying process. While suffering inevitable losses in the physical domain, the dying person may also achieve unexpected personal gains in the nonphysical domains [8]. Furthermore, to keep up their morale and stamina, the family may depend on the presence, sensitivity, and compassion of the HPs who attend the dying patient. We, the family, certainly did in Dad's case [11]. His HPs helped us cope and carry on through the ordeal [8]. And, of course, a death may prompt survivors to think about life's most substantive questions. Dad's death certainly prompted me to and eventually became a stimulus for this book.

An Overview of the Book

As these three cases indicate, I believe scientific medicine's current dominating focus on the physical domain largely overlooks the psychological, social, and spiritual domains of end-of-life care. The spiritual domain probably runs the greatest risk of being overlooked, but all three nonphysical domains deserve more attention than they get. The physical domain must not be allowed to crowd the others out.

Those convictions underlie this book. I envisioned it as an overview of current knowledge about the three nonphysical domains of end-of-life care. I began writing with hospital chaplains and social workers in mind as readers but soon decided that *all* professionals delivering terminal care could find clinically useful information here. I certainly did in researching the book. Even patients and their survivors may find some of the information useful for navigating end-of-life experiences of their own.

The main text of this book has two parts. The first part sketches the basic elements of medical care near the end of life. Chapter 2 describes the common end-of-life symptoms; Chap. 3, doctors' methods of diagnostic reasoning; Chap. 4, patient benefit/risk assessments of possible treatments; and Chap. 5, the problems of prognostication, that is, predicting outcomes, in life-threatening illnesses. Given the book's focus, this part may strike some readers as a digression into mainstream clinical medicine. I do not want to give that impression. I included these chapters for three reasons. First, these chapters describe the basic elements of all medical care. Even at the end of life, psychological, social, and spiritual care must take place in a biomedical context. Second, contemporary mainstream medical care often mystifies the medically untrained. Patients, family members, and nonmedically trained professionals cannot get full benefit from nonphysical end-of-life care unless they understand something of the physical care. And third, this review of basic elements of medical care may give even medically trained professionals helpful new perspectives on old practices.

The second part of this book addresses specific topics in the nonphysical domains of end-of-life care. Chapter 6 discusses advance care planning; Chap. 7, proxy decisions; Chap. 8, care of the family in general; Chap. 9, care of family caregivers in particular; Chap. 10, cultural influences in end-of-life care; Chap. 11, the right time to die; Chap. 12, going home in death, and the afterlife; Chap. 13, bereavement and grief; and Chap. 14, spiritual care by HPs. Chapter 15 concludes this book by summing up its major points and by suggesting new directions for end-of-life research and care.

Nearly every chapter uses the same explanatory devices. It identifies key concepts, explains them, and illustrates them with an actual case. This format emphasizes the ongoing interplay between concepts and experiences. Concepts frame our view of reality, and experiences test and sharpen those concepts. The case descriptions adhere closely to the actual cases but occasionally make minor changes to protect people's privacy, to emphasize specific points, or to promote clarity and concision. All cases involve adult patients because I draw most of the cases from my own personal contacts or adult internal medicine patients. Still, most concepts here should apply to adult and pediatric patients alike.

My commentary throughout this book takes its information from peer-reviewed articles in the professional medicine and medical ethics literatures. The articles I used report either firsthand data or the most up-to-date expert summaries. I purposefully did not rely on opinions published by professional organizations or panels (such as the recent opinions of the American Society of Clinical Oncology [16] or the Institute of Medicine [17]) to ensure my own independent assessments and to avoid undue outside influence. I occasionally offer my summary views on a topic but label them clearly. I sometimes use tables to compare findings of multiple articles on a single topic. The format of these tables will be familiar to frequent readers of the clinical literatures, but possibly not to other readers. I, therefore, explained the main points of each table in the text. Some readers may wish to skim the tables and to concentrate on the commentary in the text; other readers may wish to examine the tables directly and to draw their own conclusions. Aware of the potentially broad readership, I minimized the use of technical terms. Whenever I needed such terms, I defined them explicitly. Each chapter ends with several main lessons ("Summary Points"), a few key references ("To Learn More …"), and a list of all cited references.

By highlighting the nonphysical domains of end-of-life care, this book aims to prepare medically trained health professionals (including doctors, nurses, and others), who already provide physical end-of-life care, to give dying patients and their survivors meaningful nonphysical care, too [7, 11]. This book also aims to help other professionals (including hospital chaplains, parish-based clergy, social workers, and psychotherapists) and lay caregivers involved in end-of-life care increase the scope and effectiveness of their nonphysical care within medicine's dominant physical orientation [11]. I, therefore, believe a broad spectrum of readers will find this book helpful: dying patients themselves; their intimates, family members, and other survivors; and all caregivers—professional and nonprofessional, formal and informal.

Summary Points

1. Western Christian thought has a long, rich tradition of dying well, called *ars moriendi*.
2. Attention to the physical domain of dying has come to predominate over the nonphysical domains—the psychological, the social, and the spiritual—ever since the beginning of the scientific age about 500 years ago.
3. Each nonphysical domain in end-of-life care deserves attention comparable to that of the physical domain.

To Learn More ...

1. Block SD. Psychological considerations, growth, and transcendence at the end of life: the art of the possible. JAMA. 2001;285(22):2898–905.
2. Lo B, Quill T, Tulsky J. Discussing palliative care with patients. Ann Intern Med. 1999;130(9):744–9.
3. Meier DE, Back AL, Morrison RS. The inner life of physicians and care of the seriously ill. JAMA. 2001;286(23):3007–14.

References

1. Wilson AN. John Henry Newman: prayers, poems, medications. New York: The Crossroad Publishing Company; 1990.
2. Leget C. Retrieving the *ars moriendi* tradition. Med Health Care Philos. 2007;10(3):313–9.
3. Radey C. Knowing souls. J Clin Ethic. 1992;3(4):307–8.
4. Weissman DE. Decision making at a time of crisis near the end of life. JAMA. 2004;292 (14):1738–43.
5. Saunders C. The last stages of life. Am J Nurs. 1965;65(3):70–5.
6. Donaldson MS, Field MJ. Measuring quality of care at the end of life. Arch Intern Med. 1998;158(2):121–8.
7. Lo B, Quill T, Tulsky J. Discussing palliative care with patients. Ann Intern Med. 1999;130 (9):744–9.
8. Block SD. Psychological considerations, growth, and transcendence at the end of life: the art of the possible. JAMA. 2001;285(22):2898–905.
9. Redinbaugh E, Sullivan AM, Block SD, et al. Doctors' emotional reactions to recent death of a patient: cross-sectional study of hospital doctors. BMJ. 2003;327(7408):185–9.
10. Zinn WM. Doctors have feelings, too. JAMA. 1988;259(22):3296–8.

11. Jacobson JA, Francis LP, Battin MP, et al. Dialogue to action: lessons learned from some family members of deceased patients at an interactive program in seven Utah hospitals. J Clin Ethic. 1997;8(4):359–71.
12. Meier DE, Back AL, Morrison RS. The inner life of physicians and care of the seriously ill. JAMA. 2001;286(23):3007–14.
13. Block SD. The physician as healer. Presentation at end of life care for the hospitalized patient, University of California, San Francisco, Oct 2000.
14. Perkins HS. Commentary on "providing guidance to patients: physicians' views about the relative responsibilities of doctors and religious communities." South Med J. 2013;106 (7):407–8.
15. Perkins HS. The varied understandings of "terminal cancer" and their implications for clinical research and patient care. Med Decis Mak. 2014;34(6):696–8.
16. Smith TJ, Temin S, Alesi ER, et al. American Society of Clinical Oncology provisional clinical opinion: the integration of palliative care into standard oncology care. J Clin Oncol. 2012;30(8):880–7.
17. Institute of Medicine. Dying in America: improving quality and honoring individual preferences near the end of life. Washington, D.C.: National Academies Press; 2014.

Chapter 2
The Symptoms and Signs of Dying

Whenever a doctor diagnoses a fatal disease, the typical patient immediately asks two questions: How long will I live and will I suffer? Chapter 5 addresses the first question; this chapter, the second.

Professionals who provide any part of end-of-life care must understand something of the physical symptoms that dying patients experience. These patients often fear the symptoms of dying more than death itself. Their highest priority is symptom relief. I believe that relief is critical: Patients must be physically comfortable before their minds can focus on other aspects of care, including the psychological, social, and spiritual aspects this book describes. All professionals who attend the dying, therefore, must commit to a compassionate, effective plan to address physical symptoms.

An open discussion is a key first step in providing symptom relief. That discussion should include four elements. First, it should state explicitly the clinicians' commitment to relieve symptoms to the maximum extent possible. Making this commitment should include a sense of urgency about addressing symptoms when they arise [1]. Some clinicians, in fact, rightfully consider pain in dying "a medical emergency." Second, the discussion should mention access to expertise for treating persistent symptoms. That expertise already exists widely in numerous textbooks and journal articles. It also exists in many hospitals and healthcare systems through special palliative care teams. Third, the discussion should explain that conscientious care can significantly relieve many symptoms of dying, especially pain [2]. Common side effects of symptom relief treatments such as opioid-induced nausea and constipation can also be managed successfully. However, some symptoms such as poor appetite may yield only partly to treatment. And, fourth, the discussion should assure patients and family members that the attending professionals will never ignore patients' symptoms or neglect their care.

The attending professionals, especially the biomedically trained clinicians, must be able to respond to the concerns of patients and families as particular symptoms arise. Good care begins with good explanations. This chapter can help with its descriptions of the prevalences, presentations, and pathologies of common

H.S. Perkins, *A Guide to Psychosocial and Spiritual Care at the End of Life*,
DOI 10.1007/978-1-4939-6804-6_2

end-of-life symptoms. The chapter also briefly describes treatments for the symptoms, but limitations of space and focus prevent any extensive discussions of those treatments. Readers can find detailed discussions of them in other sources.

The symptoms of dying, of course, vary somewhat among the many diseases that can cause death. The most common fatal diseases in North America these days are (in order of descending incidence) cardiovascular diseases (including congestive heart failure and strokes), cancer, chronic obstructive lung disease (such as chronic bronchitis or emphysema), dementias, and degenerative nervous system diseases [3, 4]. Yet likely due to the public's long focus on cancer and the resulting generous funding for it most research on end-of-life symptoms concerns cancer. Relatively little research concerns other fatal diseases [5]. This chapter, therefore, follows the professional literature's lead by concentrating on cancer. I believe, though, many points about cancer apply to other fatal diseases, too. In fact, as the chapter explains later, some experts believe dying from most fatal diseases follows one "final common pathway" of symptoms [6].

The chapter starts by describing frequent symptoms in dying including pain, dyspnea (shortness of breath), fatigue, confusion (delirium), cachexia (loss of weight), anorexia (loss of appetite), and depression. The chapter then discusses three phases of symptoms as death becomes imminent. The chapter concludes by describing the various physiologic signs people use to determine for themselves a patient's time of death. A woman with brain cancer illustrates many of the chapter's points.

The Case

Mrs. W. an elderly Euro-American woman, lives with her husband, Mr. W. in a retirement center. She suffers increasing weakness in her legs over several weeks. Her mobility declines from standing and walking independently to just lying in bed. Her ability to speak also declines to the point she rarely speaks at all.

One day a bilateral paralysis suddenly strikes Mrs. W. from the neck down, and she cannot respond. Mr. W. calls an ambulance, which transports her to the nearest hospital. A work-up there reveals a large brain cancer. Treatment with maximal steroids and irradiation shrinks the tumor, reducing its encroachment on nearby normal brain tissue. The oncologist informs Mr. W. that no further treatment exists, and the tumor will eventually grow back, worsening her symptoms again. At the oncologist's suggestion Mr. W. authorizes Mrs. W.'s transfer back to the retirement center for long-term care.

Mrs. W.'s symptoms and signs distress the family over the next weeks. She regains some ability to respond but only intermittently. She fluctuates unpredictably between responsive good days and unresponsive bad days. Mr. W. feels whipsawed. Even when Mrs. W. does respond, she is confused and does not recognize family and close friends. She often complains of a headache, and groans and grimaces. Her family interprets these signs to mean Mrs. W. is suffering pain despite her intensive opioid regimen. The retirement center's primary care doctor adjusts the opioid doses with some improvement in Mrs. W.'s alertness and apparent comfort.

Mrs. W. also does not eat well and has lost considerable weight since hospital admission. The local nutritionist introduces high-calorie, high-protein foods into Mrs. W.'s diet. Her intake improves slightly. The primary care doctor explains to Mr. W. that weight loss may be inevitable with Mrs. W.'s cancer. Mr. W. also worries that his wife fatigues easily and occasionally appears short of breath. The doctor says he cannot help the fatigue much, but he does prescribe supplemental oxygen for the shortness of breath. All the while, Mr. W. wonders whether the doctor has told him "the whole story" about Mrs. W.'s condition.

Pain

Pain is the unpleasant sensory and emotional experience associated with tissue damage [7]. All pain travels along peripheral nerves to one central spinal cord tract, the spinothalamic tract, and then on to the brain. But different pain sensations travel along different fibers within those nerves. "Fast" fibers carry the transient, well-localized pains that serve a protective function. Pain from noxious pressures or extreme temperatures on the skin travels along these fast fibers. When activated, these fibers initiate an automatic reflex that instantly withdraws the affected body part from the noxious stimulus. "Slow" fibers, in contrast, carry the long-term, diffuse pains that often arise deep in the body and appear to serve no biological function. Those pains occur with many internal diseases and after surgeries.

Pain characteristics, of course, vary by site of origin [8]. Pain from bones, muscles, or joints ("somatic pains") localizes to the general area of injury and presents as dull, throbbing, or achy. Bone fractures or tissue wounds have those pains. In contrast, pain from internal organs such as heart or bowels ("visceral pains") localizes poorly and presents as cramping, squeezing, or gnawing. Heart attacks ("myocardial infarctions") or bowel obstructions have those pains. Finally, pain from nerves ("neuropathic pains") localizes either well or poorly and presents as shooting, burning, tingling, stabbing, or electrical shock-like. Tumor invasion into nerves or post-amputation "phantom" limbs have those pains [9].

Patients and their families fear pain more than any other symptom in dying [1]. Although palliative care specialists believe current treatments can alleviate pain for most dying patients, many still suffer with it regardless of their particular fatal diseases. Various studies report high prevalences of end-of-life pain with cancer (35–96 %); acquired immunodeficiency syndrome, or AIDS (63–80 %); various serious heart diseases (41–77 %); COPD (34–77 %); end-stage kidney failure (47–50 %) [7]; and end-stage liver failure, or cirrhosis (30–40 %) [10].

Why does pain still plague dying patients? Several possibilities exist. First, patients may not report their pain. Although the decreased quality of life associated with pain should prompt patients to report it, counterbalancing considerations may act as obstacles to doing so. One study asked cancer patients to rate eight attitudes

Table 2.1 Cancer patients' ratings of possible obstacles to reporting pain

Possible obstacle	Mean ratings*	
	Patients with controlled pain	Patients with uncontrolled pain
Fear of opioid addiction	1.7	2.6
Fear that increasing pain means disease is progressing	1.8	2.5
Fear of medication side effects	1.8	2.2
Fear of injections	1.8	2.1
Belief that taking opioids early in a cancer's course will make them ineffective later	1.2	2.2
Belief that treating pain distracts clinicians from fighting the cancer	0.9	1.7
Belief that "good" patients do not complain of pain	0.9	1.4
Belief that pain is inevitable	0.9	1.4
Average score for all items	1.4	2.0

Adapted from Ward et al. [11], with permission
*The rating scale ranges from 0 ("do not agree at all") to 5 ("agree very much"). 2.5 is the neutral point between the disagree and agree ratings

that might act as obstacles to reporting pain (Table 2.1) [11]. The researchers contrasted the ratings of patients with and without uncontrolled pain. Patients with controlled pain tended to doubt that *any* of the eight attitudes creates obstacles to reporting pain. Yet patients with uncontrolled pain tended to believe more (or doubt less) that *all* of these attitudes create such obstacles. Specifically, compared to patients with controlled pain, patients with uncontrolled pain tended to see as reporting obstacles the fear of opioid addiction, the fear that increasing pain means the disease is progressing, the belief that taking opioids early in a cancer's course will make them ineffective later, and the belief that treating pain distracts clinicians from fighting the cancer. Furthermore, women more than men tended to believe medication side effects create obstacles to reporting pain, the elderly and the poor more than the young and the well-to-do tended to believe "good" patients do not complain about pain, and the poorly educated more than the better educated tended to believe pain is inevitable in dying.

Second, many clinicians may lack the knowledge to provide sufficient pain relief. Professional medical organizations have responded by sponsoring educational initiatives to increase that knowledge and to encourage its use. These initiatives stress the World Health Organization's stepped approach [12] to analgesia: Nonopioids such as acetaminophen (e.g., Tylenol) or nonsteroidal anti-inflammatories (e.g., aspirin, Motrin, and Aleve) are the Step I drugs; the weak opioids such as tramadol (e.g., Ultram) or codeine are the Step II drugs; and the strong opioids such as morphine or meperidine (e.g., Demerol) are the Step III drugs. Many somatic and visceral pains respond adequately to analgesics alone when used in a stepped fashion. Yet many neuropathic pains require supplemental "coanalgesics," drugs used together with analgesics to increase their pain relief effect. Coanalgesics include antidepressants such as nortriptyline (e.g., Pamelor), desipramine (e.g., Norpramin), duloxetine (e.g., Cymbalta) or venlafaxine (e.g., Effexor), and anticonvulsants such as gabapentin (e.g., Neurontin) or pregabalin (e.g., Lyrica).

Third, because opioids occasionally depress respiration, some doctors hesitate to dose them adequately for fear of causing severe respiratory depression and death. (I firmly believe this fear arises mostly from doctors' concerns over the life-threatening harm to patients rather than over medical-legal liability for themselves.) Palliative care specialists counter that dosing opioids carefully but adequately often relieves pain completely, and, if respiratory depression occurs, special antagonist drugs can reverse the opioids' effects. Doctors must simply be prepared to use those drugs quickly if necessary. Palliative care specialists conclude that all patients deserve optimum pain relief, and doctors have the duty to provide it [1].

The old ethical doctrine of double effect justifies that duty best. Whenever an act will have simultaneous, inseparable good and bad effects, this doctrine permits the act if the total expected good effects outweigh the total expected bad effects and the actor intends only the good effects. The doctrine, therefore, permits escalating opioid doses sensibly as needed to relieve a patient's pain despite the simultaneous, inseparable risk of respiratory depression. The doctor's single-minded intention to provide adequate pain relief is the key ethically permissive feature. But, in case questions might arise afterward, the doctor should always document that intention in the medical record.

Fourth, although inpatient and outpatient palliative care services are increasingly available, doctors sometimes hesitate to use them [13]. One reason is that many doctors misunderstand palliative care as applying only to end-of-life, hospice-like situations. In fact, palliative care applies to *any* illness—fatal or not—requiring a "focus on quality of life and the alleviation of symptoms" [8]. Furthermore, doctors may underuse hospice even for dying patients, and the palliative care that goes with it. The requirements for hospice may be partly to blame [3]. To reimburse for hospice care, Medicare requires a doctor's prediction that the patient has less than six months to live. As Chap. 5 points out, many prolonged, fatal diseases including most cancers defy precise prediction of remaining life span up to only a few days before death. Doctors balk at making such uncertain predictions due to concerns about misinforming patients and family members. Medicare also requires adequate direct caregiving. While relatives or close friends may voluntarily provide that care for some dying patients say, at home, many others have no one to give it. And, while Medicare does not officially require it, one other condition often precedes hospice referral: Patient, family, and attending doctors must all agree the patient is dying and should receive no further aggressive care for disease control, only care for comfort. Achieving that agreement can be difficult, sometimes delaying or even preventing hospice referral.

How can clinicians help overcome the obstacles to getting dying patients the pain relief they need? Clinicians can certainly educate themselves about pain relief, legal and moral factors related to it, and hospice use. Clinicians should also actively educate patients and families about pain relief. The education need not take much extra time: It can often take place at the bedside as clinicians conduct their usual patient care. And they can encourage patients to report their pain and can emphasize, repeatedly if necessary, that conscientious efforts usually achieve adequate relief [2, 14].

Clinicians often need to address explicitly people's fears about opioids. Defining addiction, tolerance, and physical dependence is important in doing so [1]. "Addiction" refers to the persistent misuse of drugs—legal or not—despite significant damage to one's physical health, personal relationships, or work performance. "Tolerance" refers to increasing dosing requirements over time to achieve the same drug effects. And "physical dependence" refers to having certain symptoms and signs when a medication is withdrawn too quickly.

Having defined these terms for patients or family members, clinicians should then explain that pain relief with opioids as needed is safe and beneficial: Addiction occurs never; tolerance, rarely; and physical dependence, only occasionally. Such pain relief does not damage patients' physical health, personal relationships, or work performance but promotes instead overall fulfillment during remaining life. The absence of pain facilitates participating in relationships and functioning in other ways. Furthermore, patients in pain request only the opioid doses they need, no more. Almost no one encounters tolerance. The typical opioid regimen for adequate pain relief reaches a dosing plateau and goes no higher. But even in the rare case of true tolerance the ever-increasing doses have little significance: The opioid regimen is usually short term, limited by the patient's death. Furthermore, physical dependence presents only an occasional problem because opioids are always tapered gradually when decreased, and almost never stopped before death. (If withdrawal symptoms do arise during a taper, clinicians can temporarily increase the doses to eliminate symptoms and then taper again more slowly than before.) Above all, clinicians must stress to patients and family members that adequate pain relief is an essential part of end-of-life care. All patients deserve that relief.

Ethnic Similarities and Differences in Pain Experiences

Recent studies suggest American ethnic groups share certain pain characteristics but differ on others [15, 16]. All groups share roughly the same pain threshold (the lowest pain intensity a person detects), the same pain tolerance (the maximum pain a person will voluntarily endure), and the same willingness to use mainstream medical methods for relief. But the groups differ in heat/cold pain tolerance, which decreases steadily from Euro-Americans (EAs) to Hispanic Americans (HAs) to African Americans (AAs) [15]. The groups also differ in what prompts them to seek pain relief. EAs tend to act on high pain intensity and interference with activities; HAs, on frustration and worry; and AAs, on frustration and high numbers of pain sites. Moreover, the groups differ in their use of nonmainstream methods for relief [16]. Unlike EAs, HAs and AAs tend to use conscious rest, social support, and prayer or other spiritual techniques.

Clinicians who attend dying Americans might keep these differences in mind when assessing pain and planning pain relief measures. EAs, for example, might bear more pain than others without complaining. EAs may need frequent, explicit encouragement to report their pain. Clinicians might also need to ask patients of *all* ethnic groups for suggestions about how to use their cultural or spiritual practices to achieve holistic pain relief.

The Case

Systemic steroids, oral opioids, and maximal radiation to the tumor provide considerable relief from Mrs. W.'s dull, visceral-like headaches during the first weeks of treatment. Visits from friends, trips outside in a wheelchair, and gentle physical therapy in her room help distract her from the residual pain.

But the doctor knows Mrs. W.'s headaches will worsen again as her tumor regrows. He also knows that, as an EA, Mrs. W. may underreport her pain. He makes a mental note to ask her periodically about it. He emphasizes to the W.s that treatment now aims primarily to relieve pain, and he urges Mrs. W. to report it so he can treat it. The doctor explains what addiction, tolerance, and physical dependence are and why they pose no problem for her. He insists the opioids will help her function better, not worse. And he warns her about possible side effects including nausea, decreased appetite, and constipation. He assures the W.s he can treat those symptoms if they arise.

Three Other Common Symptoms: Dyspnea, Fatigue, and Confusion

Dying patients may fear pain most but often suffer other symptoms, too. Three symptoms—dyspnea (shortness of breath), fatigue, and confusion [2, 6]—occur especially commonly before death from such varied fatal diseases as atherosclerotic heart disease, congestive heart failure, cancer, COPD, AIDS, and kidney failure. The documented end-of-life prevalences of these symptoms range widely from 10 to 95 % for dyspnea; from 32 to 90 % for fatigue; and from 6 to 93 % for confusion. All three symptoms increase in prevalence as death nears [17].

Dyspnea (shortness of breath) is a feeling of difficult, but not necessarily rapid, breathing. It has many known causes including disturbances in the chest wall's bellows that expand and contract the lungs, in the lungs' inner membranes where oxygen diffuses into the blood, and in the receptors that monitor the blood's oxygen and carbon dioxide concentrations [18]. Multiple causes often operate at once. Curiously, up to 25 % of dying patients have dyspnea without detectable abnormalities in chest anatomy, lung function tests, or arterial oxygen or carbon dioxide concentrations [17]. Some authors attribute those dyspneas to cultural or psychiatric factors.

Whenever dyspnea arises or worsens, a doctor should perform a thorough examination to determine cause and treatment. However, a few simple measures can provide some immediate relief. Those measures include elevating the head of the bed, directing a fan gently across the patient's face, and providing supplemental low-flow oxygen by short nasal tubes [19]. Some dyspneas, of course, require more sophisticated treatments such as special breathing techniques, low-dose opioids, thoracentesis (a procedure that removes obstructing air or fluid from the space around the lungs), and tracheal intubation with mechanical respiration.

Fatigue is an extreme tiredness, exhaustion, weariness, or lethargy that impairs activity [7]. Excessive physical activity, insufficient rest, dehydration, and even poor diet can cause temporary fatigue in healthy people. That fatigue protects the body from overexertion and responds to rest. The persistent fatigue in dying patients, however, appears to serve no protective function and does not respond to rest. The pathology of that fatigue remains unknown: It can exist independent of infections, anemias, and other plausible causes. Some authors speculate that advanced cancer or its chemotherapy causes fatigue by disrupting the body's immune and inflammatory systems, but little research has tested the idea. The lack of a tested, credible clinical explanation for this end-of-life fatigue prevents effective assessment and treatment of it.

Confusion is a broadly dysfunctional mental state causing disturbed comprehension and bewilderment. The two most important end-of-life confusions are delirium and dementia. They afflict approximately 28 and 11 %, respectively, of dying patients. While these confusions share some presenting features such as short-term memory deficits and mood lability, effective treatment and prognostication depend on distinguishing the two [7, 20].

Delirium, also called "acute confusional state" or "terminal restlessness," is a sudden impairment of consciousness with cognitive, affective, and physical manifestations. Risk factors include advanced age, male gender, prior illicit drug or alcohol abuse, underlying dementia, malnutrition, brain tumor, kidney failure, and poor pain control. Delirium causes primarily inattention, that is, poor concentration. Problems with perception and reasoning may also occur. Other common features are fluctuating consciousness [9] (typically more awareness in the morning and less at night), delusions (the inability to distinguish reality from imagination), hallucinations (disordered visual, auditory, or olfactory sensations), and speech which may be louder, faster, or slower than normal. Delirium often causes either baseline drowsiness, hyperalertness, or a mixture of the two [21].

Many conditions can trigger delirium: acute illnesses such as hip fracture, infection, and kidney failure; dehydration, low blood pressure, or abnormally high or low body temperatures; abnormal concentrations of sodium, calcium, glucose, or other substances in the blood; psychoactive drugs including opioids, antidepressants, and corticosteroids; sensory impairments; constipation or stool impactions; liver or kidney failure; and physical restraints [9]. Multiple triggers often work together simultaneously [22].

Treatment can reverse many of these triggers and the resulting delirium. When an opioid causes the delirium, for example, either reducing the dose or changing the opioid can improve the delirium. Maximal improvement may require 3–4 days. Approximately 50 % of deliriums occurring a month or more before the end of life resolve within a week. Unfortunately, 88 % of deliriums occurring in the last two weeks of life do not resolve before death [22].

A patient's delirium, of course, often distresses patient, family, and caregivers. When it does, education helps [21]. Clinicians might use the preceding explanation as a guide. They should give a brief description of delirium in plain language and identify any triggers that might be reversed in the case. Clinicians might also instruct others in simple interventions that may help calm the patient: providing adequate but not glaring light, displaying pictures of family members at the bedside, eliminating extraneous noise, speaking in soothing tones, reading aloud to the patient, caressing parts of the patient's body, or holding his or her hand. If the delirium defies treatment, clinicians might explain that despite appearances to others the delirium (say, hallucinations depicting long-dead relatives) may actually be easing the patient's transition from life to death. After all, delirium may be the brain's last coping mechanism [20].

Many people confuse delirium and dementia. The importance in distinguishing the two lies in the different treatment approaches and prognoses. In contrast to delirium, dementia is a slowly and steadily progressive deterioration of memory with at least one simultaneous other cognitive problem. Both problems together must be severe enough to impair social or vocational function [7, 23]. Risk factors for dementia include advanced age, female gender, low educational attainment, and dementia in the family. Memory deficits are the usual presenting complaint and predominant feature of dementia, but the associated cognitive problems range from being unable to remember words, perform previously learned tasks, or recognize familiar objects to getting lost in familiar surroundings, losing impulse control, or having problems with planning and judgment [24]. Treatment for dementia involves frequent reorienting and other supportive care. Table 2.2 contrasts delirium and dementia.

The Case

Although Mrs. W. has no obvious dyspnea, she does have persistent fatigue in her first several weeks back on the retirement center's nursing ward. She tires quickly with simple activities such as eating a meal or sitting up briefly in bed. She may sleep for a whole day after a visit from well-wishers.

She also has new confusion. She cannot recognize close family and friends and often becomes disoriented to time and place. She sometimes sees visions of her long-dead parents, waves her hands aimlessly above her head, or makes stitching motions in the air. Her strange visions and motions improve in the morning and worsen in the evening. She cannot concentrate on conversations, however brief. And, when she talks, her speech is unusually loud and slow. Naturally, her confusion distresses the family.

Table 2.2 Distinguishing delirium and dementia

Characteristic	Delirium	Dementia
Onset	Sudden (over hours to days)	Gradual (over months to years)
Identifiable triggers	Frequent	Infrequent
Predominant deficit	Inattention/poor concentration	Memory lapses
Consciousness	May be clouded but fluctuates.	Alert until late stages
Speech patterns	Often loud and unusually fast or slow	Difficulty in finding words, articulating thoughts
Hallucinations (disordered visual, auditory, tactile, or olfactory sensations)	Common	Uncommon
Delusions (the inability to distinguish reality from imagination)	Common	Uncommon
Course in a day's time	Fluctuates. Typically better in the morning, worse at night.	Stable
Duration	Most resolve in days to weeks. Average duration is one week.	Worsens steadily over years.

Adapted from Bookbinder and McHugh [7, Table 6, p. 312], with permission

The doctor believes Mrs. W. has a delirium. Her underlying risk factors include advanced age and the brain tumor. The doctor tries to identify specific, treatable triggers. He can prove or disprove some with blood and urine tests. But he can only suspect others: her psychoactive drugs such as the opioids and steroids, her confinement to bed much of the time with resulting dehydration and constipation, and her low sensory stimulation in an unfamiliar, single-occupancy ward room.

The urine and blood tests return normal. The doctor tells the family he can still treat other possible triggers. He reassures the family that most deliriums respond to such multifaceted "shotgun" treatment plans even when no one specific cause is identified. The doctor lowers the opioid doses but makes sure Mrs. W. stays comfortable. He hydrates her with oral and intravenous fluids and increases her time out of bed. He starts a laxative. The doctor also encourages the family to orient her frequently to time and place during their visits, to place family pictures around her room, and to use a lot of therapeutic touch with her.

His treatment plan works. Mrs. W.'s delirium subsides over the next few days. She regains her ability to recognize visitors and to carry on a coherent conversation. Her speech returns to normal. The strange visions and motions stop. The family is relieved.

Cachexia (Loss of Weight) and Anorexia (Loss of Appetite)

A medical school classmate of mine once naively asked a world-renowned oncology professor what ultimately kills cancer patients. The question stumped the professor. Today some oncologists believe cachexia—the severe, progressive weight loss occurring with long-term anorexia (loss of appetite)—does so [25]. But cachexia and anorexia accompany not only fatal cancers (especially those of the internal solid organs such as the lungs, stomach, and ovaries), but also other fatal diseases (such as advanced tuberculosis, AIDS, and heart or liver failure).

Two theories suggest mechanisms for cachexia [26]. One theory says that certain diseases prompt one kind of white blood cells (macrophages) to secrete special proteins (cytokines) that cause widespread inflammation and destruction of body fats and proteins [27]. The other theory says those diseases excite nerves that stimulate secretion of natural steroids or chemical transmitters (specifically, serotonin) in the brain. One or both then stimulate the breakdown of fats and proteins by some still unknown processes. These two theorized mechanisms, if valid, may operate simultaneously.

How does cachexia differ from the other common kind of weight loss, starvation? The distinction is critical for end-of-life care (Table 2.3) [27, 28]. Starvation—prolonged food deprivation in a healthy person—prompts no abnormal inflammatory response and shifts the body into survival mode to conserve fat and protein. Starvation generally maintains blood concentrations of albumin (a major body protein) and cholesterol. Starvation also increases body sensitivity to insulin and uses energy sources efficiently. All these metabolic changes reverse with careful refeeding. Furthermore, body weight during starvation does not predict survival. Voluntary dieting, of course, is the most common example of starvation.

Cachexia, in contrast, occurs in chronically ill people and may prompt the abnormal inflammatory response the one theory suggests. Cachexia persistently and indiscriminately destroys body fats and proteins. Blood concentrations of albumin and cholesterol decrease early and remain low. Cachexia also decreases body

Table 2.3 Distinguishing starvation and cachexia

Characteristic	Starvation	Cachexia
Overall body condition	Healthy	Sick
Pathologically increased inflammation present?	No	Yes
Total body fat and protein	Conserved	Destroyed
Blood albumin	Low only late	Low from early on
Blood cholesterol	May remain normal	Low from early on
Energy use	Efficient	Inefficient
Insulin sensitivity	High	Low
Does refeeding reverse the condition?	Yes	No
Does body weight predict mortality?	No	Yes

Adapted from Thomas [28], with permission

sensitivity to insulin and uses energy sources inefficiently. These metabolic changes do not reverse with refeeding. Furthermore, body weight in cachexia *does* predict survival: Reduction to below 66 % of ideal body weight signals imminent death [26].

The anorexia that accompanies cachexia understandably worries family and caregivers because food has such important emotional and social meanings [26, 29]. Human instincts about food are primitive, deep seated, and powerful. We associate food with love, nurturance, and care. Providing and consuming food during meals provide a supportive, communal experience; emotional comfort; and culturally significant pleasures. Understandably, then, when a dying patient stops eating, loved ones fret. They may mistakenly label the patient as "starving to death" and cannot bear that prospect. Such a fate seems to them uncomfortable and inhumane. Intending to help the patient live, they try to force him or her to eat.

Recent research on starvation and cachexia strongly refutes the common assumption that they are uncomfortable [30]. Healthy people who voluntarily starve themselves in fasts may experience little discomfort after the first day. I myself have fasted a day or two at a time on many occasions and have had little discomfort. I do not feel hunger, in fact, until I resume eating. Studies on dehydration and cachexia in dying cancer patients have shown similar results. One study showed that, as fluid intake increases after dehydration, so does thirst [31]. And another showed that despite requesting and receiving little or no food or fluids 63 % of the terminal study patients never felt hunger, another 34 % felt hunger only briefly, and 62 % never felt thirst. Furthermore, patients' symptoms with thirst usually required only ice chips by mouth or lubrication on the lips for relief [32]. Cachexia may even cause a mild euphoria in dying. One doctor describes this euphoria in his 85-year-old mother as she died [33]. Suffering from chronic depression, progressive debilitation, and an acute pneumonia, his mother decided to die by not eating. The doctor-son supported her wish. Once his mother stopped eating, she smiled for the first time in months. Friends visited, and she chatted animatedly with them. But she weakened steadily after some days and became ever more sleepy. She finally did not wake up at all and died on the sixth day. She never complained of hunger or thirst.

Such evidence, systematic and anecdotal, argues strongly that limited nutrition and hydration for the dying do *not* have the ill effects many people think [30, 34]. The actual effects may be beneficial: maintaining comfort; preventing nausea, vomiting, and other troublesome problems; and producing some euphoria [9]. Monitoring patients carefully, of course, is important to ensure the good effects outweigh any bad effects. Cachexia may be one way nature eases terminal patients' transition into death.

With that idea in mind, I have changed my practice in recent years. I no longer aggressively urge dying patients to eat or drink. Instead, I offer them food and water and allow them to take what they want. I support whatever choices they make. With the approval of patients (if conscious), family members, and other attending professionals, I sometimes stop all nutrition and hydration for patients who obviously have only a few hours to live.

The Case

Mrs. W. loses considerable weight during her hospitalization. When she returns to the retirement center, she has little appetite and eats only a few bites off each meal tray. Mr. W. worries that "starvation" will keep her from recovering. He asks the doctor to prescribe "appetite pills" or feeding tubes to increase Mrs. W.'s intake. The doctor responds by explaining how starvation and cachexia differ and why Mrs. W. is probably not suffering from her anorexia. He knows Mrs. W.'s cancer may eventually cause additional weight loss, but he urges patience for the time being. The doctor also tries several simple ways to boost Mrs. W.'s intake. He decreases her opioid doses, asks the dietician to put Mrs. W.'s favorite foods on every menu, and urges Mr. W. to sit with Mrs. W. during meals to provide companionship. The doctor stresses not "force feeding" her. He also asks the nurses to give her ice chips by mouth and lubrication for the lips.

Mrs. W.'s appetite improves over the next few weeks. She remains comfortable despite the lower opioid doses. She especially enjoys the ice cream on her trays and eats better when Mr. W. joins her for meals. Her weight stabilizes but at a lower level than when she was healthy.

Depression

Psychiatric disorders affect many people, including 15 % of the general public and 12–30 % of primary care patients. These disorders are predictably even more common among patients with fatal diseases. One survey of patients with newly diagnosed cancers, for example, found psychiatric diagnoses in 47 %, including adjustment disorders in 32 % and depression in 6 % [35]. The prevalence of depression increased with the duration of the cancers.

These data raise a vexing question for professionals who attend the dying patient: Is a patient's sadness "pathologic" depression or "normal" grief? [36] Depression and grief, of course, share the characteristic of sadness. Sadness is certainly not surprising in end-of-life situations. The many losses involved—losses of health, physical abilities, roles, relationships, and anticipated futures—invariably create sad feelings. The unwelcome and sometimes unpredictable losses may frighten the patient or erode his or her sense of wholeness, continuity, and control. Depression and grief share other features, too, including sleep problems, anorexia, fatigue, and poor concentration [37].

But the different treatment strategies involved—formal psychotherapy and medications for depression, and brief formal psychotherapy or informal support groups for grief—require clinicians to distinguish the two disorders. A prominent cancer psychiatrist, Susan D. Block, MD, draws useful contrasts between the two disorders (Table 2.4) [38]. The depressed dying patient, unlike the merely grieving one, has constant sadness; feels hopeless, helpless, worthless, or guilt-ridden [37];

Table 2.4 Distinguishing "pathologic" depression from "normal" grief in dying patients

Characteristic	Pathologic depression	Normal grief
Constancy of sad feelings?	Yes	No, episodic only
Feelings of hopelessness, helplessness, worthlessness, or guilt?	Yes	No
Capacity to experience pleasure in the present?	No	Yes
Capacity to imagine a positive future?	No	Yes
Preferred involvement in death	Sometimes active (i.e., suicidal)	Always passive
Suggested management	Formal psychotherapy and/or medications	Sometimes formal psychotherapy or support groups

Adapted from Block [38], with permission

and has lost the capacity to experience pleasure in the present or to imagine a positive future. Furthermore, the depressed patient, but not the merely grieving one, may consider active steps such as suicide to cause his or her death.

Dr. Block urges attending professionals to respond actively and positively to depression, grief, and other psychiatric disorders in dying patients. She decries fatalistic attitudes revealed in such comments as "If I were dying, I'd be depressed, too." or "The patient's depression doesn't matter. The patient will die soon anyways." Dr. Block believes health professionals should take instead an active, optimistic approach to soothing the psychological and spiritual traumas of dying, even late on. Keeping up frequent contact with the patient and the family, maintaining a reassuring presence, and providing explicit emotional support help. Empathetic listening, though emotionally difficult at times, can ease patients' despair and provide important professional satisfactions. Attending professionals need only patience and courage to do it. And promising to continue contact with patient and family through the patient's death can calm any fears of abandonment.

Dr. Block also emphasizes the need for attending professionals to follow up any patient statements about suicide. She suggests responding promptly with questions such as "Would you like to talk about that?" or "What causes you to think about suicide now?" If patients do not answer directly, professionals might probe specifically for the worries that many Oregon patients who have requested physician-assisted suicide express: progressive, serious debility; inability to care for themselves; loss of control over their deaths; and burdensome medical bills and physical care for their families. Chapter 11 elaborates on these worries in its section on physician-assisted suicide.

Finally, Dr. Block encourages attending professionals to nurture the inner strengths and supportive external relationships of *all* their dying patients whether depressed, grieving, or not. She suggests four ways to do so. First, professionals can encourage dying patients to conduct life reviews and to tell especially significant memories to others. Patients can thereby remember important accomplishments, fill important gaps in their life histories for others, and leave a legacy of values and life lessons. Second, professionals can suggest patients allow family members the privilege of contributing to their care. Taking part in that care at such a meaningful time can strengthen bonds of intimacy perhaps for the last time. Third, professionals can urge dying patients to "get their affairs in order" by making testamentary wills, funeral arrangements, and plans for body disposition. Tying up life's "loose ends" in this way renews patients' sense of control over their lives and gives the satisfaction of having prepared their worldly affairs properly for survivors. And fourth, professionals can encourage patients to explore with clergy or others the big transcendent questions patients may have about life and death. Dying patients surely wonder about these questions and can use the impetus of dying to grow through them.

The Case

When healthy, Mrs. W. enjoyed playing sports, entertaining, and keeping house. But the neurologic complications from her brain tumor now make those activities impossible. She wants to go home, but her nursing care necessitates her staying permanently in a spare single room on a skilled nursing ward. She looks sad, never laughs, eats little, sleeps a lot, and mentions fatigue. She never actually complains or expresses suicidal wishes. Her doctor wonders whether she is depressed, quietly grieving her lost prior lifestyle, or simply recovering as the tumor shrinks with treatment. He chooses to monitor her mood closely.

One day Mrs. W. suddenly perks up without specific psychiatric treatment. She eats better, begins watching sports events on television, converses (though sparingly), smiles, and laughs. She enjoys visitors again and invites them to "come back soon." Both responses indicate a renewed capacity to experience pleasure and to imagine a positive future. The doctor concludes Mrs. W. has not been depressed but has either been grieving her new physical limitations or improving mentally after the tumor treatments. He can make no definite determination. Regardless, Mrs. W.'s upbeat mood continues for some weeks.

Symptoms and Signs at the Very End of Life

Some palliative care experts believe many long-term eventually fatal diseases follow a final common pathway of symptoms and signs [6, 14]. As conceived, that pathway lasts days to months and exhibits many of the symptoms and signs this chapter discusses. Cancers and some major-organ-failure diseases such as end-stage congestive heart failure, COPD, and cirrhosis may take this pathway, but the existing research documents it most strongly for cancers (Table 2.5) [39–41].

Table 2.5 Prevalence of symptoms in end-stage cancer patients in palliative care

Symptoms	1–2 months before death (%)	1 week before death (%)	1–2 days before death (%)
Pain	54	30–99	51
General malaise/debilitation or “unhealthy” feeling	58	82	–*
Dyspnea/shortness of breath	17–70	46–47	22
Anorexia/poor appetite	8	80	–
Nausea/vomiting	12–62	13–71	14
Dry mouth	–	70	87
Death rattle	–	56	56–92

Adapted from Morita [39], with permission
* The dash means no data available.

Different diseases, of course, will present variations on the general pattern depending on particular pathologies, body locations, timelines, and treatments [41]. While not all palliative care experts accept the concept of a final common pathway, I believe it provides a clinically useful framework for anticipating many patients' end-of-life experiences.

I imagine the pathway divided into three phases—the early decline, the later decline, and the very end—based on the typical progression of symptoms and signs. Readers should realize, however, that patients may not progress neatly from phase to phase in order. Pinpointing even approximate transition times between phases may be difficult [6]. Symptoms and signs wax and wane, patients may simultaneously show characteristics of more than one phase, and different clinicians may interpret patients' clinical pictures differently. Furthermore, predicting a patient's remaining life span much before death is often impossible [5, 42]. The phases, therefore, represent a general pattern of decline, not a rigid one.

The Early Decline

Increasing pain and dramatically deteriorating function characterize this phase [5,44]. It may last a few months for cancer but only a few weeks for other prolonged, fatal diseases. Individualized, intensive analgesia based on World Health Organization guidelines usually provides adequate pain relief [14]. Yet the patient's steadily increasing debilitation requires ever more help with the basic Katz Activities of Daily Living (i.e., bathing, dressing, grooming, eating, toileting, walking, and transferring). Increasingly frequent serious complications from the disease require increasingly frequent hospitalizations. Unfortunately, the American healthcare system, fragmented among specialist caregivers and between in- and outpatient settings, manages this phase poorly [14]. Family members must often take time off work to give the care that the patient needs but that the health system does not provide or the family cannot pay for. Family members may suffer financially with lost workdays and unreimbursed, out-of-pocket care expenses [10]. Their savings may dwindle [45].

A particularly curious phenomenon, called "taking to bed," [46] deserves mention here. Apparently healthy, elderly nursing home residents may fall and then complain of vague body pains or show a small weight loss (about 5 lb). They go to bed and quickly lose their appetites, functional independence, and mobility. The mortality can be high. One study reported the outcomes of such elders a year later: Only 36 % had regained the ability to walk, 6 % remained bed bound, and 58 % had died. Somewhat surprisingly, the lack of localizing medical findings predicted death best.

Taking to bed may occur as often as 1.3 times per 100 resident-months in a typical nursing home population. The exact mechanism for the sudden, unexpected decline is unknown. Some experts speculate it is progressive apathy and loss of the will to live stemming from accumulated physical and personal losses over a lifetime. Yet these elders demonstrate few features of classic depression and do not improve with antidepressants.

The Later Decline

Some describe this phase as "agonal," a word derived from the Greek noun, *agonia*, meaning a struggle or an anguish. Lasting a few days to a few weeks, the later-decline phase encompasses a progressive shutdown of whole organ systems. That process, some people believe, is nature's or God's way of helping the body disengage from life.

The patient often suffers multiple increasing symptoms at once, typically 6-10 symptoms altogether including pain, fatigue, confusion, dyspnea, dry mouth, and anorexia (Table 2.6) [2, 4, 6, 39, 40, 47]. Pain that diminished with prior intensive treatment may increase again [10, 42]. Some 40–65 % of cancer patients end up with pain "most of the time" in their last days or weeks [43, 48]. The gynecology, head-and-neck, and prostate cancers produce the most pain late on [40]. This resurgence of pain often prompts last-minute referrals to hospice for expert palliation [13].

Such chronic, unremitting pain demoralizes the terminal patient and distresses the family. For that reason palliative care specialists believe it should be treated as

Table 2.6 Levels of consciousness for terminal cancer patients at various times before death

Level of consciousness	1 week before death (%)	1 day before death (%)	6 h before death (%)
Awake	56	26	8
Drowsy	44	62	42
Comatose	0	12	50

Adapted from Morita et al. [42], with permission

an emergency. Clinicians should use adequate opioid doses, titrating them methodically upward to achieve maximal relief [9]. The ideal achievement is complete relief. Despite the highly reported prevalences of terminal pain, palliative care specialists insist skilled analgesia can relieve pain completely for 98 % of patients [2, 40]. The side effects should be minimal. Any sedation caused by the opioids usually subsides within several days after doses stabilize. And excessive sedation or respiratory depression is rare and reverses immediately with opioid antagonist medications [9]. Still, clinicians should anticipate family concerns about addiction and tolerance and address them with explicit reassurance at the time of treatment. Even if pain relief requires high opioid doses, [42] compassion dictates giving those doses for the short time until the patient's death.

For the rare cases when no other regimen achieves sufficient pain relief, induced sedation to unconsciousness is a last resort. This technique uses medication to cause the patient to sleep through the pain of his or her last days [29, 49, 50]. Such a drastic step, of course, requires careful informed consent beforehand.

Confusion presenting as hallucinations also may occur in this phase [21]. The hallucinations may be visual, auditory, or tactile. They may involve out-of-body experiences or encounters with dead relatives or significant religious or cultural symbols such as Jesus or angels [51]. A man in one of my studies described being "dead" in a prior coma and standing "on the other side of a mountain (from Jesus). I kept asking Him to take me. I was ready (to die)" [52]. Hallucinations may comfort or distress patients but usually distress family members. Clinicians should reassure them about the common occurrence of such hallucinations near death [22].

The Very End

Waning consciousness characterizes this final phase of dying, which may last just hours (Table 2.6). Patients move progressively from being awake (arousable, able to converse and to follow commands), to being drowsy (arousable with difficulty, unable to converse or to follow commands) to being comatose (unarousable) [42]. If the patient has already become drowsy or comatose and survivors have not yet said their final goodbyes, the attending professionals should urge survivors to do so. The patient, though drowsy or unresponsive, may still be able to hear.

The imminently dying patient often shows other physical signs, too (Table 2.7) [42]. The "death rattle" is a noise caused by vibrating patches of saliva or mucus when the weak and drowsy terminal patient can no longer clear those secretions from the mouth or upper airways [9, 37]. Like hallucinations the death rattle often distresses observers at the bedside [53]. Clinicians can minimize the distressing sound by restricting the patient's fluids and giving drying medications [9, 54]. Clinicians should avoid deep suctioning, which appears to irritate patients. Other physical signs of imminent death include lower jaw movements with respiration, a blue color (cyanosis) on the arms and legs, and the absence of a radial artery pulse at the wrist [42]. The dying patient may sometimes show Cheyne–Stokes breathing,

Table 2.7 Duration from onset to death for various clinical signs in terminal cancer patients

Sign	Duration from onset to death (in hours)	
	Mean	Median
"Death rattles" with respiration	57	23
Lower jaw movements with respiration	8	3
Blue color (cyanosis) of the arms and legs	5	1
Absence of a pulse at the outer wrist (i.e., over the radial artery)	3	1

Adapted from Morita et al. [42], with permission

a pattern characterized by frequent respiratory pauses for up to 20–30 s. Urine production decreases, and the patient may become incontinent of urine and feces [9]. The dying patient may also spontaneously grimace, groan, and rub or scratch specific body areas. Medically untrained observers may interpret these actions as signs of distress [55]. Attending clinicians should explain such actions are common in dying [9]. They may indicate restlessness but probably not conscious distress. Nonetheless, clinicians should pledge continuing efforts to keep the patient as comfortable as possible [9]. In this last phase of dying, clinicians' demonstrated competence, attentiveness, respectfulness, and compassion toward the patient attest best to high-quality end-of-life care.

The Case

As I originally draft this chapter, Mrs. W. is still alive three months after the diagnosis of her brain tumor. Her Karnofsky Performance Score (KPS) varies day-to-day between K40 (moderate disability, needs special assistance) and K30 (severe disability, needs institutional care) (see Table 5.2 for the full KPS scale [56]). Because her KPS falls below K50, Mrs. W. probably does not have long to live, maybe only months [40].

Nonetheless, she does not yet show signs of entering the terminal phases of her illness. Treatment with steroids and radiation has improved her alertness and function. Unlike during her hospitalization she now can sit up for a while each day in a wheelchair. She also follows simple commands. Mr. W. performs daily range-of-motion exercises with her in the room and takes her for wheelchair rides outside. Mrs. W. feeds herself. Her appetite and weight have stabilized. Mrs. W. denies any significant pain on her current opioid doses, but she fatigues easily and sleeps a lot. She wakes up for visitors, enjoys their company, and holds brief conversations with them. The W.s' children visit frequently to support both their parents.

Mr. W. provides constant companionship for Mrs. W. except when he returns to their independent living apartment to sleep or do household chores. He believes Mrs. W. will make a miraculous recovery. The primary doctor warns him gently not to get his hopes up: Mrs. W.'s improvement in function and symptoms is only temporary. The doctor urges Mr. W. to savor his interactions with her now. The primary doctor continues to see Mrs. W. several times weekly, monitoring her for signs that she is deteriorating again and becoming terminal.

Epilogue to Mrs. W's Case

Mrs. W.'s improvement holds for about 8 weeks. She then deteriorates again as she enters the early-decline phase of her illness. She can no longer sit in the wheelchair and remains all day in bed. She complains of constant, severe headaches. The doctor recommends hospice care for pain relief, and Mr. W. agrees. Hospice's titration of the opioid doses improves the headaches somewhat but never completely eliminates them. More symptoms appear as Mrs. W. enters the later-decline phase. She loses her appetite, becomes lethargic, and complains of shortness of breath. Her circumstances confuse her. She no longer understands why she is not living in the W.s' apartment. She sees visual hallucinations of family members who died long ago. And she criticizes Mr. W. uncharacteristically often.

Mrs. W. finally becomes drowsy and loses consciousness completely as she enters the last phase of her illness, "the very end." She moans and grimaces periodically. Family members at the bedside hear death rattles and notice a blue color creeping up her legs. The doctor assures the family that despite these understandably distressing observations she is not consciously suffering. Mrs. W. dies peacefully only hours later with Mr. W. and a daughter at the bedside.

Which Signs Define the Time of Death?

The last important signs in dying indicate the time the patient dies. That time has significant implications for survivors because it determines in part their memories of the death, their visitation and funeral plans, their entitlements (such as Social Security and perhaps life insurance benefits), and their possible liability for healthcare charges. That time also has significant implications for attending professionals because it affects their immediate duties and workloads, and for other patients because it may affect their access to healthcare resources (possibly including transplantable organs).

Official medical, legal, and financial procedures take that time to be whatever the attending doctor determines. He or she typically uses the clinical examination to make the determination by cessation of either effective cardiorespiratory function or total brain function. The absence of spontaneous chest wall motions, of prominent pulses (such as at the carotid arteries in the neck), and of heart and respiratory sounds through a stethoscope imply cessation of effective cardiorespiratory function. The absence of adequate respirations, of spontaneous responses or movements (such as to painful stimuli), and of brainstem reflexes imply cessation of total brain function. Hypothermia (excessively cold body temperature) and deep sedation from drugs—two conditions that can mimic cessation of total brain function—must be absent. A standardized protocol, called "the apnea test," proves inadequate respirations. All required deficits for cessation of total brain function must persist for six hours to three days, depending on the cause of brain injury. Tests for brain circulation or brain waves are never required, merely confirmatory, for determining death.

Despite these well-established medical and legal procedures, disagreements over the time of death can arise because the medically untrained may determine that time for themselves according to various other physiologic signs. Colleagues and I once asked an ethnically diverse group of elders how they recognize a person's time of death [52]. The elders mentioned several different signs; no one sign predominated. Some elders mentioned the traditional signs of cardiorespiratory death: cessation of heartbeat and visible respirations. But some mentioned other signs such as turning cool; being unable to recognize others, speak, or move; making gurgling or gasping sounds; and turning the eyes back into the head. One difference occurred by ethnic group: Some Mexican Americans and Euro-Americans, but no African Americans, recognized time of death by cooling of the body [52].

Such differences may create a difficult problem at the end of life: A moribund patient may be simultaneously alive to some people and dead to others. For example, an observer who recognizes time of death by cooling of the body may consider the patient alive at the same time another observer who recognizes time of death by cessation of respirations considers the patient dead. Such differences may complicate postmortem procedures including organ harvesting, autopsies, and transfers to the morgue or funeral home. I think that, whenever a patient meets standard medical criteria for death, clinicians should ask family members, "Do you agree or not that the body is already dead?" Only when (within reason) everyone agrees the body is dead, should clinicians stop all treatments and initiate postmortem procedures.

Summary Points

1. Pain, dyspnea (shortness of breath), fatigue, confusion, cachexia (weight loss), and anorexia (loss of appetite) occur in many prolonged, fatal illnesses.
2. Patients often do not report their symptoms, especially pain. Clinicians must ask about them, repeatedly if necessary.
3. Nearly all the pains of fatal illnesses respond well to treatment until perhaps very late. Other common symptoms do not respond consistently so well.
4. The general public may rely on physiologic signs other than cessation of heartbeat and breathing to determine for themselves when a patient dies.

To Learn More ...

1. Block SD. Psychological considerations, growth, and transcendence at the end of life: the art of the possible. JAMA. 2001;285(22):2898–905.
2. McCarthy EP, Phillips RS, Zhong Z, Drews RE, Lynn J. Dying with cancer: patients' function, symptoms, and care preferences as death approaches. J Am Geriatr Soc. 2000;48(5 Suppl):S110–21.
3. Morita T, Ichiki T, Tsunoda J, Inoue S, Chihara S. A prospective study on the dying process in terminally ill cancer patients. Am J Hospice Palliat Care. 1998;15(4):217–22.
4. Morita T, Tsunoda J, Inoue S, Chihara S. Contributing factors to physical symptoms in terminally ill cancer patients. J Pain Symptom Manage. 1999;18 (5):338–46.
5. Rabow MW, Pantilat SZ. Palliative care and pain management. In: McPhee SJ, Papadakis MA, eds. 2012 current medical diagnosis & treatment. New York: McGraw-Hill Medical, 2012. p. 74–92.
6. Swetz KM, Kamal AH. In the clinic: palliative care. Ann Intern Med. 2012;156 (3):ITC-1–14.

References

1. Rabow MW, Pantilat SZ. Palliative care and pain management. In: McPhee SJ, Papadakis MA, editors. 2012 current medical diagnosis & treatment. New York: McGraw-Hill Medical; 2012. p. 74–92.
2. Lynn J, Teno JM, Phillips RS, et al. Perceptions by family members of the dying experience of older and seriously ill patients. Ann Intern Med. 1997;126(2):97–106.
3. Council on Scientific Affairs, American Medical Association. Good care of the dying patient. JAMA. 1996;275(6):474–8.
4. Stapleton RD, Curtis JR. End of life considerations in older patients with lung disease. Clin Chest Med. 2007;28(4):8901–7.
5. Claessens MT, Lynn J, Zhong Z, et al. Dying with lung cancer or chronic obstructive pulmonary disease: insights from SUPPORT. J Am Geriatr Soc. 2000;48(5 suppl 1):S146–53.
6. Solano JP, Gomes B, Higginson IJ. A comparison of symptom prevalence in far advanced cancer, AIDS, heart disease, COPD, and renal disease. J Pain Symptom Manage. 2006;31 (1):58–69.
7. Bookbinder M, McHugh NE. Symptom management in palliative care and end of life care. Nurs Clin North Am. 2010;45(3):271–327.
8. Swetz KM, Kamal AH. In the clinic: palliative care. Ann Intern Med. 2012;156(3):ITC-1–14.
9. Pantilat SZ, Isaac M. End-of-life care for the hospitalized patient. Med Clin North Am. 2008;92(2):349–70.
10. Roth K, Lynn J, Zhong Z, Borum M, Dawson NV. Dying with end-stage liver disease with cirrhosis: insights from SUPPORT. J Am Geriatr Soc. 2000;48(5 Suppl):S122–30.
11. Ward SE, Goldberg N, Miller-McCauley V, et al. Patient-related barriers to management of cancer pain. Pain. 1993;52(3):319–24.
12. Zech FJ, Grond S, Lynch J, Hertel D, Lehmann KA. Validation of World Health Organization guidelines for cancer pain relief: a 10-year prospective study. Pain. 1995;63(1):65–76.

13. Happ MB, Capezuti E, Strumpf NE, et al. Advance care planning and end-of-life care for hospitalized nursing home residents. J Am Geriatr Soc. 2002;50(5):829–35.
14. Morrison RS, Siu AL, Leipzig RM, et al. The hard task of improving the quality of care at the end of life. Arch Intern Med. 2000;160(6):743–7.
15. Rahm-Williams F, Riley JL, Herrera D, et al. Ethnic identity predicts experimental pain sensitivity in African Americans and Hispanics. Pain. 2007;129(1–2):177–84.
16. Hastie BA, Riley JL, Fillingim RB. Ethnic differences and responses to pain in healthy young adults. Pain Med. 2005;6(1):61–71.
17. Reuben DB, Mor V. Dyspnea in terminally ill cancer patients. Chest. 1986;89(2):234–6.
18. American Thoracic Society. Dyspnea: mechanisms, assessment, and management: a consensus statement. Am J Respir Crit Care Med. 1999;159(1):321–40.
19. Tice MA. Managing breathlessness: providing comfort at the end of life. Home Healthcare Nurse. 2006;24(4):207–10.
20. Blanchette H. Assessment and treatment of terminal restlessness in the hospitalized adult patient with cancer. MEDSURG Nurs. 2005;14(1):17–22.
21. Lawlor PG, Fainsinger RL, Bruera ED. Delirium at the end of life: critical issues in clinical practice and research. JAMA. 2000;284(19):2427–9.
22. Lawlor PG, Gagnon B, Mancini IL, et al. Occurrence, causes, and outcome of delirium in patients with advanced cancer. Arch Intern Med. 2000;160(6):786–94.
23. Johnston CB, Harper GM, Landefeld CS. Geriatric disorders: dementia. In: McPhee SJ, Papadakis MA, editors. 2012 current medical diagnosis & treatment. New York: McGraw Hill Medical; 2012. p. 60–3.
24. Aminoff MJ. Kerchner GA. Nervous system disorders. In: McPhee SJ, Papadakis MA, editors. 2012 current medical diagnosis & treatment. New York: McGraw Hill Medical; 2012. p. 789–90.
25. Caput P. Cancer cachexia. Nutr Rev. 1988;46(11):385–8.
26. Hamerman D. Molecular-based therapeutic approaches in treatment of anorexia of aging and cancer cachexia. J Geront: Med Sci. 2002;57A(8):M511–8.
27. Giacosa A, Frascio F, Sukkar SG, Roncella S. Food intake and body composition in cancer cachexia. Nutrition. 1996;12(1 suppl):S20–3.
28. Thomas DR. Distinguishing starvation from cachexia. Clin Geriatr Med. 2002;18(4):883–91.
29. Loewy EH. Terminal sedation, self-starvation, and orchestrating the end of life. Arch Intern Med. 2001;161(3):329–32.
30. Byock I. Patient refusal of nutrition and hydration: walking the ever-finer line. Am J Hosp Palliat Care. 1995;12(2):8–13.
31. Burge FI. Dehydration symptoms of palliative care cancer patients. J Pain Symptom Manage. 1993;8(7):454–64.
32. McCann RM, Hall WJ, Groth-Juncker A. Comfort care for terminally ill patients: the appropriate use of nutrition and hydration. JAMA. 1994;272(16):1263–6.
33. Eddy DM. A conversation with my mother. JAMA. 1994;272(3):179–81.
34. Rabow MW, Pantilat SZ. End-of-life care: nutrition & hydration. In: McPhee SJ, Papadakis MA, Rabow MW, editors. 2012 current medical diagnosis & treatment. New York: McGraw-Hill Medical; 2012. p. 89–90.
35. Derogatis LR, Morrow GR, Fetting J, et al. The prevalence of psychiatric disorders among cancer patients. JAMA. 1983;249(6):751–7.
36. Chochinov HM, Wilson KG, Enns M, Lander S. Prevalence of depression in the terminally ill: effects of diagnostic criteria and symptoms threshold judgments. Am J Psychiatry. 1994;151(4):537–40.
37. Pantilat SZ. End-of-life care for the hospitalized patient. Med Clin North Am. 2002;86 (4):749–70.
38. Block SD. Psychological considerations, growth, and transcendence at the end of life: the art of the possible. JAMA. 2001;285(22):2898–905.
39. Morita T, Tsunoda J, Inoue S, Chihara S. Contributing factors to physical symptoms in terminally ill cancer patients. J Pain Symptom Manage. 1999;18(5):338–46.

40. Mercadante S, Casuccio A, Fulfaro F. The course of symptom frequency and intensity in advanced cancer patients followed at home. J Pain Symptom Manage. 2000;20(2):104–12.
41. Vainio A, Auvinen A with the members of the Symptoms Prevalence Group. Prevalence of symptoms among patients with advanced cancer: an international collaborative study. J Pain Symptom Manage. 1996;12(1):3–10.
42. Morita T, Ichiki T, Tsunoda J, Inoue S, Chihara S. A prospective study on the dying process in terminally ill cancer patients. Am J Hospice Palliat Care. 1998;15(4):217–22.
43. McCarthy EP, Phillips RS, Zhong Z, Drews RE, Lynn J. Dying with cancer: patients' function, symptoms, and care preferences as death approaches. J Am Geriatr Soc. 2000; 48(5 suppl 1):S110–21.
44. Zomogyi-Zalud E, Zhong Z, Lynn J, Hamel MB. Elderly persons' last six months of life: findings from the hospitalized elderly longitudinal project. J Am Geriatr Soc. 2000;48 (5 suppl 1):S131–9.
45. Lynn J, Ely EW, Zhong Z, et al. Living and dying with chronic obstructive pulmonary disease. J Am Geriatr Soc. 2000;48(5 suppl 1):S91–100.
46. Clark LP, Dion DM, Barker WH. Taking to bed: rapid functional decline in an independently mobile older population living in an intermediate-care facility. J Am Geriatr Soc. 1990;38 (8):967–72.
47. Conill C, Verger E, Henriquez I, et al. Symptom prevalence in the last week of life. J Pain Symptom Manage. 1997;14(6):328–31.
48. Cleeland CS, Gonin R, Hatfield AK, et al. Pain and its treatment in outpatients with metastatic cancer. N Engl J Med. 1994;330(9):592–6.
49. Sanft T, Hauser J, Rosielle et al. Physical pain and emotional suffering: the case for palliative sedation. J Pain. 2009;10(3):238–42.
50. Quill TE, Lo B, Brock DW, Meisel A. Last-resort options for palliative sedation. Ann Intern Med. 2009;151(6):421–4.
51. Houran J, Lange R. Hallucinations that comfort: contextual mediation of deathbed visions. Percept Mot Skills. 1997;84(3 Pt 2):1491–504.
52. Perkins HS, Cortez JD, Hazuda HP. Patients' diverse beliefs about what happens at the time of death. J Hosp Med. 2012;7(2):110–6.
53. Jacobs LG. Managing respiratory symptoms at the end of life. Clin Geriatr Med. 2003;19 (1):225–39.
54. Kompanje EJO. "Death rattle" after withdrawal of mechanical ventilation: practical and ethical considerations. Intensive Crit Care Nurs. 2006;22(4):214–9.
55. Bruera E, Sweeney C, Willey J, et al. Perception of discomfort by relatives and nurses in unresponsive terminally ill patients with cancer: a prospective study. J Pain Symptom Manage. 2003;26(3):818–26.
56. Schag CC, Heinrich RL, Ganz PA. Karnofsky performance status revisited: reliability, validity, and guidelines. J Clin Oncol. 1984;2(3):187–93.

Chapter 3
Decisions at the End of Patients' Lives

All patient care requires decisions addressing certain clinical questions. What is the patient's most likely diagnosis? What is the most dangerous possibility? Which laboratory tests will clarify the diagnosis? How should they be sequenced? Which treatments will serve the patient best? How should the treatments be administered? Each large decision encompasses many smaller ones. End-of-life care with the intense emotions, complex technical treatments, and stark life-or-death implications involved presents some of the most difficult clinical decisions of all.

Despite medicine's recent efforts to explain options for care and to consider patients' wishes, the methods doctors use to make clinical decisions still baffle most patients, family members, and others without medical training. Unfamiliarity with the professional medical culture in general and with basic medical concepts in particular causes some of the bafflement. Yet even doctors sometimes do not recognize the decision-making methods they use.

To help all parties involved, this chapter describes clinical decision-making methods. It does so by addressing topics especially relevant to end-of-life care: How do doctors judge medical futility? Who should participate in end-of-life care decisions? Which clinical decision-making methods do doctors use most often? and How might negotiations resolve decisional conflicts? The complicated case of an advanced liver-failure patient illustrates both general clinical decision-making methods and specific decisions common in end-of-life care.

The Case

Mr. T., a 63-year-old African American, suffers from advanced, long-term alcoholic liver failure with cirrhosis. His illness involves frequent accumulations of abdominal fluid hindering his breathing, an overactive spleen that destroys blood cells, and

H.S. Perkins, *A Guide to Psychosocial and Spiritual Care at the End of Life*,
DOI 10.1007/978-1-4939-6804-6_3

occasional stomach bleeds causing disordered thinking and even obtundation. At home Mr. T. often ignores his low-salt diet and skips his medications. He also smokes. Mr. T.'s next of kin, his daughter, cares for him as best she can although she lives with her own family some miles away.

The daughter brings Mr. T. to the emergency room one night after finding him bloated with abdominal fluid, "asleep" on the floor, and unresponsive. She believes Mr. T. stopped his medications again. The doctors admit him immediately to one of the hospital's wards.

How Do Doctors Make Clinical Decisions?

Doctors' decisions often seem opaque to the medically untrained. The reasons for this opaqueness include the complexities of human biology, the ever-present factual uncertainties of clinical situations, the changes in patients' circumstances with time, the varied clinical backgrounds and idiosyncratic decision-making styles of individual doctors, and their infrequent or unclear explanations. I believe most clinical decisions are sound despite their seeming opaqueness. Yet the unclear decision-making process can frustrate patients and family members and confuse other health professionals. It also can undermine patient or family trust in doctors and even raise suspicions that they withhold important information. To avoid these problems, doctors must understand their own decision-making processes and be able to explain those processes to others.

Any explanation, however, must address the common misperception that clinical medicine is an exact science. This misperception portrays clinical medicine as using basic concepts or research data to derive the one sure, correct solution for each clinical problem. Twentieth-century medical education promoted that view, and its implication was obvious: By mastering medical science the doctor can solve each clinical problem with the single correct solution.

But clinical medicine is *not* an exact science like physics or chemistry. Clinical medicine does not have ever precise, objective observations; invariant rules of deductive logic; and single correct solutions to problems. Instead, as Jerome Kassirer (the former editor of the *New England Journal of Medicine*) says, clinical medicine is "not a science (but is) based on science." [1]. It is like other applied disciplines. Its "basic" concepts are many, variably applicable, and occasionally even conflicting, and its applications are sometimes inexact, even "messy."

To help everyone understand the clinical decision-making process, the following discussion describes some of the basic concepts of the process. The descriptions fall into two parts according to the two questions a doctor must answer whenever a patient presents a medical problem: What is causing the problem? and How should it be treated? I group concepts under the question most closely related to them. But I reemphasize that none of the concepts is invariant, absolute, or all-determinative. Rather, an essential part of the doctor's "art" has always been knowing when to adhere to basic concepts and when to deviate from them.

What Is Causing the Patient's Problem?

To answer this question, the doctor must first define the patient's problem in clear, clinically useful terms. The patient's original chief complaint often suffices but sometimes requires reformulation to permit further diagnostic analysis. This reformulation involves distilling the patient's presentation into its most clinically important features. For example, a chief complaint of "dizziness" requires reformulation and distillation into either vertigo, syncope, or disequilibrium—each implying different diagnoses and treatments.

The doctor then applies one of four methods—hypothesis generation based on differential diagnosis, Bayesian probabilities, pathophysiology, or pattern recognition—to narrow down the diagnostic possibilities. The first three methods are deductive, applying general concepts to the patient's specific case. Deductive thinking is typically slow, conscious, logical, and less prone to errors. Doctors usually apply deductive thinking in difficult or atypical cases. The fourth method is inductive, extrapolating knowledge from similar past cases to the new case of this patient. Inductive thinking is typically quick, mostly subconscious, intuitive, and more prone to errors. Some authors believe inductive thinking evolved to help humans survive by reacting quickly to danger. Doctors usually apply inductive thinking in common or familiar cases [1].

Experienced doctors often make diagnoses without realizing how they do so. Yet patients, family members, and professional colleagues need some explanation of doctors' methods to understand the resulting diagnostic conclusions. Doctors should be able to recognize the methods they use, and professional colleagues should have a basic familiarity with the possibilities. To help all involved, I briefly describe all four methods here.

Hypothesis Generation Based on the Diagnostic Possibilities

This method, first articulated by Arthur Elstein, involves generating diagnostic hypotheses and testing them against the patient's clinical data [2]. The doctor creates a mental list of diagnostic possibilities for the patient's problem ("the differential diagnosis"); distills the most important features of the case presentation from the history, physical examination, and laboratory results; and tests each diagnostic possibility by weighing the case's features for and against it. The logic behind the testing goes as follows: A patient with a particular disease should show the symptoms and signs that characterize the disease. If the patient does not show many of these symptoms and signs, then he or she probably does not have that disease. This logic can rule out many diagnoses. The diagnostic possibilities surviving this culling process then guide further workup or treatment. Generating and testing hypotheses, of course, mimic the scientific method. But the "weights" of the presentation features and the thresholds for ruling a diagnosis in or out are purely subjective. They differ among doctors, sometimes yielding different diagnostic conclusions in the same situation.

The doctor might use the hypothesis generation method to evaluate a patient presenting with a cough. The differential diagnosis for cough includes common diseases (such as asthma, acute and chronic bronchitis, gastroesophageal reflux ("heartburn"), upper respiratory tract infections and allergies, and side effects of some drugs). The differential diagnosis also includes rarer but more dangerous possibilities (such as lung cancer or tuberculosis). The clinician might investigate the diagnostic possibilities by reviewing the patient's drug list, asking about pets and allergies, checking the patient's vital signs (including body weight), and ordering a chest X-ray.

Bayesian Probabilities

Thomas Bayes, an eighteenth-century English clergyman, pioneered a rigorous method for using new information to revise up or down initial probability estimates about uncertain conditions or events [1]. This method, called Bayesian analysis, has application in clinical medicine as a way to adjust probability estimates for a disease according to positive or negative diagnostic test results [3]. A diagnostic "test" is any information about the patient's condition: a historical detail, an observation on physical examination, or a laboratory test. A positive diagnostic test result typically increases the probability estimate for the disease; a negative diagnostic test result decreases that estimate. Specifically, the disease's "pretest probability" (initially the prevalence of the disease in a population) is multiplied by the test's unique "likelihood ratio" (a number which characterizes the test's ability to discriminate patients with and without the disease) to produce the disease's "posttest probability."

Though arithmetically simple in concept, Bayesian analysis in its strictest form has significant disadvantages for clinical use. Research has not determined most likelihood ratios (doctors must guess at them), people cannot reason easily with probabilities, and multiple simultaneous test results require long Bayesian terms to calculate an exact posttest probability. The process quickly becomes too complex for clinical use.

Fortunately, a simpler, more intuitive approach captures the essence of Bayesian analysis if not its precision. This approach involves a qualitative scale of probabilities for a disease: very unlikely (less than 10 %), unlikely (10–33 %), uncertain (34–66 %), likely (67–90 %), and very likely (more than 90 %). A pretest probability estimate in the middle three categories, but not the two end categories, ordinarily requires further testing to rule a disease in or out. A positive test result boosts the disease probability up one category while a negative test result drops that probability down one category. Unencumbered by detailed arithmetic calculations, doctors have been able to use this simplified Bayesian analysis in their everyday diagnostic work.

I believe this simplified Bayesian analysis can also help the medically untrained understand medical tests results. A doctor, for example, might use Bayesian probabilities to justify an exercise stress test for a middle-aged woman with atypical chest pain and a family history of early atherosclerotic heart disease. A positive test might increase the doctor's estimate of significant atherosclerotic heart disease from "uncertain" to "likely." Most patients would likely understand this conclusion more easily if expressed as a qualitative disease probability (i.e., "uncertain" or "likely") than as a quantitative one (i.e., "34–66 %" or "67–90 %").

Pathophysiology

Probably the least used of all decision-making methods, pathophysiology involves reasoning from a known pathologic cause to a predictable physiologic effect. This method inspired the dramatic twentieth-century medical education reforms that made medical students first into scientists and then into clinicians. As a result, students for many years have studied first the basic medical sciences such as anatomy and physiology before starting actual clinical training. Such pathophysiologic reasoning, though rarely invoked explicitly today, might still help the doctor diagnose, say, a particular heart valve's constriction (aortic stenosis) in a "dizzy" patient with a characteristic heart murmur.

Pattern Recognition

Seasoned doctors probably use this method most often. It involves comparing the important features of a new case to familiar disease prototypes. Beginning medical students make these comparisons slowly and consciously based on "classic" textbook descriptions of diseases as comparators. But more experienced clinicians make comparisons faster and less consciously while simultaneously recognizing subtle variations in disease presentation. "Anchoring" is a danger, however: Features noted early in an evaluation tend to influence the diagnostic conclusion more than do features noted later on. Still, some highly experienced doctors using pattern recognition can simply glance at a patient and instantly make a correct diagnosis. A doctor might use pattern recognition to diagnose a facial rash by quickly identifying the characteristic features of seborrheic dermatitis (scaly patches, indistinct borders, an underlying red color, and rare itching), psoriasis (scaly plaques, sharp borders, a dark or violaceous color, and frequent itching), or rosacea (small nodules, some with pus, indistinct borders, a flushing red color, and no itching).

I emphasize that no doctor makes all diagnostic decisions by only one method. A doctor may use different methods in the same case, that doctor may use different methods in different cases, and different doctors may use different methods in the same case. Therefore, the doctor trying to explain diagnostic decisions to the medically untrained should be able to identify the method or methods used in the patient's case and the contribution each method makes to an overall diagnostic conclusion.

The Case

The ward doctors use all four diagnostic decision-making methods to identify the likely cause for Mr. T.'s "sleepiness." The doctors immediately relabel the problem as "mental status changes," a clinically meaningful category, and list the diagnostic possibilities including advanced liver disease; kidney failure; a stroke, concussion, or bleed in the head; chemical disorders of the blood including low sodium or high calcium; sedative ingestion; and serious infection (the hypothesis-based-on-differential-diagnosis method). The doctors believe advanced liver disease most likely because Mr. T. has presented similarly for past exacerbations of it (pattern recognition). Initial laboratory tests of liver, kidney, and blood coagulation functions and blood counts support that diagnosis. Those results boost the doctors' probability estimate for it from a pretest "likely" to a posttest "very likely" (Bayesian probabilities). At the same time, various other negative blood chemistries, urine toxicology screenings, a head scan, and bacterial cultures drop their probability estimates from a pretest "uncertain" to a posttest "unlikely" for other diagnostic possibilities (such as low sodium, high calcium, sedative ingestion, a head mass problem, and a serious infection). In addition, the doctors reason that only advance liver disease with its attendant brain swelling, abdominal fluid accumulation, and spleen overactivity can explain all the features of Mr. T.'s clinical presentation (pathophysiology).

Mr. T.'s doctors, therefore, decide to treat him for an exacerbation of his liver disease. Doctors typically plan treatment focused on the highest priority diagnostic possibilities, that is, the most likely and the most dangerous ones. Mr. T.'s initial workup just happens to reveal his advanced liver disease as both the most likely and the most dangerous diagnostic possibility.

How Should the Patient's Problem Be Treated?

Having whittled down the "working diagnoses" to only one or a few possibilities, doctors usually initiate treatment by medicine's traditional "try-and-see" approach. Medicine's poor understanding of diseases through much of history made treatment effects often unpredictable and sometimes harmful. "Try and see" made sense under the circumstances. Doctors would try a treatment and see its effects. If they had little or no benefits (or even harms), doctors would stop that treatment and try another. Even as the medical science revolution has produced more consistently effective treatments, the "try-and-see" approach has persisted due to individual variability in responses among patients.

The "try-and-see" approach requires constant monitoring of treatment results to determine the ongoing balance of benefits and harms for each patient. The doctor conducts this monitoring by following the course of a few "sentinel variables" specific to each disease or treatment. These variables indicate whether treatment "is working," that is, providing a favorable balance of benefits over harms for the patient. If the treatment is "working," the doctor continues it. If it is not, the doctor stops it. Doctors, patients, and others must all commit time and patience to the "try-and-see" treatment process. Unfortunately, many patients and family members do not understand it. They expect quick, sure results. When the results are not like that, these patients or family members may become frustrated and angry. Doctors respond to them best by acknowledging the frustration and anger, describing the use of sentinel variables, and estimating (if possible) the timeline for assessing treatment results definitively. Doctors may also have to remind patients and family members that waiting patiently for a properly timed treatment assessment is one of the hardest parts of being ill.

The Case

Realizing that Mr. T. may also have some stomach bleeding (a well-known complication of advanced liver disease), the ward doctors begin treating him in the usual "try-and-see" manner with standard, first-line treatments. They order salt restriction, modest dose diuretics, suction to the stomach, abdominal fluid aspirations, and a bowel purgative. Mr. T.'s alertness and certain blood tests serve as the "sentinel variables" for monitoring his progress. Some of these variables including his alertness improve at first; others do not. When Mr. T. slips back into a coma on hospital day 3 and all the sentinel variables deteriorate again, the doctors conclude that the standard treatment is not working. They must find out why.

Repeat blood tests confirm that Mr. T. is bleeding intermittently into his stomach and his liver function is declining. He appears to be teetering on the brink of complete liver failure. The ward doctors treat him urgently as best they can. The doctors consult the liver specialists about Mr. T.'s chances for a liver transplant. The specialists explain they have followed Mr. T. for more than a year. He now has seventh place on their local transplant priority list. They predict he must wait a month or more for a donated liver unless he appears to be dying and his liver failure score increases. Then he would move to the top of the priority list.

Mr. T. deteriorates further over the next day. The liver specialists now predict he will die shortly without a transplant. They tell the ward doctors that Mr. T.'s liver failure score will rise enough to make him eligible for a transplant only on transfer to the intensive care unit (ICU). The liver specialists promise to pursue a transplant immediately for Mr. T. if the ward doctors make that transfer.

The treatment choice is clear: Sending Mr. T. to the ICU gives him a chance, though only a small one, at receiving a transplant and living a meaningful life for some more months or years. In contrast, keeping him on the ward gives him no chance at a transplant and ensures he will die within days. The ward doctors consult Mr. T.'s daughter by telephone. She waffles on the decision. One minute she wants Mr. T. to go to the ICU and to have a chance at a transplant. The next minute she wants him to stay on the ward and to receive comfort care there. In effect, the ward doctors must decide what to do. And they must decide quickly.

Yet they are split. The most senior ward doctor, an attending, believes that Mr. T. is irreversibly dying and no treatment can save him. This doctor favors comfort care for him on the ward. A more junior ward doctor, the resident, believes Mr. T. deserves a chance at fully aggressive ICU care and a life-saving transplant. The resident bases his opinion on having seen doctors refuse at first to treat his grandmother after a stroke. They relented when the family pressed for treatment. His grandmother lived several more happy years afterward. The anxious, most junior ward doctors—the interns—can muster no opinion. But a ward nurse with ICU experience warns, "I've seen many patients like Mr. T. go to the ICU. They all die there, sometimes miserably." The ward doctors face an obvious futility judgment about ICU treatment and a liver transplant for Mr. T.

How Do Doctors Judge Medical Futility?

Clinical medicine has always required futility judgments [4]. In fact, such judgments predominated over the centuries when medicine had no scientific grounding and few effective treatments. The treatments of those times produced few benefits and many harms. The Hippocratic oath, therefore, warned doctors against causing unnecessary suffering by treating patients when illness had "overmastered" them [5]. Many doctors even tried to protect their reputations as good healers by completely refusing to attend the irreversibly dying.

A History of Modern Medical Futility Judgments

The bioscientific revolution of the nineteenth and twentieth centuries transformed clinical medicine dramatically, yielding curative or palliative treatments for many formerly untreatable illnesses. But a backlash began around 1970. People worried that a "technologic imperative" had come to dominate clinical practice. They defined this imperative as the irresistible temptation to use technologic treatments indiscriminantly even with little realistic expectation of benefit to patients. People wondered whether doctors instituted such treatments oblivious to their low chance for success and high chance for harm and then could not stop the treatments once they proved ineffective or harmful. Of course, desperate patients and family members sometimes fell victim to that imperative, too. While the backlash against technologic medicine started slowly, it soon erupted full force as Americans became convinced about needless patient suffering and uncontrolled medical costs. At the same time the new hospice philosophy created an attractive alternative to such "high-tech" medicine at the end of life.

This backlash helped fuel a vigorous debate about medical futility. The debate raged from the mid-1980s into the early 2000s and focused on disagreements over subjective aspects of an operational definition of the futility concept; specific illogical, inconsistent, or frankly discriminatory applications [6]; potentially deleterious general effects on the traditional doctor–patient relationship (such as seeming to abandon the sickest patients); and conceptual confusions between futility and rationing [4, 5, 7–9]. The debate eventually lost most of its steam, leaving some conceptual and practical problems unresolved. Still, doctors face futility judgments in everyday patient care [5, 9]. Those judgments deserve the clearest possible guidance.

Judging Medical Futility in Patient Care

I believe one particular article provides the best currently available guidance for judging futility [10]. This article combines good intuition, conceptual rigor, and clinical usefulness. Authors Schneiderman, Jecker, and Jonsen begin the article by defining futility as a quality of an action: uselessness for achieving a defined goal. They then expand on the widespread intuition that futility has two independent dimensions—one quantitative, one qualitative. Futility can occur by either dimension. The quantitative one involves the conclusion drawn from the attending doctor's experience, other doctors' experience, or empirical studies that a medical treatment has not succeeded in the last 100 similar cases. Critics decry this threshold success rate of less than 1 % as merely subjective, and it certainly is. But I believe it falls well within most health professionals' intuitive idea of futility thresholds (running typically from 1 to 5 %) and is low enough to protect against many unjustified, discriminatory judgments.

The qualitative dimension of futility, a somewhat more controversial idea, involves the assessment that a treatment offers no expectation of progress toward a justified treatment goal, that is, a benefit the patient can appreciate. The authors explain the idea by distinguishing effects and benefits. Effects are changes merely in biological variables or processes that patients usually cannot appreciate. Examples include changes in blood chemistries or blood pressure. Because patients do not appreciate them, effects cannot serve as adequate treatment goals. In contrast, benefits are changes in symptoms or body functions that patients *can* appreciate for making their lives better. Examples of benefits include relieving pain or increasing the distance one can walk. Because patients *do* appreciate them, benefits *can* serve as adequate treatment goals. The authors conclude their argument by ruling out as possible benefits merely permanent states of unconsciousness, overwhelming suffering, an all-consuming struggle to survive, or dependence on the ICU. The authors believe patients cannot appreciate such results. Thus, the treatments that create only them are necessarily futile.

Schneiderman, Jecker, and Jonsen grant doctors final authority to decide medical futility but urge prior consultation with patients, proxies, and other health professionals (especially the specialists most experienced with similar cases). I support that position with some elaboration. I believe patients (or their proxies when patients lack decision-making capacity) deserve the greater say in choosing overall *treatment goals* because patients know best what they will appreciate as beneficial outcomes. But, because patients and proxies may sometimes misunderstand the limitations of treatments, I believe doctors ordinarily deserve the greater say in choosing the best *treatment means* to achieve patients' goals. Naturally, patients can provide important input for decisions about treatment means [5, 11]. A patient, for example, may prefer to limit interventions to only "high-yield" treatments or to entertain some "long-shot" treatments. Such preferences might reasonably influence a doctor's futility judgment about treatment means. I emphasize again, however, that a doctor should take great steps to prevent futility judgments from becoming unilateral power plays. Ideal medical decision-making seeks consensus from patients (or proxies) and doctors alike on treatment goals and means. The burden of proof should always fall most heavily to those who favor withholding standard treatments on futility grounds.

Schneiderman, Jecker, and Jonsen also allow for some variability in futility judgments. Because reasonable, well-meaning people may reach different conclusions in the same case, the authors are willing to accept some range of futility judgments as valid. They are also willing to accept for good purposes the rare use of treatments ordinarily considered futile but not entailing excessive suffering for patients.

The Case

The ward doctors must decide about the futility of immediate ICU transfer and a possible liver transplant for Mr. T. His circumstances do not appear to meet criteria for qualitative futility. If intensive care keeps Mr. T. alive until he receives a transplant, he might return to a life he appreciates (perhaps living independently and seeing his daughter and her children). But his circumstances may or may not meet criteria for quantitative futility. Mr. T. presents several drawbacks for transplantation. He has not shown that he understands his liver disease or treats it conscientiously. For instance, he often does not follow its basic treatment regimen at home [12]. He also smokes, putting him at increased risk for dying unexpectedly from another disease such as a heart attack or stroke. The liver specialists, however, do not completely write off his chances for benefiting from a transplant. As one doctor says, "We've transplanted patients as sick as Mr. T. before. Some have lived for two years or longer. Mr. T. might surprise us and do well." Furthermore, Mr. T. submitted to long-term treatment by the liver specialists and allowed himself placement on the transplant waiting list. The ward doctors reason he must, therefore, have wanted a transplant or at least been willing to have one.

Who Should Participate in End-of-Life Medical Decisions?

Critical treatment decisions these days usually prompt doctors to consult patients about their preferences. Mr. T. himself might have been able to give his transplantation preferences only days before hospitalization or even briefly when he regained alertness during hospitalization. Fortunately, today's medical ethic encourages doctors to consider patients' preferences about these decisions, but that ethics obviously has not always held sway. Doctors historically made all medical decisions for patients. Doctors' arcane medical knowledge, their benevolent professional ethics, and patients' passivity due to illness and to limited education seemed to justify this paternalism for centuries. But medical decision-making changed drastically in the twentieth century as patients became more educated, more aware of their rights, and more assertive than ever before. Patients began campaigning for a voice in their care, especially to refuse unwanted treatment. As they established that voice, patient autonomy became central to modern medical practice.

Still, incorporating patient autonomy into everyday practice has proven more complicated than expected. Many patients have welcomed their new decision-making authority. But many others have not, preferring instead to let their families or doctors still make medical decisions for them. How can doctors know which patients wish to exercise their decision-making rights, and which do not? Whenever possible, of course, doctors should ask each patient's preferences about participating in medical decisions. But can research help doctors anticipate those preferences?

To answer this question, I examined North American studies about patients' preferences and perceptions about participating in decisions about their care. (I assumed these study patients, almost all of whom were inpatients, generally suffered from critical or terminal illnesses.) I identified three published American studies, three published Canadian studies, and one unpublished American study addressing patients' *preferences* for such participation (Table 3.1). Large majorities of respondents (62 % or greater) in six studies [13–18] and a near majority (47 %) in the seventh [19] preferred participating in care decisions either by having final control themselves or sharing it with their doctors. Though not performed across all the studies, subgroup analyses suggested that more female, Euro-American, unmarried, and better educated respondents than others preferred to participate in those decisions [13–15, 18].

I also identified two published American studies, one published Canadian study, and one unpublished American study addressing patients' *perceptions* of their actual participation in decisions about their care (Table 3.2). Mere minorities of respondents (48 % or less) in all four studies reported such participation [17–20]. Of the four studies just two performed subgroup analyses, and they produced conflicting results by age, gender, and ethnic subgroup [18–20].

This limited review suggests that, although most North Americans want to participate in decisions about their care, too few do so. To a greater extent than they may realize, doctors may still control care decisions for critically or terminally ill

Table 3.1 Patients' preferences for participation in medical decisions

Study	Patients	Preferences for autonomy	
		Whole-sample analyses	Subgroup analyses
Cassileth, 1980, USA*[13]	Mixed cancer inpatients and outpatients. Various ethnicities.	62.5 % wanted to participate; 37.5 % did not.	Euro-American, younger, and more educated respondents wanted to participate. No gender differences.
Blanchard, 1988, USA [14]	Cancer inpatients. No ethnicities reported.	69 % wanted to participate; 31 % did not.	Female, younger, and unmarried respondents wanted to participate.
Kelner, 1995, Canada [15]	Elderly inpatients with chronic respiratory or neurologic diseases. No ethnicities reported.	71 % wanted to participate; 29 % did not.	Unmarried, more educated, and higher socioeconomic respondents wanted to participate. No age or gender differences.
Hofmann, 1997, USA [19]	Intensive care unit patients. Various ethnicities.	47 % wanted to participate.	No analyses reported.
Deber, 1996, Canada [16]	Cardiac angiogram inpatients. No ethnicities reported.	26 % wanted doctor control; 29 %, patient control; 45 %, shared control.	No analyses reported.
Stewart, 2004, Canada [17]	Coronary care unit patients. "Mostly" Euro-American.	34 % wanted doctor control; 29 %, patient control; 37 %, shared control.	No analyses reported.
Perkins, no date, USA [18][a]	Ward patients with potentially fatal diagnoses. Various ethnic groups.	36 % wanted doctor control; 40 %, patient control; 22 %, shared control.	African Americans and men wanted doctor control; other ethnic groups and women wanted patient control.

[a]In preparation
*USA = United States of America

patients. Doctors may need to redouble their efforts to include patients (or proxies) in those decisions.

The Case

If Mr. T. were able to communicate, his doctors would surely consult him about the transplantation decision. His input might have tipped the decision definitively one way or the other. But the obtundation prevents his input at the crucial moment.

Table 3.2 Patients' perceptions of participation in medical decisions

Study	Patients	Perceptions of autonomy	
		Whole-sample analyses	Subgroup analyses
Hofmann, 1997, USA* [19]	Intensive care unit patients. Various ethnicities.	23 % had discussed CPR[a] wishes with doctor; 12 % had discussed ventilator wishes.	No analyses reported.
Stewart, 2004, Canada [17]	Coronary care unit patients. "Mostly" Euro-Americans.	53 % perceived doctor control; 24 %, patient control; and 23 %, shared control.	No gender difference.
Happ, 2007, USA [20]	Intensive care unit patients on ventilators. Various ethnicities.	40 % of patients perceived their participation in treatment decisions.	No ethnic or gender differences. The young participated in even everyday decisions; the elders, mainly only in withholding decisions.
Perkins, no date, USA[b] [18]	Ward patients with potentially fatal diagnoses. Various ethnic groups.	33 % perceived doctor control; 22 %, patient control; and 26 %, shared control.	Minorities and men perceived mostly doctor control. Euro-Americans and women perceived mostly patient control.

[a]*CPR* cardiopulmonary resuscitation
[b]In preparation
*USA = United States of America

Unfortunately, other possible sources of information about Mr. T.'s wishes do not decide the matter, either. The liver doctors themselves do not recall Mr. T.'s expressing to them any clear wishes about transplantation. Of course, the doctors may have never asked. They may have just assumed that, because Mr. T. did not object, he agreed to the sequence of treatments for advanced liver disease including transplantation. As a result, the doctors may have missed their opportunity to learn explicitly Mr. T.'s overall treatment wishes and life goals. The observation of one end-of-life expert might apply here: We doctors focus "so much on fixing the physiological wreckage that we (forget about) the lives our patients (wish to) lead." [21].

The doctors then consult Mr. T.'s daughter about any wishes he previously expressed about transplantation. She remembers none. In fact, she says Mr. T. never talked much about his medical problems or hinted at what he wanted for treatment.

Finally, the doctors consult research about treatment wishes among African Americans. Unfortunately, as Chap. 10 suggests, the limited data about African American men do not argue definitively for one decision or the other. Arguing for a transplant, some data show African Americans, both men and women, generally request more life support than do other American ethnic groups [22–24] and are willing to live with greater disabilities than are other groups [25, 26]. But perhaps arguing against a transplant, other data suggest that African American men, unlike

African American women, feel disempowered in the medical care system and fear harm from it [27]. The conflicting data leave the doctors still wondering which decision would serve Mr. T. best.

Conflicts About End-of-Life Care

Conflict characterizes much of modern American life. In fact, decisions about end-of-life care almost guarantee conflicts among the parties affected because of the high, life-or-death stakes; the burdensomeness of many life-prolonging treatments; their uncertain results; and the intense emotions involved. Even the ward doctors in Mr. T.'s case encounter their own conflict over which treatment decision would serve him better: ICU transfer and possible transplant or continued ward care aimed at comfort. Many end-of-life conflicts turn acrimonious but they need not be. The doctors in Mr. T.'s case worked through their conflict respectfully.

Table 3.3 lists factors that can fuel these end-of-life conflicts. Sources for the factors include patients and families, health professionals, and the healthcare system. I find that a few factors from each source commonly fuel many of the conflicts: from among the patient and family factors, poor understanding of the medical situation, and disruptive family dynamics; from among the health professional factors, poor communication skills, differences over diagnosis, treatment or

Table 3.3 Factors that fuel conflicts in end-of-life care

Source	Factor
Patient and family	Poor understanding of the medical situation
	Knowledge-base differences
	Disruptive family dynamics (especially long-simmering conflicts)
	Underlying impatience or guilt feelings
Health professionals	Poor communication skills (e.g., unclear explanations, discomfort with discussing death, insensitivity to others' emotions)
	Differences among health professionals over diagnosis, treatment, or prognosis
	Unfamiliarity with the patient's values or treatment goals
	Differences between patient and health professional values
	Inadequate symptom control
	Excessive hesitation or haste in using life-support treatments
	Seeing the patient's death as a defeat
	Fatigue
	Excessive fear of retaliatory lawsuits
The healthcare system	Overly routine use of "high-tech" life-support treatments
	Overly restrictive policies and regulations
	Costliness of care

prognosis, and unfamiliarity with the patient's values or treatment goals; and from among the health system factors, the overly routine use of "high-tech" life-support treatments (with the attendant psychological difficulties in stopping them). Because these factors can be easily overlooked, health professionals who find themselves mired in end-of-life conflicts might review Table 3.3 for possible unrecognized sources of conflict.

In Mr. T.'s case the factors playing the most important roles in the conflict over treatment arise from health professional factors. They include the ward doctors' different predictions about the possibility and success of a liver transplant, and their unfamiliarity with Mr. T.'s general life philosophy and goals for treatment.

Negotiating Resolutions to End-of-Life Conflicts

When a conflict over end-of-life care arises, the lead doctor attending the patient should take responsibility for finding a patient-centered resolution. Ideally, that doctor serves as moderator for discussions about the conflict. But doing so requires effective leadership.

Whoever the moderator is should arrange the time, place, and participant composition of the initial discussion. Its timing should allow for maximum participation by the most important parties affected and for unhurried, private exchanges. I find scheduling discussions after regular work hours has the advantage of minimizing interruptions. The discussion may occur at the bedside (if there are few participants and privacy can be maintained) or in ward conference rooms otherwise. In either case the place should be quiet, comfortable, and convenient for all. Whenever possible, the patient should attend and be permitted to choose the family and friends who attend. The lead doctor, any key consulting doctors, and a nursing representative, chaplain, or social worker should also attend. The moderator may have to exercise judgment in limiting the number of participants. I find six to be about the maximum. Table 3.4 lists stages and steps in the negotiating process.

The moderator might begin the discussion by reminding everyone that its primary purpose is to serve the patient's interests. That is, all the care decisions that result from the discussion should produce not merely "effects" (improvements in physiologic variables), but "benefits" (improvements in symptoms, function, or longevity that the patient will appreciate).

The moderator should then summarize the patient's case history. The moderator may need to have reviewed beforehand the rationale for past critical decisions and to have delineated what about the case can be known with certainty and what cannot. The moderator might conclude with a best-guess prediction about the patient's future. However uncomfortable doctors feel about such predictions, patients and families want them. Of course, in making a prediction, the doctor should always acknowledge the possibility of unforeseen intervening events that might change the prediction. Throughout this summary the moderator should speak plainly, avoiding the "doctor speak" of specialized concepts and terms.

Table 3.4 Negotiating resolutions to end-of-life conflicts

Stage of the negotiation	Suggested steps
Preparing for the discussion	Find a quiet, comfortable, convenient place for discussion.
	Invite all appropriate parties to attend.
Beginning the discussion	Remind all about the primary purpose of the discussion: to serve the patient's interests best.
	Outline the patient's case history in plain language.
	Define the conflict and identify its source from your viewpoint: the goals of care, the means of care, or both.
Encouraging respectful exchange of differing views	Listen closely.
	Repeat back comments to ensure accurate understanding.
	Express empathy. Let people know you have heard them.
	Correct misunderstandings about healthcare practice, law, or policy.
Identifying and implementing solutions	Brainstorm solutions.
	Decide on a solution.
	Share responsibility for the decision(s).
	Promote transparency by making sure to inform all affected parties about the decisions (unless the patient objects).
	Plan follow-up for implemented decisions.
Responding if no solution arises	Remember that the passage of time resolves many conflicts.
	Keep communications open whenever possible.
	Encourage healthy coping strategies for all.
	Suggest referral to a hospital ethics consultant or committee.
	Always protect the patient's interests.

Data from Quill [2]

After that case review the moderator should try to articulate the core conflict and indicate whether it concerns the goals of care, the means of care, or both. This whole "setup" for further discussion should take only a few minutes.

The moderator can then begin the negotiations in earnest. He or she should encourage key participants, starting with the patient, to express their views about the subject of the conflict. Strong emotions may run through these views. The moderator may, therefore, occasionally need to remind participants that respectful listening holds the key to resolution, and that all should be allowed to speak freely. A full initial airing of views and emotions may set the stage for cool, rational deliberations later. The moderator or other health professionals present may also need to acknowledge the expressed emotions, to anticipate any unexpressed emotions, and to express empathy. Such empathetic expressions let people know their emotions have been heard [2]. When each speaker finishes, the moderator should confirm an accurate understanding by briefly summarizing that person's views.

The moderator should eventually invite suggestions for resolution of the conflict. He or she should facilitate the brainstorming without dominating it [2]. For the sake of progress, the moderator may sometimes need to take an active role in guiding the discussion. One way to do so is to elicit specific information depending on whether the conflict concerns the goals or means of care. If the goals of care, the patient's wishes about treatment outcomes (especially function, relief of symptoms, and survival) should carry the greatest weight in any resolution. The patient should have the chance to express those wishes explicitly if possible. (Otherwise, well-informed, trustworthy proxies should try to do so.) Because articulating wishes about treatment outcomes unnerves many patients, I recommend the moderator elicit those wishes with specific questions. He or she might ask the patient, for example, What do you enjoy doing these days? What are you still looking forward to doing? or What makes life worthwhile for you now? In addition, the moderator might ask the patient about trade-offs between survival and comfort. For example, given the choice, would you prefer longer survival with some discomfort or shorter survival with little discomfort? Answers to these questions can provide critical information about which functional outcomes the patient considers benefits (that is, treatment goals) and how much comfort the patient is willing to give up to achieve prolonged survival (that is, the comfort-survival trade-off).

If the conflict involves the means of care, the patient's wishes about specific treatments should carry the greatest weight in the resolution. Most expressed treatment wishes are refusals. No treatment, of course, can be forced on informed, aware patients. The United States Constitution and voluminous case law protect patients' bodily privacy, permitting them to refuse even life-saving treatments. To elicit treatment refusals in conflict negotiations, the moderator might ask the patient, Have you ever been seriously ill before? What treatments did you receive? and Would you refuse any of those treatments again (or others you know about) even if your life depended on them? Any discussion about the means of care should also include input from health professionals knowledgeable about the relevant treatments. The health professionals should be able to explain the potential advantages and disadvantages of those treatments. I offer one warning, however: Patients should not be allowed to demand just any treatment. Health professionals need not —and sometimes must not—provide treatments that violate law, personal conscience, or established professional practice.

The health professionals participating in conflict negotiations must have a sound grasp of healthcare practice, law, and policy. Misunderstandings in all three areas are common. In healthcare law, for instance, common misunderstandings include that forgoing life support for decisionally incapacitated patients requires prior evidence of their wishes, that withholding fluids or nutrition is illegal, that hospital risk management must approve all life-support terminations, and that administering high doses of analgesics to relieve pain invites prosecution. Another common misunderstanding involves the hospital attorney's role: It is solely to protect the *hospital*'s interests. Therefore, whenever the hospital's interests conflict substantially with the patient's, health professionals may have the legal duty *not* to follow the hospital attorney's advice.

Negotiating to a single resolution may sorely test the moderator's patience and understanding of human beings. Keys to success are inclusiveness, openness to suggestions, good listening, and respectful responses. I find that most conflict discussions have a natural time limit, between a half hour and an hour. In concluding the discussion, the moderator may need to exercise firm but fair direction in suggesting a solution based on what participants have said. The moderator should acknowledge his or her share of responsibility for the decision, make sure any centrally affected parties not present are informed (unless the patient objects on solid grounds), and plan follow-up for implementation of the solution. The moderator should also report the discussion, its conclusion, and any dissents in the patient's medical record. I believe most conflicts can be resolved in this matter.

Some negotiations, of course, will not achieve a resolution, and health professionals should not always expect one. Participants in those cases should remember that just the passage of time resolves many conflicts. Meanwhile, the moderator of the conflict discussion should try to maintain communications with the disagreeing parties and encourage everyone in healthy coping strategies. Furthermore, the moderator might suggest referral to a clinical ethics consultant or committee. Rarely, though, an ongoing impasse threatens a patient's vital interests perhaps by prolonging unnecessary suffering. Those situations, I believe, justify doctors in exercising rightful professional authority to implement a carefully considered, definitive, patient-centered resolution despite the lack of consensus.

Implementing Decisions to Withdraw Life Support

Withdrawing life support and allowing a patient to die are heart-wrenching for everyone. Such actions, once fully considered and agreed to as promoting the patient's interests, demand courage from the patient's lead doctor. He or she should not expect nurses or other health professionals to stop the respirator or withdraw the nasogastric feeding tube. Rather, the lead doctor should take responsibility for performing the withdrawal. He or she should set a convenient time to do it and invite the family to attend. At the designated time, the doctor should briefly review the decision and its rationale and perform the withdrawal. Afterward he or she should make reassuring visits or calls to the family and the staff well into the grieving period after the patient dies.

The Case

The ward doctors demonstrate some of the best features of conflict negotiation in Mr. T.'s case. They explicitly acknowledge to one another at the start that they all aim to promote the best for him. They also acknowledge that the conflict involves both the goals of care (prolonged survival versus comfort) and the means of care

(ICU transfer and possible transplantation versus continued ward care and comfort measures).

The doctors invite Mr. T.'s daughter to join the negotiations over the telephone. They express empathy to her about her distress over Mr. T.'s dire medical condition and ask her views about his likely wishes under the circumstances. Unfortunately, she gives them no clear, consistent answer. She then declines her decision-making prerogative and asks the doctors to do what they think best. With the management decision falling squarely to them, the doctors promise to notify the daughter when they reach one.

The doctors briefly review Mr. T.'s hospital course and then brainstorm about management options. They listen carefully to each other's views: the attending doctor to the resident doctor's proposal for ICU transfer and an attempt at transplantation; the resident doctor to the attending doctor's proposal for continued ward care aimed at comfort. Each doctor repeats back the other's view to ensure an accurate understanding. The doctors note the pros and cons of each option and agree no "halfway" compromise exists. Management must be all one option or all the other. Furthermore, the attending ward doctor does not "pull rank" to get his own way. He insists the different management options rest on their merits, not on the hierarchical status of their proponents.

In addition, the ward team consults the liver specialists to ensure an accurate understanding about the technical aspects of transplantation (e.g., Mr. T.'s initial position on the priority list, the time he would likely have to wait for a donated liver, the factors that would move him up on the priority list, the liver specialists' prior treatment experience in similar situations, their willingness to try a transplant with Mr. T., and Mr. T.'s predicted life expectancy of 1–2 years with a successful transplant).

Finally, the doctors apply the futility criteria of Schneiderman, Jecker, and Jonsen to the transplantation option [10]. The doctors agree that a successful liver transplantation would provide Mr. T. significant life benefits such as continuing to enjoy his children and grandchildren—not merely prolonging permanent unconsciousness, extreme suffering, or dependence on the ICU. Furthermore, the liver specialists believe the success rates of transplants for patients such as Mr. T. are low but higher than 1 %. This reasoning suggests that transplantation for Mr. T. meets neither the qualitative nor quantitative definitions of futility.

Time for the debate quickly runs out as Mr. T.'s condition deteriorates further. Forced into a decision, the doctors opt for one last chance to save Mr. T.'s life or at least to give his daughter time to visit him before death. They, therefore, transfer Mr. T. to the ICU for immediate transplantation preparations. The transfer maximizes his priority for a liver transplant. However, when the liver specialists evaluate Mr. T., they note even further deterioration in his liver function. They then declare him too sick for transplantation and remove him from the eligibility list. They order comfort care only, notify Mr. T.'s daughter, and liberalize visiting hours for her. Mr. T. dies the next day.

An Epilogue to Mr. T.'s Case

I presented Mr. T.'s case to two audiences of more than 70 doctors each. The case presentation stopped at the point when the ward doctors had to decide between ICU transfer and continued ward care. I asked the audiences to vote by secret ballot for one option or the other. Both audiences split exactly evenly between the two options.

Summary Points

1. Doctors must answer two fundamental questions in patient care: What is causing the patient's problem? and How should that problem be treated?
2. The diagnostic process, addressing the question What is causing the patient's problem? may use one or more of the following approaches: hypothesis generation based on diagnostic possibilities, Bayesian probabilities, pathophysiology, or pattern recognition. Treatment typically uses a "try-and-see" approach.
3. Medical futility has both quantitative and qualitative dimensions. The quantitative dimension defines futility as the treatment's lack of success in the last 100 applications; the qualitative dimension defines futility as the treatment's being unable to achieve any benefits the patient would appreciate.

To Learn More ...

1. Kassirer JP. Teaching clinical reasoning: case-based and coached. Acad Med. 2010;85(7):1118–24.
2. Quill TE. Recognizing and adjusting to barriers in doctor-patient communication. Ann Intern Med. 1989;111(1):51–7.
3. Schneiderman LJ, Jecker NS, Jonsen AR. Medical futility: its meaning and ethical implications. Ann Intern Med. 1990;112(12):949–54.

References

1. Kassirer JP. Teaching clinical reasoning: case-based and coached. Acad Med. 2010;85(7):1118–24.
2. Quill TE. Recognizing and adjusting to barriers in doctor-patient communication. Ann Intern Med. 1989;111(1):51–7.
3. Riegelman RK. Studying a study and testing a test: how to read the medical literature. Boston: Little, Brown and Company; 1981. p. 127.
4. Council on Ethical and Judicial Affairs, American Medical Association. Medical futility in end-of-life care. JAMA. 1999;281(10):937–41.
5. Kasman DL. When is medical treatment futile? A guide for students, residents, and physicians. J Gen Intern Med. 2004;19(10):1053–6.

6. Alpers A, Lo B. When is CPR futile? JAMA. 1995;273(2):156–8.
7. Truog R, Brett AS, Frader J. The problem with futility. N Engl J Med. 1992;326(23):1560–4.
8. Jecker NS, Schneiderman LJ. Futility and rationing. Am J Med. 1992;92(2):189–96.
9. Schneiderman LJ, Faber-Langendoen K, Jecker NS. Beyond futility to an ethic of care. Am J Med. 1994;96(2):110–4.
10. Schneiderman LJ, Jecker NS, Jonsen AR. Medical futility: its meaning and ethical implications. Ann Intern Med. 1990;112(12):949–54.
11. Helft PR, Siegler M, Lantos J. The rise and fall of the futility movement. N Engl J Med. 2000;343(4):293–6.
12. Friedman LS. Liver, biliary tract and pancreas disorders. In: McPhee SJ, Papadakis MA, Rabow MW, editors. 2012 current medical diagnosis and treatment. New York: McGraw Hill Medical; 2012. p. 663, 672.
13. Cassileth B, Zupkis RV, Sutton-Smith K, March V. Information and participation preferences among cancer patients. Ann Intern Med. 1980;92(6):832–6.
14. Blanchard CG, Labrecque MS, Ruckdeschel JC, Blanchard EB. Information and decision-making preferences of hospitalized adult cancer patients. Soc Sci Med. 1988;27 (11):1139–45.
15. Kelner M. Activists and delegators: elderly patients' preferences about control at the end of life. Soc Sci Med. 1995;41(4):537–45.
16. Deber RB, Kraetschmer N, Irvine J. What role do patients wish to play in treatment decision making? Arch Intern Med. 1996;156(13):1414–20.
17. Stewart DE, Abbey SE, Shnek ZM, Irvine J, Grace SL. Gender differences in health information needs and decisional preferences in patients recovering from an acute ischemic coronary event. Psychosom Med. 2004;66(1):42–8.
18. Perkins HS, Cortez JD, Hazuda HP. Differences by ethnic group and gender in patients' perceptions and preferences about control over treatment decisions in critical or terminal illness. Unpublished manuscript, March 2012.
19. Hofmann JC, Wenger NS, Davis RB, et al. Patient preferences for communication with physicians about end-of-life decisions. Ann Intern Med. 1997;127(1):1–12.
20. Happ MB, Swigart VA, Tate JA, Hoffman LA, Arnold RM. Patient involvement in health-related decisions during prolonged critical illness. Res Nurs Health. 2007;30(4):361–72.
21. Lynn J. Death foretold: prophecy and prognosis in medical care. BMJ. 2001;322(7283):43 (17 Feb 2001).
22. Blackhall LJ, Frank G, Murphy ST, et al. Ethnicity and attitudes towards life sustaining technology. Soc Sci Med. 1999;48(12):1779–89.
23. Crawley LM, Marshall PA, Lo B, Koenig BA. Strategies for culturally effective end-of-life care. Ann Intern Med. 2002;136(9):673–9.
24. Tulsky JA, Cassileth BR, Bennett CL. The effect of ethnicity on ICU use and DNR orders in hospitalized AIDS patients. J Clin Ethics. 1997;8(2):150–7.
25. Cykert S, Joines JD, Kissling G, Hansen CJ. Racial differences in patients' perceptions of debilitated health states. J Gen Intern Med. 1999;14(4):217–22.
26. Braun UK, Beyth RJ, Ford ME, McCullough LB. Voices of African American, Caucasian, and Hispanic surrogates on the burdens of end-of-life decision making. J Gen Intern Med. 2008;23(3):267–74.
27. Perkins HS, Cortez JD, Hazuda HP. Advance care planning: does patient gender make a difference? Am J Med Sci. 2004;327(1):25–32.

Chapter 4
Risks, Benefits, and Patients' Treatment Decisions at the End of Life

For centuries patients made no treatment decisions for themselves while a doctor-knows-best paternalism dominated medical care. Patients knew little of medical science and assumed doctors' treatment decisions served patients' interests well.

As I mentioned before, attitudes in the USA began to change in the mid-twentieth century. The rising level of the general public's education and the simultaneous assertion of personal rights prompted patients to demand more involvement than ever before in treatment decisions. Patients wanted both more information and more decision-making authority. Numerous studies have since documented this shift in attitudes. Many patients want some control, partial or full, over treatment decisions; only a few patients want to maintain sole doctor control [1–5]. The shift in attitudes, however, is not limited to the USA but has spread to most other Western countries, too. Western medicine has changed as a result: Patient autonomy has eclipsed the old medical paternalism and become the new dominant ethic for making treatment decisions.

The doctor–patient relationship has changed accordingly. Changes in patients' decision-making role have changed doctors' caregiving role. Doctors still have their superior technical knowledge but no longer make unilateral treatment decisions with it. Instead, they use that knowledge to inform patients about treatment options, benefits, and burdens. Doctors then consciously encourage patient choice. That choice often determines end-of-life treatment especially when standard medical care offers few benefits and considerable harms.

Under their new decision-making autonomy, patients have had to accept new responsibilities to inform themselves and to make decisions in a process called informed consent. However simple that informed consent idea seems, implementing it in the complicated world of critical or terminal care often proves difficult [6, 7]. Patients who ordinarily grasp technical information easily and make firm, clear decisions with it can become befuddled with their lives at stake. Patients must learn unfamiliar medical concepts quickly under stress; define and express core personal values; assess the physical, psychological, social, and financial implications of

H.S. Perkins, *A Guide to Psychosocial and Spiritual Care at the End of Life*,
DOI 10.1007/978-1-4939-6804-6_4

various treatment options; and make a choice—occasionally an irreversible, life-or-death one. Family and attending health professionals (HPs) may sometimes disagree with patients' choices, complicating the decision-making process. Perhaps for those reasons some patients relinquish decision-making authority back to doctors whenever the illnesses are chronic and serious (such as with diabetes or heart failure [8]), or life-threatening (such as with cancer or atherosclerotic heart disease) [2, 9, 10].

Still, today's patients command great leverage over treatment decisions in end-of-life situations, especially with life-support refusals [11, 12]. To engage patients effectively in the decision-making process, HPs must expand their perspectives of patients beyond the physical body. HPs must take account of patients' psychological, social, and spiritual circumstances. Those circumstances create the backdrop for patients as they weigh treatment options, make decisions, and live with the consequences. And those circumstances can exert considerable influence over the decision making process itself [13]. To understand patients' decisions, then, HPs must understand the values that emerge from patients' circumstances and that affect their treatment assessments.

This chapter addresses those values. It focuses its discussion on seriously or terminally ill patients by first describing how doctors might break bad news, say, of life-threatening situations, to them. This chapter then discusses specific values patients often consider in balancing treatment benefits and burdens to make their choices. Next, it critiques the death-with-dignity idea as it relates to such choices. Finally, this chapter discusses possible changes in patients' treatment choices over time, common benefit-burden trade-offs for certain ethnic groups and genders, and the key distinction between established treatments and experimental interventions. The case of a severely burned young man who adamantly refuses life-sustaining treatment illustrates many of the points.

The Case

Mr. C. is a smart, handsome young man. He is also a college graduate, rodeo athlete, avid outdoorsman, and former military pilot who has worked in his father's real estate business since military discharge. One day he and his father are inspecting some farmland for development. Unknown to them, a gas line on the property is leaking. When the men start their car nearby, they accidentally ignite a fiery explosion, and both suffer deep, widespread burns. Mr. C.'s face, for example, is instantly burned beyond recognition, and his hands reduced to stumps. He survives the ambulance ride to the hospital, but his father does not.

Mr. C. clings precariously to life as he arrives at the hospital. The staff begin aggressive life support and transfer him to the intensive care unit (ICU). To everyone's amazement Mr. C. survives, but he undergoes many operations over the next months including skin graftings, removal of one eye, and partial amputations of both hand stumps. He also endures daily, painful antibiotic immersion baths to prevent infections. The doctors believe Mr. C.'s life depends critically on continuing these treatments.

However, Mr. C. eventually becomes so despondent that he demands a stop to all treatments and a discharge home to die. Desperately wanting him to live, his mother pleads with Mr. C.'s doctors not to grant his wish. She is a deeply religious woman and hopes Mr. C. will yet embrace the Christianity he abandoned years before. Meanwhile Mr. C.'s doctors interpret his refusal of further life-saving treatments as a sign he is irrationally depressed. They conclude he can no longer make good decisions for himself and should be declared "legally incompetent." They want a guardian appointed to make decisions for him.

The doctors, therefore, consult Dr. W., a psychiatrist, for help with the guardianship. Dr. W. interviews Mr. C. as part of his clinical assessment, hears Mr. C.'s arguments for refusing further treatment, and finds them surprisingly rational and coherent. Mr. C. says emphatically in summary, "I do not want to live as a nearly blind and severely crippled man." Dr. W. concludes he himself would feel the same way in Mr. C.'s position. Dr. W. decides to assist Mr. C. in the only ways possible—by helping him make a terminal care plan and by engaging a lawyer to press it before the hospital's administrators and doctors. Mr. C., Dr. W., and the lawyer eventually propose the following plan: If a court upholds Mr. C.'s right to refuse treatment, he may stop all life support, remain in the hospital, and receive comfort care there until he dies.

The case never reaches court. To avoid formal adjudication, the hospital's administrators and doctors agree to the proposal. Nonetheless, having won recognition for his right to refuse life-sustaining treatment, Mr. C. unexpectedly changes his mind. He decides not to exercise his newly acknowledged right and chooses instead to continue all treatments as before. He finally returns home after months of hospitalization and makes a new life for himself. He earns a law degree, gets married, and launches a lecturing career based on his hospital experiences.

Mr. C. admits years later that his intervening life has been worthwhile and that he is glad to have received those life-saving treatments in the hospital. He still insists, however, he would make the same demand now as then to stop all of them if he landed in the same situation again. He asks rhetorically, "What gives a doctor the right to keep alive a patient who wants to die?" Dr. W. also observes correctly that had Mr. C. been so severely burned only a few years earlier, he "would have simply died." The powerful new medical technologies that became available in the interim and then saved his life have created a new controversy about control over them. Though starting with Mr. C.'s case, that controversy about control over life-sustaining technologies rages through many subsequent medical and legal cases. It eventually resolves in patients' favor [14].

Breaking Medical Bad News

Health professionals often equate "medical bad news" with some new or recurrent fatal disease. Mr. C.'s burns certainly qualify as bad news in that way: they could kill him. But I agree with one expert who expands the definition of *medical bad news* to include "any (medical) information that seriously and adversely affects one's view of the future" [15]. Under this expanded definition Mr. C. frequently faces more medical bad news even after he survives the immediate risk of death. His grim future prospects include prolonged hospitalization, daily painful burn treatments, severe disabilities caused by near blindness and widespread body scarring, and radically altered self-image and life plans. Such suffering and disabilities surely qualify as medical bad news and deserve as careful and sensitive a handling as does any fatal disease.

Disclosing the bad news of fatal or disabling diseases is an important obligation of medicine, and the public expects doctors to fulfill it [16]. In fact, as they have disclosed to patients ever more in recent years, doctors have discovered that medical bad news, though sad, rarely harms patients permanently. Rather, the open communication involved often provides unexpected benefits: It can correct misunderstandings, promote trust, encourage realistic treatment plans, and help calibrate expectations for the future. Many HPs now suspect that patients already know their medical bad news before their doctors divulge it. Thus, the therapeutic privilege—the old practice of withholding information from a patient solely on the suspicion it would harm him or her—no longer holds much credence. Today's HPs firmly believe patients deserve to hear bad news unless they specifically request not to.

Patients themselves almost never initiate discussions about medical bad news. They expect their doctors to do so [17]. So how might a doctor deliver medical bad news clearly and sensitively? Table 4.1 lists six steps in the process according to the mnemonic SPINES. The table also lists specific tasks within each step [15]. First, the doctor *sets* an overall approach. The senior attending doctor primarily responsible for the patient's care should ordinarily take the lead in planning, conducting, and following up the bad news discussion. Before starting it, this doctor-moderator should review the patient's past medical course and current condition and discuss them both as necessary with the patient's other attending doctors to determine points of agreement and disagreement [18]. This doctor-moderator should also mentally rehearse delivering the bad news, imagine how the patient might react emotionally, and prepare responses for any distraught reactions.

Second, the doctor *plans* the logistics. The doctor should ask the patient beforehand who should attend important discussions about the patient's diagnosis and treatment. The doctor should also locate a quiet, private, and accessible meeting place and agree on a time that allows for lengthy, uninterrupted discussion. Naturally, the doctor should dress professionally for the meeting [19] and bring a supply of tissues to offer tearful participants.

Third, the doctor starts the bad news discussion by *introducing* himself or herself and briefly explaining (as necessary) his or her role in the patient's care. The doctor

Table 4.1 Steps in breaking medical bad news: the SPINES model[a]

Step	Specific tasks
1. *Set* an overall approach.	a. Determine who should deliver the bad news. (It should usually be the senior principal doctor attending the patient.)
	b. Review beforehand the patient's past medical course and present condition.
	c. Determine points of agreement and disagreement among the attending doctors. Avoid surprise disagreements in front of the patient or family!
	d. Rehearse mentally how to deliver the news.
	e. Imagine how the patient and others may react emotionally. Rehearse how to respond.
2. *Plan* for the discussion.	a. Ask the patient ahead of time which family members and friends should be present.
	b. Identify a quiet meeting room.
	c. Set a time that allows for lengthy, uninterrupted discussion. If time constraints exist, ensure the room has a clock placed where you can watch it unobtrusively.
	d. Dress professionally.
	e. Bring handkerchiefs or tissues for teary participants.
3. *Introduce* yourself (if necessary) and *inquire* about the patient's understanding of the illness.	a. Introduce yourself and explain your role in the patient's care.
	b. If you have time constraints, mention them at the outset.
	c. Use body language to create an atmosphere of trust (e.g., sit near the patient, make eye contact, or hold the patient's hand as appropriate).
	c. Ask what the patient already understands about the situation. Correct misunderstandings respectfully.
	d. Ask what the patient foresees for his or her condition, its treatment, and its possible outcomes.
	e. Ask what specific wishes the patient has for other aspects of his or her future.
4. Deliver the bad *news*.	a. Tailor the approach to "where the patient is." Try to use the patient's own words, concepts, and wishes to let the patient know you have *really* heard him or her.
	b. Warn the patient that bad news is coming.
	c. Avoid jargon, euphemisms, acronyms, or overly technical concepts.
	d. Give the news in small doses, checking periodically for participants' understanding.
	e. Respond to questions about remaining life span with time ranges, not particular times. Couch those ranges as best-guess estimates.

(continued)

Table 4.1 (continued)

Step	Specific tasks
5. *Empathize.*	a. Observe the patient and others for emotions.
	b. Note emotions as they arise and mentally categorize them for later recall.
	c. Encourage participants to express their emotions.
	d. Acknowledge the emotions explicitly. Say something like "I know this is not the news you wanted to hear. I wish it were different."
	e. Validate the emotions with empathic statements such as "I imagine, if I were in your shoes, this news would make me discouraged, afraid, or angry. Are you feeling any of those emotions?"
	f. Consider expressing your own feelings of sadness.
6. *Summarize* the discussion and *set* a follow-up action plan.	a. Summarize the patient's circumstances briefly.
	b. Avoid saying, "There is nothing more I can do."
	c. Propose therapeutic goals for future care.
	d. Pledge ongoing involvement (including visits to the patient).
	e. Give realistic hope (even if it must be directed at comfort, not cure).
	f. Give contact telephone numbers and invite the patient and family to call if necessary.
	g. Promise to review matters soon (always within a day or so).

[a]Modified from the SPIKES model. Baile et al [15], with permission

should also elicit the names and relationships to the patient of those in attendance. If time constraints exist, the doctor should mention those constraints at the outset [19]. (I recommend the time-constrained doctor face a clock on the wall or on a nearby table to be able to monitor the time unobtrusively.) If time runs out, the doctor should negotiate with participants another time to address any remaining concerns. The doctor should, of course, use good body language throughout the discussion to convey a caring attitude. The doctor might sit down next to the patient and make direct eye contact with him or her. The nonverbal part of the discussion is just as important as the verbal part.

Then the doctor *inquires* about what the patient already knows about the illness. Specifically, the doctor might ask, "What do you know about your condition?" or "What is it like for you to be ill?" A recent study catalogues the responses of 209 cancer, congestive heart failure (CHF), or chronic obstructive pulmonary disease (COPD) patients to a similar question, What do you understand about your illness? [20] Patients usually address some of five topics—naming the disease or describing its pathology, recounting an illness history, describing symptoms or functional limitations due to the illness, ascribing a cause to the disease, and predicting a

Table 4.2 Prevalences of responses to the question, What is your understanding of your illness?

Themes	Cancer patients[a] (%)	Congestive heart failure patients[b] (%)	Chronic obstructive pulmonary disease patients[b] (%)
Naming a diagnosis	77	39	41
Giving an illness history	55	43	33
Describing symptoms or functional limitations due to the illness	4	30	26
Ascribing a cause to the disease	1	7	23
Predicting a prognosis	27	5	7

[a] N = 69; [b] N = 70; Condensed from Morris et al. [19], with permission

prognosis. The content of patients' responses, however, varies by disease (Table 4.2). Many patients with cancer name their diagnoses and give illness histories, often relating detailed treatment chronologies. Few, however, describe their symptoms or functional limitations, and almost none explicitly attribute their cancers to specific causes such as smoking (even when these patients have smoking-related cancers). Some cancer patients mention their prognosis in general terms such as "I don't have a long future." Unlike the patients with cancer, only a few patients with CHF or COPD name their diagnoses or give illness histories. Many of these patients describe instead their symptoms or functional limitations, and some COPD patients attribute their diseases specifically to smoking. Almost no CHF or COPD patients mention prognosis in any way. The content of responses also varies by patients' education: Patients with more overall education tend to mention diagnosis, pathology, and prognosis more often than do patients with less education. To form as complete a picture as possible of the patient's understanding, the doctor may need to prompt the patient about those topics the patient does not spontaneously mention.

The doctor, of course, must always listen attentively [21]. The doctor must be fully "present," blocking out all mental distractions. Furthermore, the doctor should periodically try to summarize what the patient has said. Coulehan suggests using patient-centered comments such as "Let's see if I have this right …," "It sounds as though you are telling me …," or "This is what I think I heard from you. Did I miss anything you especially wanted me to understand?" [22]. The doctor should use the patient's own words and concepts whenever possible. Accurate summaries show the doctor's close attention to what the patient has said and increase the patient's trust in the doctor. Summaries also help check for important knowledge gaps that still need to be addressed.

Fourth, the doctor delivers the *news*, preparing the patient by first warning that bad news is coming. The doctor should use ordinary, nontechnical language throughout. Specifically, he or she should avoid jargon, euphemisms [18], acronyms, and excessively technical concepts [21]. If technical concepts are necessary, the doctor should define them. But the doctor should realize that medically untrained people may not grasp those concepts with the first explanation. The doctor may need to give several explanations, using slightly different words each time. The doctor should also check periodically for understanding by simply asking "What questions do you have?" or "May I go over any points again?" But people may hesitate to ask questions at first. They may have misguided fears about bothering the "busy" doctor or looking ignorant, or may be overwhelmed at the bad news and simply forget questions they ordinarily would ask. The doctor, therefore, must invite questions several times during the discussion as the bad news "sinks in" [16]. The doctor should also anticipate questions that might occur to the patient or family later and address those questions up-front.

Fifth, the doctor *empathizes* with the patient, the family, and the other HPs caring for the patient. Many doctors find this empathizing to be the hardest part of breaking bad news: confronting heart-wrenching reactions is exhausting. Still, empathizing forms the compassionate bedrock of the best bad news discussions. Because bad news delivers a serious, life-changing blow, those affected often need immediate emotional support to be able to cope. The doctor leading the discussion should begin to provide it. Participants' spontaneous reactions to the news can guide how to do so. Two experts recommend categorizing reactions into one of the three categories as an aid to later recall: physiologic (including the adrenaline-fueled fight-or-flight response), cognitive (including denial, blame, excessive intellectualization, and disbelief), and emotional (including anger, fear, anxiety, sadness, despondency, guilt, and helplessness) [23]. At an appropriate time before the discussion ends, the lead doctor should acknowledge any potentially distressing reactions and offer to address them. The doctor must judge on the spot whether to address those reactions right then with the whole group or later with specific participants in private.

Several emotionally difficult reactions of discussion participants may present challenges for the lead doctor. He or she should prepare beforehand to manage them. Most participants, of course, feel sadness at learning medical bad news. But, if anyone shows abject despondency, the doctor might say something like "I know this news makes you very, very sad. It makes me sad, too. I wish I could change matters, but I can't. I will, however, stand by you through them." In addition, participants may feel frank anger and turn it back on the doctor, thereby "shooting the messenger" who has delivered the bad news. They may accuse the doctor of misdiagnosing the patient, misleading them about the patient's progress, or mishandling the patient's treatment. The doctor must never reflect that anger back onto the accusers or others. Reflecting it only fuels it more. Instead, the doctor should analyze it, acknowledge it openly, and respond to it calmly and constructively. Participants may also suddenly experience fears for the future. One expert urges responding to such fears explicitly with supportive but realistic reassurances [16]. I believe the lead doctor can always provide realistic reassurances—even in the grimmest of situations—if only to promise his or her continued involvement. The doctor can promise, for example, "to see this situation through with you—wherever it takes us." Furthermore, the lead doctor (and other clinical HPs in attendance)

Table 4.3 How health professionals might respond to emotionally stressful reactions from patients or family members[a]

Reactions	Possible responses[b]
Feeling despondent at receiving very bad news	"I wish the news were better. It's disappointing for us all."
Expressing unrealistic hope	"Sadly, medicine has limits on what it can achieve. Your particular hope is probably not realistic. So let's focus on other goals we *can* realistically achieve to make matters better."
Expressing feelings of loss, grief, or hopelessness	"I wish things had turned out differently. I imagine this has been a terrible blow for you."
Expressing severe disappointment in the physician or in medicine	"I know you are disappointed in the way things have turned out. I am, too. One of the hardest things for me is reaching the limits of what medicine can do. But despite those very real limits we can still do things to help."
Expressing demands for excessively aggressive treatment aimed at cure	"I can imagine how disappointed you are in the way things are turning out. Because aggressive treatment at this point does not offer ['you' or the patient's name] any curative benefit, we should channel our efforts into offering what we can—conscientious comfort care."
Reacting to severe medical complications	"This setback is hard for you. It's hard for us all. But we now need to do the best we can with the hand we've been dealt."

[a]Adapted from Quill et al. [24], with permission
[b]I have modified the wording of some of the authors' original comments to conform to my own communication style.

should liberally enlist the help of clergy, psychiatrists, or social workers to help follow up the most severe emotional reactions.

Some experts warn health professionals generally against "I'm sorry" responses. These experts believe people misinterpret such responses as pitying, as admitting a clinical error (even if there is none), or as prematurely cutting off discussion of a problem [23]. Other responses prove better. Table 4.3 presents some for when patients or others react in emotionally stressful ways [24]. Health professionals should, of course, not repeat these responses rotely but modify them to suit circumstances and personal communication styles. Above all, health professionals should remember that patients and survivors appreciate spontaneous, genuine expressions of compassion more than stock statements.

Especially while inquiring about what the patient knows, delivering the bad news, and empathizing (Steps 3, 4, and 5), the doctor must keep his or her own insecurities, biases, and fears from interfering with giving the patient and the family emotional support. Two common threats to that support are time urgency and action orientation. Medical training inculcates both of those values, and they serve doctors well in many other patient care situations. But, while breaking medical bad news, doctors must resist them due to the resulting temptation to control the discussion or to stop it prematurely. I agree with one expert who emphasizes letting the bad news discussion follow its natural course. "Do not fear tears or silences," this expert suggests. "Allow (the) tears … And don't … leap to fill awkward silences. Give both the time they need. They eventually end" [25]. Of course, just being present, waiting patiently, and remaining attentive through those raw moments convey caring [21]. And, if compassion prompts such gestures, holding the patient's hand or laying one's hand gently on the patient's shoulder can have a powerfully soothing effect without using words.

Sixth, the doctor *summarizes* the discussion, *sets* an action plan for future care, and documents both in the patient's medical record. The doctor should never say, "There is nothing more I can do." Even when cure is impossible, the patient always hopes for comfort, and the doctor can take steps to accomplish it. The doctor can also plan for therapies to maintain the patient's optimal function. Furthermore, the doctor can implement activities to bolster patient and family spirits. He or she can make follow-up calls or visits. The doctor can also provide contact telephone numbers and invite the patient and family to call whenever necessary [21]. If the patient is a "take-charge" person, the doctor can even assign the patient specific tasks such as informing family members and close friends not present at the bad news discussion; scheduling and attending specialist consultations; screening prospective palliative or hospice care agencies; and making a will or funeral preparations. Such an action plan dispels the patient's sense of helplessness.

Some patients and families may retain little information from the first "bad news" discussion. The doctor should, therefore, schedule follow-up discussions for later [16, 21]. Those discussions can review the patient's interim illness history, answer questions that have cropped up, address new problems, and recast hope as circumstances change. One reassuring lesson I have learned: The doctor need not worry about conducting the "perfect" bad news discussion. Patients and family readily forgive missteps if the doctor's aftercare is compassionate and dependable.

Of course, creating an end-of-life care plan requires different treatment decisions than those in ordinary curative care, and patients' treatment assessments play a central role in them. The next sections describe the values patients often use to balance the benefits and burdens of treatments near the end of life.

Patients' Treatment Assessments

Important Values to Consider

Expected outcomes drive treatment decisions. Patients seek a doctor's treatment, hoping to improve their illness outcomes. Historically, doctors responded out of the Hippocratic oath's paternalistic directive to treat "for the benefit of the sick" [26]. Doctors chose from among the possible treatments and outcomes those the doctors felt would benefit patients most. The new patient autonomy, however, has upended that old paternalism. Now, patients assess treatments and outcomes for themselves [27, 28]. The resulting individualized treatment plans require doctors and other HPs to keep a strong focus on individual patients' values.

I believe everyone has a unique *collection* of values related to health and health care, but the *individual* values are not all unique. People share some of them, especially ones related to end-of-life care. Recent research has identified a few of those shared end-of-life values. Knowing them can help health professionals anticipate how dying patients might assess treatment options and outcomes.

Table 4.4 presents the end-of-life care values that have earned consistent endorsement by doctors or patients in recent studies. I have classified each value in the table under one of the six care domains this book emphasizes. Five of those domains—the physical, the psychological, the social, the spiritual, and the miscellaneous others—address care of the patient; the other addresses care of the family. As I describe the values in each domain, I illustrate them with patient or HP quotes from the medical literature.

The reader should note that normal curative acute care focuses on the physical and psychological domains of the patient. Between the two domains the physical predominates. It prompts more patient visits, commands more attention, and has a greater impact on treatment than does the psychological. Curative acute care typically concerns itself with saving lives, relieving physical symptoms, and improving bodily function. Mr. C.'s treatment on arrival at the hospital follows this pattern. To save his life, the staff institutes immediate, aggressive burn care including intensive care unit admission, contact isolation, broad-spectrum intravenous and topical antibiotics, and extensive wound debridement. But shortly afterward the staff must contend with Mr. C.'s psychologically driven protests against the painful antibiotic baths.

Yet, whenever cure will not occur and the patient is dying, the focus of care shifts. End-of-life care takes some focus off the patient's physical and psychological domains and trains it on his or her social and spiritual domains and on the care-of-the-family domain. Mr. C.'s demand to stop all life-saving treatments signals the possibility he will need end-of-life care soon. Such care would affect his social domain (relations with his mother), perhaps his spiritual domain (his lapsed Christianity), and the care-of-the-family domain (the physical and emotional toll on his mother if he were to be discharged home to die). Common end-of-life values come to the fore.

Mr. C.'s caregivers might consult the list of values in Table 4.4 under those circumstances. But I would caution the caregivers about four points. First, the list is only preliminary. Some important end-of-life care values may still not be identified. And others, though identified, may still lack the necessary widespread, documented endorsement for inclusion in the table. Second, the reader should understand that a dash in the table does not mean the particular consensus statement or study considers the value unimportant. Instead, the dash means only that the statement or study *does not address* the value. The cross-cultural studies by my colleagues and me, for example, do not explicitly address many of the values in the table, but we still consider them all important in end-of-life care. Third, I used a somewhat

Table 4.4 Consensus values in domains of terminal care

Values	Endorsed by doctors?				Endorsed by patients?			
	JAGS [35] (doctors)[a]	IOM [36][a]	Emanuels [37]	Steinhauser [38] (doctors)	JAGS [35] (patients)	Singer [32]	Steinhauser [38] (patients)	Perkins [31]
Physical domain								
Relieve symptoms.	Yes	Yes	Yes	Yes	Yes	Yes	Yes	–[b]
Promote self-care and other physical abilities.	Yes	Yes	–	–	Yes	–	–	–
Psychological domain								
Relieve symptoms.	Yes	Yes	Yes	Yes	Yes	–	Yes	–
Promote emotional health.	Yes	Yes	Yes	–	Yes	–	–	–
Trust professionals' clinical and interpersonal skills.	Yes	–	–	Yes	Yes	–	Yes	Yes
Facilitate patients' thinking and communication.	Yes	–	Yes	Yes	Yes	–	Yes	–
Achieve some patient control in dying.	Yes	–	–	–	Yes	Yes	–	Yes
Social domain								
Nurture significant relationships.	Yes	–	Yes	Yes	Yes	Yes	Yes	–
Spiritual domain								
Promote spiritual health.	Yes	Yes	Yes	Yes	Yes	–	Yes	Yes
Miscellaneous other values relating to the patient								
Fulfill financial, legal, and ethical duties.	Yes	–	–	Yes	Yes	–	Yes	–
Avoid inappropriate prolongation of dying.	Yes	–	–	–	Yes	Yes	–	Yes

(continued)

Table 4.4 (continued)

Values	Endorsed by doctors?				Endorsed by patients?			
	JAGS [35] (doctors)[a]	IOM [36][a]	Emanuels [37]	Steinhauser [38] (doctors)	JAGS [35] (patients)	Singer [32]	Steinhauser [38] (patients)	Perkins [31]
Family care domain								
Prevent physical and financial burdens on others.	Yes	–	Yes	Yes	Yes	Yes	Yes	–
Prepare family for the death.	–	Yes	–	Yes	–	–	Yes	–
Comfort grieving family.	Yes	Yes	–	–	Yes	–	–	–

[a]JAGS = Journal of the American Geriatric Society. IOM = Institute of Medicine.
[b] A dash means that little or no data exist for the item.

arbitrary cutoff to label a value "endorsed." The table summarizes four studies with data from doctors and four with data from patients. I considered a value "endorsed" by a group whenever two or more studies for that group designated the value as important in end-of-life care. And, fourth, although each of these values is endorsed by at least one study of doctors and one of patients, a few values (such as "promoting the patient's emotional well-being" and "preparing the family for the patient's death") are endorsed by more doctor studies than patient studies. Despite my usual patient-centered approach, I include those values in the table because I believe doctors must have seen firsthand the importance of the values in the care of dying patients. I also believe patients would not object to including those values in a terminal care plan and might even endorse them if asked.

The Physical Domain

All four doctor studies and three of the four patient studies cite relief of physical symptoms as important in end-of-life care. Two of the doctor studies also cite promoting physical abilities, especially self-care, as important. Both of those values have played central roles in revolutionizing end-of-life care and now receive as much emphasis in end-of-life care as in acute curative care. In fact, the Preamble to the National Hospice and Palliative Care Organization's Standards of Practice alludes to them both. "When cure is no longer possible," the Preamble states, "hospice recognizes that a peaceful and comfortable death is an essential goal of health care … and that intensive palliative care focuses on … enhanced quality of life for the terminally ill." That quality of life certainly depends on optimizing bodily function as much as on relieving physical symptoms.

The Psychological Domain

All four doctor studies and two of the four patient studies cite relief of psychological symptoms as important. Yet three of the four doctor studies also cite *promotion* of psychological health as important. Thus, in both the physical and psychological domains, doctors and other HPs may actually orient more toward health promotion at the end of life than do patients. (Patients may just not think about health promotion at such times.) Accordingly, one HP who is not a doctor described his or her efforts to keep dying patients from despondency. "I want to do everything I can to encourage (their) living long and healthy (sic)," this HP said. "(I try not) to have an air of pessimism (but to) give optimism" [29]. Furthermore, two doctor studies and three patient studies cite as important the trustworthiness of health professionals' clinical and interpersonal skills. Dying patients and their families need to believe such care will be technically skilled, dependable, and compassionate. They also need to believe the dying patient will be treated respectfully and never abandoned to face death alone. Furthermore, three doctor studies and two patient studies cite as important in end-of-life care facilitating patients' thinking and communication. Addressing HPs' good interpersonal skills and communication with patients, one family member recalled the care for her dying father. "The (medical) residents always approached my father as if he was a person," she said. "There weren't any divisions between them .... They would sit on his bed, hold his hand, talk about their families, his family, ..." [30].

Doctors and patients, however, may differ over the value of patient control in dying. Only one doctor study but three patient studies cite such control as important. One of the three patient studies asked the question, "Are the doctors more likely to do what you want if you sign an advance directive?" Many patients in that study believed their doctors would follow those wishes. One patient said, "Of course, your doctor is going to do what you want ... because that's (his) duty ... He does what his patient wants" [31]. And another said, Doctors "honor a man's agreement and his will" [31]. Similarly, a patient in a different study said, "That's my life .... It's very, very important to me ... that I make choices for myself" [32].

The Social Domain

This domain acknowledges contacts with family or friends as helping to sustain patients emotionally through the difficult dying process. Accordingly, three of the four studies for doctors and patients alike recognize nurturing significant relationships as important in end-of-life care. A patient dying with the human immunodeficiency virus (HIV) said in one study, "I (had) never told anyone in my family that I was HIV-positive. (But) to complete my living will, I had to tell (my brother) ... (It) was quite a challenge for me. And I did tell him, and everything has just worked out fine. He's 100 % supportive, ... we were close before ... but now we're really close" [32]. And a hospice volunteer remembered a similar relationship-nurturing experience. "(This hospice patient) liked ... (car) rides," the volunteer said. "He couldn't walk around very well, but he invited another patient to come out. She was very debilitated, too. So the three of us would drive around the community. As debilitated as some patients get, they're still capable of helping someone else ..." [30].

The Spiritual Domain

All four doctor studies and three of the four patient studies explicitly cite promoting spiritual health as important in end-of-life care. Spirituality for most people, of course, involves religion and God. A social worker said in one study, "When terminally ill patients discuss death ... religion and spirituality almost always come up. Patients ... want to share what their experience (with God) has been on earth ... (But) sometimes you see fear of the unknown. (Their) comfort in discussing it depends on how strong their faith (is) ..." [29]. Also in a religious vein a patient in another study explained, "We're here on borrowed time. When God tells you (that) you gotta leave ... that's when you die. He is going to take your soul" [33]. And still another patient said, God "prepares a place for you. He takes good people to live with the angels" [34]. A few people, though, described their spirituality without mentioning God or religion. A family member remembered a recently dead relative whose spirituality focused on nature alone. "(When) he got home," this family member said, "(the attendants) got him out of the ambulance. I remember his saying, 'Can I wait (here) just a minute to remember the sunshine?'" [30]. And another patient described heaven as being without God. It is, he said, "a classroom where we can learn something (new)—a different dimension, a new type of being" [34].

Miscellaneous Other Values Relating to the Patient

This domain contains two values. One is patients' fulfilling particular financial, legal, and ethical duties at the end of life. These duties might include filling out an advance directive, drawing up a testamentary will to distribute property posthumously, arranging other kinds of legacies, and planning for a funeral. Two of the four studies for doctors and patients alike cite the importance of this value. "I've already made a will because I'm a single parent," said one study patient. "I didn't want my children to go through making decisions that would be difficult … (I want) to take the burden off them" [29]. And another study patient said, "I had the opportunity to sit down with my parents and talk about their advance directives, talk about their funeral arrangements, and spend the whole afternoon thinking about death and what would happen in this situation and that situation" [29]. The other value in this domain is avoiding inappropriate prolongation of dying. Just as no one wishes to die prematurely, no one wishes to languish long before death. Many people express such wishes as dying at the right time. (Chapter 11 addresses that topic in detail.) As one study patient said, "I wouldn't want life supports if I'm going to die anyway. There's no dignity in it. It's just a guinea pig thing" [32]. And yet another said, "I don't want to be kept alive artificially forever just to die later on and (to) suffer (in the meantime) …" [32].

The Family Care Domain

This domain acknowledges the family as an important focus for end-of-life care. The domain comprises at least three values. One is preventing physical and financial burdens on others. Three out of the four studies each for doctors and patients explicitly endorse this value. As a patient said in one study, "I'd (want to) die here (in the hospital), not at home. I wouldn't put the (care) burden on my family" [32]. And another patient explained, "I hope to stop myself from becoming a burden to (my children). Looking after somebody either takes a lot of money, … or you have to do it yourself. I do not wish for my children to be in the position of having to do that (for me). I would rather die faster than later" [32].

A second value here is preparing the family for the death. Two of the four doctor studies endorse this value. One study quotes a nurse who described past experiences with families who took patients home to die. "When the last breath came," this nurse said, "the families panicked. They brought the patient into the emergency room and went though (CPR). Preparing the family, assessing what they actually know, and figuring out what you have to teach them is essential" [30]. Also, one woman recalled HPs' willingness to answer her questions during her husband's terminal illness. "Whatever you ask," this woman said, "they tell you whether you want to hear it or not. That's good. I asked, 'Why is his urine brown?' (The hospice volunteer) told me, 'His kidneys are shutting down.' After(wards) my husband would ask (me what that person had said) … and I would tell him…. We'd talk about it" [29]. The treatment plan for another man, the same study reports, apparently called for stopping nutrition and hydration. His "children would say, 'Daddy is starving to death.' (His wife, the stepmother,) had never experienced a death before. But because the hospice (personnel had) prepared her, she was able to say to the (children), 'No, he's not (starving). His body's shutting down, and we have to realize that'" [29].

And a third value in this domain involves comforting and supporting the family in their grief. Two of the four doctor studies endorse this value. Having discussed beforehand the family's beliefs about what happens after death can suggest what health professionals might say and do at a patient's death to act on this value. One woman insisted that when a friend died, she stayed in her body "15 or 20 min, long enough to be warm (for her husband) to hold (her) for a while and to say good-bye" [34]. A HP witnessing such a death might encourage this man and other survivors to hold the newly dead woman if they wish and to tell her their good-byes at the same time. Other families may believe death releases patients from their suffering so they can finally rest. Recalling his dead sister's long, difficult fight against leukemia, one man remembered the family's heart-wrenching decision to stop her ventilator. "It was sad," he explained. "But she (was then) no longer suffering. Her soul went … to heaven, and she's resting (now)." Of course, many grievers hope to reunite someday with dead relatives and friends in heaven [33, 34]. Even when not sure about grieving survivors' beliefs, health professionals might offer comfort by expressing those widespread hopes. HPs might say something like "Many of us see death as a release from suffering. We hope that to be so with your (dead) loved one. We can also hope that you will see (him or her) again someday in the hereafter."

The Case

Table 4.4 serves as a checklist for evaluating Mr. C.'s care *as the end-of-life care he wants* [35–39]. The intensive burn treatments certainly keep him alive, but they often miss the mark as compassionate end-of-life care. The frequent skin-grafting surgeries, meticulous wound debridements, and antibiotic baths cause him great physical pain. He lives in an infection-control isolation room, cut off from most people. His social interactions, therefore, are brief and infrequent. And he can do little for himself including feeding, bathing, and toileting himself. Miserable, he soon begs to stop all treatments and to be allowed to die. Perhaps he is really pleading for the multifaceted physical, psychological, social, spiritual, and familial care that comes with state-of-the-art end-of-life care.

When Mr. C. expresses his wish to stop treatments, his doctors understandably suspect a depression. They automatically and not necessarily correctly assume Mr. C. is irrational for wanting to die. And they do not ask his reasoning behind the wish and, thus, do not perceive the importance to him of regaining some control over his life.

They do, however, consult Dr. W., the psychiatrist. He provides the care Mr. C. needs most—someone to listen. Dr. W. uses what he hears to broaden treatment's focus beyond the narrow physical domain. Specifically, Dr. W. listens to Mr. C.'s complaints and concludes they are reasonable. Dr. W. then helps Mr. C. reformulate those complaints into a plan for his terminal care.

Although I have little direct information about the planning process, I imagine here how Dr. W. would approach Mr. C.'s case. Dr. W. would quickly realize that promoting Mr. C.'s treatment refusal requires simultaneous attention to all five domains of end-of-life care. Dr. W. would also see that Mr. C.'s despondency blinds him to his mother's own despondency. Dr. W., therefore, would remind Mr. C. that his mother (like any other mother) badly wants her child to live. Fulfilling Mr. C.'s wish to stop all treatments and to return home to die will exact too terrible a toll on her. His wish will grieve her, she will exhaust herself providing the physical end-of-life care he needs, and she may be left with huge medical bills afterward. Furthermore, his mother may resent his decision to stop treatment and to die because she still hopes he will embrace his Christian faith again. Dr. W. would explain that, if Mr. C. stops life-sustaining treatments, he will bear some responsibility for supporting his mother through the dying process.

Dr. W. would also advise Mr. C. to make legal, financial, and ethical preparations for dying. Mr. C. can state his treatment refusals in an instructional advance directive and appoint a trusted proxy to carry them out if he ever becomes unable to make refusal decisions for himself. (Dr. W. may not serve as Mr. C.'s proxy due to legal prohibitions, and Mr. C.'s mother should not due to her conflict of interests over Mr. C.'s wishes.) Mr. C. can also draft a testamentary will, plan his funeral, and arrange for body disposition after death. Mr. C. should complete these preparations before he stops the life-sustaining treatments.

Whatever his actual approach, Dr. W. then empowers Mr. C. by engaging a lawyer to press his case. Mr. C. can finally help himself. He, Dr. W., and the lawyer draft a terminal care plan, which Mr. C. believes prevents the inappropriate prolongation of his death. The plan calls for Mr. C. to stop all life support and then to continue comfort care in the hospital until he dies. Mr. C.'s mother will not have to bear the burden of providing Mr. C.'s terminal care at home. Dr. W.'s unconventional care strategy transforms Mr. C.'s helplessness into a valid hope for taking back some control over his life. Dr. W. thereby earns Mr. C.'s trust. Mr. C. eventually gains solid recognition for his right to refuse life-sustaining care.

Table 4.5 Contrasting methods for assessing treatment outcomes

Characteristic	Proportionality	Trade-offs
Focus of benefits-to-burdens assessments	Individual treatments	Whole outcome states after one or more treatments
Underlying conceptual framework	Biomedical	Biopsychosocial
Main benefits considered	Narrow: typically only survival, physical symptoms, and bodily function	Expansive: psychosocial values as well as survival, physical symptoms, and bodily function
Form of assessment	Qualitative	Quantitative

Methods for Balancing Competing Values

Patients' treatment choices depend on medical circumstances, underlying personal values (as in Table 4.4 for terminal patients), and the balance of treatments' benefits and burdens. I discuss here two methods for assessing that balance. One method compares benefits and burdens among individual treatments and the other, among overall health-state outcomes [40–43]. The two methods differ in the focus of their benefits-to-burdens assessments; their underlying conceptual framework; the main benefits considered; and the form, qualitative or quantitative, of the assessments (Table 4.5).

Assessing Individual Treatments

This method suits best choosing one treatment among several options when the final treatment choice will dominate much of subsequent care. Examples include choosing among surgery, irradiation, and chemotherapy to treat certain cancers, or between medical clot dissolution and surgical clot removal to treat new strokes. The typical informed consent process uses this method. The benefits-to-burdens assessments focus on each treatment option individually. The underlying conceptual framework is usually biomedical, and the main benefits considered are physical—survival, symptoms, and bodily function. The form of assessment is rough, intuitive, and qualitative: comparing each treatment option against all others for the desirability and likelihood of benefits and burdens. The right treatment choice offers the greatest likely balance, or "proportion," of benefits over burdens for the patient. This comparative balancing is called "proportionality."

How do patients make such benefits-to-burdens assessments for individual treatments? Cardiopulmonary resuscitation (CPR) provides a good illustration. It is probably the most frequently assessed individual treatment due to its brief, self-contained, intense, and potentially life-saving nature. CPR burst dramatically into clinical medicine in the early 1960s when it revived patients who suffered sudden death during surgery. Because those patients would have certainly died before, mainstream medicine quickly adopted CPR and made it the standard of care for all cardiorespiratory arrests.

Yet the generalized use of CPR has not fulfilled initial hopes. The realized benefits have actually been quite disappointing: Only 1–22 % of people who have CPR in or out of the hospital survive to hospital discharge. And many patients discharged alive suffer severe, permanent neurologic damage. Unwitnessed cardiorespiratory arrests, patient age over 70, and underlying diagnoses of cancer or widespread infection ("sepsis") consistently predict unlikely survival and poor function after CPR [44]. Only 10.5 % of elderly resuscitated cancer patients, for example, survive to hospital discharge, and even less, only 2.3 %, survive to hospital discharge when they start out confined to bed for more than half the time [45]. Furthermore, patients who do survive to hospital discharge live a median of only an additional 5 months and endure poor physical function in the meantime.

At the same time CPR involves great burdens. They often include intensive care unit admission; mechanical ventilation; and invasive, uncomfortable procedures such as central vein catheter or tracheal breathing tube placements. Complications may include rib fractures (from the chest compressions) and damage to the brain and other major organs (from prolonged oxygen deprivation).

Unfortunately, patients may choose to have CPR without realizing its low survival rates, poor functional results, and potential complications. Television and movie depictions may be responsible, giving people an overly favorable impression of CPR outcomes [16]. Yet, when informed realistically about those outcomes, many patients who would ordinarily choose CPR refuse it [46].

The Case

At admission the doctors make Mr. C. eligible for CPR as part of their all-out effort to save his life. Mr. C. himself may actually agree at first. Only with full-CPR status will the intensive care burn unit admit him, and will he be allowed to undergo his many later surgeries. Yet Mr. C. soon changes his mind as he decides to stop all life-sustaining treatments. I imagine Dr. W. and the other doctors thoroughly explain to Mr. C. the potential benefits and burdens of CPR as he is making his decision. Eventually his CPR refusal becomes an important part of the end-of-life proposal Mr. C., Dr. W., and the lawyer present to the hospital and its doctors.

Health-State Outcomes

The other assessment method focuses on the benefits and burdens of overall health-state outcomes rather than of individual treatments. A health-state outcome may result from primarily one treatment or many treatments used simultaneously or in sequence.

Several techniques fall under this method [47]. The rating-scale technique is probably the simplest, most intuitive, and clinically most useful. The rating scale is a line with the worst and best possible health-state outcomes, usually death and return of full health, at the ends as anchors. Demarcations at equal intervals run along the line, say, from 0 to 10. Death is usually 0; return of full health, 10. The demarcations allow an arithmetic comparison of the overall values of different health-state outcomes. The patient marks on the line the value of an anticipated health-state outcome compared to the two anchor states [47]. The distance of the mark from the lower anchor point serves as the measure of that outcome's value to the patient. Of course, if the patient considers some live health-state outcomes worse than death, the rating-scale line must extend beyond death into the negative range. Several studies, for example, suggest Americans consider chronic coma, severe dementia, and other severely impaired states worse than death [48, 49].

Another technique, time trade-off, is conceptually harder. With it, the patient evaluates a fixed longer survival with a specific chronic illness against systematically varied shorter survivals in perfect health. The fixed survival time with the chronic illness may be arbitrary (perhaps, say, 1 or 5 years) or have grounding in data (perhaps a median survival time for such diseases as heart failure, emphysema, or Alzheimer's dementia). Time trade-off, in effect, asks the patient to imagine an exchange or "trade-off" between survival time and symptoms or disabilities. When the patient exactly equates the value of shorter survival in perfect health to longer survival with symptoms and disabilities of the chronic illness, the ratio of the two survival times represents the patient's evaluation of living with the chronic illness.

Still another technique, willingness-to-pay, resembles time trade-off except the patient buys freedom from symptoms or disabilities with money. The patient estimates the maximum percentage of family income he or she would be willing to pay for a cure. That percentage is the patient's evaluation of the cured state; 100 % minus that percentage is the patient's evaluation of the ill state.

One other technique, the standard gamble, requires explicit probabilistic reasoning and has had some use in carefully designed research. But, because most people have difficulty thinking in terms of probabilities, the technique has found little use in patient care. For that reason I do not describe the standard gamble here.

The Case

If Mr. C. tries to make an assessment early in his hospitalization, he would rate his health state as extremely poor. His consistently expressed wish to die rather than to continue treatments indicates he considers his state worse than death, that is, below zero by the rating-scale technique. And, miserable as he is, he would surely trade his severely disfigured and disabled condition even for a short time with complete cure by the time trade-off technique or would spend a high percentage of family income for a cure by the willingness-to-pay technique. Both results would show great dissatisfaction with his ill state.

However, Mr. C.'s assessment of his health state improves once he gains recognition for his right to refuse the burn treatments. To everyone's surprise he stops refusing them and claims he wants to live after all. He finishes the treatments and survives to return to a somewhat normal life. He earns a law degree, marries, and begins a successful lecturing career. If he re-evaluates his health state then, the evaluation would be more positive than at any other time since the accident. Mr. C. would probably rate his health state in the positive range, that is, better than death. And he would likely trade only a relatively long survival without disfigurement and disabilities for his current condition (by the time trade-off technique) and would sacrifice only a lower percentage of family income for a complete cure (by the willingness-to-pay technique). Both results would indicate improved satisfaction with his health state.

Death with Dignity: A Comment

Mr. C.'s severely disfiguring and disabling burns; his prolonged, painful treatments; and his resulting demands to be allowed to die bring to mind "death with dignity." Some readers would surely recommend it to him under the circumstances.

Many people believe today's high-tech care (as in Mr. C.'s case) erodes dying patients' sense of worth and steals from them any control over their lives. This belief has prompted widespread calls for changes in end-of-life care to ensure "death with dignity." Proponents of the idea seek to "dignify" dying through honoring patients' wishes and respecting patients' intrinsic worth and life accomplishments [50]. One author has even proposed "dignity-conserving" end-of-life care along illness-, person-, and community-specific lines [51].

Yet one old but insightful article takes issue with the "death-with-dignity" idea. The author of this article claims that, though well intentioned, the idea lacks sufficient clarity to provide much guidance for care [52]. He identifies four possible competing meanings for the idea: (1) letting "nature" take its course and avoiding mechanical support that merely maintains an unconscious or uncommunicative existence, (2) taking steps (such as physician-assisted suicide) to avoid the outrages of critical illness, (3) coping maturely, and (4) maintaining self-esteem while dying.

The author concludes that "death with dignity" has diverse, confusing, and often unrelated meanings. He also doubts the idea, regardless of the particular meaning taken, captures the complexities of human dying. "Death with dignity," therefore, may just be an overly simple slogan.

This author then offers four ideals as alternatives to "death with dignity" for guiding end-of-life care—respect for persons, reverence for the human body, human worth based on inclusion in the human community, and an ennobling destiny. I agree: Those ideals address a depth of human dying that the "death-with-dignity" idea does not. They provide a touchstone for humane, compassionate end-of-life care. In fact, I believe the end-of-life care values in Table 4.4 derive from them. But, while endorsing the four ideals generally, I admit they need further specification for direct applicability in clinical practice.

How Patients' Treatment Assessments Change with Time

Patients' treatment assessments may change with time and circumstance, complicating decisions about their care. Mr. C. apparently changes his assessments twice. He accepts life-sustaining burn treatments shortly after hospital admission. Then he changes his mind and demands to stop all treatments and to go home to die. Finally, after gaining recognition for his right to refuse, he changes his mind again and resumes treatments as before. He eventually completes the treatments, goes home, and creates a new personal and professional life for himself despite ongoing handicaps. Mr. C. says later during some lectures that he would refuse life-sustaining treatments again in the same situation, but he is glad that he had them when he did and lived.

As Mr. C.'s case shows, patients' treatment assessments may change dramatically. The changes—some as sudden and unexpected as Mr. C.'s—may frustrate HPs' attempts to match treatments to patients' current wishes. Those changes, for instance, undercut the fundamental assumption behind advance directives that a patient's past treatment wishes accurately predict his or her future wishes. A well-known phenomenon, called response shift, may play a key role in causing the changes. According to response shift, changing circumstances cause people to change the values by which they make benefits–burdens assessments of treatments or health-state outcomes. The assessments themselves may change, too, as a result [53]. I find that phenomenon common with patients. A patient who, when well, has always insisted he or she would refuse cancer chemotherapy accepts it willingly upon receiving a cancer diagnosis. The same phenomenon may explain Mr. C.'s own change in treatment assessments. The treatments he finds unacceptable when they seem forced on him early become acceptable to him later when he gains the power to reject them.

The response shift phenomenon has stimulated considerable research into the stability of patients' treatment wishes over time. Unfortunately, the results are mixed. Patients' wishes concerning life support show modest stability by the percentages (67–88 %) over 1–24 months [54–57]. But the associated kappa statistics—measures of agreement beyond chance—deteriorate from fair or good after one month (range 0.35–0.70 on a scale of −1.00 to +1.00) [58] to very poor after 5–16 months (range 0.01–0.29) [59] (see Chap. 7 for more explanation of the kappa statistic).

Three factors, in particular, may affect the stability of treatment wishes. One factor, elapsed time, produces instability with many kinds of treatment wishes: The longer the elapsed time between elicitations, the more changes occur [60]. The other two factors—recent hospital discharge [61] and severe depression—create instability specifically with wishes about life support. Recent hospital discharge tends to decrease patient wishes for life support at first, but those wishes increase again in as little as three months [61]. Furthermore, severe, but not mild or moderate, depression tends to decrease patient wishes for life support, but those wishes increase again after successful antidepressive treatment [55]. Age, gender, ethnicity, and education have no consistent effect on the stability of treatment wishes. Still, some treatment wishes show more stability than others: refusals (compared to acceptances) [61], wishes about the riskiest or most invasive treatments such as CPR, and wishes (usually refusals) expressed in instructional advance directives [57].

Patient assessments of health-state outcomes can also change [57]. One study examined people's ratings of seven health-state outcomes with varying undesirability (coma; inability to reason; inability to communicate; severe, uncontrollable pain; confinement to bed; legal blindness; and functional deafness). Subjects rated each outcome twice some 10–11 months apart. Overall, subjects rated coma the most undesirable outcome at both times. They also rated two outcomes with high initial undesirability ratings (inability to reason and inability to communicate) and another with a low initial undesirability rating (legal blindness) more stably than the three other outcomes with intermediate initial undesirability ratings (severe, uncontrollable pain; confinement to bed; and functional deafness) [59]. Therefore, wishes pertaining to future health-state outcomes having extremely high or low undesirability ratings vary little compared to wishes pertaining to future health-state outcomes having intermediate undesirability ratings. Wishes pertaining to current health states also vary little [57].

These stability studies can provide some help in deciding treatment for the currently uncommunicative patient who earlier refused certain life-support treatments or health-state outcomes. Overall, respect for persons requires honoring their expressed wishes for themselves. But, when the patient cannot express current wishes but did express prior ones, HPs face the possibility that the patient's prior wishes might have changed in the interim. Prior refusals create the most perplexing situations. What should HPs do?

The stability studies suggest considerations for deciding whether to honor past refusals. Considerations, often strong, *for* honoring such refusals exist whenever the patient has consistently refused (preferably in writing) any similarly risky, invasive treatments or whenever treatments will likely produce a coma or other highly undesirable health-state outcome (such as one involving little ability to think or communicate). Considerations, often weak, *against* honoring such refusals exist whenever the patient's immediate life-threatening crisis presents a significant, unexpected change from the patient's usual past health state (so the patient's prior refusals might no longer apply); whenever other intervening events could reasonably have changed those refusals; or whenever the patient expressed them during a severe untreated depression or soon after a hospital discharge. In general, unless solid evidence strongly suggests a patient's prior refusals no longer remain valid, HPs should honor them even though the patient cannot express them at the moment. But HPs should always consider life support the default treatment in highly uncertain situations.

The Case

When Mr. C. refuses life support, he might be suffering from a situational depression. Without knowing the details of Mr. C.'s treatment regimen, I assume that, if he needs antidepressive treatment, he receives it from Dr. W. and that the treatment does not change his refusal. I, therefore, think the hospital and its doctors eventually decide correctly to honor his refusal. They probably also continue to offer the life-sustaining treatments for his burns but do not force those treatments on him. Thus, I also think, when Mr. C. changes his mind and accepts those treatments again, the hospital and its doctors decide correctly to reinstitute them.

How Patients' Assessments Differ by Demographics

Four recent studies associate ethnic group and gender with varying patient assessments of life-support treatments and health-state outcomes. Overall, Euro-Americans (EAs) want life support least, African Americans (AfAs) want it most, and Korean Americans (KAs) and Hispanic Americans (HAs) fall in between [62–65]. Specifically AfAs typically want various life-support treatments two to six times as often as do EAs; and KAs and HAs, about one and a half times as often as do EAs (Table 4.6).

Two of these four studies also address gender. They suggest that, overall, American women want life support about 40 % less than do American men [64]. This gender difference applies across EAs, AfAs, and HAs [65]. A study of Britons shows the same gender difference but examines possible reasons for it. Perhaps due to their own prior caregiving burdens, British women tend to refuse life support for themselves because they wish not to impose the accompanying caregiving burdens on others. British men, in contrast, tend to accept life support because they trust its effectiveness and do not think so much about the caregiving burdens on others [46].

Only one study reports assessments of health-state outcomes by ethnic group. That study suggests EAs value disabled states less than do AfAs. Subjects

Table 4.6 Treatment wishes of other American ethnic groups relative to Euro-Americans

Study	Subjects	Treatments	Odds ratios of wishes[a]		
			Korean Americans	Hispanic Americans	African Americans
Sehgal [62]	Hemodialysis patients	Hemodialysis if demented	–	–	3.9
McKinley [63]	Cancer patients	Intensive care unit monitoring	–	–	5.8
		Life support in general	–	–	2.8
		Cardiopulmonary resuscitation	–	–	3.1
		Nasogastric feedings	–	–	2.3
		Mechanical ventilation	–	–	3.4
Blackhall [64]	Community-dwelling seniors	Life support in general	1.2	1.8[b]	2.1
Barnato [65]	Medicare beneficiaries	Hospitalization until death	–	2.2	2.3
		Life-prolonging drugs	–	1.2	1.9
		Mechanical ventilation	–	1.4	2.1

[a]An odds ratio here results from dividing the odds that members of another ethnic group want a particular treatment by the odds that Euro-Americans want it. (As the referent group, Euro-Americans always have an odds ratio of 1.0.) The ratio shows how much more likely subjects from the other ethnic group want the treatment than do the Euro-American subjects. For example, according to the Sehgal study, African American hemodialysis patients are 3.9 times more likely to want hemodialysis continued if they become demented than are Euro-American hemodialysis patients.

[b]The term "Hispanic Americans" generally refers to anyone from a Latin American country (e.g., Cuba, Puerto Rico, El Salvador, or Mexico). The Hispanic Americans in this study, however, came only from Mexico.

rated the desirability of three disabled states—inability to do basic self-care (such as bathing, dressing, and grooming), inability to walk more than a few steps, and permanent confinement to a nursing home. EAs rated each state only a third as desirable as did AfAs [66]. No studies report assessments of health-state outcomes by gender.

Health professionals, of course, must apply such research results cautiously in patient care. The studies are few and the data are sparse. Furthermore, because individuals in no demographic group share all the same values or beliefs, health professionals must avoid overgeneralizing about the whole group. Mr. C.'s case illustrates the dangers. He is an EA man. According to the studies, *as an EA*, he should tend not to want life support—a conclusion consistent with his intermediate wishes but not his later ones. But, *as a man*, he should tend to want those treatments —the opposite of his intermediate wishes. Anticipating Mr. C.'s wishes from these demographics, therefore, would certainly confuse his caregivers. I, therefore, believe (as Mr. C.'s case shows) enough variability exists within any ethnic or gender group that HPs must not try to deduce a patient's values just from his or her demographics. Rather, the surest way to learn a patient's assessments of treatments or health states is, of course, to ask. Perhaps Mr. C.'s burn doctors need to do exactly that not only to hear his treatment refusals but also to understand the reasons behind them.

The Therapeutic Misconception

Despairing over Mr. C.'s "pathetic plight," Dr. W. might question how long treatment can sustain "life in (such) a wrecked body" [14]. Yet a clinical research project might be exploring at the very same time new approaches to such extensive burns, and Mr. C. might qualify to participate. Although he refuses established treatments, he just might accept experimental interventions. Other patients in similar circumstances surely would accept *any* intervention, even unproven and experimental, that seems to offer increased chances at survival.

Systematic drug research has improved medical treatment vastly in recent decades. A rigorous, stepwise testing protocol has been developed and now plays a central role in vetting new drugs. Once finished with animal trials, a promising new drug undergoes four phases of human trials. Phase 1 trials establish the drug's toxicity and dosing range in humans. Based on the milligram-per-kilogram doses in the animal trials, doses in the Phase 1 human trials begin low and escalate rapidly until unbearable toxicities occur. Those toxicities fix the upper limits of the drug's doses. Phase 2 trials continue to monitor the drug's toxicities but primarily assess its benefits. These trials use relatively few participants under strictly regulated clinical conditions, which exist, say, on a special drug-testing ward at a university hospital. If the drug continues to show promise, Phase 3 trials expand the participant base to large numbers of patients in "real-world" clinical practices where drug use may differ from the ideal conditions of Phase 2 trials. Participants in Phase 3 trials, for example, may have varying comorbidities (i.e., coexisting diseases other than the target disease the drug aims to treat) or varying adherence (the extent to which subjects take the drug as prescribed). Favorable results from these three trial phases usually then convince the Food and Drug Administration to approve the drug for use as an established treatment against the target disease. Phase 4 trials occur later and test a previously approved drug for diseases other than the original target disease. "Postmarketing" data collection also continues to monitor for side effects. This stepwise protocol has yielded great success over many years for distinguishing the few clinically useful new drugs from the many others that fail due to excessive harms or inadequate benefits.

Because drug trials often occur in hospitals, clinics, and similar treatment settings, patients often confuse established drug treatment and experimental drug research. Several authors first recognized this confusion in psychiatric drug trials of the early 1980s. Many patients who participated in those trials never understood that random chance, not a doctor's therapeutic judgment, determined whether they received the experimental drug or a placebo. The authors called this confusion "the therapeutic misconception." They defined it as the lack of understanding that research protocols impose certain restrictions (such as randomization and blinding), which clinical care never does and which may sometimes actually harm participants [67]. I modify the authors' original idea by calling this "the process kind of therapeutic misconception" because it involves the research process.

The idea of the therapeutic misconception, I think, has expanded over the years to include still another kind, which I call "the outcome kind of therapeutic misconception." This kind involves the research outcomes for the participant. Many patients for whom all established treatments have failed may think that clinical research, most often drug trials, still offers a chance for cure. Their attending HPs sometimes share this belief. In fact, one study of Phase 1 cancer drug trials found that more than 70 % of participants believed the experimental drugs "would control their cancer." Their optimism did not correlate with age, gender, or education but *did* correlate with higher expected benefits from medical interventions and higher self-rated optimism, physical health, quality-adjusted life expectancy, strength of religious convictions, and willingness to take risks [68].

In reality, only about 5 % of participants benefit from Phase 1 drug trials [69]. That small percentage convinces me the informed consent process before a patient's entry into research must clearly differentiate the aims of established treatment and clinical research. Established treatment aims to benefit the particular patient. Considerable prior research and/or clinical experience supports as reasonable the expectation that the patient will benefit directly. Treatment aligns the intentions of patient and doctor. In contrast, clinical research aims to produce "generalizable" knowledge. Only genuine prior uncertainty about whether the experimental intervention will work (called "clinical equipoise") justifies the research. The researcher, the drug manufacturer, and future patients may benefit from it, but the research participant almost certainly will not. The research may even harm the participant. Thus, research often divides the intentions of patient-participant and researcher. I, therefore, believe the patient who is considering participation in clinical research must understand those differences clearly.

Because the two kinds of therapeutic misconception may have subtle but powerful effects on decisions, I urge a HP whom the patient trusts to be present whenever a researcher asks a critically ill patient to participate in clinical research. (The patient's primary care doctor often fulfills this advisory role best, but the consulting Dr. W. might fulfill it best for Mr. C.) That HP must ensure the patient understands accurately the possible benefits and burdens of research participation. To do so, the advising HP must listen closely to the informed consent discussion between researcher and patient and ask key questions that reflect the patient's most important values.

As the primary care doctor, I once hospitalized a long-time patient of mine for a newly diagnosed, life-threatening leukemia. I knew this elderly grandmother loved most being at home and having her grandchildren visit her there. But her leukemia was already far advanced at diagnosis. The consulting oncologist told me he could offer no established chemotherapies, only a Phase 1 experimental drug. I asked to sit in when the oncologist discussed the drug research with the patient.

After the oncologist's initial description the patient said she wanted the experimental drug. But she appeared to have the outcome kind of therapeutic misconception: She believed the drug was her only chance to live, and she did not grasp the potentially serious complications involved. Specifically, she would not likely benefit and might suffer significant toxicities from the escalating drug doses. I, therefore, asked the oncologist a few questions in front of the patient to clarify the drug's possible benefits and burdens to her. The oncologist answered candidly roughly as follows:

Question: Is there a reasonable expectation the drug will benefit the patient by either improving symptoms or lengthening survival?
Answer: No, there is too little experience with the drug to expect this patient will benefit directly. Future patients, however, might benefit from the knowledge we gain from this patient's participation in the research.
Question: What serious side effects might occur?
Answer: Life-threatening infections or bleeding, loss of appetite and weight, and severe fatigue.
Question: How long (out of a projected remaining lifespan of only several months) will the patient be hospitalized to receive the drug?
Answer: At least several weeks but longer if serious complications occur.
Question: Might the patient die sooner with the drug than without it?
Answer: Yes, especially if serious infections or bleeding occurs.
Question: Once the patient joins the research, can she withdraw early if she wishes?
Answer: Each participant must commit to completing the study. Otherwise, the participant undertakes research risks and produces little useful data.

After all these questions the oncologist emphasized that the patient should join the study only for the right reason: to try to help future patients, not to benefit herself. He also reminded her that she might have only a small chance at benefit, and it would almost certainly not be a cure. She might also experience serious symptoms and life-threatening complications. The experimental drug might even shorten her life. The patient reconsidered her initial choice. She said she wanted to maximize her time at home and the visits with her grandchildren. She, therefore, finally chose not to participate in the research but entered hospice instead and returned home to die. Other patients in her situation might, of course, choose freely and knowingly to participate in the research. I can accept that choice as long as the informed consent process addresses forthrightly the possibility of the therapeutic misconception.

Summary Points

1. The SPINES mnemonic helps structure a discussion to deliver medical bad news: Set an overall approach, Plan for the discussion, Introduce yourself and Inquire about the patient's understanding of the situation, deliver the bad News, Empathize, and Summarize the discussion and Set a follow-up action plan [15].
2. Doctors and patients alike perceive relieving physical and emotional symptoms, nurturing significant relationships, promoting spiritual health, and preventing physical and financial burdens on others as important values in dying patients' assessments of treatments or outcomes.
3. In addition, patients more than doctors perceive achieving some control over the dying process and trusting health professionals' clinical and interpersonal skills as additional important values in those assessments [31–39].
4. Generally, Euro-Americans want life support least, African Americans want it most, and Korean Americans and Hispanic Americans fall in between [62–65].
5. Doctors should clearly explain to patients who are considering participation in research the differences in aims and procedures between established treatment and clinical research.

To Learn More …

1. Baile WF, Buckman R, Lenzi R et al. SPIKES—a six-step protocol for delivering bad news: application to the patient with cancer. Oncologist. 2000;5 (4):302–11.
2. Barnato AE, Anthony DL, Skinner J, Gallagher PM, Fisher ES. Racial and ethnic difference in preferences for end-of-life treatment. J Gen Intern Med. 2009;24(6):695–701
3. Froberg DG, Kane RL. Methodology for measuring health-state preferences—II. Scaling methods. J Clin Epidemiol. 1989;42(5):459–71.
4. Quill TE, Townsend P. Bad news: delivery, dialogue, and dilemmas. Arch Intern Med 1991; 151(3):463–8.
5. Vanderpool HY. The ethics of terminal care. JAMA. 1978;239(9): 850–2.

References

1. Blanchard CG, Labrecque MS, Ruckdeschel JC, Blanchard EB. Information and decision-making preferences of hospitalized adult cancer patients. Soc Sci Med. 1988;27 (11):1139–45.
2. Mazur DJ, Hickam DH. Patients' preferences for risk disclosure and role in decision making for invasive medical procedures. J Gen Intern Med. 1997;12(2):114–7.
3. Kraetschmer N, Sharpe N, Urowitz S, Deber RB. How does trust affect patient preferences for participation in decision-making? Health Expect. 2004;7(4):317–26.
4. Flynn KE, Smith MA, Vanness D. A typology of preferences for participation in health-care decision-making. Soc Sci Med. 2006;63(5):1158–69.

5. Ainslie N, Beisecker AE. Changes in decisions by elderly persons based on treatment description. Arch Intern Med. 1994;154(19):2225–33.
6. Gattellari M, Voigt KJ, Butow PN, Tattersall MHN. When the treatment goal is not cure: Are cancer patients equipped to make informed decisions? J Clin Oncol. 2002;20(2):503–13.
7. McNutt RA. Shared medical decision making: problems, process, progress. JAMA. 2004;292 (20):2516–8.
8. Arora NK, McHorney CA. Patient preferences for medical decision making: Who really wants to participate? Med Care. 2000;38(3):335–41.
9. Degner LF, Aquino Russell C. Preferences for treatment control among adults with cancer. Res Nurs Health. 1988;11(6):367–74.
10. Deber RB, Kraetschmer N, Irvine J. What role do patients wish to play in treatment decision making? Arch Intern Med. 1996;156(13):1414–20.
11. Keating NL, Landrum MB, Arora NK, et al. Cancer patients' roles in treatment decisions: Do characteristics of the decision influence roles? J Clin Oncol. 2010;28(28):4364–70.
12. Barry MJ, Edgman-Levitan S. Shared decision making—the pinnacle of patient-centered care. N Engl J Med. 2012;366(9):780–1.
13. Sullivan M. The new subjective medicine: taking the patient's point of view on health care and health. Soc Sci Med. 2003;56(7):1595–604.
14. White RB. Case studies in bioethics, case no. 228: a demand to die. Hast Cen Rep. 1975;5 (3):9–10.
15. Baile WF, Buckman R, Lenzi R, et al. SPIKES—a six-step protocol for delivering bad news: application to the patient with cancer. Oncologist. 2000;5(4):302–11.
16. Clayton JM, Butow PN, Tattersall MHN. When and how to initiate discussion about prognosis and EOL issues with terminally ill patients. J Pain Symptom Manage. 2005;30 (2):132–44.
17. Girgis A, Sanson-Fisher RW. Breaking bad news: consensus guidelines for medical practitioners. J Clin Oncol. 1995;13(9):2449–56.
18. Fox E, Arnold R, Brody B. Medical ethics education: past, present and future. Yale curriculum in ethical and humanistic medicine. Acad Med. 1995;70(9):761–9.
19. Morris DA, Johnson KS, Ammarell N, et al. What is your understanding of your illness? A communication tool to explore patients' perspectives of living with advanced illness. J Gen Intern Med. 2012;27(11):1460–6.
20. Coulehan JL, Platt FW, Egener B, et al. "Let me see if I have this right …": Words that help build empathy. Ann Intern Med. 2001;135(3):221–7.
21. Wenrich MD, Curtis RJ, Shannon SE, et al. Communicating with dying patients within the spectrum of medical care from terminal diagnosis to death. Arch Intern Med. 2001;161 (6):868–74.
22. Balaban RB. A physician's guide to talking about end-of-life care. J Gen Intern Med. 2000;15 (3):195–200.
23. Quill TE, Townsend P. Bad news: delivery, dialogue, and dilemmas. Arch Intern Med. 1991;151(3):463–8.
24. Quill TE, Arnold RM, Platt F. "I wish things were different": expressing wishes in response to loss, futility, and unrealistic hopes. Ann Intern Med. 2001;135(7):551–5.
25. Fallowfield L. Giving sad and bad news. The Lancet. 1993;341(8843):476–8 (20 Feb).
26. The Hippocratic Oath, fourth century, B.C.
27. American Medical Association. Principles of Medical Ethics, 1980.
28. Pearlman RA, Jonsen AR. The use of quality-of-life considerations in medical decision-making. J Am Geriatri Soc. 1985;33(5):344–52.
29. Steinhauser KE, Christakis NA, Clipp EC, et al. Preparing for the end of life: preferences of patients, families, physicians, and other care providers. J Pain Symptom Manage. 2001;22 (3):727–37.
30. Steinhauser KE, Clipp EC, McNeilly M, et al. In search of a good death: observations of patients, families, and providers. Ann Intern Med. 2000;132(10):825–32.

31. Perkins HS, Geppert CMA, Gonzales A, et al. Cross-cultural similarities and differences in attitudes about advance care planning. J Gen Intern Med. 2002;17(1):48–57.
32. Singer PA, Martin DK, Kelner M. Quality end-of-life care: patients' perspectives. JAMA. 1999;281(2):163–8.
33. Perkins HS, Cortez JD, Hazuda HP. Patients' diverse beliefs about what happens at the time of death. J Hosp Med. 2012;7(2):110–6.
34. Perkins HS, Cortez JD, Hazuda HP. Diversity of patients' beliefs about the soul after death and their importance in end-of-life care. South Med J. 2012;105(5):266–72.
35. Lynn J. Measuring quality of care at the end of life: a statement of principles. J Am Geriatr Soc. 1997;45(4):526–7.
36. Field MJ, Cassel CK, editors. Approaching death: improving care at the end of life. Washington, D.C: National Academy Press; 1997.
37. Emanuel EJ, Emanuel LL. The promise of a good death. Lancet. 1998;351(Suppl 2):21–9.
38. Steinhauser KE, Christakis NA, Clipp EC, et al. Factors considered important at the end of life by patients, family, physicians, and other care providers. JAMA. 2000;284(19):2476–82.
39. Perkins HS, Cortez JD, Hazuda HP. Cultural beliefs about a patient's right time to die: an exploratory study. J Gen Intern Med. 2009;24(11):1240–7.
40. Slevin ML, Stubbs L, Plant HJ, et al. Attitudes to chemotherapy: comparing views of patients with cancer with those of doctors, nurses, and general public. BMJ. 1990;300(6737):1458–60 (02 June).
41. Matsuyama R, Reddy S, Smith TJ. Why do patients choose chemotherapy near the end of life? A review of the perspective of those facing death from cancer. J Clin Oncol. 2006;24 (21):3490–6.
42. Mitera G, Zhang L, Sahgal A, et al. A survey of expectations and understanding of palliative radiotherapy from patients with advanced cancer. Clin Oncol. 2012;24(2):134–8.
43. Weeks JC, Catalano PJ, Cronin A, et al. Patients' expectations about effects of chemotherapy for advanced cancer. N Engl J Med. 2012;367(17):1616–25.
44. Taffet GE, Teasdale TA, Luchi RJ. In-hospital cardiopulmonary resuscitation. JAMA. 1988;260(14):2069–72.
45. Vitelli CE, Cooper K, Rogatko A, Brennan MF. Cardiopulmonary resuscitation and the patient with cancer. J Clin Oncol. 1991;9(1):111–5.
46. Arber S, Vandrevala T, Daly T, Hampson S. Understanding gender differences in older people's attitudes towards life-prolonging medical technologies. J Aging Stud. 2008;22 (4):366–75.
47. Froberg DG, Kane RL. Methodology for measuring health-state preferences—II. Scaling methods. J Clin Epidemiol. 1989;42(5):459–71.
48. Pearlman RA, Cain KC, Patrick DL, et al. Insights pertaining to patient assessments of states worse than death. J Clin Ethics. 1993;4(1):33–41.
49. Patrick DL, Starks HE, Cain KC, Uhlmann RF, Pearlman RA. Measuring preferences for health states worse than death. Med Decis Mak. 1994;14(1):9–18.
50. Hendin H. Selling death and dignity. Hast Cen Rep. 1995;25(3):19–22.
51. Chochinov HM, Hack T, Hassard T, et al. Dignity in the terminally ill: a cross-sectional cohort study. Lancet. 2002;360(9350):2026–30 (21/28 Dec).
52. Vanderpool HY. The ethics of terminal care. JAMA. 1978;239(9):850–2.
53. Shultz DS, Carnevale FA. Engagement and suffering in responsible caregiving: on overcoming maleficence in health care. Theor Med. 1996;17(3):189–207.
54. Danis M, Garrett J, Harris R, Patrick DL. Stability of choices about life-sustaining treatments. Ann Intern Med. 1994;120(7):567–73.
55. Ganzini L, Lee MA, Heintz RT, Bloom JD, Fenn DS. The effect of depression treatment on elderly patients' preferences for life-sustaining medical therapy. Am J Psychiatry. 1994;151 (11):1631–6.
56. Rosenfeld KE, Wenger NS, Phillips RS, et al. Factors associated with change in resuscitation preference of seriously ill patients. Arch Intern Med. 1996;156(14):1558–64.

57. Ditto PH, Smucker WD, Danks JH, et al. Stability of older adults' preferences for life-sustaining medical treatment. Health Psych. 2003;22(6):605–15.
58. Everhart MA, Pearlman RA. Stability of patient preferences regarding life-sustaining treatments. Chest. 1990;97(1):159–64.
59. Lockhart LK, Ditto PH, Danks JH, Coppola KM, Smucker WD. The stability of older adults' judgments of fates better and worse than death. Death Studies. 2001;25(4):299–317.
60. McParland E, Likourezos A, Chichin E, Castor T, Paris BEC. Stability of preferences regarding life-sustaining treatment: a two-year prospective study of nursing home residents. Mt Sinai J Med. 2003;70(2):85–92.
61. Ditto PH, Jacobson JA, Smucker WD, Danks JH, Fagerlin A. Context changes choices: a prospective study of the effects of hospitalization on life-sustaining treatment preferences. Med Decis Making. 2006;26(4):313–22.
62. Sehgal A, Galbraith A, Chesney M, et al. How strictly do dialysis patients want their advance directives followed? JAMA. 1992;267(1):59–63.
63. McKinley ED, Garrett JM, Evans AT, Danis M. Differences in end-of-life decision making among black and white ambulatory cancer patients. J Gen Intern Med. 1996;11(11):651–6.
64. Blackhall LJ, Frank G, Murphy ST, et al. Ethnicity and attitudes toward life sustaining technology. Soc Sci Med. 1999;48(12):1779–89.
65. Barnato AE, Anthony DL, Skinner J, Gallagher PM, Fisher ES. Racial and ethnic difference in preferences for end-of-life treatment. J Gen Intern Med. 2009;24(6):695–701.
66. Cykert S, Joines JD, Kissling G, Hansen CL. Racial differences in patients' perceptions of debilitated health states. J Gen Intern Med. 1999;14(4):217–22.
67. Appelbaum PS, Roth LH, Kidz C. The therapeutic misconception: informed consent in psychiatric research. Inter J Law Psych. 1982;5(3–4):319–29.
68. Weinfurt KP, Castel LD, Li Y, et al. The correlation between patient characteristics and expectations of benefit from Phase 1 clinical trials. Cancer. 2003;98(1):166–75.
69. Daugherty CK. Impact of therapeutic research on informed consent and the ethics of clinical trials: a medical oncology perspective. J Clin Oncol. 1999;17(5):1601–17.

Chapter 5
The Challenges of Prognosis Near the End of Life

"How long do I have to live, doctor?" the patient often asks as a panic-stricken first response to a new fatal diagnosis. Emotional turmoil grips the patient. Life as usual no longer stretches out endlessly. The future seems instead palpably short and unpredictable. Matters peculiar to the end of life, which the patient may have ignored before, inject themselves with urgency. The patient faces momentous treatment decisions, end-of-life financial and legal plans, and arrangements for body disposition and a funeral. He or she may also have to say important good-byes and otherwise attend to meaningful relationships; plan purposefully for special events such as trips, weddings, or reunions; or close out a life's work. The patient may simultaneously search for answers to such pressing existential or spiritual questions as How will I die? What legacy will I leave? Is there life after death? and If so, what will that life be like for me?

Doctors typically dread the survival question [1]. It heralds not only tumultuous changes in the patient's life but also dramatic changes in the doctor–patient relationship. Much of how the patient responds to the diagnosis depends on the doctor's answer. Yet the doctor faces a difficult challenge because the patient usually wants two potentially conflicting qualities in the answer—honesty and optimism [2]. The doctor asks him- or herself, How can I accurately disclose a fatal prognosis without also dashing many of the patient's hopes for the future? In any case the doctor may foresee a long, uncomfortable, emotionally draining follow-up discussion. No wonder many doctors, though acknowledging the importance of informing patients about fatal prognoses, actually prefer to avoid the topic of survival. Nearly half of American internists in a large nationwide survey [3] said they do not volunteer survival predictions. Rather, they make those predictions—and quite reluctantly—only if patients ask. In fact, the more experienced the internists, the more likely they wait for patients to ask explicitly for such predictions. Some doctors surely hope patients never ask.

So exactly what accounts for doctors' wish to avoid the survival question? That national survey of American internists suggests some reasons. First, most respondents [3] felt poorly prepared to address the question. They wanted better training

H.S. Perkins, *A Guide to Psychosocial and Spiritual Care at the End of Life*,
DOI 10.1007/978-1-4939-6804-6_5

for it. Second, even more respondents admitted that delivering a survival prediction stresses them, and nearly all respondents considered making an accurate survival prediction more stressful than making an accurate diagnosis [3]. One respondent said, "The course of (an) illness is much more difficult to pin down (than a diagnosis). You're talking about what's going to happen in the future!" Respondents also insisted that increased experience with the question does not reduce the stress involved. Third, nearly as many respondents believed patients "expect too much certainty" from such a prediction. "Patients routinely request prognoses regarding conditions which (are) highly variable from patient to patient," explained a respondent. "I find this aspect of clinical practice … the most difficult (part) …. (It's) the 'art of medicine.'" And exactly half of respondents claimed a prognostic error causes patients (and perhaps colleagues) to lose confidence in a doctor's clinical abilities. Of course, the survival question also raises for everyone the sobering, even frightening reality of death. Most people, including doctors, prefer not to think about it.

Still, clinical reality makes the survival question almost impossible to avoid. The same national survey showed that American general internists, pulmonologists or critical care specialists, and hematologists or oncologists face the question roughly 6, 25, and 100 times a year, respectively, in their practices (Table 5.1) [3]. Still, difficult as it is, addressing that question may actually benefit dying patients and their doctors. It may summon latent courage that neither patients nor doctors knew they had. The question may also begin an especially meaningful dialogue between them. That dialogue may help patients clarify their end-of-life goals, cull out the unrealistic ones, and plan ways to accomplish the others. The dialogue may also help patients and doctors together anticipate decisions about withholding life-support treatments and making hospice referrals (Table 5.1). And it may provide doctors important professional satisfactions for guiding patients compassionately through the dying process.

By explicitly discussing end-of-life prognoses, or clinical predictions, this chapter aims to reduce their fearsomeness to all. The chapter begins by discussing

Table 5.1 Internists' median annual encounters with the survival question, decisions to withhold or withdraw life support, and hospice referrals

Kinds of internists	Median times the question How long do I have to live, doctor? arises in clinical practice annually	Median annual number of decisions about withholding or withdrawing life support	Median annual number of hospice referrals
General internists	6	5	5
Pulmonologists or critical care specialists	25	16	10
Hematologists or oncologists	100	25	38

Data from Christakis and Iwashyna [3], with permission

the anxiety that often causes patients and family members to ask doctors the survival question in the first place. The chapter then reviews people's attitudes about prognoses. It next addresses the accuracy of doctors' subjective clinical predictions, considers potentially helpful additional predictive factors, and describes several common functional trajectories to death. The chapter closes with suggestions of steps for discussing fatal prognoses with patients. The case of a man newly diagnosed with colon cancer illustrates many of the points.

The Case

Mr. F., an accomplished 73-year-old architect, has never been seriously ill in his life. He lives with his wife, who is beginning to show memory deficits characteristic of dementia, and works full time at his profession. Mr. F. enjoys an especially close relationship with his youngest granddaughter, who is eight. He picks her up from school every weekday and helps her with her homework. Mr. and Mrs. F. sometimes prepare dinner for her when the girl's parents work late. Mr. F. says he looks forward to seeing his granddaughter grow up.

Mrs. F. persuades Mr. F. to undergo his first-ever screening colonoscopy. (A colonoscopy involves inserting a lighted, flexible, roughly yard-long tube into the rectum to inspect the colon higher up.) A gastroenterologist, a bowel specialist, performs the procedure and spots a "suspicious" mass just up from the rectum. He biopsies the mass. The pathology reading comes back several days later: cancer. The doctor delivers the bad news to Mr. F. at a follow-up office visit. The news catches Mr. F. completely off guard, "stunning" him. He asks the doctor, How long do I have to live?

Anxieties About the Future with a Potentially Fatal Illness

A new cancer diagnosis, of course, sets any patient reeling. While the immediate anxiety often prompts the question Mr. F. asked, How long do I have to live? other questions invariably follow. Can my cancer be treated? Can it be cured, or is it "a sure death sentence?" How will the cancer and its treatments affect my daily life? Will I be able to fulfill my usual responsibilities at home and at work? Will I eventually have to depend on others for my personal bodily care? Will the cancer and its treatments burden my family physically, emotionally, or financially? [1, 2] And, if the cancer becomes terminal, the patient may ask still other questions. What symptoms will I have now? Will I suffer? Will I ever need life-support machines? How will I die? Will I have control over any part of my death? Will I be able to die at home? How will my family react to seeing me die? [1, 2, 4, 5].

The particular questions vary from patient to patient. Unfortunately, doctors cannot predict which questions to address with which patients. Fear of appearing ignorant, bothering busy doctors, or getting more bad news [6] prevents many patients from asking their questions outright. And no sociodemographic or clinical characteristics indicate which questions particular patients want answered. The doctors must guess. Often specialists, they tend to anticipate only technical questions in their specialties. These doctors assume—usually incorrectly—that patients will ask other questions if they want or need other information. As a result, the informing process succeeds only partially. Patients are often satisfied with the technical information they receive but not with the nontechnical information they receive about such matters as future symptoms, physical function, and care needs.

Thankfully, doctors have allies to help with addressing patients' questions and anxieties. Nurses are especially good at it, but other health professionals (HPs) can be, too. I, therefore, believe doctors should explicitly encourage other HPs to contribute to the overall informing process. Good reasons support such a shared approach: Questions and anxieties may arise as these other HPs care directly for patients, the patients may feel most comfortable voicing their questions or anxieties to these caregivers, the caregivers can begin responding to patients' needs on the spot, and involvement of the other HPs may unburden the doctors.

Whoever addresses dying patients' questions and anxieties should explicitly invite them if patients do not raise them spontaneously. Caregivers might say, for example, "Many patients in your situation worry with questions about their futures. Do you have any questions like that? If so, I can try to answer them. If I cannot, I can ask someone else (perhaps your doctor) to try." If the patient volunteers no questions or anxieties, the caregiver might suggest some. He or she might say, "Some people wonder, What will my symptoms be like later on? What care needs will I have? or How will my family react as they see me decline? Do you have any questions like those?" Of course, all doctors and other caregivers who address a dying patient's questions and anxieties should answer well within their knowledge of the case and inform the other attending caregivers, say, through progress notes, about the patient's ongoing fears and informational needs.

The Case

When Mr. F. asks the survival question, the doctor squirms. He is obviously uncomfortable answering. He does not explore the anxieties behind Mr. F.'s question. Rather, the doctor just says he cannot answer without doing more tests to determine the cancer's spread. He then schedules several scans and a return visit. The ordinarily unflappable Mr. F. leaves the office in a mental "fog."

This "fog" envelops Mr. F. on his drive home. Troubling questions pummel his mind. Will I never see my granddaughter grow up? What will my wife do without me as her dementia progresses? Should I abandon my partially completed architectural plans for my one daughter's new home? Should I immediately redraft my will and plan my funeral? Should I agree to treatment or just take to bed and die?

When Mr. F. arrives home, he tells Mrs. F. his diagnosis. Upset by the news, Mrs. F. peppers Mr. F. with questions. He cannot answer most of them. His lack of answers only upsets her more. Mr. F. finally realizes that nearly everything the doctor said after "colon cancer" went "in one ear and out the other." Mr. F. thinks, in fact, that he did ask other questions besides the survival question and that the doctor probably answered them. But he—Mr. F.—did not hear the answers. Finally, Mr. F. tells Mrs. F. he must go for more tests. Mrs. F. cuts him off. "You don't need all those tests," she says angrily in her denial. "Those doctors are only trying to make big profits off of you."

A Recent History of Prognosis in Clinical Medicine

In my own hospital ethics consultations I repeatedly sensed doctors' hesitancy to give or use prognoses in practice. The famous SUPPORT study confirmed that sense. It showed that doctors often hesitate to use authoritative prognoses even when given to them. Specifically, doctors in that study received "accurate predictive information on future functional ability (and) survival" for particular, seriously ill patients of theirs. Fifty-nine percent of the doctors acknowledged receiving the prognostic information, but only 15 % said they discussed it with the patients. Data analysis detected no impact of the information on any patient outcomes including prevalence or timing of do-not-resuscitate (DNR) orders, days spent in the intensive care unit (ICU) or on mechanical ventilators, or use of hospital resources [7].

Such hesitancy to give or use survival prognoses may arise from worry about possible prediction errors or about patients' tendency to consider such prognoses either too literally or not at all. One researcher, however, takes an intriguingly different view: He sees prognostication as a recently neglected clinical skill. He thinks twentieth-century medical education progressively emphasized diagnosis and treatment and deemphasized prognosis. He supports his view with comparative readings of medical textbooks from the 1890s to the 1980s [8]. He took pneumococcal pneumonia as a representative disease and compared the textbook space allocated to its diagnosis, treatment, and prognosis. The textbooks, he found, devoted ever more space over time to the diagnosis and treatment of pneumococcal pneumonia while devoting ever less space to its prognosis. By the 1980s, prognosis received almost no space at all. The researcher concluded that diagnosis and treatment had come to overshadow prognosis completely, and thus, doctors got out of the habit of using it clinically.

This researcher has suggested a historical explanation for the progressive deemphasis of prognosis. Before 1900, medical practice focused on determining survival. Doctors believed outcomes from illness depended on patients' unchangeable underlying physical constitutions, ranging from healthy and robust to sickly and weak. A doctor's main task involved first observing the symptoms, signs, and complications of an illness to identify the patient's underlying physical constitution. Then the doctor would predict whether that constitution would support the patient enough to allow him or her to survive the illness. Patients wanted to know whether they would live.

But, as clinical medicine increasingly embraced its new science early in the twentieth century, medical practice began to focus on diagnosing disease, that is, the malfunction of body parts or processes. Medical thinking evolved to believe disease actually changes a patient's underlying physical constitution from naturally healthy to unnaturally ill. The doctor's main task shifted from speculating about survival to identifying specific diseases. Patients wanted to know what caused their symptoms.

By the late twentieth century, medical science had evolved still further. It had discovered many new treatments that could actually cure or control disease. Those treatments included insulin, antibiotics, vaccines, life-support techniques, immunologic modifiers, transplantation, antidepressants, and many others. The new treatments could convert a patient's ill physical constitution back to a healthy "natural" one. With such a significant treatment armory, medicine boldly adopted the language of war against disease. Terms such as "the war on cancer," "battling disease," "the therapeutic armamentarium," and "magic bullet" treatments illustrated medicine's confident fighting spirit. Medical practice changed focus again, this time to determining treatment. Many treatments proved so successful that doctors' main task became picking and administering the treatment that would cure the patient. Making an accurate diagnosis became merely a stepping stone to choosing an effective treatment. Patients wanted to know how their diseases could be treated. Patients also began to believe medicine could cure almost anything. Prognosis, therefore, diminished steadily in clinical importance throughout all this history.

Yet, with the successful cure of many acute diseases, chronic diseases have risen to dominate clinical practice at the start of the twenty-first century. Medical practice must now focus on delaying the progression of chronic diseases such as atherosclerosis, the rheumatic diseases, and the dementias [5]. Some of those diseases eventually cause death; nearly, all cause long-term disability. Now, patients with those diseases want doctors' predictions about survival and function once again. Prognosis, I believe, is reemerging as an important part of clinical practice.

The Case

Mr. F.'s mental "fog" persists for about four weeks after the gastroenterologist discloses the cancer diagnosis. But both Mr. and Mrs. F. finally begin to accept it. Mr. F. undergoes the additional tests the doctor orders. The results indicate the cancer has spread only to the closest lymph nodes. The doctor recommends surgery, describing it as "definitive" but says other, unspecified treatments may be necessary, too. He casually mentions the possibility of a colostomy. The doctor refers Mr. F. to a colorectal surgeon, a "radiation cancer doctor," and a "chemical cancer doctor."

Still not informed about his prognosis for survival, Mr. F. does not understand why he needs two cancer doctors in addition to the surgeon. Like many other patients in that situation, he thinks the treatment involves simply "cutting out the cancer." He expects to have the surgery immediately, be cured, get "back on (his) feet within a few weeks," and resume his normal life.

Unfortunately, the actual treatment regimen is not so simple. Mr. F. undergoes 2 months of daily preoperative radiation and intravenous chemotherapy. Extreme fatigue, loss of appetite, and weight loss set in. "I'm vanishing," an alarmed Mr. F. thinks. "My body is disappearing!" Still, the doctors insist at first he is "doing well." When his weight loss reaches 50 lb and Mr. F. can barely eat at all, his doctors change their minds and hospitalize him for extensive nutrition before the surgery.

Doctors' Subjective Predictions for Survival in Potentially Fatal Illnesses

Potentially fatal illnesses cause patients to worry most about two kinds of prognoses —survival time and interim function. The greater worry concerns survival time. Expectations for survival, of course, affect patients' treatment preferences. For example, as I mentioned in Chap. 4, most of the general public initially request cardiopulmonary resuscitation (CPR) for cardiac arrest because they believe CPR offers high survival rates. But those rates are actually quite low: Survival to hospital discharge runs only about 20 % after in-hospital arrests and only about 5 % after out-of-hospital arrests. When informed about those low survival rates [9–13], many people change their minds to refusing CPR [14]. A similar phenomenon may occur with palliative treatment for advanced cancer. Many patients may take such treatment, whether radiation or chemotherapy, because they mistakenly believe the treatment offers a chance for cure [2, 15–17]. The treatment does not, and many patients might refuse it if they knew that.

With outcome expectations influencing their treatment preferences so much, patients depend heavily on doctors' predictions. How accurate, then, are doctors' subjective predictions of survival? I reviewed the literature about them. I found it small and focused mostly on cancer patients (perhaps because cancer dominates the popular American image of fatal diseases) [18]. The accuracy of such predictions even for advanced cancer is only fair. Overoptimism accounts for most of the inaccuracy. One summary of eight studies reports that doctors' median survival prediction for their advanced cancer patients was six weeks while the patients' median actual survival was only four weeks [19]. Furthermore, the doctors' survival predictions fell within 1 week, 2 weeks, and 4 weeks of the patients' actual survivals only 25, 43, and 61 % of the time, respectively. Doctors' accuracy improved, however, as they made repeated predictions for the same patient and as patients neared their deaths. That second phenomenon, increased predictive accuracy as an event nears, is called "the horizon effect" and occurs commonly in weather forecasting [20].

Doctors' excessive optimism occurs especially commonly in survival predictions for hospice patients. Most of those patients, of course, have cancer. One study found 80 % of those predictions were inaccurate. Overestimates of survival were nearly four times more common than underestimates. Overall, doctors predicted survivals averaging more than five times longer than patients' actual survivals [21]. A similar study showed hospice doctors predicted their patients would survive a median of 75 days and told the patients a median of 90 days. The patients actually survived a median of only 29 days. Predictive inaccuracies did not differ among hospice doctors, nurses, and social workers [20, 22].

The overoptimism of doctors' survival predictions may be either *intentional* or *unintentional*. One study explored the possibility of *intentional* biasing among hospice doctors. Twenty-eight percent of doctors surveyed admitted they intentionally give patients overly optimistic survival predictions; 12 %, intentionally overly pessimistic ones; and 37 %, strictly accurate ones. Twenty-three percent of doctors said they give no survival predictions at all [23]. Two reasons may explain the tendency to give patients *intentionally* overly optimistic survival predictions. One reason may be the natural human reluctance to give bad news. Doctors, like most others, prefer to please people with good news rather than to upset them with bad news. Good news leaves people encouraged; bad news leaves them discouraged. Therefore, when news is bad, doctors may intentionally couch it as positively as possible, trying to soften the blow. Another reason for intentionally overly optimistic predictions might be the internal and external pressures on doctors to maintain hopefulness among patients and family members. People often say they want honesty from their doctors. But, when the survival news might be bad, people may really want a shading toward optimism more than strict honesty. Sensitive to those wishes, doctors may be unable to resist the temptation to cast bad news intentionally toward the optimistic.

Two reasons may also explain doctors' giving *unintentionally* overly optimistic survival predictions. One of those reasons might be the "planning fallacy," the bias that, according to one author, helps people cope with death by planning to fight it actively rather than to accept it passively. This bias tends to exaggerate the benefits and minimize the burdens of the struggle [24]. The other reason might be the emotional investment doctors make in their patients. Doctors hope patients do well. Making the right diagnosis, giving the best treatment, and producing an outcome the patient values give doctors professional satisfaction. A patient's death, in contrast, may give only disappointment or the sense of professional failure. Doctors may even grieve for their patients who die [25]. Doctors may, therefore, subconsciously pin their expectations more on good outcomes than bad ones.

Some commentators from within medicine judge doctors' predictive inaccuracies harshly, especially the tendency to overoptimism about survival. These commentators often blame that optimism specifically for the growing rates of chemotherapy use in far advanced cancer [26], the relatively low rates of timely hospice referrals [27, 28], and the persistently high rates of in-hospital deaths for patients with end-stage cancer [29]. One commentator says that doctors do not recognize when the end of life has begun [30]; another, that they cannot "recognize the nearness of death." [29] One commentator simply calls doctors' survival predictions "poor" [29] and another calls them "notoriously inaccurate." [1] Taking a less negative view, yet another commentator suggests doctors' predictions simply "reflect the finitude of the possible." [31].

I believe doctors' subjective survival predictions will always play an important role in end-of-life care despite their inaccuracies. Most patients and family members want those predictions, treatment decisions depend on them, and Medicare requires them for assessing hospice eligibility. The challenge is to make those predictions as accurate as possible while delivering them as humanely as possible.

Conceptual Aids to Doctors' Subjective Predictions

However important already, doctors' subjective outcome predictions can certainly use improvement. Partly for that reason an outcomes research movement has grown up in recent years. That research follows particular diseases and treatments to determine their outcomes. The data can then be organized into conceptual aids to doctors' subjective predictions about not only survival but also interim function.

Conceptual Aids for Predictions About Survival

Data useful for making survival predictions can be either qualitative or quantitative. One kind of qualitative data, disease staging, has proven helpful with cancers, heart failure, and other potentially fatal diseases. Cancer staging, for example, typically categorizes a cancer in one of four stages: confined to the original organ site (stage 1), contiguously extended outside that organ (stage 2), seeded to regional lymph nodes (stage 3), or spread widely throughout the body (stage 4). These stages help determine treatment. They also provide a crude survival prognosis. In general, survival decreases the wider the cancer has spread.

Several kinds of quantitative data pertaining to survival prognoses exist, too. One kind expresses survival as the percentage of patients who typically live for a certain time, say 5 years, after the diagnosis. For example, 40–65 % of patients who like Mr. F. have colon cancer with local lymph node spread and who take full treatment can expect to live at least five more years [32]. Another kind of quantitative data expresses survival as a median, that is, half of patients live less than the specified time and half live more. For example, the median survival for patients with heart failure is 5 years; the median for patients with severe emphysema, 4 years [33, 34]. And still another kind of quantitative data expresses survival prognoses as a time range. For example, patients with Alzheimer's dementia typically live 3–15 years after the diagnosis [35]. Such data can be helpful to patients and family members but only if HPs explain the data in everyday language.

I believe these data can also help doctors make and communicate survival predictions in two ways. First, both the qualitative and quantitative kinds of data can anchor doctors' predictions for purposes of limiting excessive optimism or pessimism. Second, the quantitative data—especially when expressed as means, medians, or ranges—convey the important idea that survival predictions for an individual patient are never precise, only approximate [36]. They are merely best guesses. Still, that idea may frustrate some patients and family members who expect from medicine specific, accurate survival time predictions. When such expectations arise, doctors might address them by explaining the inherent inexactness of all clinical predictions. Definitive knowledge in a particular situation often requires a wait-and-see approach.

Efforts to improve survival predictions have not been limited to guessing survival times directly. Parallel efforts have tried to guess those times indirectly from patients' interim function. The rationale here is simple: The breakdown of the body's physiological and psychological resources creates ever more functional impairments the closer one comes to death. Certain functional changes, therefore, should reflect how close one is to death [37].

Early research about survival times and function focused on the well-established Karnofsky Performance Status Scale (KPSS) (Table 5.2) [38]. First published in the late 1940s, the KPSS grades patient function from KPSS-100 % (full function: lives at home, provides self-care, performs other normal activities, and can work; no evidence of disease) through KPSS-50 % (partial function: lives at home; can provide some self-care but cannot work; and requires considerable medical care and assistance) to KPSS-0 % (no function: dead). Doctors initially used the KPSS to

Table 5.2 Karnofsky Performance Status Scale (KPSS)

Overall condition	Performance status (%)	Level-specific descriptions
Lives at home. Needs no care assistance. Performs normal activities and can work.	KPSS-100	Has no symptoms or signs of disease.
	KPSS-90	Has only minor symptoms or signs of disease.
	KPSS-80	Has significant symptoms or signs of disease. Must exert effort to perform normal activities.
Lives at home. Mostly able to care for self but sometimes needs assistance. Cannot work.	KPSS-70	Cares for self. Cannot perform normal activities.
	KPSS-60	Cares for self most of the time. Needs occasional assistance.
	KPSS-50	Needs considerable medical care and assistance.
Cannot care for self: needs institutional-level assistance. Disease may progress rapidly.	KPSS-40	Moderately disabled. Needs significant assistance.
	KPSS-30	Severely disabled. Needs hospital or other institutional care although death is not imminent.
	KPSS-20	Severely disabled. Needs intensive institutional care.
	KPSS-10	Moribund. Fatal disease progresses rapidly.
	KPSS-0	No function: dead.

Adapted from Doyle et al. [38], with permission

predict cancer patients' ability to care for themselves [4], their need for hospitalization [28, 39], and their response to chemotherapy [39]. Some doctors then tried to apply the scale to predicting survival, too. The KPSS proved only moderately useful for that task. Low KPSS scores accurately predicted short survivals: Any patient with a score below KPSS-50 % did not survive longer than 6 months. KPSS scores higher than 50 %, however, did not necessarily predict longer survivals possibly because a dying patient's function may deteriorate suddenly, drastically, and irreversibly [4]. The KPSS also proved only modestly discriminating at low scores. The clustering there of many poorly functioning, advanced cancer patients limited the KPSS's power to discriminate the varying survivals among those patients [40]. Furthermore, the KPSS had low interrater reliability, that is, low reproducibility of ratings across observers. Some commentators, therefore, criticized the scale for excessive subjectivity [37]. For all those reasons the KPSS has been inadequate as a survival prediction aid for clinical purposes [28].

Despite the failure of the KPSS as a survival prediction aid, researchers recognized the intuitive sense of tying survival to function in prognostic schemes. Many studies subsequently validated that intuition [30]. Researchers, therefore, kept the KPSS' basic functional template but added to it two other recently identified predictor variables, nutritional intake and level of consciousness, to create the new Palliative Performance Scale (PPS) (Table 5.3) [18, 41]. The functional part of the PPS, like the KPSS, tracks basic patient functions such as walking, providing self-care, and participating in other activities. Overall, the PPS has eleven grades from PPS-100 % (mobile and active) to PPS-0 % (dead) [18, 30, 39]. Those grades parallel various extents of disease. Function is the most powerful predictor variable in the PPS. Nutrition, defined as either poor appetite or weight loss, is the next most powerful predictor variable, and level of consciousness is the least powerful [5, 42, 43]. The sequence of steps for using the PPS takes account of the three variables' different predictive powers in determining a patient's PPS score.

Despite the similar functional templates of the KPSS and the PPS, some commentators see the PPS as more structured and more objective than the KPSS [44]. The limited follow-up data, which focus on cancer [18, 45] and palliative care patients [40, 42], support that impression. Unlike the KPSS the PPS has proven its discriminatory predictive power for survival all along its scale except in one or two short mid-scale intervals [18].

Trying to improve further on the PPS, some researchers have tested additional predictor variables, but those variables have made only small differences in accuracy for predicting survival. Symptoms (including shortness of breath, dry mouth, and swallowing difficulties) and physical signs (including mental changes) have proven modestly predictive [20, 30, 39, 46, 47]. Laboratory data (including white blood cell count and lymphocyte count) have proven only slightly predictive [20]. Social and psychological variables have not proven predictive at all [42]. The Palliative Prognostic Index, another scale for predicting the survival of terminal cancer patients, uses three of the same variables as the PPS (functional status, oral intake, and presence or absence of delirium) and adds two other variables, presence or absence of shortness of breath at rest and of swelling [30].

Table 5.3 Victoria Hospice Society Palliative Performance Scale, Version 2 (PPSv2)[a]

PPS level	Ambulation	Activity and evidence of disease	Self-care	Intake	Conscious level
100 %	Full	Normal activity and work No evidence of disease	Full	Normal	Full
90 %	Full	Normal activity and work Some evidence of disease	Full	Normal	Full
80 %	Full	Normal activity with effort Some evidence of disease	Full	Normal or reduced	Full
70 %	Reduced	Unable normal job/work Significant disease	Full	Normal or reduced	Full
60 %	Reduced	Unable hobby/house work Significant disease	Occasional assistance necessary	Normal or reduced	Full or confusion
50 %	Mainly sit/lie	Unable to do any work Extensive disease	Considerable assistance required	Normal or reduced	Full or confusion
40 %	Mainly in bed	Unable to do most activity Extensive disease	Mainly assistance	Normal or reduced	Full or drowsy ± confusion
30 %	Totally bed bound	Unable to do any activity Extensive disease	Total care	Normal or reduced	Full or drowsy ± confusion
20 %	Totally bed bound	Unable to do any activity Extensive disease	Total care	Minimal to sips	Full or drowsy ± confusion
10 %	Totally bed bound	Unable to do any activity Extensive disease	Total care	Mouth care only	Drowsy or coma ± confusion
0 %	Death	–	–	–	–

PPS = Palliative Performance Scale.
[a]Accessed at www.npcrc.org/files/news/palliative_performance_scale_PPSv2.pdf, 24 August 2016
[b]Instructions for use of PPS (see also definition of terms)
1. PPS scores are determined by reading horizontally at each level to find a "best fit" for the patient, which is then assigned as the PPS % score.
2. Begin at the left column and read downward until the appropriate ambulation level is reached and then read across to the next column and downward again until the activity/evidence of disease is located. These steps are repeated until all five columns are covered before assigning the actual PPS for that patient. In this way, "leftward" columns (columns to the left of any specific column) are "stronger" determinants and generally take precedence over others.
Example 1: A patient who spends a majority of the day sitting or lying down due to fatigue from advanced disease and requires considerable assistance to walk even for short distances but who is otherwise fully conscious level (sic) with good intake would be scored at PPS 50 %.
Example 2: A patient who has become paralyzed and quadriplegic requiring total care would be scored at PPS 30 %. Although this patient may be placed in a wheelchair (and perhaps seem initially to be at 50 %), the score is 30 % because he or she would be otherwise totally bed bound due to the disease or complication if it were not for caregivers providing total care including lift/transfer. The patient may have normal food intake and be with full conscious level.

Example 3: However, if the patient in example 2 is paraplegic and bed bound but still able to do some self-care such as feed themselves (sic), then the PPS would be higher at 40 or 50 % since he or she does is (sic) not "total care".
3. PPS scores are in 10 % increments only. Sometimes, there are several columns easily placed at one level but one or two that seem better at a higher or lower level (sic). One then needs to make a "best-fit" decision. Choosing a "half-fit" value of PPS 45 %, for example, is not correct. The combination of clinical judgment and "leftward precedence" is used to determine whether 40 or 50 % is the more accurate score for that patient.
4. PPS may be used for several purposes. First, it is an excellent communication tool for quickly describing a patient's current functional level. Second, it may have value in criteria for workload assessment or other measurements and comparisons. Finally, it appears to have prognostic value.
Definition of Terms for PPS
As noted below, some of the terms have similar meanings with the differences being more readily apparent as one reads horizontally across each row to find an overall "best fit" using all five columns.
1. Ambulation
The items "mainly sit/lie," "mainly in bed," and "totally bed bound" are clearly similar. The subtle differences are related to items in the self-care column. For example, "totally bed bound" at PPS 30 % is due to either profound weakness or paralysis such that the patient not only can not get out of bed but is also unable to do any self-care. The difference between "sit/lie" and "bed" is proportionate to the amount of time the patient is able to sit up versus need (sic) to lie down.
"Reduced ambulation" is located at the PPS 70 % and PPS 60 % level. But using the adjacent column, the reduction of ambulation is tied to inability to carry out their (sic) normal job, work occupation or some hobbies, or housework activities. The person is still able to walk and transfer on their (sic) own but at PPS 60 % needs occasional assistance.
2. Activity and Extent of Disease
"Some," "significant," and "extensive" diseases refer to physical and investigative evidence which shows degrees of progression. For example, in breast cancer, a local recurrence would imply "some" disease, one or two metastases in the lung or bone would imply "significant" disease, whereas multiple metastases in lung, bone, liver, or brain, hypercalcemia, or other major complications would be "extensive" disease. The extent may also refer to progression of disease despite active treatments. Using PPS in AIDS, "some" may mean the shift from HIV to AIDS, "significant" implies progression in physical decline, new or difficult symptoms and laboratory findings with low counts. "Extensive" refers to one or more serious complications with or without continuation of active antiretrovirals, antibiotics, etc.
The above extent of disease is also judged in context with the ability to maintain one's work and hobbies or activities. Decline in activity may mean the person still plays golf but reduces from playing 18 holes to 9 holes, just a par 3, or to backyard putting. People who enjoy walking will gradually reduce the distance covered although they may continue trying sometimes even close to death (e.g., trying to walk in halls).
3. Self-Care
"Occasional assistance" means that most of the time patients are able to transfer out of bed, walk, wash, toilet, and eat by their own means, but that on occasion (perhaps once daily or a few times weekly), they require minor assistance.
"Considerable assistance" means that regularly every day the patient needs help, usually by one person, to do some of the activities noted above. For example, the person needs help to get to the bathroom but is then able to brush his or her teeth or wash at least hands and face. Food will often need to be cut into edible sizes, but the patient is then able to eat of his or her own accord.
"Mainly assistance" is a further extension of "considerable." Using the above example, the patient (sic) now needs help getting up but also needs assistance washing his face and shaving, but can usually eat with minimal or no help. This may fluctuate according to fatigue during the day.

"Total care" means that the patient is completely unable to eat without help, toilet or do any self-care. Depending on the clinical situation, the patient may or may not be able to chew and swallow food once prepared and fed to him or her.
4. Intake
Changes in intake are quite obvious with "normal intake" referring to the person's usual eating habits while healthy. "Reduced" means any reduction from that and is highly variable according to the unique individual circumstances. "Minimal" refers to very small amounts, usually pureed or liquid, which are well below nutritional sustenance.
5. Conscious Level
"Full consciousness" implies full alertness and orientation with good cognitive abilities in various domains of thinking, memory, etc. "Confusion" is used to denote presence of either delirium or dementia and is a reduced level of consciousness. It may be mild, moderate, or severe with multiple possible etiologies. "Drowsiness" implies either fatigue, drug side effects, delirium, or closeness to death and is sometimes included in the term stupor. "Coma" in this context is the absence of response to verbal or physical stimuli; some reflexes may or may not remain (sic). The depth of coma may fluctuate throughout a 24-h period.

Far less research has focused on survival prediction aids for other potentially fatal diseases. However scarce, the data on these diseases consistently identify prior patient function as the most powerful predictor of survival [48–50]. Other important predictors include patient age [49, 51], the particular disease [51], and certain physiologic variables such as severity-of-illness scores [48, 51].

Some newer prediction aids take the form of equations. These equations, however, usually have narrow derivation conditions that only occasionally apply precisely to the clinical situations at hand. In addition, the equations incorporate many variables, require extensive clinical assessments, and involve complicated arithmetic computations. For all these reasons, such aids have typically proven too cumbersome to use in everyday clinical practice [51].

Nonetheless, I believe a few simple concepts underlying all these aids *can* help doctors make survival predictions. For example, with a cancer the doctor might first tether a preliminary prediction of the patient's survival to the general survival data for the type and stage of cancer. The doctor might next intuitively modify that prediction up or down depending mostly on the patient's functional abilities (walking and self-care) and nutritional indicators (appetite and weight loss). The PPS can prove helpful at this point. Then, the doctor might modify the prediction in minor ways by considering certain symptoms (shortness of breath, dry mouth, and swallowing difficulties) and signs (mental changes). With other potentially fatal diseases the doctor might base his or her survival predictions on some intuitive mix of published data about the disease, his or her own direct experience with it, and the patient's disease severity and age. But, to emphasize again, the doctor must state plainly (and perhaps repeatedly) to patients, families, and others that all survival predictions are mere guesses and that unforeseen events may intervene at any time to change those predictions.

Conceptual Aids for Predictions About Function

Fatally ill patients worry not only about how long but also about how well they will live until death [52]. They dread a poor quality of life at the end. When asked to describe components of such a quality of life, many patients mention functional decline, changing social roles, symptoms (especially pain), loss of control over dying, financial burdens, stress on the family, and ambivalences over wanting to know medical details and fearing bad news [2]. Functional decline may be the most important of all these components because it can demoralize patients by itself and exacerbate so many of the other components.

Modern American culture values privacy and independence highly. Perhaps for that reason many dying American patients worry most about losing the ability to perform basic personal care functions such as dressing, eating, and toileting. These self-care dependencies and the increasing care-receiving roles that accompany them can damage patients' self-image and burden their families. Dying patients and their families may, therefore, need some forewarning to prepare themselves emotionally for the functional declines ahead. If patients and family members do not broach the subject themselves, doctors or other HPs may have to [36].

The results of one clever study, conducted as part of a broader study, can help with this informing process [53]. The broader study interviewed a large, representative sample of American elders annually over many years. One question in the interviews asked about people's ability to perform seven "activities of daily living" (ADLs): bathing, grooming, dressing, eating, toileting, transferring between bed and chair, and walking a short distance (such as across a small room). The researchers identified the fatal diseases that killed decedents in the study over one particular year and grouped those diseases into five major categories—sudden death, cancer, major organ failure, progressive frailty, and miscellaneous others. Each category represented about 20 % of deaths. The researchers then plotted the average number of ADLs decedents in each fatal disease category could perform by the number of months between the decedents' last annual interviews and their deaths. Each disease category included enough decedents interviewed in each of the 12 months before death to allow plotting the whole group's ADL functional decline over the year.

Four of the five disease categories showed distinct patterns of decline (see Fig. 5.1). Sudden death showed full function until the simultaneous loss of all ADLs at death. Massive blood clots to the brain or lungs (i.e., strokes or pulmonary emboli, respectively) illustrate this disease category. Cancer (identified as 'terminal illness' in the figure) showed high initial function, either full or decreased by only one ADL over many months. But then as many as four ADLs disappeared quickly during the last 3 months before death. Solid organ cancers such as cancers of the colon or lungs illustrate this category. Major organ failure started the last year with a modest functional deficit of about two ADLs. A steady overall loss of another two ADLs followed. Meanwhile acute exacerbations and partial recoveries superimposed their waxing and waning pattern over the general downward functional trend until one exacerbation suddenly caused death. Late-stage heart failure and

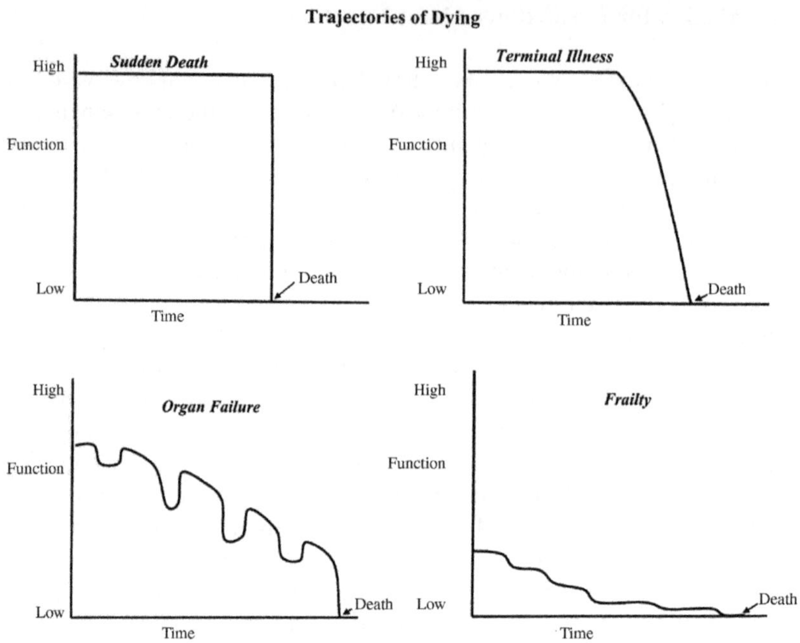

Fig. 5.1 Functional trajectories of dying. The ordinate shows the average number out of seven "activities of daily living"—bathing, grooming, dressing, eating, toileting, transferring between bed and chair, and walking a short distance (such as across a small room)—elders could perform at the time of their last annual study interviews. The abscissa counts down in months the time intervals between those interviews and the elders' deaths within the year. The *rightmost point* in each graph designates time of death for all elders in the particular fatal disease category. (From Lunney et al. [64], with permission)

emphysema illustrate this disease category. Progressive frailty started the last year at a low functional level. Decedents had already lost three or more ADLs. Function declined another three or so ADLs through the year in a downward, stair-step pattern. Temporary plateaus of stable function interrupted the otherwise steady declines. The progressive dementias common among nursing home residents illustrate this category. The unclassifiable, "miscellaneous other" diseases showed no distinct pattern. Long-term atherosclerotic heart disease illustrates this disease category and comprised nearly half of its diseases.

These graphs of functional decline may explain why more cancer patients than any others populate hospices. Of all five fatal disease categories, only cancer shows a sharp, predictable final functional downturn that leaves sufficient time for a hospice referral. Unfortunately, recognizing the exact time when that irreversible downturn toward death occurs is hard, perhaps explaining the late chemotherapies and late hospice referrals that many advanced cancer patients receive [29].

The Case

Like many medically untrained patients Mr. F. begins his cancer treatment with overly simple expectations. They arise partly due to the gastroenterologist's inadequate explanations about the disease and its treatment. He explicitly mentions only surgery. Mr. F., therefore, thinks "they just cut out your cancer, and you get back on your feet in two to three weeks. You go back to living your usual life. You're cured." Mr. F. does not anticipate the several months of arduous preoperative radiation and chemotherapy. Those treatments eventually cause him severe fatigue, loss of appetite, and weight loss. When Mr. F. must be hospitalized as a result, he thinks the treatments themselves are killing him. "If someone had told me what this whole process would involve," he says, "I would have thought twice about starting it."

Mr. F. is not afraid to die but does not want to do so just yet. He badly wants to see his youngest granddaughter grow up. This deeply human desire would surely fuel "a planning fallacy" about treatment: In Mr. F.'s mind the anticipated benefits (a prolonged life enjoying his granddaughter) would always outweigh the adverse effects (specifically the side effects of radiation and chemotherapy before the surgery and the fatigue and colostomy after the surgery). I believe Mr. F. would have chosen full treatment even if he had always understood its harsh reality. As he tells Mrs. F. at one point, "I either take the treatments the doctors tell me to take, or I die." Foreknowledge of the treatment side effects likely would not have changed his final decision.

But Mr. F. struggles with the frustration that "no one tells you what to expect, how you're doing. You're just told to go to the specialists." The gastroenterologist rightfully bears some responsibility for this frustration. The doctor should use the prognostic aids available to him to explain to Mr. F. the possible benefits and burdens of treatment. The doctor could mention the 40–65 % 5-year survival rate with full treatment. He could explain the likely duration and side effects of preoperative radiation and chemotherapy. And he could describe the typical functional trajectory of cancer and assure him about long-term postoperative function especially concerning basic personal care. Furthermore, the doctor should engage a specially trained nurse to visit Mr. and Mrs. F. preoperatively to prepare them for colostomy care later. Above all, the doctor should reassure Mr. F. that he will not burden his family with his personal care in the foreseeable future or perhaps ever. Such an informative approach with predictions concerning survival and interim function might go far in correcting Mr. F.'s misguided expectations and calming his fears.

Patients' Attitudes About Doctors' Clinical Predictions

Some doctors hesitate to make clinical predictions out of concern that the medically untrained public will not understand those predictions correctly. The public in general, of course, does not have the clinical sophistication of trained HPs. But that fact does not necessarily mean the public takes doctors' predictions too literally, dismisses them too readily, or does not appreciate the uncertainties behind them [31].

I believe, in fact, the public *does* take a remarkably perceptive view of clinical predictions. People do appreciate, for example, the potential inaccuracy involved. One study found that 88 % of proxies of intensive care unit patients expect and accept such inaccuracy [54]. The public attributes that inaccuracy often to incomplete human knowledge about medical matters but occasionally to God's working unexpected miracles in human affairs. I also believe the public views doctors' clinical predictions not simply as baseless speculation or perfect foreknowledge but as something more complicated: valuable "best guesses" about the future based on doctors' past clinical experiences. The public's confidence in those guesses increases, of course, when the predictions of multiple doctors agree or when the predictions come true [54]. Whatever inaccuracy in the details, doctors' honest, well-reasoned, best-guess predictions surely can help patients and family members prepare mentally for the future. For instance, those predictions can help patients and family members brace themselves for hardships ahead [54]. Professionals of all kinds who attend the dying should, therefore, be able to discuss clinical predictions honestly and supportively with patients and families.

The Case

After his surgery Mr. F. repeatedly asks his surgeon, the nurses, and other HPs how he is doing. None answer him straightforwardly. For example, Mr. F. asks the surgeon immediately after the cancer resection, "How do we know whether the operation worked?" The surgeon evades the question. "You were pretty sick," he says. "I cut out a lot of tissue. You'll be recovering for months. I will see you once after discharge but then won't need to see you again for a year." Such answers frustrate Mr. F. He feels "stonewalled." Yet, as long as he responds "well" to standard treatments, he tolerates the evasiveness.

While he cannot get a clear, specific prediction for himself, Mr. F. takes an appreciative but realistic view toward doctors' clinical predictions in general. He sees them as valuable educated guesses, which sometimes may be wrong. He also says he would not question a doctor's competence if the doctor admits to uncertainty about the future or makes a prediction that proves wrong. "But I hate being kept purposely in the dark," Mr. F. explains. "I'm an impatient patient. The doctors should educate me, tell me what is going on. I don't like the feeling that I'm the only one who doesn't know what will happen to me."

Prognostic Discussions with Seriously or Terminally Ill Patients

Timing, Barriers, and Benefits

Patients with potentially serious or terminal illnesses should be offered candid prognostic discussions at some point for the benefits they can provide. Still, doctors who undertake these discussions must consider suitable timing and potential barriers for them.

Timing for Prognostic Discussions

Timing for prognostic discussions especially near the end of life presents a challenge simply because no time ever seems right. Nonetheless, doctors may look for certain signs as cues for such discussions. The cues may include increasing disease severity, decreasing responsiveness to treatment, increasing frequency or duration of hospitalizations, chronically diminished appetite or weight loss, the patient's loss of the will to live, or just the patient's request for the discussion [55–58]. Perhaps the most common cue is whatever expected survival time intervals define terminality for the immediate attending doctors. Those time intervals, however, may vary widely. Many doctors use the time interval of 6 months, Medicare's definition of terminality for hospice eligibility. Still, a nationwide survey revealed that American internists consider expected survival time intervals anywhere from 1 to 75 weeks as defining terminality for them. The more experienced internists tend to favor the shorter intervals [3]. The wide range and arbitrariness of these intervals suggest that no single one will ever be acceptable to all doctors [3].

The Case

Mr. F.'s case, of course, does not present many of the cues that typically prompt doctors to have prognostic discussions with seriously or terminally ill patients. Mr. F. does not have increasing disease severity, decreasing responsive to treatment, or increasing frequency of hospitalizations. Specifically, he shows no signs of residual cancer after the surgery and does not need follow-up hospitalizations for any reason. Furthermore, he regains his appetite and some of his lost weight. He also firmly states he wants to live and participates again in activities meaningful to him. He even resumes his architectural practice and the afternoon homework sessions with his granddaughter. And with Mr. F.'s good chance, 40–65 %, to live at least another 5 years, no one would consider him "terminal" by his expected survival time.

Nonetheless, by asking the question, "How do we know whether the operation worked?", Mr. F. is obviously requesting a prognostic discussion. That moment is a suitable time to begin the discussion.

Barriers to Prognostic Discussions

Whatever cues doctors use to initiate them, prognostic discussions near the end of life should probably be more frequent and more substantive than they are. Two studies support this conclusion. One study of terminal cancer patients examined medical records for evidence of such discussions. Only 38 % of records documented any prognostic discussions, and only 9 % documented estimates of expected survival times [59]. The other study asked elders seriously ill with cancer, heart failure or chronic lung disease, and their doctors whether they had ever discussed prognosis with each other. The patients and their doctors agreed only about 20 % of the time, a poor level of agreement beyond chance, that such discussions had occurred [60]. Both of these studies, therefore, suggest end-of-life prognostic discussions occur far too infrequently.

Why do they occur so infrequently? I believe significant barriers exist for all parties who have a stake in the discussions—doctors, patients, families, and institutions. But I focus here on barriers for doctors for two reasons: First, these discussions cannot realistically occur without the doctors' taking the lead in promoting them and, second, society expects the doctors to do so [22].

Table 5.4 lists some common barriers to prognostic discussions for doctors [3, 6, 14, 36, 58, 60, 61]. Three of the barriers relate to timing. One is doctors' reluctance to discuss a bad prognosis unless firmly established by clinical data [8]. Whenever the data are uncertain (and they often are), doctors tend to say nothing rather than to risk giving inadvertently misleading information. Another barrier is the false belief that patients will ask for what they want to know. This belief causes many doctors to wait to disclose a bad prognosis until patients ask for it. But many patients wait simultaneously for the doctors to initiate the disclosure. As each side waits for the other to start the disclosure, too often no one does so. A third barrier is the belief that patients are either too sick or not sick enough at the moment to hear a bad prognosis. Because patients may *always* be perceived as either too sick or not sick enough for the discussion, it may never occur. All three of these barriers, then, discourage end-of-life prognosis discussions. One author calls the phenomenon "a collusion" between doctors and patients to postpone the discussion forever [57].

Potential adverse reactions of patients and family members may also create for doctors significant barriers to prognostic discussions [1]. Doctors may anticipate that patients and family members may not understand a bad prognosis, may not retain it accurately for long, or may not know how to respond constructively. Doctors may also worry that disclosing a bad prognosis may cause patients or family members either an immediate emotional breakdown or some longer-term mental or physical harms, any of which may prove difficult to control [1]. Furthermore, doctors may encounter families who ask doctors to compromise their professional ethics by keeping prognoses from patients altogether [6].

The emotional stresses on doctors themselves create additional barriers to prognostic discussions. Doctors may not know for certain what and how much information patients or family members need or want. Doctors may also lack confidence in their ability to deliver a fatal prognosis accurately and

Table 5.4 Barriers for doctors to having prognostic discussions with seriously or terminally ill patients

Category of barriers	Specific barriers	References
Timing of discussion	Reluctance to discuss a bad prognosis if not yet firmly established by the clinical data	Bradley et al. [59], Hancock et al. [6], Helft [57], Lamont and Christakis [23]
	Belief that doctor should wait for patient to ask first about prognosis	Hancock et al. [6], Helft [57]
	Belief that patient is either too sick or not sick enough for a prognosis discussion	Hancock et al. [6], Jackson et al. [61]
Adverse reactions of patients or family	Doubts about whether patient or family will understand prognosis correctly	Hancock et al. [6], Jackson et al. [61], Kiely et al. [36], Murphy et al. [14]
	Doubts about whether patient or family will retain discussion content accurately for long	Fried et al. [60], Jackson et al. [61]
	Worry that patient or family may not know how to respond constructively	Hancock et al. [6]
	Fears of an immediate, uncontrollable emotional breakdown for patient or family	Hancock et al. [6]
	Fears about long-term mental or physical harms for patient or family	Bradley et al. [59], Hancock et al. [6]
	Worry that family might ask doctor to compromise professional ethics by keeping the prognosis from patient altogether	Hancock et al. [6]
Emotional stresses for doctors	Uncertainty about what and how much information patient or family needs or wants	Jackson et al. [61]
	Lack of confidence in one's ability to communicate a bad prognosis accurately and compassionately	Bradley et al. [59], Hancock et al. [6], Lamont and Christakis [23]
	Discouragement in the face of a fatal disease without curative treatment	Hancock et al. [6]
	Temptation to give a more hopeful prognosis than warranted	Christakis and Iwashyna [3]
	Grief for the dying patient	Hancock et al. [6]

compassionately [6]. They may even become discouraged at having to acknowledge a fatal disease without being able to offer curative treatment. And among the barriers I mentioned earlier, doctors may struggle with the temptation to give better, more hopeful prognoses than warranted [23, 30]. Many doctors understandably just prefer to give seriously ill patients happy rather than sad news. And doctors may grieve for dying patients whom they know well or have worked hard to save.

Table 5.5 Benefits to having prognostic discussions with seriously or terminally ill patients

Specific benefits	References
Opportunity to fulfill a doctor's duty to give patients (or proxies) important prognostic information	Hancock et al. [6]
Prompt for doctors to formulate predictions to guide further care	Bradley et al. [59], Perkins [62]
Prompt for patient to rethink priorities for his or her remaining life	Kiely et al. [36]
Emotional preparation of patient when death is imminent	Kiely et al. [36], Lynn [63], Perkins [62]

Benefits to Prognostic Discussions

Despite the potential barriers to end-of-life prognostic discussions, common sense suggests they can have significant benefits for doctors and patients alike (Table 5.5) [6, 36, 59, 62, 63]. Such discussions fulfill a doctor's duty, now firmly embedded in medical ethics, to disclose prognoses to patients as an important part of informed consent [6]. These discussions also prompt doctors to formulate as working hypotheses explicit tentative outcome predictions [62] for use in guiding end-of-life care [59, 62]. Of course, such "working hypothesis" predictions may require changes as patients' circumstances change. The discussions may also prompt patients to revise priorities for their remaining lives. Patients may decide to relive meaningful past events (say, through photographs or calls to family or close friends), arrange for significant legacies, repair damaged relationships, say important good-byes, address end-of-life financial or legal affairs, or enroll in medical research [36]. And, when predictions point to imminent death, prognostic discussions may help prepare patients emotionally for it [36, 62, 63]. For example, those discussions may reorient patients mentally from curative to palliative care and lead them to accept timely hospice referral.

Guidelines to Discussing Prognosis with a Seriously Ill Patient

How might doctors or other HPs discuss a bad prognosis, perhaps a fatal one, with a seriously ill patient? One article gives especially clear steps for doing so [61]. Some of those steps resemble the steps in delivering other kinds of medical bad news such as a worrisome diagnosis or a failed treatment. (The reader might consult Chap. 4 for a broad overview of delivering medical bad news.) For simplicity's sake I assume here that the principal doctor managing the patient's care leads the prognostic discussion. Some circumstances, however, may require other HPs do that. If so, those other HPs should coordinate their planned disclosures with the principal doctor to avoid saying anything that unnecessarily contradicts prior disclosures or causes confusion.

Table 5.6 outlines the steps in the discussion process and suggests questions for introducing some of the steps. The principal doctor should review the patient's

Table 5.6 Steps in discussing a bad prognosis

Step	Substeps	Description of step	Sample introductory questions (if relevant)
1		Review the patient's illness course. Read up on the particular disease. Understand the rationale for the diagnostic and therapeutic plan.	N/A[a]
2		Assess the patient's readiness to discuss prognosis.	
	2a	Assess the patient's current understanding of the illness.	"How do you think you are doing now?"
	2b	Ask whether the patient has ever thought about future complications.	"Have you ever thought what it would be like if you become more ill?"
	2c	Determine timing for the prognostic disclosure.	"Should I discuss prognosis now or wait? If I wait, when would a better time come? Why would it be better?"
3		Determine an approach to the discussion based on the patient's readiness to discuss prognosis and on the urgency of the situation.	
	3a	The patient prefers to discuss prognosis right away whatever the urgency of the situation. Begin with general, open-ended introductory questions.	"What is your body telling you about what is ahead?"
	3b	The patient hesitates to discuss prognosis, and there is no immediate urgency to do so. Mention the benefits of discussing prognosis, but do not force the discussion. Offer the opportunity for a discussion later.	"May we schedule a later time to discuss what might be ahead for you? How about (name a time soon)?"
	3c	The patient hesitates to discuss prognosis, and there is an immediate urgency to do so. Introduce the subject deftly but firmly.	"We need to talk about what lies ahead for you. It is important. May we do it now?"
4		Give prognostic information tailored to the patient's wishes and needs. Many patients consider predictions of survival time and function most important; others consider prediction of symptoms and strategies of care most important. Always use everyday language and explain key terms.	N/A
5		Document the discussion in a progress note. Include the time, date, and place; the participants involved; the information disclosed: any questions the patient or others asked; the answers given; and the informational areas that still need addressing.	N/A

Data from Jackson et al. [61]
[a]N/A = not applicable.

illness course before the discussion (Step 1). If that doctor does not specialize in the patient's particular disease, he or she might read up about it in, say, general medical textbooks. (Nondoctors might consult general nursing textbooks or reliable medical Internet sites such as WebMD.) The doctor should consult knowledgeable colleagues about any important lingering questions. The doctor should initiate the prognostic discussion with the patient only when he or she thoroughly understands the rationale for the diagnostic and therapeutic plan. This preliminary review often suggests how urgent the prognostic discussion is.

The doctor should approach the discussion by assessing first the patient's readiness for it (Step 2). This step might include three substeps. The doctor might initially ask the patient's current understanding of his or her illness (Step 2a). A simple question such as "How do you think you are doing?" might suffice to open the conversation. The doctor might then assess whether the patient has already anticipated future complications (Step 2b). The doctor might ask, "Have you thought about what it would be like if you ever become more ill?" I find asking that question in a normal conversational tone avoids alarming the patient. The patient's reply may then help the doctor decide on timing the prognostic disclosure (Step 2c). The doctor might ask him- or herself, "Should I discuss prognosis with this patient now or wait for another time? If not now, when would be a better time, and why would that time be better?" The doctor should ordinarily have sound reasons for delaying. But it might be justified, for example, if a definitive diagnosis is not yet firmly established, if a crucial therapy has not yet declared itself a success or failure, or if key family members have not yet arrived to give the patient support. The doctor, of course, must not delay the discussion excessively. Setting a reasonable time limit for initiating it and adhering to that time limit can help prevent such unwise delay.

Patient preferences and the urgency of the situation often lead the doctor to one of the three conclusions at this point (Step 3). First, the patient prefers to discuss prognosis right away whatever the urgency of the situation (Step 3a). The doctor might act on this conclusion by asking such questions as "What is your body telling you now about what is ahead?" Second, the patient hesitates to discuss prognosis, and no urgency exists to do so (Step 3b). The doctor might act on this conclusion by sensitively acknowledging the fearsomeness of discussing the future of an illness and also by explaining the possible benefits of doing so. Those benefits include allowing time to digest the prognostic information, to develop good coping tactics, to reevaluate treatment goals, and perhaps to make a timely hospice referral [18, 30, 39, 57, 61]. Of course, the doctor should never force the discussion on such a patient. Rather, the doctor might offer the opportunity for a discussion later. Third, the patient hesitates to discuss prognosis, and some urgency for the discussion exists (Step 3c). Causes for such urgency might include uncontrollable, increasingly distressing symptoms; failing treatments aimed at cure; a life-threatening downturn in the patient's condition; patient or family requests for life support likely to be ineffective; or imminence of death [6, 39]. The doctor should act on this conclusion by pressing deftly but firmly to have the discussion. He or she might say to the patient, "We need to talk about what lies ahead for you. It's important. We should do so as soon as possible. May we do it now?"

In all cases the doctor must tailor the content of the prognostic discussion to the information the patient wants or needs (Step 4). Unfortunately, no clinical or demographic factors reliably help the doctor know beforehand what that information is. Society may expect the patient to die in bed, at home, and with his or her life business complete [63], but that ideal occurs only occasionally. In fact, expectations about dying vary among patients, and not all of them want society's ideal. Therefore, the doctor must ask what each one already understands and still wants to know. The doctor must then listen carefully to learn how to meet each patient where he or she is.

Many patients consider survival time and functional course the most important parts of a terminal prognosis. The doctor might describe to those patients the best, worst, and most likely dying scenarios based in part on the range of expected survival times [36]. Such an approach may avoid many overly optimistic predictions. Other patients, however, may rather learn about future symptoms, comfort interventions, personal control in dying, and "dignity" measures [1, 39, 63]. The doctor might emphasize to those patients strategies of care. Such an approach would aim to ease worries about prolonged, painful deaths. The doctor should, of course, always use everyday language and explain key terms. The doctor should only rarely confront the patient with prognostic information he or she really does not want to hear—and then only when absolutely necessary.

Finally, the doctor conducting the prognostic discussion must document it in the patient's medical record (Step 5). That documentation informs other HPs what information the doctor, the patient, and the family members have already discussed and how the various participants reacted. The documenting progress note should give the time, date, and place of the discussion; the people attending; the information disclosed; any questions the patient or others asked; the answers the doctor or other HPs gave; how participants took the news; and the informational areas that still need addressing.

The Case

Mr. F.'s illness presents numerous opportunities for prognostic discussions. An intelligent, motivated patient, Mr. F. asks the doctors and nurses repeatedly about his prognosis. But the HPs fumble every opportunity for a substantive discussion. As a result Mr. F. never gets key information about his expected survival time or future function.

The prognostic communication failures begin at the office visit when the gastroenterologist first tells Mr. F. he has colon cancer. Mr. F. asks the typical question, "How long do I have to live?" Although the gastroenterologist surely knows the 5-year survival data for colon cancer, he does not answer the question. He simply mentions surgery. Mr. F. concludes the cancer will be cut out, he will be cured, and his normal life will resume again. But the treatment is not so simple. The gastroenterologist refers Mr. F. without explanation to a "radiation cancer doctor" and a "chemical cancer doctor" for preoperative treatment. Those doctors start the treatment but do not clearly explain its possible side effects. The treatment sickens Mr. F. severely. Extreme exhaustion sets in, his appetite vanishes, and his weight drops 50 lb in just a few months. "I'm melting away," Mr. F. thinks, but he never complains to the doctors. They always tell him he is doing well. "I didn't know the right questions to ask," he explains later. "I trusted the doctors to know what they were doing." Mrs. F. decides meanwhile that Mr. F. does not need the preoperative treatments after all. She urges him to stop them. She tells Mr. F., "The doctors are only trying to profit from you." Finally, Mr. F.'s severe weight loss catches his doctors' attention. They hospitalize him, stop the preoperative radiation and chemotherapy, and start treatments to boost his nutritional status.

Discussions about prognosis do not improve in frequency or substance after the surgery excising the colon cancer. Mr. F. asks his surgeon immediately afterward, "How do we know whether the operation worked?" The surgeon replies only tangentially with "You're pretty sick right now. I did major surgery on you. I cut out a lot of colon." A specialized hospital nurse then visits Mr. F. several times over the next few days to teach him colostomy care. Mr. F. asks her how he will do after discharge. "You'll do just fine," the nurse replies. Finally, Mr. F. asks the surgeon again at discharge how he—Mr. F.—will do at home. "You'll do well," says the surgeon tersely. "I won't need to see you again for a year." Mr. F. concludes he will recover over a few weeks and then go back to living life as before, including full-time architectural practice.

Unfortunately, his recovery does not proceed that way. Mr. and Mrs. F. immediately struggle with the practical problems of colostomy care at home. The F.s feel totally unprepared. They have difficulty finding colostomy supplies, changing the bags, and identifying the most compatible foods for Mr. F. to eat. "You don't know what care you need at home until you need it. And then there are so few people to talk to about your problems," he says. The surgeon's office staff provide little practical help. Mr. F. finally asks his accountant brother, a longtime diabetic, for advice. His brother suggests home health care. Mr. F. contacts such an agency, and the staff there assure him they can help. But the agency must await the surgeon's order for service. That order arrives about two weeks later, and the agency at last begins its help for Mr. F. The help goes well. One day when the colostomy care has become nearly routine, Mr. F. asks the home healthcare nurse, "What needs to happen now?" "Your colostomy is doing fine, and you look better. Continue managing it the way you are," she says. "But you'll have to ask your doctor about the colon cancer." Mr. F. still cannot get straight prognostic information about his underlying disease.

Now several years after the surgery, life for Mr. F. has yet to return to what it was before. He is still thin and monitors his diet carefully to avoid diarrhea. He tires easily and must take long naps each morning and afternoon. But he works a few hours a day and helps his granddaughter with her homework. At this point the surgeon might review with Mr. F. the 5-year survival statistics, that is, 40–65 %

survival for patients with his stage of colon cancer at resection. The surgeon might also review Mr. F.'s current Palliative Performance Scale (PPS) score (Table 5.3). Mr. F. ambulates fully and is working with some effort at his architectural practice but has no further evidence of cancer. He needs no help with personal care, has normal or decreased nutritional intake (with a stable but low body weight), and has full consciousness. His PPS score, therefore, is 80 %. The surgeon should tell Mr. F. this number and explain its implications: Mr. F. should do well for the foreseeable future. However, the doctor must remember the high PPS score does not ensure against an eventual downturn in Mr. F.'s health due to the cancer. The surgeon might also explain that the good function Mr. F. now enjoys is the most powerful predictor for an overall good prognosis.

Unfortunately, Mr. F. gets none of this information and is understandably frustrated. He hates being ill-informed about his cancer. As he says, his architectural business requires him to ask specific questions; to get "straight" answers from building contractors; to make timely decisions and clear plans; and to anticipate results in detail. He wonders why HPs do not behave similarly. "I'm the patient, and I know less about what's going on with me than anybody else. I don't like the feeling that I'm the only one who does not know what will happen to me. I don't like being in the dark. So talk to me even if you're not certain. I want to take part in the conversation about me." He then adds, "I realize I have to learn how to be a patient. I don't know what to expect or how to react. And I'm impatient. So educate me." For sure, the healthcare system has not met Mr. F.'s information needs about prognosis.

The system can and must do better about informing patients with potentially fatal illnesses. Important end-of-life plans for patients and survivors alike depend on timely, candid, and thorough prognostic information. Whichever patients want that information should be able to get it.

Summary Points

1. Patients with fatal diseases often worry about survival time and interim function. Though aware of the uncertainties inherent in predictions, patients and family members typically want doctors' predictions about those two variables.
2. Function, both prior and current, is the most important predictor of survival with many fatal diseases. Current nutritional status is the second most important predictor with cancer.
3. Four of five main categories of fatal diseases—sudden death, cancer, organ system failure, and progressive frailty—have distinct downward functional trajectories during the last year of life. Knowing the trajectory of the pertinent disease may help patient, family members, and attending professionals anticipate a patient's course.

To Learn More …

1. Christakis NA. The ellipsis of prognosis in modern medical thought. Soc Sci Med. 1997;44(3):301–15.
2. Jackson VA, Jacobsen J, Greer JA et al. Cultivation of prognostic awareness through the provision of early palliative care in the ambulatory setting: a communication guide. J Palliat Med. 2013;16(8):894–900.
3. Lamont EB, Christakis NA. Prognostic disclosure to patients with cancer near the end of life. Ann Intern Med. 2001;34(12):1096–1105.
4. Lunney JR, Lynn J, Foley D, Lipson S, Guralnik JM. Patterns of functional decline at the end of life. JAMA. 2003;289(18):2387–92.

References

1. Barnett MM. Does it hurt to know the worst? Psychological morbidity, information preferences and understanding of prognosis in patients with advanced cancers. Psycho Oncol. 2006;15(1):44–55.
2. Kutner JS, Steiner JF, Corbett KK, Jahnigen DW, Barton PL. Information needs in terminal illness. Soc Sci Med. 1999;48(10):341–52.
3. Christakis NA, Iwashyna TJ. Attitude and self-reported practice regarding prognostication in a national sample of internists. Arch Intern Med. 1998;158(21):2389–95.
4. Yates JW, Chalmer B, McKegney FP. Evaluation of patients with advanced cancer using the Karnofsky performance status. Cancer. 1980;45(8):2220–4.
5. Teno JM, Weitzen S, Fennell ML, Mor V. Dying trajectory in the last year of life: Does cancer trajectory fit other diseases? J Palliat Med. 2001;4(4):457–64.
6. Hancock K, Clayton JM, Parker SM, et al. Truth-telling in discussing prognosis in advanced life-limiting illnesses: a systematic review. Palliat Med. 2007;21(6):507–17.
7. The SUPPORT Principal Investigators. A controlled trial to improve care for seriously ill hospitalized patients: the study to understand prognoses and preferences for outcomes and risks of treatments (SUPPORT). JAMA. 1995;274(20):1591–8.
8. Christakis NA. The ellipsis of prognosis in modern medical thought. Soc Sci Med. 1997;44 (3):301–15.
9. Council on Ethical and Judicial Affairs, American Medical Association. Guidelines for the appropriate use of do-not-resuscitate orders. JAMA. 1991;265(14):1868–71.
10. Blackhall LJ, Ziogas A, Axen SP. Low survival rate after cardiopulmonary resuscitation in a county hospital. Arch Intern Med. 1992;152(10):2045–8.
11. Ebell MH, Becker LA, Barry HC, Hagen M. Survival after in-hospital cardiopulmonary resuscitation: a meta-analysis. J Gen Intern Med. 1998;13(12):805–16.
12. Ehlenbach WJ, Barnato AE, Curtis JR, et al. Epidemiologic study of in-hospital cardiopulmonary resuscitation in the elderly. N Engl J Med. 2009;361(1):22–31.
13. Tian J, Kaufman DA, Zarich S, American Heart Association National Registry for Cardiopulmonary Resuscitation Investigators, et al. Outcomes of critically ill patient who received cardiopulmonary resuscitation. Am J Respir Crit Care Med. 2010;182(4):501–6.
14. Murphy DJ, Burrows D, Santilli S, et al. The influence of probability of survival on patients' preferences regarding cardiopulmonary resuscitation. N Engl J Med. 1994;330(8):545–9.
15. Elkin EB, Kim SHM, Casper ES, Kissane DW, Schrag D. Desire for information and involvement in treatment decisions: elderly cancer patients' preferences and their physicians' perceptions. J Clin Oncol. 2007;25(33):5275–80.

16. Weeks JC, Catalano PJ, Cronin A, et al. Patients' expectations about effects of chemotherapy for advanced cancer. N Engl J Med. 2012;367(17):1616–25.
17. Mitera G, Zhang L, Sahgal A, et al. A survey of expectations and understanding of palliative radiotherapy from patients with advanced cancer. Clin Oncol. 2012;24(2):134–8.
18. Lau F, Downing GM, Lesperance M, Shaw J, Kuziemsky C. Use of palliative performance scale in end-of-life prognostication. J Palliat Med. 2006;9(5):1066–75.
19. Glare P, Virik K, Jones M, et al. A systematic review of physicians' survival predictions in terminally ill cancer patients. BMJ. 2003;327(7408):195–8 (26 July).
20. Chow E, Harth T, Hruby G, et al. How accurate are physicians' clinical predictions of survival and the available prognostic tools in estimating survival times in terminally ill cancer patients? A systematic review. Clin Oncol. 2001;133(3):209–18.
21. Christakis NA, Lamont EB. Extent and determinants of error in doctors' prognoses in terminally ill patients: prospective cohort study. BMJ. 2000;320(7233):469–73 (19 February).
22. Forster LE, Lynn J. Predicting life span for applicants to inpatient hospice. Arch Intern Med. 1988;148(12):2540–3.
23. Lamont EB, Christakis NA. Prognostic disclosure to patients with cancer near the end of life. Ann Intern Med. 2001;34(12):1096–105.
24. Smith TJ, Longo DL. Talking with patients about dying. N Engl J Med. 2012;367(17):1651–2.
25. Redinbaugh E, Sullivan AM, Block SD, et al. Doctors' emotional reactions to recent death of a patient: cross sectional study of hospital doctors. BMJ. 2003;327(7408):185–9 (26 July).
26. Earle CC, Neville BA, Landrum MB, et al. Trends in the aggressiveness of cancer care near the end of life. J Clin Oncol. 2004;22(2):315–21.
27. Christakis NA, Escarce JJ. Survival of Medicare patients after enrollment in hospice programs. N Engl J Med. 1996;335(3):172–8.
28. Harrold J, Rickerson E, Carroll JT, et al. Is the palliative performance scale a useful predictor of mortality in a heterogeneous hospice population? J Palliat Med. 2005;8(3):503–9.
29. Smith JL. Commentary: Why do doctors overestimate? BMJ. 2000;320(7233):472–3 (19 February).
30. Lamont EB. A demographic and prognostic approach to defining the end of life. J Palliat Med. 2005;8(Suppl 1):S12–21.
31. Lynn J, Teno JM, Harrell FE. Accurate prognostications of death: opportunities and challenges for clinicians. West J Med. 1995;163(3):250–7.
32. McQuaid KR. Ailmentary tract cancers. In: McPhee SJ, Papadakis MA, Tierney LM, editors. 2008 Current medical diagnosis and treatment. New York: McGraw Hill Medical; 2008. p. 1413–4.
33. Chesnutt MS, Murray JA, Pendergast TJ. Pulmonary disorders. In: McPhee SJ, Papadakis MA, Tierney LM, editors. 2008 Current medical diagnosis and treatment. New York: McGraw Hill Medical; 2008. p. 221.
34. Bashore TM, Granger CB, Hranitzky P. Heart disease. In: McPhee SJ, Papadakis MA, Tierney LM, editors. 2008 Current medical diagnosis and treatment. New York: McGraw Hill Medical; 2008. p. 350.
35. Johnston CB, Harper GM, Landefeld CS. Geriatric disorders. In: McPhee SJ, Papadakis MA, Tierney LM, editors. 2008 Current medical diagnosis and treatment. New York: McGraw Hill Medical; 2008. p. 59.
36. Kiely BE, Stockler MR, Tattersall MHN. Thinking and talking about life expectancy in incurable cancer. Semin Oncol. 2011;38(3):380–5.
37. Mor V, Laliberte L, Morris JN, Wiemann M. The Karnofsky performance scale: an examination of its reliability and validity in a research setting. Cancer. 1984;53(9):2002–7.
38. Doyle D, Hanks G, MacDonald N. Oxford textbook of palliative medicine. New York: Oxford University Press; 1993. p. 109.
39. Head B, Ritchie CS, Smoot TM. Prognostication in hospice care: Can the palliative performance scale help? J Palliat Med. 2005;8(3):492–502.
40. Bruera E, Miller MJ, Kuehn N, MacEachern T, Hanson J. Estimate of survival of patients admitted to a palliative care unit: a prospective study. J Pain Symptom Manage. 1992;7(2):82–6.

41. www.npcrc.org/files/news/palliative_performance_scale_PPSv2.pdf
42. Reuben DB, Mor V, Hiris J. Clinical symptoms and length of survival in patients with terminal cancer. Arch Intern Med. 1988;148(7):1586–91.
43. den Daas N. Estimating length of survival in end-stage cancer: a review of the literature. J Pain Symptom Manage. 1995;10(7):548–55.
44. Morita T, Tsunoda J, Inoue S, Chihara S. Validity of the palliative performance scale from a survival perspective. J Pain Symptom Manage. 1999;18(1):2–3.
45. Maltoni M, Nanni O, Pirovano M, et al. Successful validation of the palliative prognostic score in terminally ill cancer patients. J Pain Symptom Manage. 1999;17(4):240–7.
46. Pirovano M, Maltoni M, Nanni O, et al. A new palliative prognostic score: a first step for the staging of terminally ill cancer patients. J Pain Symptom Manage. 1999;17(4):231–9.
47. Morita T, Tsunoda J, Inoue S, Chihara S. Survival prediction of terminally ill cancer patients by clinical symptoms: development of a simple indicator. Japanese J Clin Oncol. 1999;29 (3):156–9.
48. Mayer-Oakes SA, Oye RK, Leake B. Predictors of mortality in older patients following medical intensive care: the importance of functional status. J Am Geriatr Soc. 1991;39 (9):862–8.
49. Wu AW, Damiano AM, Lynn J, et al. Predicting future functional status for seriously ill hospitalized adults: the SUPPORT prognostic model. Ann Intern Med. 1995;122(5):342–50.
50. Levenson JW, McCarthy EP, Lynn J, Davis RB, Phillips RS. The last six months of life for patients with congestive heart failure. J Am Geriatr Soc. 2000;48(5 Suppl):S101–9.
51. Knaus WA, Harrell FE, Lynn J, et al. The SUPPORT prognostic model: objective estimates of survival for seriously ill hospitalized adults. Ann Intern Med. 1995;122(3):191–203.
52. Somogi-Zalud E, Zhong Z, Lynn J, Hamel MB. Elderly persons' last six months of life: findings from the hospitalized elderly longitudinal project. J Am Geriatr Soc. 2000;48(5 Suppl):S131–9.
53. Lunney JR, Lynn J, Foley D, Lipson S, Guralnik JM. Patterns of functional decline at the end of life. JAMA. 2003;289(18):2387–92.
54. Zier LS, Burack JH, Micco G, et al. Doubt and belief in physicians' ability to prognosticate during critical illness: the perspective of surrogate decision makers. Crit Care Med. 2008;36 (8):2341–7.
55. Pfeiffer MP, Mitchell CK, Chamberlain L. The value of disease severity in predicting patient readiness to address end-of-life issues. Arch Intern Med. 2003;163(5):609–12.
56. Aabom B, Kragstrup J, Vondeling H, Bakkesteig LS, Stovring H. Defining cancer patients as being in the terminal phase: Who receives a formal diagnosis, and what are the effects? J Clin Oncol. 2005;23(30):7411–6.
57. Helft PR. Necessary collusion: prognostic communication with advanced cancer patients. J Clin Oncol. 2005;23(13):3146–50.
58. Mack JW, Cronin A, Keating NL, et al. Association between end-of-life discussion characteristics and care received near death: a prospective cohort study. J Clin Oncol. 2012;30 (35):4387–95.
59. Bradley EH, Hallemeier AG, Fried TR, et al. Documentation of discussions about prognosis with terminally ill patients. Am J Med. 2001;111(3):218–23.
60. Fried TR, Bradley EH, O'Leary J. Prognosis communication in serious illness: perceptions of older patients, caregivers, and clinicians. J Am Geriatr Soc. 2003;51(10):1398–403.
61. Jackson VA, Jacobsen J, Greer JA, et al. Cultivation of prognostic awareness through the provision of early palliative care in the ambulatory setting: a communication guide. J Palliat Med. 2013;16(8):894–900.
62. Perkins HS. The varied understandings of "terminal care" and their implications for clinical research and patient care. Med Decis Mak. 2014;34(6):696–8.
63. Lynn J. Serving patients who may die soon and their families: the role of hospice and other services. JAMA. 2001;285(7):925–32.
64. Lunney JR, Lynn J, Hogan C. Profiles of older medicare decedents. J Am Geriatr Soc. 2002;50(6):1108–12

Chapter 6
Advance Care Planning and Preparing to Die

Advance planning is the process of making decisions about one's future medical, legal, or financial affairs in case one becomes incapable of making those decisions later [1–4]. Such planning often addresses end-of-life affairs such as testamentary bequests, organ donation, disposition of the dead body, and funeral arrangements. This chapter focuses only on advance *care* planning (ACP), the kind of advance planning that addresses premortem medical care. ACP commonly tries to answer such questions as What roles should patient, proxies, and doctors take in decision-making? When should aggressive, potentially curative treatments be used? How should their outcomes be evaluated? Should the patient ever enter clinical research trials? When should purely comfort measures begin? and Where does the patient want to die?

Advance care planning appeals intuitively to Americans [1, 5–7]. They know that days, months, or even years of mental incapacity may precede a modern death, and they fear both overtreatment and undertreatment in those interim periods. For that reason most Americans want some control over their end-of-life treatment even if they cannot decide it for themselves at the time [8]. They want to express their wishes early to guide their treatment later.

ACP has taken three forms over its history [8]: instructional legal documents, proxy-designating legal documents, and more informal, dedicated discussions. The two kinds of legal documents are called advance directives (ADs). Instructional ADs give the patient's explicit instructions for treatment in times of future incapacity; proxy-designating ADs, in contrast, name one or more people to make treatment decisions at those times based on the incapacitated patient's previously expressed or currently presumed wishes. Both kinds of ADs have gained legal status throughout the USA [9–11]. Still, many Americans sign neither but simply express that information verbally to their families or doctors. These informal, dedicated discussions are also a kind of ACP.

The relationship of do-not-resuscitate (DNR) orders to ACP deserves special clarification here. Some authors consider DNR orders a kind of ACP [12]; I disagree [13]. DNR orders resemble ACP superficially because they issue an

H.S. Perkins, *A Guide to Psychosocial and Spiritual Care at the End of Life*,
DOI 10.1007/978-1-4939-6804-6_6

instruction about one possible future treatment: never use cardiopulmonary resuscitation (CPR). In contrast, ACP often issues instructions about many other possible future treatments, too, and may designate a proxy decision-maker. More particularly, though sometimes grounded in instructional ADs or ACP discussions, DNR orders do not meet the definition of either (Table 6.1). They need not reflect the patient's explicit or presumed treatment wishes (such as when CPR would be futile or, worse, harmful for the patient) and do not require patient decision-making capacity at signing or patient decision-making *in*capacity at implementation. Furthermore, DNR orders require a doctor's signature, not the patient's, to become clinically and legally valid and never require formal witnessing under law. (They may, however, require cosigning by the patient or by a valid proxy under some hospitals' policies.) I, therefore, do not consider here DNR orders a kind of ACP.

This chapter begins by discussing cues that should prompt ACP. The chapter then describes the three kinds of ACP in the rough chronological order of their widespread acceptance: instructional ADs first, proxy-designating ADs next, and broad ACP discussions last. The descriptions address the origin, content, advantages, and disadvantages of each kind of ACP. Each kind improved on the prior one, but none is perfect. This chapter focuses mostly on dedicated discussions, the kind of ACP I prefer and call "broad ACP." These discussions address a patient's relevant life history, personal values, and goals of care [14]; the mental preparation of patient and survivors for the coming end-of-life experience (including its difficulties and uncertainties); and the doctor's pledge to attend patient and family faithfully throughout. This chapter also discusses Physicians' Orders for Life-Sustaining Treatment (POLST), an AD that suits currently terminally ill patients well and has widespread popularity [15, 16]. The chapter closes by outlining a stepwise approach to ACP and by describing the "negotiation of death," a process by which the dying patient, survivors, and attending clinicians plan together how the patient should die [17]. Notably, many of this chapter's references date from the 1990s when much of the most important ACP research occurred. The case of Mrs. G., the same patient as in Chap. 7, illustrates key concepts.

Table 6.1 Contrasting do-not-resuscitate orders and advance care planning

Characteristic	Do-not-resuscitate orders	Advance care planning
Kinds of directive	One treatment instruction only	Often multiple treatment instructions and/or a proxy designation
Legal meaning	A doctor's order	A patient's expressed wishes
Signer	Doctor	Patient
Witnessing necessary under law?	Never	Always
Patient has decision-making capacity at signing?	Not necessarily	Always
Patient has decision-making capacity at implementation?	Not necessarily	Never

The Case

Eighty-four-year-old Mrs. G. has suffered back pain for many years. Moderate doses of opioids relieve it partially, but Mrs. G. refuses higher doses due to the drowsiness, nausea, and constipation they cause. She has also experienced two recent, potentially life-threatening problems with her lungs—a blood clot and a pneumonia. She recovered from both with intensive inpatient treatment. Although each illness presented the opportunity for formal ACP with her doctors, it never took place. That omission complicates management of her next medical crisis.

Still, Mrs. G. conducts her own informal ACP discussion after she learns a distant relative survived a cardiac arrest and stroke only to endure mechanical ventilation and tube feedings before dying years later. Wanting to avoid such a fate, Mrs. G. expresses her end-of-life wishes to her husband and two daughters during a family conversation. Most of her expressed wishes concern property bequests, but a few concern medical treatments. Unfortunately, her treatment wishes are often vague or inconsistent. For example, Mrs. G. says at different times both "Treat me if I have any chance to live," and "I don't want to live if I cannot participate in life. I don't want to be a vegetable." She never reconciles those potentially contradictory statements. She also instructs the family never to let her care bankrupt them. "Don't sell the house," she says, "just to pay my medical expenses."

During that family conversation the elder daughter asks specifically for Mrs. G.'s wishes about mechanical ventilation. Mrs. G. not only describes her horror at the distant relative's experience with the ventilator but also mentions a friend's more hopeful one. Doctors placed this friend on a ventilator and then decided she could not survive. They then weaned her progressively off the ventilator and expected her to die. To everyone's surprise the friend began breathing spontaneously and lived on for many years. "You just never know," Mrs. G. concludes. When the elder daughter presses again for her wishes, Mrs. G. quickly ends the conversation by saying only, "You and your father can decide about that ventilator when the time comes. You'll know what to do." While this informal ACP discussion identifies Mr. G. and the elder daughter as Mrs. G.'s preferred proxies, it merely postpones critical decisions about mechanical ventilation. The discussion leaves the family confused about what Mrs. G. really wants. A systematic ACP process, especially if guided by a doctor, might have helped.

Mrs. G. subsequently suffers a cardiac arrest at home, is resuscitated there, and is transferred emergently to a hospital. She is placed on a ventilator in an intensive care unit (ICU) but successfully weans off after a few days. A cardiologist, a neurologist, and a pulmonologist take care of her at first. She improves. But then her heart condition deteriorates again when she can no longer communicate. A second cardiologist urges placing her back on the ventilator, but her family has reservations about the idea. Relying on her sketchy prior instructions, they have to decide what to do. Fortunately, a skilled ethics consultant eventually joins the case to help everyone work through the impasse. Mrs. G. finally dies with hospice care in the hospital but without the ventilator.

Advance Care Planning

The Origins of Advance Care Planning

Amazing new life-saving treatments including CPR, mechanical ventilation, and hemodialysis first appeared in the 1950s and 1960s. These treatments revolutionized medicine. They saved the lives of patients who would have died from their illnesses before. CPR quickly became standard treatment for cardiac arrest [18]. Intensive care units (ICUs) cropped up to use those new treatments expertly for serious heart damage and rhythm disturbances, pneumonias and other causes of respiratory failure, and kidney failure. Critical care became its own specialty.

Yet experience with the new treatments eventually showed they do not always produce good results. A few lucky patients revived miraculously to resume good mental and physical function, but many did not. For example, as I noted before, overall survival rates to hospital discharge after cardiac arrest and CPR rarely exceeded 22 % [18, 19]; those rates for patients over 70 and for cancer patients rarely exceeded 10 % [20, 21]. Some patients who *did* survive cardiac arrest and CPR often suffered severe brain damage, even permanent unresponsiveness. Some mechanically ventilated patients remained forever dependent on their ventilators. Even many hemodialysis patients attained only modest survivals—three to five years on average for all such patients and less for diabetics—and they lived functionally diminished lives on dialysis [22]. Americans began to rethink their initial enthusiasm for the new treatments. That enthusiasm soon gave way to a more sobering view. Many people finally decided they did not want the new treatments for themselves.

Americans, of course, have long embraced their common law right to refuse treatments, even life-saving ones [23–26]. But, for such refusals to be legally secured, patients historically had to voice them on the spot in the crisis. The possibility of becoming too weak or brain-damaged to voice on-the-spot refusals led many people to demand ways to make treatment decisions in advance.

Beginning in the late 1960s, ACP arose to meet this demand. The Euthanasia Education Council, later renamed Concern for Dying, published the first instructional AD in the 1970s [27]. Called A Living Will [7], it instructed caregivers, "If … (t)here is no reasonable expectation of my recovery from physical or mental disability, I request that I be allowed to die and not be kept alive by artificial means or 'heroic measures.' I do not fear death itself as much as the indignities of deterioriation, dependence, and hopeless pain…." A Living Will requested comfort measures under such circumstances. It, however, had no official legal standing partly due to the unfortunate name. Lawyers could not understand a "will" to be executed *before, not just after,* a person's death [28]. Nonetheless, signers tried to persuade caregivers to comply with A Living Will anyway, and for that reason it closes with the plea, "I hope you who care for me will feel morally bound to follow (this) mandate" [27].

Two court cases subsequently gave ACP important legal support. The Karen Quinlan case in 1976 recognized for the first time the authority of proxies to refuse life support for an incompetent patient. Specifically, the court allowed the parents of permanently comatose Karen Quinlan to stop her ventilator and to allow her to die if she could never regain a "cognitive, sapient" life. The court reasoned that Quinlan herself would not want life support under those conditions. Then the Nancy Cruzan case in 1990 gave a boost specifically to instructional ADs [29]. In that case the United States Supreme Court let stand a Missouri law permitting proxies to stop life support for an incompetent patient only with "clear and convincing evidence" of the patient's wishes to do so. Many scholars interpreted the Cruzan decision to mean instructional ADs provide the necessary clear and convincing evidence that Missouri and some other states demand in those situations [30].

Advance care planning in its various forms has since become a familiar feature of modern American life. Considerable professional and popular literature about ACP has accumulated, addressing such questions as: Which circumstances should cue ACP? How do the advantages and disadvantages of each kind of ACP balance out? and Which kind produces the best results? The next several sections take up those questions.

Cues for Conducting Advance Care Planning

Doctor–patient ACP discussions about end-of-life care have recently gained widespread support [3, 26, 31, 32]. Patients in general want them [6, 33–35], and doctors favor them for the many patients who are rapidly deteriorating [36], admitted to ICUs, unlikely to survive CPR, or terminally ill [37]. Professional medical societies also favor such discussions. Oncology societies, for example, encourage them for all cancer patients likely to die within a year [36, 38]. Unfortunately, actual performance of these discussions falls far short of such goals [39, 40]. Only 23–30 % of critically ill patients [41, 42], 18–38 % of terminally ill patients [37], and 27–40 % of end-stage cancer patients [36, 43] report actually having the discussions with their doctors. This disconnect between performance and goals [34, 35] raises the question, What cues should trigger doctor–patient ACP discussions about end-of-life care? [44].

The discussions, of course, should occur at the optimal time for patients [45], but patients and doctors differ over when that time is. Patients generally want ACP discussions earlier in the doctor–patient relationship or earlier in a serious illness than when doctors ordinarily conduct them [46]. But patients do not consistently specify one time they consider optimal. Doctors must guess at it.

Still, clinical experience suggests reasonable cues for ACP discussions about end-of-life care. Table 6.2 lists the cues one expert uses [47]. The table separates those cues into cues that should prompt immediate ACP and cues that should prompt near-term, but not necessarily immediate, ACP. The reader should note that diagnosing a potentially fatal disease requires breaking the bad news in a timely way. Breaking bad news differs fundamentally in purpose from conducting ACP: Breaking bad news *informs* the patient about a serious diagnosis while ACP *plans*

Table 6.2 Suggested cues for conducting advance care planning

Cues for immediate advance care planning
• The doctor foresees an imminent death.
• The patient talks about wanting to die.
• The patient suffers intolerable, overwhelming, and uncontrollable symptoms.
Cues for near-term advance care planning
• The doctor would not be surprised if the patient dies within 6–12 months.
• The patient has a potentially fatal illness (especially if it prompts hospitalization).
• The patient chooses a potentially life-prolonging treatment with only a small chance of success.
• The patient initiates discussions about advance directives or end-of-life care (including hospice).

Adapted from Quill [47], with permission

with the patient for future care. The two often benefit from different timing and approaches. (I discuss approaches for breaking bad news in Chap. 4.) Cues for conducting ACP immediately include the doctor's foreseeing the patient's death as imminent and the patient's expressing a wish to die or suffering intolerable, overwhelming, and uncontrollable symptoms [48]. Cues for near-term but not necessarily immediate ACP include the doctor's not being surprised if the patient dies within 6–12 months, or the patient's having a life-limiting illness (especially when it precipitates hospitalization), choosing a potentially life-prolonging treatment with only a small chance of success [43], or initiating discussions about ADs or end-of-life care [47]. Doctors may use other cues that also make sense such as low Palliative Performance Scale scores (see Chap. 5), the presence of multiple serious coexisting diseases, or a history of multiple failed chemotherapies [43].

If the doctor does not initiate the ACP discussion when the patient or family want it, the patient or family should initiate it themselves [49]. The best care in today's American health system often requires such assertiveness from patients and families. The doctor should be willing to go with it.

The Case

Strong cues for conducting ACP occur both before and during Mrs. G.'s last hospitalization. Beforehand either her blood clot or her pneumonia would have justified near-term ACP because each was a potentially fatal illness prompting hospitalization. Furthermore, if ever revealed to her doctors, Mrs. G.'s AD (which she signed in the interim) and informal ACP discussion with her family would also have justified near-term ACP because she initiated those end-of-life care conversations herself.

Other, equally strong cues for ACP occur during her hospitalization. The cardiorespiratory arrest certainly qualifies as a potentially fatal disease that prompts hospitalization. In addition, the doctors might reasonably expect her to die imminently (as the first cardiologist thinks) or within 6–12 months. Despite her successful resuscitation the full cardiorespiratory arrest outside the hospital gives the elderly Mrs. G. more than a 90 % likelihood of dying before hospital discharge. This likelihood only increases as her condition deteriorates later.

Notably, Mrs. G. shows at times early in the hospitalization that she can understand conversations and respond appropriately. Though sometimes groggy, she consistently opens her eyes and moves her arms to the neurologist's commands. She also communicates meaningfully with her family. Mr. G., for example, once asks her, "Do you want one of my 'famous kisses'?" Mrs. G. nods and puckers up. Mr. G. then kisses her and asks, "Did you like that?" She nods yes again. Such incidents suggest Mrs. G. can perform some ACP at that time. But her doctors do not try it.

Unfortunately, as the hospitalization continues, Mrs. G.'s responsiveness diminishes. It finally disappears completely as the question of reventilating her arises. Because Mrs. G. can no longer express her wishes, her doctors must rely on her written AD and the prior family discussion for guidance in making that decision. Even then, the decision-making process proves difficult.

Barriers to Advance Care Planning

Westerners have long considered illness abnormal and health normal. Only the ill, not the healthy, have presented for medical care. Thus, Western medicine evolved to diagnose, treat, and cure diseases in the few who were ill [50]; it paid little attention to preventing diseases in the many who were healthy. Preventative measures traditionally fell to public health programs. Yet by the mid-twentieth century the emergence of vaccines, antibiotics, and other powerful preventative measures against infections forced Western medicine to expand its focus to include such measures. ACP fits that preventative approach because people typically use ACP to *prevent* end-of-life treatments they do not want. Unfortunately, ACP like many other preventative measures has struggled to gain a foothold in cure-oriented Western medicine. ACP still faces barriers, many unintentional, from patients and families, doctors, and healthcare institutions alike. Those barriers often produce what one author calls a powerful, unwitting "collusion" to avoid meaningful ACP altogether.

Patient and Family Barriers

Talking about death is hard because it raises the specter of losing everything, even existence itself [51]. For that reason the seventeenth-century writer La

Rochefoucauld observed, “One can no longer look steadily at death than at the sun” [52], and the modern-day end-of-life care expert, Ira Byock, M.D., says, Death “constantly threatens to (awaken) us from the dream of life” [53].

Understandably, few patients can muster the courage to face their fears about death and to plan for it. Most patients just avoid the subject. Some have no wishes about future care and do not want to formulate any [37]. Others may never be able to broach the subject with their families or doctors. These patients may wait for others to broach the subject and secretly hope that no one does. These patients may even kid themselves into believing that, if no one else mentions end-of-life ACP, it must not be important. Still other patients may wait for the ideal time to conduct ACP but always be just too sick or not sick enough to do so. The ideal time, therefore, never arrives.

Both patients and family members may use other strategies, too, to block or delay ACP. Some people may deny fatal diagnoses or the limits of curative medicine [45, 50, 54]. About a third of advanced cancer patients, for instance, do not believe they are dying [55, 56], and many request aggressive chemotherapy up until only days before death [57]. Some patients may even conceal symptoms they fear will signal irreversible terminal illness or will end their treatments for cure [50]. Patients and family members may also delay ACP by endlessly requesting more tests, second opinions, or long-shot treatments [38]. The desperate and impressionable may even look outside mainstream medicine for the cure they are convinced exists (and others enthusiastically promise). These people may believe mainstream medicine is hiding the cure from them for the sake of experimentation or profit [58].

Patients and family may also avoid mentioning ACP for fear of upsetting the doctor. They realize the doctor wields great power over their lives. Patients and family feel vulnerable [50]. They, therefore, hesitate to suggest anything (such as ACP) they think might anger the doctor. (Unfortunately, some doctors *do* react angrily to such suggestions and thereby stifle further meaningful ACP discussions [50].) Patients and family may even fear the doctor will retaliate by prematurely withholding or withdrawing any needed life support [54].

Doctor Barriers

Doctors present their own barriers to ACP [51]. Many feel ill-prepared to perform it [54]. Recent research shows those feelings are justified: Doctors receive little training in end-of-life communication and fall short of standards in their attempts at ACP [14]. Doctors may also believe ACP dashes people’s hopes [38, 46, 54], leaving patients and families despondent. Research, however, consistently disproves that idea. Doctors may also fear that ACP unleashes patient emotions that are difficult to witness, require a lot of time to assuage, and leave the doctor exhausted. Finally, some doctors may see no role for ACP at all in medical practice. These doctors may believe they must constantly fight death, their foremost enemy, with a maximal effort to save patients' lives [46], or that every patient death otherwise risks legal or professional censure.

Institution Barriers

Like patients, family members, and doctors, healthcare institutions also present barriers to ACP. The fragmentation of care that now characterizes the American health system creates fragmentation of ACP, too [54]. No one professional or institution has responsibility for ensuring ACP is done in a timely, effective way. As a result ACP often happens in an uncoordinated, piecemeal way [59]. Furthermore, information about a patient's prior ACP can easily get lost in a voluminous electronic medical record or during complicated information transfers across institutions. Finally, data-driven, bottom-line-oriented health insurance policies do not specifically reimburse for ACP [54]. That lack of reimbursement certainly discourages routine ACP [37]: If the doctor performs it, he or she must do it for free. Of the three institutional barriers to ACP mentioned here, I believe the last is conceptually easiest to fix: performing ACP deserves reimbursement.

Important Communication Skills for Advance Care Planning

Good doctor communication skills, I believe, can overcome many of the barriers to ACP [60]. Some of those skills such as creating rapport with patients, asking questions insightfully, and listening carefully to patients' answers should underlie all health professional–patient interactions. But certain skills become especially important in ACP: thinking positively, fostering trust, and using good communication techniques likely to serve the purpose. The need for such communication skills in ACP seems obvious. But, because many doctors' ACP discussions may fall short of accepted standards [14, 61, 62], I discuss here the three skills I believe are especially important in ACP. I talk in terms of doctors' conducting the ACP process, but other clinicians such as nurses may also do so.

Thinking Positively

Preexisting attitudes can have a powerful influence on how human endeavors work out. Thinking positively about ACP, therefore, may go a long way to realizing its benefits. Unfortunately, many doctors dislike the process, feel ill-prepared to conduct it, and try to avoid it. Such negative attitudes can only diminish the success of ACP.

Therefore, optimal outcomes depend on a positive attitude. And good reasons justify such an attitude. For one, ACP gives patients the rare opportunity to take stock of their lives. ACP challenges patients to examine their core personal values and goals in life. Some patients have never done so before. ACP may even prompt important changes in activities or attitudes for whatever time remains. ACP also offers rewards for doctors including the chance to get to know patients on a deeply meaningful level. During ACP, patients may reveal their most heartfelt thoughts about living and dying, making themselves emotionally vulnerable. Tender hopes and troubling fears may surface. The doctors, in turn, have the opportunity to respond out of humaneness and compassion. For that significant opportunity they should consider ACP more a privilege than a burden [47]. Of course, this privilege should inspire doctors to improve constantly on their ability to conduct ACP.

Fostering Patients' Trust

The vulnerability that comes with illness makes patients enter any doctor–patient relationship by trusting their doctors. That baseline trust is an advantage for doctors: Patients begin by assuming their doctors are well trained and conscientious in their care.

Doctors then have the opportunity to build on that trust whenever patients suffer life-threatening illnesses and need ACP [13]. One way to build trust involves slowing down the usual clinical pace and exercising the patience necessary to do ACP. Doctors' typical time urgency can make patients feel hurried, even harried. Patients may need time for the difficult end-of-life decisions to "percolate." Doctors should allow patients that time to the extent possible. I find those decisions often happen eventually at the "right" time and pace [63]. Clinical urgency, however, occasionally requires doctors to press for decisions before patients feel ready to make them, but doctors should do so only when absolutely necessary [13].

Another way to build trust involves listening attentively to patients' views and responding thoughtfully. Using patients' own words helps. Such responses tell patients the doctor has heard them and cares about what they think. Still another way to build trust is assuring patients explicitly that they may safely express even their rawest feelings such as frustration, discouragement, and despair. Doctors should reply with empathy. They might say, for example, "Many patients in similar circumstances tell me they have those feelings" or "I would feel the same way in your situation." Such empathy "validates" or "legitimizes" those feelings for patients [13, 54]. And still another way to build trust is showing respect for past decisions by patients, proxies, or other clinicians. Denigrating those decisions only reflects badly on the doctors making the judgments and may paradoxically undermine patients' respect for them [13].

One final way to build trust involves pledging good-faith efforts to follow a patient's wishes as closely as possible in critical illness. Doctors, however, must be honest: Some circumstances, perhaps unforeseen beforehand, may not permit strict adherence to a patient's wishes. Doctors should explain that expressing wishes during ACP does not guarantee they can be followed later.

Using Effective Communication Techniques

Some techniques, though conceptually simple, can have considerable impact on ACP results. One such technique is conducting ACP with patients frequently, even routinely. Here, the old adage, practice makes perfect, applies. But research suggests doctors conduct ACP too infrequently, too late, or not at all [43]. Seventy percent of doctors in one study said they never conduct end-of-life care discussions with patients while only 20 % said they "routinely" do so. I suspect the cues listed in Table 6.2 justify ACP with many more patients than get it now.

Each ACP session also deserves adequate doctor–patient face time, at least 20–30 min for initial ACP, I believe, to allow patients to talk liberally. Yet research indicates the typical session lasts only 5.6 min, surely too little [62]. Doctors must allow more time for ACP to make it worthwhile.

Certain general communication techniques can help the doctor use productively whatever time exists for ACP. (A later section in this chapter discusses techniques specific to a stepped approach for conducting ACP.) The doctor should give a rationale for ACP at the outset to allay any false patient fears, say, about an impending new terminal diagnosis. The doctor might say to the patient, "I have thoroughly explained your current medical condition. But I think we should talk about your overall goals for care so that I can know them for the future. I want to try to honor those goals even if you become very sick and can't communicate your goals to me directly" [64]. The doctor might then add, I have this discussion with … "every patient regardless of his or her current condition," "every patient whose symptoms we are still trying to control," "every patient hospitalized for this problem," or "every sick patient even if we cannot tell how long he or she will live," depending on the doctor's usual practice and the patient's current condition.

Of course, the doctor should always try to engage patients "where they are." This technique requires something hard for doctors: stepping out of the typical doctor mind-set and giving patients some control over the conversation [47]. The doctor must encourage patients to steer the ACP conversation at times where they want it to go. Open-ended questions—that is, questions without prespecified answers such as just yes or no—go a long way to giving patients that encouragement. Examples of open-ended questions used commonly in ward rounds include "What is bothering you most today?" and "How do you feel now?" Open-ended follow-up probes such as "Tell me more" or "Go on" invite patients to elaborate on prior responses.

These open-ended questions require the doctor to listen a lot. Unfortunately, the doctor usually talks 70 % of the time and the patient only 30 % of the time in the typical brief ACP session [62]. Patient-centered ACP surely requires reversing those percentages. Careful listening, though, has its rewards. It enables the doctor to learn how patients understand their illnesses. That understanding in the case of life-threatening illnesses often touches the psychological, social, and spiritual domains as well as the pathophysiologic. Perhaps for that reason careful listening by the doctor can increase the satisfaction with end-of-life care for both patient and doctor.

Of course, engaging patients "where they are" also means making ACP as relevant as possible to them by tailoring discussions to patients' understanding [37, 65].

One well-known way to maintain relevance is to use the ask–tell–ask method in giving information. The doctor asks what information patients want, gives that information, and asks what patients then understand. The doctor explains more if necessary [56].

The doctor should also be careful not to overload patients with information. He or she must remember that few patients have had prior medical training. To understand medical information accurately, most need small doses, time, repetition, and perhaps alternate explanations. The doctor should always keep his or her comments brief, simple, and specific [54]. The doctor should also avoid vague terms (such as "poor prognosis," "doing everything," and "being a vegetable") and confusing medical jargon (such as "comatose," "metastatic," and "end-stage") [14]. Of course, an accurate understanding of the medical situation occasionally requires using a technical term. If so, the doctor should define it briefly and simply. But so accustomed are clinicians to using technical medical language that they seldom realize when they are using it with patients. Clinicians must conscientiously guard against that mistake because technical language effectively excludes the medically untrained from discussions.

I realize conducting *any* kind of good communication in practice takes time and faces obstacles. The greatest obstacle is perhaps lack of reimbursement. Unfortunately, quality ACP cannot become standard practice until health insurance reimburses it adequately.

Kinds of Advance Care Planning

Three kinds of ACP have evolved over the years. First came instructional ADs, then proxy-designating ADs, and finally broad ACP discussions. I think of them as first-generation, second-generation, and third-generation ACP, respectively. The three kinds have different advantages and disadvantages. Each kind tries to correct for the disadvantages of earlier kinds, but none is perfect yet. Therefore, no one kind completely supplants the others, and people may use combinations of the three.

I describe here the three kinds in chronological order. I begin with an overview of ADs. Then I sequentially take up instructional ADs, proxy-designating ADs, and broad ACP discussions. Mrs. G.'s own AD and the ACP discussions with her family illustrate each kind.

Advance Directives: An Overview

Table 6.3 compares nine prominent ADs. Eight of the nine contain both treatment instructions and proxy designations. The lone exception is the original A Living Will. It pioneered the concept of written instructional ADs [27]. (Because A Living

Table 6.3 Characteristics of selected advance directives

	A Living Will [109]	The Medical Directive [110]	Your Life, Your Choices [111]	Texas Advance Directive Act [112]	Five Wishes [113]
Publication date	1978	1990	1996 and 2001	1999 and 2004	2000
Author or publisher	Concern for Dying	LL Emanuel, EJ Emanuel	RA Pearlman et al. VHA[a]	Office of the Attorney General, State of Texas	Aging with Dignity
Area of legal validity	Unclear	Unclear	Unclear	Texas	36 states and the District of Columbia
Additional information	*Questions and Answers about the Living Will*; other articles, case studies, and films	Accompanying professional journal article	52-page workbook	5-page *Senior Texans: Advance Care Planning* supplement at www.oag.state.tx.us/AG_publications/txts/Advance_Care.shtml (accessed, November 20, 2006)	"Five Wishes" video
Eligible signers	Unspecified	Adults of sound mind	Any nonpregnant person capable of willful, voluntary decisions	Nonpregnant adults may sign all 3 documents; proxies or 2 doctors of incompetent patients may sign out-of-hospital DNR[a] order.	Any person older than 17 years
Length and format	2 pages and a wallet card	4 pages but no wallet card	4 pages and a wallet card	8 pages but no wallet card	12 pages and a wallet card
Conditions for effecting	No recovery expected from "extreme physical or mental disability"	4 mental incompetence states with varied survivals and disabilities: coma, vegetative state, and others	Coma or vegetative state, stroke, dementia, and terminal illness	Illness likely to kill patient within 6 months or an irreversible illness that leaves the patient unable to "care for self or make decisions and (patient likely to die) without life support"	Imminent death, severe brain damage (such as permanent coma), or other illnesses the patient describes
Treatment choices	Refuse "artificial means" or "heroic (life-support) measures."	Choose or refuse life-support and other measures. May choose "undecided" or trials of treatment.	Choose or refuse life-support measures and hospice care. May choose "don't know."	Choose or refuse individual or all life-support measures.	Choose or refuse all life-support measures, or let doctor decide.

(continued)

Table 6.3 (continued)

	A Living Will [109]	The Medical Directive [110]	Your Life, Your Choices [111]	Texas Advance Directive Act [112]	Five Wishes [113]
Choices about cadaver	Tissue donation	Organ or body donation for transplantation, education, or research	Organ donation, autopsy, other research on body; plus wishes about burial or cremation	None	Organ donation, and burial or cremation
Special features	• Is most familiar early directive, now outdated. • Is very brief. • Allows request for home care if it would "not jeopardize … recovery to a meaningful and sentient life or impose an undue burden" on family.	• Is brief. • Lists life-support and other measures. • Poses questions to help patient define illnesses that do not justify prolonging life. • Allows personal statement of wishes. • Acknowledges need for proxy's judgment when patient's wishes unclear, patient is undecided, or situation differs from the 4 states as described in the directive. • States priority between patient's instructions and proxy's decisions if they conflict. • Emphasizes doctor's ongoing duties to diagnose, educate, give prognoses, and make recommendations.	• Has easy-to-read layout with excellent graphics. • Is comprehensive. • Workbook defines key concepts simply and gives case examples. • Emphasizes patient's grasp of concepts. • Explains in everyday English some common serious illnesses and their treatments. • Keys patient wishes to explanations in grid format. • Helps patient define relevant spiritual beliefs, hopes for recovery, and acceptable risks.	• Combines laws about instructional directives, proxy directives, and out-of-hospital DNR orders. • Emphasizes that spouses often do not know patients' wishes. • Distinguishes powers of attorney for medical care and property management. • Directives in English and Spanish; out-of-hospital DNR orders in English only. • Requires doctor to transfer patient if doctor cannot fulfill patient's wishes. • Allows setting an expiration date. • Authorizes out-of-hospital DNR identification band. • Explains hospice care.	• Uses simple language. • Addresses personal, emotional, and spiritual needs as well as medical ones; claims to be "the Living Will with a Heart and Soul." • Gives instructions for revoking prior directives. • Lists traits of a good proxy. • Lists decision-making powers of proxy. • Allows patient to write additional instructions. • Includes funeral wishes.

(continued)

Table 6.3 (continued)

	A Living Will [109]	The Medical Directive [110]	Your Life, Your Choices [111]	Texas Advance Directive Act [112]	Five Wishes [113]
			• Workbook includes directives; website gives parallel references to VHA directives. • Gives the patient tips for end-of-life discussions with family and doctors. • Cautions about ambiguities of common terms such as "pulling the plug." • Allows instructions about how strictly to follow patient's wishes. • Includes funeral wishes. • Uses divider tabs, sidebar summaries, and prompts for easy reference. • Gives completion checklist. • Gives contact information for relevant organizations.		

(continued)

Table 6.3 (continued)

	A Living Will [109]	The Medical Directive [110]	Your Life, Your Choices [111]	Texas Advance Directive Act [112]	Five Wishes [113]
Drawbacks	• Does not define key concepts. • Gives vague instructions that are hard to interpret clinically.	• Uses word-dense layout. • Explains treatments too simplistically. • Allows mutually exclusive treatment choices.	• Workbook is long. • States some hopes as facts such as directives "prevent confusion and … ease the burden on families." • 1996 workbook directives use legal language and key only to Washington State's law.	• Gives sketchy definitions of life-support measures. • Makes unrealistic promises to "prevent arguments and bad feelings at the end of life," to achieve a "gentle" or "good end of life," and to help give survivors "a more peaceful bereavement."	• Copyrighted. • Costs $1–$5. • Does not define life-support measures. • Makes unrealistic promises such as to give patient "control (over) … how you are treated if you get seriously ill," to keep patient "fresh and clean" always, and to protect family from guessing "what you want" and from making "hard choices."

Table 6.3 (continued)

	Advance Health Care Directive [114]	VHA Long-Term Care Advance Care Proxy Planning [115]	VHA Advance Directives [116]	Medical Orders for Life-Sustaining Treatment [117]
Publication date	2000–2006	2002	1998 and 2003	2005
Author or publisher	California Medical Association	Volicer et al., VHA National Ethics Committee	VHA National Center for Ethics in Health Care	Rochester Health Commission
Area of legal validity	California	Unclear: not an "official" form but intended as a model for facilities developing their own forms	Veterans Health System	New York
Additional information	6-page *Advance Health Care Directive Kit* and *Deciding To Forego Treatment: Advance Directives*	Accompanying professional journal article	7-page VHA handbook defining staff duties	20-page *Guidebook to MOLST*
Eligible signers	Any legal adult California resident	Proxies and professional caregivers of incapacitated veteran patients	Veteran patients	A nurse or other health professional completes, and a doctor signs on behalf of patient or proxy.
Length and format	8 pages and a wallet card	2 pages but no wallet card	6 pages but no wallet card	4 pages but no wallet card
Conditions for effecting	Kit says any illness in which "you are unable to make your own decisions," but directive specifies only death within months or permanent life-support if there is dependence with inability to make decisions.	Illnesses that require hospitalization and life support	Irreversible unconsciousness or a terminal illness	Patient's current illness(es). Other illnesses not specified
Treatment choices	Choose or refuse a proxy, a doctor, a care facility, life-support measures, and release of medical information.	Choose or refuse acute care hospitalization, life-support measures, artificial hydration and nutrition, and comfort care.	Refuse all life-support measures.	Choose or refuse CPR, intubation and mechanical ventilation, nutrition and hydration, antibiotics, and comfort care. May choose trials of treatment.

(continued)

Table 6.3 (continued)

	Advance Health Care Directive [114]	VHA Long-Term Care Advance Care Proxy Planning [115]	VHA Advance Directives [116]	Medical Orders for Life-Sustaining Treatment [117]
Choices about cadaver	Organ donation, autopsy (for education or research), and body disposal	None	None	None
Special features	• Uses simple language. • Offers DNR form for patients and preferred-intensity-of-treatment form for doctors to complete on behalf of patients in long-term care facilities. • Kit and DNR form in English and Spanish. • Kit answers common questions succinctly. • Kit gives space to list people with copies. • Allows personal statement about treatment wishes. • Reminds proxy to use patient wishes to guide proxy decisions. • Gives option to allow proxy to make decisions before patient loses decisional capacity. • Requires signatures of 2 witnesses, notary public, or nursing home ombudsman. • Allows setting an expiration date. • Encourages use of pre-hospital DNR form. • Replaces Natural Death Act Declaration and Durable Power of Attorney for Health Care.	• Covers incompetent nursing home residents who have no advance directives. • Suggests participants for patient's care-planning conference. • Reminds signers to update with every year, transfer, or significant change of medical illness.	• Commits explicitly to respecting patient's wishes. • Urges outpatient advance care planning. • Offers tools to aid discussions. • Urges patient education and shared doctor–patient decisions. • Defines key concepts simply (including decision-making capacity). • Allows for verbal instructions when patient cannot sign. • Reminds patient to communicate wishes and to distribute directive. • Assigns to facility staff the responsibility for conducting the process. • Gives instructions for filing and implementing. • Requires review with every year or hospitalization. • Allows health professional to transfer patient to another health professional if directive violates original professional's conscience. • Outlines dispute resolution procedures.	• Adopted from Oregon's model document, called POLST. • Aims to "translate current treatment preferences into physician orders." • Summarizes but does not replace New York Living Will or New York Health Care Proxy directive. • Provides an additional document for special patients such as those lacking capacity to consent. • Requires review at every hospitalization or every transfer, at changes in wishes or in medical illness, and at least every 60 days for nursing home residents and every 90 days for others. • Requires doctor to fill out parts about nonCPR treatments based on "clear and convincing" evidence of patient's wishes. • Specifies full treatment as default for uncompleted parts. • Uses bright pink paper for easy identification.

(continued)

Table 6.3 (continued)

	Advance Health Care Directive [114]	VHA Long-Term Care Advance Care Proxy Planning [115]	VHA Advance Directives [116]	Medical Orders for Life-Sustaining Treatment [117]
Drawbacks	• Copyrighted. • Kit and forms cost \$1–\$5. • Gives sketchy definitions of life-support measures.	• Makes exaggerated claim that patients "can continue to direct their medical care" when they are no longer competent. • Is specific for VHA; may have limited validity elsewhere.	• Does not effect parts of state-authorized directives that conflict with VHA policy. • Keeps health professionals from overruling proxy decisions. • Is specific for VHA; may have limited validity elsewhere.	• Is copyrighted. • Specifies no other conditions for effecting a patient's wishes except the patient's current illness. • Requires burdensomely frequent reviews.

Adapted from Perkins [4], with permission

[a]*VHA* = Veterans Health Administration; *DNR* = do not resuscitate; *CPR* = cardiopulmonary resuscitation; *POLST* = Physicians' Orders for Life-Sustaining Treatments.

Will was the only existing AD for years, many people developed the habit of calling *all* instructional ADs "living wills." To avoid ambiguity here, I use "A Living Will" to mean only that original instructional AD and not later ones.) A Living Will was a laudable first step for ACP but had one insurmountable obstacle involving its core instruction, "If … I can no longer take part in decisions for my own future, (and) … if there is no reasonable expectation of my recovery from physical or mental disability, I request that I be allowed to die and not be kept alive by artificial means or 'heroic measures.'" The vagueness of that instruction bewildered clinicians as they tried to implement it in clinical care. They soon concluded that so sweeping are the interpretations required that the instruction gives virtually no useful guidance. Many more clinically sophisticated instructional ADs have appeared since A Living Will. For that reason I recommend against using it. I believe it has only historical significance today.

The other eight ADs listed in Table 6.3 appeared after 1989. They represent a huge leap forward in sophistication. Specific states drafted three of the eight, but all states and the District of Columbia now recognize ADs legally [60]. Explanatory workbooks, professional journal articles, or even videos accompany these ADs. Adult patients, especially if not pregnant, may sign six of the eight. Clinicians must sign the other two—the Veterans Health Administration Long-Term Advance Care Proxy Planning and the Medical Orders for Life-Sustaining Treatment—on behalf of patients. Effecting the documents usually requires either irreversible, severe brain damage or terminal illness. All eight ADs allow signers to accept or refuse life-support treatments. Three allow signers to request hospice or comfort care explicitly, and four allow them to permit organ donation. The length of these ADs ranges from 1 to 12 pages. Some, but not all, of the ADs assume or encourage proxy designation; try to explain concepts briefly and simply; and require periodic review of patients' wishes.

Unfortunately, several of the ADs exaggerate their benefits. Five Wishes, for example, promises to allow the patient to keep control over "how you are treated if you get seriously ill." Similarly, the AD under the Texas Advance Directive Act promises to "prevent arguments and bad feelings at the end of life," to give the patient "a good end of life," and to help survivors achieve "a more peaceful bereavement." I cannot find empirical substantiation for these promises.

The Case

Mrs. G. signed her AD at her attorney's office probably when she was preparing her testamentary will [66]. Many attorneys now include ADs in a standard package of end-of-life documents for clients to sign.

Though Mrs. G. used none of the ADs listed in Table 6.3, her AD shares many of their features (Table 6.4). Sections 1–4 and 7 designate her proxies and their powers. Section 1 appoints Mr. G. as "Primary Advocate" and the elder daughter as "Successor Advocate." Section 2 empowers her advocates to make treatment

Table 6.4 Mrs. G.'s instructional directive and designation of the person(s) to have the durable power of attorney for health care*

1. *Designation of advocates*: I appoint my husband as Primary Advocate with my durable power of attorney for health care. I appoint my elder daughter as Successor Advocate if my husband cannot serve as Primary Advocate.
2. *Invocation of these powers*: My advocate may exercise this durable power of attorney for health care after my attending doctor and one other doctor or mental health professional have determined that I cannot participate in making decisions about my medical or psychological affairs.
3. *Authority of my advocate*: A. Before making any decision my advocate will try to discuss with me the matter at hand. B. If my advocate cannot determine my wishes on the matter, he or she will try to follow my relevant prior instructions whether communicated orally or in writing. C. If my prior instructions do not resolve the matter or I have left no relevant prior instructions, then my advocate will decide the matter based on my best interests. D. I give my advocate the following specific authorizations subject to any restrictions from my prior instructions: i. To arrange for my admission to any health care facility ii. To arrange for my discharge from any health care facility iii To consent to any treatment for me [including cardiopulmonary resuscitation (CPR)] iv. To refuse any treatment for me (including CPR) v. To read my health care records and to disseminate their information as my advocate considers best for me vi. To donate to a needful recipient my blood during my lifetime or my organs after my death E. My advocate may authorize or refuse any treatment to relieve pain for me even if that treatment may produce physical damage, addiction, or a hastened death. My advocate may not intentionally cause my death.
4. *Release from liability*: On behalf of me and my estate, I release from liability all my caregivers so long as they try to follow my instructions in good faith.
5. *Statement of wishes* A. Any written or oral wishes I express after signing this document will supersede it in case of any conflict. B. I do not wish any life-sustaining treatment if … 1. I have a condition that my attending doctor believes is incurable or irreversible and will result in a reasonably quick death if he or she does not implement life-sustaining treatment; or 2. I have a coma or vegetative state that my attending doctor believes is irreversible. C. By "life-sustaining treatment" I mean any medical intervention that uses machines, drugs, or other nonnatural means to maintain a vital bodily function and that merely prolongs the dying process for someone with a terminal illness. I consider feedings through tubes into the stomach, intestines, or veins as "life-sustaining treatments."

(continued)

Table 6.4 (continued)

D. I ask to live my last days at home or in a hospice rather than in a hospital if such a disposition does not put at risk my chances for recovery and does not create excessive burdens for my family.	
6. *Revocation of prior directives*: I revoke any prior instructional directives or designations of people to have powers of attorney for health care matters.	
7. *Release of medical information*: My advocate may obtain, use, and disseminate information about my health condition as he or she believes will accomplish my wishes or promote my best interests.	
Patient's signature	Witnesses' signatures
Date	Date
*Modified and condensed from Mrs. G.'s original (see case discussion)	

decisions for Mrs. G. whenever her attending doctor and a comparable clinician determine Mrs. G. cannot make those decisions for herself.

Section 3 describes the decision-making authority and procedures Mrs. G. expects her advocates to use. She directs them to try first to discuss the matter with her directly. If that proves impossible, she asks them to decide her treatment based primarily on her relevant prior instructions and only secondarily on the proxies' guesses about her best interests. She also authorizes her advocates to arrange admissions and discharges from healthcare institutions, to accept or refuse any treatments (including CPR and pain treatments), to read and disseminate her medical information as needed, and to donate her blood during her lifetime and her organs after death. Mrs. G. specifically prohibits her advocates from intentionally causing her death. Section 4 releases Mrs. G.'s advocates from legal liability as long as they exercise good faith in trying to follow her instructions.

Section 5 gives Mrs. G.'s specific instructions about treatment. She asks her advocates not to use life-support treatments (meaning any "machines, drugs, (various kinds of tube feedings) or other nonnatural means that maintain a vital bodily function … and merely prolong the dying process") if she ever becomes irreversibly comatose or terminally ill. She requests instead care at home or in a hospice if that care will not threaten her recovery or overburden her family.

Section 6 revokes any prior ADs. Section 7 repeats the advocates' authority to "obtain, use, or disseminate" Mrs. G.'s medical information as needed to fulfill her wishes or to promote her best interests.

The First Generation of Advance Care Planning: Instructional Advance Directives

Instructional ADs provided the logical first step in ACP's evolution. People wanted a way to refuse invasive life-sustaining treatments such as CPR or mechanical ventilation and to request conscientious pain management during a future permanent coma or similar illness. The solution seemed simple: People should record their treatment instructions for future reference.

Despite the simplicity of the idea, instructional ADs give four reasons for caution [4, 67]. First, patients using these ADs have to decide about sophisticated life-sustaining treatments they often understand poorly [29, 68]. Many patients, for example, badly underestimate the harms and overestimate the benefits of such treatments. Typical lay estimates for survival after cardiac arrest and CPR, for example, run around 70 % even though actual survival runs much lower [4, 68].) In response, experts have tried mightily to write into instructional ADs descriptions of common life-sustaining treatments. I believe none of these descriptions have succeeded adequately at being brief, accurate, and colloquial enough to ensure understanding. Second, patients (and their families) often expect their AD instructions to be followed exactly. Yet medical crises are often complex and unpredictable. Unforeseen complications arise, frustrating even the best intentions to follow patients' prior instructions [69].

Third, the instructions, necessarily general to cover a wide range of crises, do not speak for themselves. They require extensive interpretation in specific clinical circumstances. Illnesses may, therefore, play out far differently than anyone anticipates beforehand. Fourth, most instructional ADs require terminal illnesses or persistent comas to activate their instructions. Yet comas must last a long time, usually 1–3 months, to be called "persistent," and many other illnesses such as deliriums, late-stage dementias, and even severe psychiatric illnesses that leave patients unable to make decisions are not strictly covered under these ADs. All these four reasons lead one expert to conclude that instructional ADs have had little influence over the use of CPR [13] and other life-support treatments [67, 70–73]. I generally agree, but I describe one exception shortly.

The Case

Despite these drawbacks instructional ADs do have some benefits. Mrs. G.'s instructional AD has at least two. In it she refuses "any medical intervention that … maintain(s) a vital bodily function and merely prolongs the dying process …" if she ever becomes irreversibly, imminently terminal or permanently comatose or vegetative. This instruction, though too vague by itself to decide the reventilation issue that arises during her last hospitalization, provides the family encouragement as they decide it. "Mom meant these (instructions) as something (for us) to hang on to," the elder daughter explains afterward. "They made it a little easier for us" to refuse the ventilator on her behalf. The family may reason that Mrs. G. is expressing in her AD the wish to forego life-sustaining treatments, though not further specified, in extreme medical situations. The family could reasonably conclude she is permitting them to forego the ventilator in the *particular* critical situation that does occur.

The instructional AD might also console the family in one way after Mrs. G. dies [25, 71, 74]. Mr. G. and the daughters surely feel occasional doubts about whether refusing the ventilator was right for Mrs. G. [74]. Mr. G., for example, sometimes tells his daughters regretfully, "Your mother really wanted to live." Still, the family can remind themselves at those times about the wish in her AD to spend her "last days (comfortably) at home or in a hospice." The family can reassure themselves (rightfully, I believe) they fulfilled that wish as best they could.

Unfortunately, Mrs. G.'s instructional AD has many shortcomings of instructional ADs. She signs it at the attorney's office. Doing so offers the convenience of signing it with other end-of-life documents but prevents immediate access to expert clinical knowledge in case of questions. I do not know whether Mrs. G. *ever* consults her primary doctor about the instructional AD [75]. If like many other patients she does not, important questions arise about the validity of the wishes she expresses. How well does she anticipate and understand her likely future illnesses? How well does she understand life-sustaining treatments? How firm or flexible are her decisions about treatments? Would she ever change her mind about the acceptability of certain outcomes, say, living with disabilities? Furthermore, Mrs. G.'s focus on bequeathing possessions during the family ACP discussion raises doubts for me about her attention to the medical decisions and about her knowledge of them at the time. Her AD's vague treatment instructions overall (and specifically her expressed ambivalence about ventilation during the family discussion) also raise doubts for me about whether they meet legal standards of "clear, convincing, and consistent" evidence to base decisions solely on those instructions.

Even if Mrs. G. tells her doctor about her AD, he or she may not follow up with a dedicated doctor–patient ACP discussion. If the doctor does, the discussion may encounter some common problems with such discussions: an insufficient allotment of time for it; a focus on dire situations that discriminate poorly (because nearly all patients would refuse life-sustaining treatments anyway in those situations); outcomes described only as the extremes of life or death, never as intermediate outcomes of life with various disabilities; and no elicitation of relevant patient life values, goals for care, or an underlying logic for decisions. Such a discussion would provide little practical clinical guidance.

Access to Mrs. G.'s AD also proves to be a problem at hospital admission. Her emergency transfer by ambulance occurs so hurriedly that the hospital staff do not know her name on arrival and admit her only as "Jane Doe." They obviously cannot gain access to her prior medical records and AD. Furthermore, amid the chaos of the transfer, the family forget to bring their copy of her AD. Before they can produce it, Mrs.G. has already been sent to the intensive care unit (ICU) and put on a ventilator. These admission events dismay the family for what they consider excessive haste. They do not appreciate Mrs. G.'s immediate need for life support. They wonder whether the doctors have little interest in Mrs. G.'s prior treatment wishes as expressed in her AD.

Mrs. G.'s AD also does not define precise criteria for determining two crucial times later in her hospitalization. Determining both times eventually requires difficult, on-the-spot judgments by clinicians and family. One time is when Mrs. G.

can no longer "participate in making decisions about (her) medical or psychological affairs" and needs her advocates to make medical decisions for her. Mrs. G. can likely make decisions early in her hospitalization but certainly not later. Yet the exact time she loses decision-making capacity cannot be pinpointed. The doctors finesse the problem by seeking decisions from the family throughout the hospitalization as long as Mrs. G. does not object.

Determining the other critical time does not have such a simple solution. That time is when Mrs. G. becomes terminal and her terminal care instructions should take effect. The AD defines a terminal illness as "an incurable or irreversible (condition that) will result in a reasonably quick death … (without) life-sustaining treatment." The words "irreversible" and "reasonably quick" introduce the ambiguity that underlies the dispute between the second cardiologist and the family about when all "nonnatural means … merely prolong the dying process." Though difficult to determine, a clearer definition of terminal illness might resolve the dispute more quickly and allow Mrs. G. to transfer to hospice sooner than what actually happens.

Finally, Mrs. G.'s AD does not anticipate the fragmentation in today's health care. That fragmentation raises the pivotal question in Mrs. G.'s case, Who is her primary attending doctor for purposes of declaring her terminal? The second cardiologist assumes the role of primary attending doctor only because all the other attending doctors—the pulmonologist, the first cardiologist, and the neurologist—have left the case. But that second cardiologist refuses to declare Mrs. G. terminal for having "an incurable or irreversible condition that would result in a reasonably quick death." "I know what is wrong with her heart," he tells the family. "She has a rhythm problem, and I can fix it. If you do not put her back on the ventilator, she will die (needlessly)."

Some people claim that while instructional ADs may not provide much benefit, they at least do no harm. I disagree. Mrs. G.'s AD risks several possible harms that she could not anticipate. Two thankfully do not occur. The AD's prohibition against intravenous "feedings" in terminal illness could complicate comfort management if interpreted as banning intravenous fluids and morphine. The family might interpret such a ban as causing Mrs. G. to suffer "starvation," dehydration, and excessive pain at the end. Furthermore, the AD's criteria for prohibiting hospice care—jeopardizing "my chances for recovery" or creating "excessive burdens for my family"—might appear to prevent such care completely. After all, the second cardiologist believes Mrs. G. can still be cured, and hospice care at home might excessively burden Mr. G.

The one harm that does occur partly due to Mrs. G.'s instructional AD concerns the reventilation dispute between the second cardiologist and Mrs. G.'s family. That cardiologist insists on reading the AD, its definitions, and its conditions quite narrowly. He does not believe that, by it, Mrs. G. is terminal and should receive comfort care only. Rather, he is convinced she can survive, and he briefly convinces Mr. G. of that hope. Mr. G., therefore, leans at first toward the cardiologist's view of reventilating Mrs. G. The daughters, in contrast, believe Mrs. G. is irreversibly dying, and her AD clearly refuses all life support in such circumstances. Unlike the cardiologist, they favor comfort care only.

The second cardiologist unfortunately angers the daughters in subsequent discussions due to his insistence on reventilation. They believe they have already decided the issue by signing a form earlier in the hospitalization agreeing to no more mechanical ventilation. "Why is he bothering us again with this question when he should know from the chart that we already made that decision?" they ask themselves. "He should not think he can be The Big Hero just by sweeping in and fixing her heart." The daughters view the cardiologist as refusing to see Mrs. G. as a "whole person" or to honor her prior instructions. They also believe he is giving Mr. G. false hopes and driving a psychological wedge between them and their father. Eventually the daughters begin to see this cardiologist as "pompous" and "obnoxious" and to distrust his clinical judgment. Only the skillful intervention by the ethics consultant resolves the dispute.

As Mrs. G.'s case illustrates, instructional ADs do not speak for themselves. Implementation always requires interpretation in light of circumstances [76]. Good-faith interpretation uses intimate knowledge of the patient and his or her life, reasonable surmises about the patient's care wishes, and patient-specific methods of decision-making. No instructional AD provides all that. End-of-life care, therefore, necessarily involves negotiation. The last section of this chapter takes up that topic.

Physician Orders for Life-Sustaining Treatment (POLST): A Special Instructional Advance Directive

By the 1990s, state laws had long required emergency medical technicians, if called for cardiorespiratory arrests in the field, to perform CPR and then to transfer patients automatically to hospitals even if the patients were terminal and had prior DNR orders [77]. People realized these laws defeated a main goal of hospice care: letting terminal patients die at home if they wished. Therefore, clinicians and others in Oregon decided to create a new instructional AD unconstrained by such laws. The reformers called the new AD the Physicians' Orders for Life-Sustaining Treatment (POLST) [15, 16, 78].

When published in 1996 and legally approved, POLST gave patients a way to express their treatment wishes as immediately valid, "active medical orders that do not require interpretation or further activation" anywhere [79]. POLST also authorized proxies to make life-support decisions for the future for patients who had already lost decision-making capacity. POLST validated these life-support decisions permanently across admissions to the same inpatient institution, across admissions to different inpatient institutions, and even across admissions to different outpatient institutions [80]. Many states have since legalized the POLST idea although a few states call it by different names such as Medical Orders for Life-Sustaining Treatment (MOLST) in New York, or Physician Orders for Scope of Treatment (POST) in Tennessee and West Virginia. More states will likely legalize the idea in the future.

POLST in the typical form come as an easy-to-read, bright pink, single sheet of paper [80, 81]. That sheet gives patients' names and birthdates at the top, brief summaries of patients' medical conditions and the names of the informants (i.e., patients or proxies) next, and the authorizing clinicians' signatures and contact information at the bottom. POLST typically address such treatments as CPR, admission to acute care hospitals or ICUs, comfort care, antibiotics, and tube feedings. Patients (or proxies acting on incapacitated patients' behalf) pick among distinct, predetermined choices to accept or refuse each treatment:

- *cardiopulmonary resuscitation (CPR)* if the patient in crisis has no pulse or respirations: attempt or do not attempt;
- *medical interventions* if the patient has a pulse or respirations: use "comfort measures only" (transfer to the hospital only if comfort cannot be achieved at the current location), "limited additional interventions" (transfer to the hospital if needed but avoid intensive care measures), or "full treatment" (transfer to the hospital and use all measures including intensive care as needed);
- *antibiotics*: never use, decide whether to use or not when infection occurs, or definitely use whenever they will prolong life; and
- *artificially administered nutrition by tube*: never use, use only for a "defined trial period," or use long-term if necessary.

The authorizing clinicians record informants' choices as medical orders; sign those orders; and distribute copies to the medical record and to patients, proxies, and other key people. Patients or their caregivers must then post POLST sheets in prominent places around the patients' homes (such as on refrigerator doors) for ready reference. Clinicians can consult the sheets and implement the orders immediately in an emergency.

One follow-up study showed that paramedics had little difficulty finding, understanding, and implementing POLST in the field [82]. Nursing homes that have used POLST, however, have encountered predictable, mostly minor problems with them (Table 6.5). In one study, 62 % of such nursing homes reported problems with retrieving patients' POLST forms from other healthcare institutions. Furthermore, 37 % reported problems with getting doctors to fill the forms out; 34 %, with getting doctors to sign the forms; 28 %, with disagreements from families over POLST decisions; and 21 %, with interpreting the orders for implementation in an emergency [75].

The differences POLST make in actual patient care are difficult to assess, but some studies have begun to try [15]. The early results are encouraging. For example, Americans appear willing to accept and use POLST. One study measured POLST completion rates among nursing home residents. Sixty percent in geographic areas actively promoting POLST completed POLST forms, and 48 % in geographic areas *not* actively promoting POLST did so [79]. Both percentages dwarf the usual 5–30 % completion rates of traditional ADs [2, 10–12, 23, 24, 26, 75, 83–85]. Another study contrasted the prevalences of certain life-sustaining treatment orders between nursing home residents with completed POLST forms and nursing home residents without

Table 6.5 Problems in using the Physician Orders for Life-Sustaining Treatment (POLST) forms

Problem	Percentage of nursing homes reporting the problem (%)[a]
Retrieving POLST forms from other health care facilities	62
Getting doctors to complete the POLST	
• To fill out the form	37
• To sign the form	34
Implementing POLST	
• Facing family disagreements over the content	28
• Interpreting the orders	21

Adapted from Wenger et al. [79], with permission
[a]This study contrasted nursing homes in areas promoting POLST educational materials, policies and procedures with nursing homes in areas not doing so. I report the percentages for all nursing homes together because no differences between the two groups reached statistical significance.

Table 6.6 Documented orders about life-sustaining treatments for nursing home residents by POLST[a] status

Life-sustaining treatment	Percentage of residents with completed POLST forms who had orders (%)	Percentage of residents without completed POLST forms who had orders (%)
Cardiopulmonary resuscitation	98	86
Medical interventions related to balancing curative versus comfort care	97	14
Antibiotics	96	3
"Artificial" tube feedings	92	7

Adapted from Hickman et al. [81], with permission
[a]Physicians Orders for Life-Sustaining Treatment

such forms (Table 6.6). Orders concerning CPR were nearly equally prevalent between the two groups, 98 % versus 86 %. But *orders* about curative or comfort care, antibiotics, and artificial feedings were far more prevalent in the POLST group (all 92 % or more) than in the non-POLST group (all 14 % or less) [81]. Still another study surveyed participants in a comprehensive elder care program for their POLST completion rates and the consistency between their POLST and actual care. Almost all program participants (57 of 58 (98 %)) had completed POLST forms, and the care for completers in their last two weeks of life often complied with their POLST statements. Specifically, 91 % of CPR decisions, 46 % of cure versus comfort decisions (including hospitalization), 86 % of antibiotic decisions, and 94 % of feeding tube decisions did so. Overall, 39 % of completers had full compliance with their POLST, and another 51 % had majority compliance [86]. The reader must remember, however, that such cross-sectional studies cannot prove cause-and-effect. Verbal patient wishes, doctor practice habits, or the unavailability of treatments rather than POLST may have actually determined treatment decisions. Additional studies must sort out the cause-and-effect possibilities.

Even as ongoing research assesses the precise effect of POLST on life-sustaining treatment, I support their use. They surely promote ACP and improve the conformity of some care to patients' wishes. Patients who are imminently dying (including home hospice patients) or who refuse specific life-sustaining treatments for all situations will most likely benefit from POLST. Attending clinicians will also likely find POLST helpful when clinical assessment in the life-threatening emergency is simple (pulse/no pulse or respirations/no respirations) and treatment options are few and distinct (CPR/no CPR or hospital transfer/no hospital transfer). But POLST will likely have little effect in other situations, which present unclear diagnoses or complex treatment options.

The Case

POLST forms were not available in Mrs. G.'s case. Her state of residence was still debating, developing, and pilot-testing various versions of POLST at the time. In fact, the state has still not legalized any of those versions several years later.

Even if POLST had been available, they would probably not change Mrs. G.'s care. She is not clearly terminal when she suffers the cardiorespiratory arrest at home, and she expresses no universal refusal of CPR beforehand. In fact, she tells her family the opposite—that she wants "every chance to live." Furthermore, neither the paramedics nor the emergency room doctors can confidently determine Mrs. G.'s underlying diagnosis or prognosis at hospital admission. Rather, as often happens, the critical illness requires some days to "declare" itself and its course. For what anyone knows early on, Mrs. G. may still regain all her prior function and return to "participating actively in life." But she needs sophisticated life-sustaining treatments to stay alive in the meantime. By all those considerations the family and the paramedics make the right decision to perform CPR in the home and to transfer Mrs. G. to the hospital for intensive life support. POLST would not affect that decision.

The Second Generation of Advance Care Planning: Proxy-Designating Advance Directives

The early instructional ADs soon revealed serious problems with their use [12, 23]. The problems included inadequate patient understanding of medical facts and treatments at signing [68], inaccessibility of the ADs in the crisis [87], irrelevance of the instructions to the crisis [88], and the occasional lack of agreement between a patient's prior instructions and current best interests [2, 67]. Proxy-designating ADs, officially called durable powers of attorney for health matters, emerged to circumvent such problems [23].

Sound reasons justified these new ADs. Long tradition in medicine had doctors already consulting family members as unofficial proxies for treatment decisions that patients could not make themselves. Family members, often named as patients' proxies in these ADs, were assumed to know patients best and to promote their interests most faithfully. These ADs legitimized that important tradition [89, 90]. In addition, many patients no longer wanted to leave to chance the particular people who would make decisions for them in a crisis. They wanted to choose their proxies beforehand and to document those choices. Furthermore, with these ADs clinicians no longer needed to recruit stop-gap proxies at the bedside or over the telephone during the crisis. Such tasks are time-consuming, risk suboptimal proxies and proxy decisions, and divert clinicians' attention away from urgent medical matters. These ADs, in contrast, designated clearly the proxies whom patients wanted and provided readily accessible contact information for the crisis.

Other reasons supporting proxy-designating ADs have emerged with time. These ADs improve decision-making by introducing particularity and flexibility that static instructional ADs do not have. Proxies can respond to particular, immediate circumstances and need not guess at them beforehand. Proxies can also change decisions as necessary if circumstances change. Furthermore, proxies can participate in give-and-take with doctors and thereby build collaborative decision-making partnerships with them. Mutual trust is bound to increase as a result.

Proxy-designating ADs also offer an ease of completion that instructional ADs do not. Most patients already understand proxy decision-making and know their preferences for proxies [83, 91]. Patients typically need no special explanations of concepts (such as of particular diseases and treatments as with instructional ADs). Therefore, signing proxy-designating ADs is simpler for patients and quicker for doctors than signing instructional ADs. In fact, signing proxy-designating ADs easily fits into the standard 15-min patient office visit.

Though an improvement over instructional ADs, proxy-designating ADs present their own problems. Patients may never disclose their proxy choices to the proxies themselves or to other key people. Proxies unaware of their selection beforehand may not be willing or prepared to serve as proxies in the crisis. Disputes may also arise among the proxies whom the patient chooses and others who expected to be chosen. The AIDS epidemic presented that problem. Many early AIDS patients lived a secretly gay life. When they became severely ill, they often had not yet "come out" to their families but had had long-term, intimate partners. These patients usually preferred their partners, not their blood relatives, as proxies. Proxy-designating ADs clarified these patients' proxy choices but still proved awkward in some cases. Clinicians sometimes had to inform family members in the crisis that they were not the patient's preferred or official choices for proxies. Clinicians then had to negotiate treatment decisions between marginalized partners and hurt or judgmental family members [13].

The designated proxies themselves may also present problems. Proxies may lack adequate knowledge of patients' wishes [3, 92–94] or have difficulty understanding medical facts. Proxies may also lack the necessary courage to make critical life-or-death decisions. They may just not have the sturdy backbone to handle the momentousness of the decisions, the uncertainties involved, or the sometimes withering criticism from others. Furthermore, proxies may not clearly separate their own interests from patients' interests. Strong emotional attachments can conflate the two [94]. (Legally separating powers of attorney for medical and financial matters may alleviate that problem.) Thus, all clinicians who attend vulnerable, dying patients have the duty to monitor proxy decisions for reasonable agreement with patients' interests. If proxy decisions deviate significantly from those interests, clinicians must act to protect the patients. In extreme cases clinicians may have to insist on a change of proxies.

The Case

Mrs. G.'s proxy-designating AD provides clear benefits: Most importantly among them it identifies two trustworthy, accessible family members as her proxies. Mrs. G. names Mr. G. as Primary Advocate due to his deep knowledge of her. As her lifetime partner, he surely knows her better than does anyone else. And Mrs. G. names her elder daughter as Successor Advocate partly due to her excellent organizing skills. That daughter organizes much of the family response to Mrs. G.'s illness crisis including round-the-clock bedside watches. In addition, the AD helps the attending clinicians avoid having to guess about whom Mrs. G. wants as proxies.

In addition, the AD informs Mr. G. and the elder daughter about their selection as proxies. Both must sign a form accepting that role (Table 6.7). They thereby

Table 6.7 Acceptance of responsibilities by Mrs. G.'s advocates*

I accept the responsibilities of serving as Mrs. G.'s advocate under the following conditions:	
1. My authority as her advocate becomes effective only when she cannot make her own decisions about medical care.	
2. I may decide to withhold or withdraw life-sustaining treatment and to allow her to die only if she has previously expressed a clear, convincing, and consistent wish for that decision under the circumstances.	
I will act according to fiduciary standards that promote her best interests. I will presume that her treatment wishes, when known, indicate those interests.	
Primary Patient Advocate's Signature	Successor Patient Advocate's Signature
Date	Date

*Modified and condensed from Mrs. G.'s original

acknowledge empowerment as advocates only for times when Mrs. G. becomes unable to make medical decisions for herself. Mr. G. and the elder daughter promise to withhold life support from Mrs. G. only with "clear, convincing, and consistent" prior wishes from her to do so. And they vow to serve her always as fiduciaries. Moreover, when the two sign this form, they have the chance to read her instructions and to prepare themselves for possible future decisions on her behalf. The AD specifies many of those decisions. It allows decisions about institutional admissions and discharges, treatments, dissemination of medical information, and donation of blood and organs. It also allows decisions about pain relief treatments even if those treatments risk physical damage, "addiction," or hastened death. But the AD prohibits decisions that would "intentionally" cause Mrs. G.'s death.

Along with its benefits Mrs. G.'s proxy-designating AD has drawbacks, too. One drawback is the unclear division of decision-making authority among Mr. G. as Primary Advocate, the elder daughter as Successor Advocate, and the younger daughter who has no official decision-making authority. Mr. G. believes at first that, as Primary Advocate, he must shoulder the whole decision-making burden. But the two daughters also want a say. The G. family eventually work out a decision-making method that includes all three. Less harmonious families, however, might end up fighting over the proxy decision-making authority.

Mrs. G.'s AD also does not specify the circumstances that would invalidate either named proxy's decisions for her. Some of those circumstances could surely be anticipated and documented beforehand. Of course, a named proxy's decisions ordinarily deserve the benefit of the doubt for promoting the patient's wishes or interests. Yet protecting the patient occasionally justifies, even demands, clinicians overrule the proxy. Such a drastic step requires enormous diplomatic skills.

Conflicts of interest always create a threat to the validity of proxy decisions. Such conflicts may be financial, emotional, or physical. Financial or emotional conflicts are usually obvious. The less obvious physical conflict of interests often involve exhausting caregiving at home. While most proxies surely make responsible patient care decisions despite conflicts of interest, clinicians sometimes have difficulty knowing that for sure [5].

Mr. G. like many other intimates faces emotional conflicts of interest in making his proxy decisions. He may not be able to disentangle his emotional interests from Mrs. G.'s, thereby unwittingly confusing his with hers. The reventilation issue poses just such a severe conflict of interest. Because Mr. G. loves Mrs. G., he wants her to live. He pledges selflessly (and unrealistically) to care for her fully if she does live. The second cardiologist's plan to ventilate Mrs. G. fuels Mr. G.'s hope. That hope pulls Mr. G. toward reventilation. But Mr. G. also sees Mrs. G.'s suffering in her current critical illness. Reventilation and intensive care mean even more discomfort and loss of privacy for her and do not guarantee her survival. Mr. G. wonders whether, if Mrs. G. could speak, she would choose comfort care instead. Compassion for her pulls him toward comfort care.

Mr. G.'s conflictedness illustrates the wear and tear decision-making can have on proxies. Though I do not know about Mrs. G. in particular, many patients choose their proxies without thinking about how much the decision-making process may adversely affect those proxies. After Mrs. G. dies, Mr. G. certainly regrets not reventilating her. Still, I think he and his daughters make the best decision for her.

Mrs. G.'s proxy-designating AD also holds her advocates to fulfilling her best interests in situations for which she never expressed treatment wishes. This best-interests decision-making method has gained widespread acceptance lately in medical ethics and clinical medicine. Nonetheless, I believe the best-interests standard is difficult, perhaps even impossible, to meet. It is future-oriented, requiring clinical predictions. A proxy must imagine all possible outcomes, choose the best one according to the patient's presumed interests, and judge accurately that treatment will produce such an outcome. Clinical predictions are notoriously inaccurate for clinicians, surely more so for medically untrained proxies. As a result, reasonable differences of opinion arise frequently, causing conflicts among proxies, clinicians, and others.

Such a conflict arises in Mrs. G.'s case. Mr. G. and the daughters disagree initially about whether reventilation would fulfill Mrs. G.'s best interests. Persuaded by the second cardiologist that Mrs. G. can survive, Mr. G. believes initially her best interests lie in reventilation because it gives her the best chance to "participate in life" again. The daughters, however, believe Mrs. G. is irreversibly, imminently dying, and her best interests lie, as she states in her AD, in comfort care at home or in a hospice. (Even after Mrs. G. dies, no one can know for sure which decision promoted her interests better.)

I favor proxy decisions based not on the patient's presumed best interests but on the best "fit" of treatment and outcomes to the patient's life history and personal values. This method is called the constructed judgment method. Unlike the best interests method, the constructed judgment method based on a patient's life history is past-oriented, and its application open to reasoned discussion among the patient's family and friends. Proxies often find the constructed judgment method easier to use than the best-interests method because they can draw on their own memories of the patient. (I discuss the constructed judgment method in greater detail in Chap. 7.) The G. family, for example, have seen Mrs. G. suffer from back pain for many years. They have also seen her enjoy herself despite that pain. She likes her life, she has said, as long as she can move around; do basic self-care; and visit her children, grandchildren, and great grandchildren. But, once she becomes bed-bound and unresponsive late in this hospitalization, the family realizes she can no longer live a life acceptable to her. The family correctly, I believe, opt to let her die under hospice care.

Two other minor objections about Mrs. G.'s proxy designation occur to me, too. It does not specifically combine or separate proxy decision-making powers over medical and financial matters. Many people assume that having one kind of proxy authority automatically means having the other. That assumption is incorrect. In fact, separating the two kinds of powers often makes good sense as a protection for everyone involved. For that reason, whenever people claim decision-making powers for a patient, clinicians should check the official documents for an exact definition of the powers granted.

Mrs. G.'s AD also makes the confusing statement that her proxies "may not *intentionally* cause (her) death." That statement might exacerbate her family's regrets over refusing reventilation. I firmly believe, however, the family *justifiably intends* not to prolong her dying at the end but rather to promote her comfort. Many other clinicians would also agree with the family's decision. But, in case regrets might surface after Mrs. G.'s death, her clinicians should explicitly reassure the family about their support for the decision not to reventilate. The clinicians should also explicitly assume some responsibility for the decision.

The Third Generation of Advance Care Planning: Broad Discussions

Both the instructional and the proxy-designating ADs take seemingly logical approaches by trying to match future treatment decisions *exactly* to patients' prior wishes. Instructional ADs function like a movie script; AD-designated proxies, like movie directors. By using ADs the patients, or "script writers," expect to control the future action [95]. Yet voluminous, rigorous research over many years raises serious doubts about that expectation [2]. Life-threatening illness almost never follows the anticipated storyline and sometimes veers quite far off script. Furthermore, the "script" instructions are often vague, [26, 69], and the "directors" can guess patients' wishes little better than chance [2, 26, 70, 96]. Life-threatening crises with their unexpected complications invariably force impromptu responses. Some years ago I addressed this point in an article entitled "Controlling Death: The False Promise of Advance Directives" [4]. The false promise is that ADs can make critical illness yield to prior human direction. They cannot.

If these two kinds of ADs cannot extend patient autonomy to future life-threatening crises, does ACP have any value at all? I say yes but only with changes in ACP's purpose and approach. I believe that rather than trying to match future care exactly to individual patients' prior wishes, ACP should try to prepare patients (and their families) mentally and emotionally for the end of life and the decisions involved. The preparation should include understanding the *real* benefits of planning, eliciting patients' relevant values and goals for end-of-life care, readying everyone for the difficulties ahead, and securing the steady support of doctors and other HPs throughout the experience [49]. I call this approach "broad ACP."

The Value of Planning

As the Cold War escalated in late 1957, President Eisenhower addressed a planning conference of business executives recruited to serve in key government jobs during any future national security emergency. He made a seemingly curious statement at the time. He said, "I heard long ago in the Army (that) plans are worthless, but planning is everything … (An) emergency, … is not going to happen the way you (expect). So (you may have to) … take all the plans … and throw them out." But the planning, Eisenhower then added, "makes it possible to carry forward (intelligently in) … an emergency. You prepare yourselves (for the unexpected)."

Eisenhower's statement reflects the fundamental problem of instructional and proxy-designating ADs: confusing the value of the plan with the value of the planning. The real value of broad ACP is preparing people mentally for the end-of-life crisis to come [49, 90]. It creates not a detailed instruction manual for patients' end-of-life care but a supportive partnership among patients, proxies, HPs, and others for facing the end of life bravely together. Mrs. G.'s case points out some potential benefits of broad ACP.

Broad ACP should reassure people about future quality care while dispelling unrealistic expectations of prior control. Mrs. G. and her family expect her AD to control fully her care at the hospital. But, when she is admitted immediately to the ICU and placed on a ventilator despite contrary instructions in her AD, the family become frustrated and angry. They believe the doctors are intentionally ignoring Mrs. G.'s wishes in order to treat her maximally. The family feel steamrollered. Broad ACP could have prepared them beforehand for the uncertainties and complexities of such a critical illness. Unforeseeable situations arise often, requiring unanticipated decisions on the spot [97]. As Mrs. G.'s case shows, acutely life-threatening events require quick clinical responses, which may not permit consultation with patients or families. Broad ACP might have also encouraged the G. family to ask immediately about decisions they do not understand and about aspects of the situation that might lie beyond medical control.

Broad ACP can also communicate to patients and family the commitment of strong moral support in the end-of-life crisis. A patient's long-time primary care doctor historically provided that support with attentive presence and advice. But extensively fragmented care unfortunately prevents such care today. Hospital-based doctors must provide that service as best they can even when they do not know patients and families well. Of course, the doctors alone need not provide it. Other professionals including nurses, chaplains, social workers, and family ministers can help. In Mrs. G.'s case I do not know whether her primary care doctor visits her in the hospital or provides moral support in other ways. Most of the hospital-based doctors (especially the second cardiologist) certainly do not provide that support. But, as I will detail shortly, the ethics consultant does. He meets with the family several times, supports them as they decide about reventilation, and helps implement the decision afterward. The family's minister also gives critical support by meeting and praying with them as they deliberate.

Broad ACP planning should also help patients' views "mature" in their decision-making approach with time and experience [98]. Patients should learn what medical treatment can and cannot realistically accomplish. A well-informed, respectful, ongoing discussion beginning with the doctor in the clinic and continuing with the doctors in the hospital promotes this maturation best. Unfortunately, to the extent Mrs. G. does not have either well-informed or ongoing discussions before hospitalization, her ACP may not optimally promote such maturation for her or her family. The family's views, in fact, must mature quickly in the mere hours they have to decide about reventilation.

Focusing on Patients' Personal Values and Goals for End-of-Life Care

Several experts find patients' life histories, personal values, and goals for care more useful than patients' specific treatment wishes for guiding medical care [13, 39, 51, 99–101]. I certainly do. Life histories and personal values lay the conceptual foundation for defining specific functional goals for care that patients would appreciate [8]. Patients rightfully play the key role in defining those goals. But once the overall functional goals are set, patients should allow their doctors considerable leeway to choose the best treatments for achieving those goals. Mrs. G. specifies clearly her minimal functional goals for care: being able to move about somewhat independently, to give herself basic care, and to enjoy her family. Ventilation immediately upon admission makes sense when she might still regain responsiveness and achieve those goals. But reventilation later on makes less sense when she slips back into unresponsiveness and may never again achieve any of her three functional goals.

HPs might begin eliciting patients' functional goals for care by asking the high points of their lives. This exercise sets the context within which to interpret patients' functional goals later. Mrs. G. might mention meeting Mr. G., marrying him, and raising their children. HPs might then ask, "What makes life worth living for you now?" "What do you still look forward to doing?" Or "What do you enjoy most in life these days?" These questions elicit patients' particular values and functional goals at the moment. Mrs. G. might say she enjoys most being at home with Mr. G. and seeing her family. She might also say she looks forward to playing with her first great grandchild or seeing him go off to kindergarten.

Patients and doctors participating in broad ACP can then collaborate on articulating tentative patient-centered functional goals for care. Doctors might suggest some based on the prior discussion and might allow patients to accept, reject, or modify those goals. Doctors might also ask patients how much leeway they will permit their proxies and doctors for revising the goals and determining treatments to try to achieve them. Doctors should always emphasize the tentative nature of care goals, which may have to change in response to changing medical conditions.

Preparing Patients and Survivors for the End of Life

Powerful emotions, many troubling, arise for patients and their survivors as patients near death [97, 102–104]. Attending HPs should not only address those emotions when patients or survivors volunteer them but also anticipate them when patients or survivors do not. I recommend HPs do so by avoiding the usual questions, How are you doing? And How are you feeling? Both questions invite the unthinking, vacuous response, "I'm fine." Instead, I recommend more substantive questions such as What are you feeling now? or What bothers you most today? Those questions open the conversation to discussing heartfelt emotions.

Of note, doctors (or other HPs) conducting broad ACP have the chance to help people prepare to confront these emotions even *before* the end-of-life crisis strikes. HPs might introduce the topic of emotions by saying something like "Patients and their survivors may experience strong emotions toward the end of patients' lives. To help you and others later on, we might talk about those emotions now so you will recognize them if they occur later." The HPs, of course, should mention that they cannot be sure these emotions will ever come up. The HPs might just say, "Although many deaths produce similar emotions, each death is unique in its details."

One of the most common end-of-life emotions is fear about the uncertain or the unknown to come. The process of dying probably creates more of this fear than does the prospect of death itself. That very idea prompted the statement in A Living Will, "I do not fear death itself as much as the indignities of deterioration, dependence, and hopeless pain. I, therefore, ask (for) medication to alleviate (my) suffering" while I am dying.

HPs might ask directly whether patients or their survivors fear what will happen in the dying process and whether they want to talk about it. If they do, the HPs might give a general description of how terminal illness evolves. As Chap. 5 explains, patient function declines, sometimes gradually, sometimes suddenly. The underlying fatal disease determines much of the pattern of decline.

The patient's attending doctors should assume responsibility for addressing prognosis for the patient's disease. They might use the Fig. 5.1 in Chap. 5 to describe the most likely functional course for a patient's disease. The doctors might warn that such declines are hard to experience for patient and survivors alike. In Mrs. G.'s case a waxing and waning neurologic course hampers early predictions of her eventual survival. The family's hopes soar when she regains responsiveness early on but then nosedive again when she slips into permanent unresponsiveness later. Furthermore, Mrs. G.'s illness does not fit any of the four established categories for end-of-life functional decline. Her illness falls instead into the remaining miscellaneous category that defies specific functional prognosis. As I said earlier, about 20 % of American deaths do.

Doctors conducting broad ACP might also mention that the main goals of care may have to shift at some point from attempting cure to promoting comfort. When that shift happens, it may anger observers who do not agree with its timing. Different observers recognize death as imminent at different times. Those who recognize death as imminent early may believe others are making the patient suffer too long; those who recognize it as imminent only later may believe others are letting the patient die prematurely. Acrimonious conflicts may follow.

Furthermore, since the rise of hospice many people expect a symptom-free death. That expectation is only partly realistic. Pain care has improved so much in recent years that many patients can now die pain-free. But other symptoms do not yield so well to palliative treatment. Mrs. G. does not appear to have much pain during her hospitalization, but she does have other symptoms. The restlessness during her last night responds well to simple sedation, but the retained airway secretions recur throughout her hospitalization despite repeated, uncomfortable tracheal suctionings. Doctors doing broad ACP might explain the variable response of terminal symptoms to treatment. The doctors might conclude by saying, "I cannot promise complete comfort at the end of life, but I can promise my best efforts to achieve it."

Doctors conducting broad ACP should also warn people about American medicine's extensive fragmentation. It results from the explosion of technical medical knowledge and the accompanying high degree of specialization among clinicians and within institutions. Coordination of care often suffers even at the end of life [69]. Care may sometimes seem unfriendly, disorganized, rigid, confusing, and inconsistent. It can frustrate patients and families. In Mrs. G.'s case, for example, the family feels whipsawed by the doctors' inconsistent, conflicting opinions. Early on, the first cardiologist says Mrs. G. is imminently dying while the neurologist says she might recover. Later the second cardiologist appears unwilling to honor a document the family signed earlier with the pulmonologist to prevent reventilating Mrs. G. Such experiences can bewilder, even anger, the medically untrained. They feel vulnerable and may question clinicians' competence. Doctors doing broad ACP, therefore, should acknowledge the problem early. They should urge patient and family members to learn as much about the patient's care as possible and to engage it actively. The patients and family members should ask questions; take notes about important care decisions; and bring photocopies of problem lists, medication lists, and other key medical documents to meetings with doctors [13]. Such active participation can minimize the discontinuities of care and protect patients.

Importantly, HPs conducting broad ACP should remind patients and family members about some surprisingly positive aspects of the dying experience: It offers emotional benefits in addition to the emotional trials [102]. One benefit is the opportunity for personal growth. Legacies can be planned, relationships strengthened, and spiritual values explored. Mrs. G.'s terminal illness, for example, surely strengthens ties within her family and deepens their faith connections to God. Another benefit is reliving treasured memories and otherwise enjoying patients' company as never before. The G.s enjoy reliving significant memories with Mrs. G. (such as Mr. G.'s "famous kisses") while she can still communicate. Yet another benefit for some family members is gaining satisfaction from helping with direct care of the patient. The G. family, for example, maintains a presence at Mrs. G.'s bedside and gains satisfaction from helping the nurses keep her comfortable. One final benefit is the peacefulness that often accompanies death. Clinicians can reassure patients and survivors that the imminently dying often slip quietly into unconsciousness shortly before death and may not even know when they die. This peaceful interlude before death may be nature's way of easing patients from life into death.

The Health Professionals' Pledge

Broad ACP should conclude with a simple, heartfelt pledge from the HPs—the principal attending doctors most importantly—never to abandon dying patients or their survivors. The great uncertainty, distress, and sense of loss at the end of life can leave patients and survivors emotionally adrift. They need HPs' steady, supportive presence. The HPs should not just monitor dying patients physiologically but should also address the emotional burdens of patients and survivors. HPs can ease those burdens just by being willing to hear them. The nurses in Mrs. G.'s case address the family's emotional burdens particularly well. (I find, in general, nurses and chaplains are especially skilled at doing so.) In fact, the elder daughter returns to the hospital some months after Mrs. G. dies specifically to thank the nurses for the compassion they showed Mrs. G. and the family during the hospitalization.

A Final Comment About Broad ACP

Some may claim that broad ACP as I describe it here is too burdensome for busy clinicians. The process does seem like a huge smorgasbord of topics. Comprehensive broad ACP, of course, may not be possible in today's busy clinical reality. But as with other tasks clinicians should tailor their approach to the needs of specific patients and family members, choose the most important aspects to address, and follow through on them to the extent time and other resources permit. Clinicians should also recruit clergy and other HPs to help in the process.

Steps in Conducting Advance Care Planning

Because talking about dying is always awkward, no time feels quite right for ACP. Thus, patients, family members, and professionals should never expect to feel completely comfortable that the right time for ACP has come. Still, the Patient Self-Determination Act of 1990 (PSDA) requires health care institutions to conduct at least some ACP with patients on admission. Institutions must inform patients about their medical decision-making rights, ask whether patients have executed ADs, and provide information about ADs for those patients who want it [66, 87, 105]. Unfortunately, these discussions give ACP short shrift. Admission clerks, not doctors, nurses or other HPs, often conduct the discussions, and patients cannot focus on serious ACP decisions during the typically hectic admission process [2, 93]. PSDA-mandated discussions, therefore, cannot replace thoughtful, dedicated ACP discussions with HPs at other times [92].

Most patients and end-of-life care experts recommend conducting those dedicated ACP discussions during routine office visits to primary care clinicians when patients are medically stable [46, 93]. Good sense, of course, suggests waiting until clinician and patient have developed a good working rapport. One expert estimates establishing such rapport takes at least four visits [31]. Clinicians and patients generally agree the clinician bears responsibility for broaching the subject [33, 46, 93]. Most patients want their spouses and/or adult children—their usual choices for proxies—to attend the discussions [46]. If the patient is healthy, the clinician should explain that discussing ACP at that time does not mean the patient is dying [64]. ACP is merely planning for future care. Furthermore, ACP need not consume a lot of time. One study timed initial ACP discussions at 13–18 min altogether; [7, 87, 92] another study, at 30–40 min. I allot 30 min in an extended office visit for the initial substantive ACP discussion. Then I allot 5–10 min during later visits whenever questions or changes arise. Unfortunately, by not covering such discussions [87] current national reimbursement policies create a significant obstacle for busy clinicians to perform ACP at all.

Table 6.8 suggests five steps in the ACP process [7, 106]. The reader will note these five steps draw from all the generations of ACP. The table lists the skills required for each step, and the component actions to be taken [106, 107]. (This discussion assumes the patient's principal doctor, usually a primary care doctor, conducts the process, but any HP may do so as long as the principal doctor has foreknowledge of it and the HP communicates the results back to that doctor.) Step 1 introduces the idea of ACP. Step 2 provides initial information and plans for the definitive follow-up discussion. Step 3 helps the patient articulate relevant personal values, goals for care, and specific treatment wishes. Step 4 records and distributes the patient's statements in summary documents. And Step 5 reviews and updates the patient's ACP summaries later on as needed. The clinician ideally can devote one or two partial office visits to Steps 1, 2, and 5, and another visit to Steps 3 and 4 together.

Step 1. Introduce ACP. Many patients have only the vaguest knowledge about ACP before doing it. The doctor, therefore, should take a few minutes to introduce ACP. The doctor should explain its importance and the steps in doing it. The doctor should emphasize the purpose of ACP: to give family, friends, doctors, and other professionals the patient's guidance for making medical decisions on the patient's behalf if the patient ever cannot make those decisions in the future.

Step 2. Prepare for the definitive discussion. At the same time as Step 1 the doctor might give the patient a sample AD and explanatory ACP materials. The doctor should also ask the patient's preferences for proxies and set a date for the definitive Step 3 discussion. The doctor should urge the patient to read the ACP materials, discuss them with the preferred proxies and any other key people in the patient's life, and bring the preferred proxies to the Step 3 discussion.

Step 3. Help the patient articulate personal values, goals of care, and specific treatment wishes. Because many people assume ACP involves only the decision

Table 6.8 Steps in conducting advance care planning (ACP)

Step	Skills required	Component actions
1. Introduce ACP.	• Explanation • Encouragement	A. Raise the idea of ACP. B. Explain its importance. C. Explain the process.
2. Prepare for the substantive ACP discussion.	• Explanation • Encouragement	A. Provide a sample advance directive or other printed ACP materials. B. Ask the patient's preferences for proxies. C. Set a time for a definitive step 3 discussion. D. Recommend the patient discuss the ACP materials with potential proxies and other key people in the patient's life. E. Urge the patient to bring the potential proxies to the step 3 discussion.
3. Help the patient articulate relevant personal values, goals for care, and specific treatment wishes.	• Explanation • Supportive elicitation • Unhurried, intent listening	A. Initiate the ACP process with a brief explanation of its purpose. B. Ask the patient, "What makes life worthwhile for you?" "What do you still look forward to doing?" or "What do you still enjoy in life?" C. State important personal values and goals for care derived from the patient's answers in part B. D. Summarize what the patient has said and confirm an accurate understanding: "Do you want to change anything I said in summarizing your answers?" E. Describe specific life-sustaining treatments most likely to require decisions in the future when the patient cannot express his or her wishes. F. Elicit the patient's wishes about those treatments. G. Use broad ACP to prepare patient and proxies in general for end-of-life care.
4. Record the patient's statements in an ACP summary and distribute copies of it.	• Accurate written description of the patient's values, goals for care, and treatment preferences • Sensible distribution of copies of the ACP summary to the people most likely to need them for future decisions	A. Summarize the patient's statements in the patient's own words if possible. B. Have the patient review, sign, and date the ACP summary. C. Distribute copies of that summary to key people in the patient's life (e.g., proxies, principal doctors, and other significant people) and to the medical record. D. Give copies to the patient and encourage storage in a logical, accessible place for retrieval in a crisis. E. Urge the patient to tell others where the ACP summary is stored.

(continued)

Table 6.8 (continued)

Step	Skills required	Component actions
5. Review and update the ACP summary as needed.	• Conscientious, orderly tracking and follow–up • Logical storage of materials	A. Review with patient prior expressions of values, goals, and treatment wishes. B. Elicit any changes. C. Record the changes. D. Have the patient review, sign, and date any amended ACP summary. E. Destroy all outdated ACP summaries.

Data from Emanuel et al. [106], Back et al. [107], and Emanuel [7]

about CPR, they expect the doctor to ask immediately about it. The typical CPR question is "What treatments do you want if your heart or lungs ever stop working?" I find patients' answers to that question often unhelpful in clinical situations and, therefore, save it for last. The more clinically helpful information, in contrast, concerns the patient's relevant life history [8] and personal values [51]. Practical goals for care can emerge from them. For that reason I believe the definitive ACP discussion should focus on those topics.

Still, the doctor should acknowledge patients' initial expectations to talk about life-sustaining treatments. He or she might open the discussion with "I realize you may expect to talk about resuscitations, breathing machines, and similar treatments. We will talk about them a little later. But I want to make sure *all* of your care in the future—including the other kinds of treatments—is as true to you as possible even if you cannot tell me your wishes at the time. To do that, I need to know about you as a person including what is most important to you in your life."

Then the doctor might pose one of the questions aimed at learning a patient's relevant personal values and goals for care: What makes life worth living for you now? What do you still look forward to doing? or What are you enjoying most in life these days? [74] Intent listening holds the key to getting useful answers. Unfortunately, doctors tend to do more talking than listening in all doctor-patient encounters including ACP discussions. ACP requires the doctor consciously to practice good listening. A key part is dedicating a little necessary time, effort, interest, and patience to the process. The doctor, for example, might draw the patient out with probes such as "Tell me more." or "Is there anything else?" The doctor might also use the patient's own words to summarize whenever the patient finishes a reply [14]. The doctor might even invite corrections by asking, "Do you want to change anything I said in summarizing your answers?"

Next the doctor might elicit the patient's wishes about the life-sustaining treatments most likely to require decisions when the patient cannot express his or her wishes directly. Such decisions usually involve hospitalization, CPR, and mechanical ventilation. The doctor should explain the risks and benefits of each treatment and then might ask the patient's wishes from among three choices: I want, I don't want, and I am not sure. The doctor should also remind the patient that implementing treatment choices in the crisis always requires some interpretative leeway [60] based on the patient's life and values.

Finally, the doctor should try to prepare patient and proxies for end-of-life care by conducting broad ACP. The doctor might mention not only the pressures, uncertainties, frustrations, and emotions involved but also the rewards.

Step 4. Record the patient's statements in an ACP summary document and distribute copies of it. The doctor should summarize the patient's statements in a prominently placed medical progress note and attach any formal supporting documents such as ADs. The patient should review, sign, and date all the documents including the progress note. The doctor should then make copies of everything and distribute them to the patient, the proxies, and the medical record. The doctor should also urge the patient to store his or her copies in a logical, accessible location (not a safe deposit box) for easy retrieval during a crisis and to disclose the location to trusted others. Finally, the patient should distribute copies to key nonproxy family members and doctors.

Step 5. Review and update the ACP summary document as needed. Doctor and patient should review together the patient's wishes and ACP summary document whenever the patient experiences significant life changes or an acute hospitalization involving life support [24]. The doctor should record any changes. Then the patient should review, sign, and date the new summary. Doctor, patient, and others should keep copies of the new ACP summary and destroy any old ones.

Negotiating Death

End-of-life care has many aspects: partly sound scientific deduction based on data, partly patient treatment wishes, partly happenstance, and partly potential medical juggernaut [5, 17]. The various aspects can confuse patients, family members, and sometimes even clinicians themselves. But, as Eisenhower said about all emergencies, the complexity of the situation and the ever-present element of chance ensure that the end-of-life crisis never develops as expected. Broad ACP may help prepare people to confront the crisis, but much of its management requires on-the-spot negotiations.

Some theoretical and empirical research, though scarce, illuminates this important phenomenon of end-of-life negotiations. One author defines them as the process by which the various meanings people attribute to a patient's dying evolve into a single plan to deescalate life support. She calls the process the "negotiation of death." [17] This author believes different observers derive different meanings from the manner of a patient's death. For patient or family the meanings might include "giving up" on life by the patient, "giving up" on the patient by others [17], avoiding a prolonged death, honoring the patient's wishes, achieving "death with dignity," or allowing a "natural death." For the clinician the meanings might include being defeated by death, accepting the eventual failure of all life support, recognizing the illness' "declaration" that the patient is dying, honoring the patient's or family's wishes for comfort care over life prolongation, or risking legal or professional censure for the death.

Amid the various meanings attributed to a patient's dying, end-of-life negotiations aim to create the best end-of-life care plan acceptable primarily to the patient and only secondarily to interested others. Most of the negotiations occur during informal conversations with the patient or family as the doctors make their rounds or the nurses provide their bedside care. The plan that emerges involves compromise. It is also dynamic, changing to accommodate changes in the patient's medical condition or in observers' views. Still, the plan rarely achieves a total, ideal concord among all parties; some dissatisfaction usually persists.

The plan specifies which treatments will be used when [17]. It often involves "orchestrating" deliberate, staged treatment decisions, actions, and inactions [108]. The plan may continue life support treatments already in use when some people simply cannot bear to stop them. The plan may also withhold other life support treatments when people see them as invasive and ineffective. Most commonly, the consciously withheld treatments are CPR, mechanical ventilation, or late-stage cancer chemotherapies. The plan may withhold still other treatments, which might be effective in other circumstances (such as antibiotics or certain tube feedings) [76] but which cannot help the irreversibly dying patient [108]. All the while, discussions about whether the plan qualifies as a good, dignified, proper, or natural death may swirl in the background among observers. Most people hold strong views on this subject, and such views may surface prominently in the end-of-life crisis.

All professionals attending the dying patient can play an important role in negotiating the end-of-life treatment plan. Their sympathetic observer's perspective can help patient and family alike identify the meanings the death has for them, and consider the best treatment plan in light of those meanings. Multiple meanings may simultaneously exist for any one person. A wife may wish that her husband live a long time while she also wishes that he not suffer unnecessarily. Clinicians can help people sort through their feelings but must guard against hijacking with their own technical knowledge the process of others' finding meaning in the experience [17].

The Case

If Mrs. G. as an outpatient had discussed her AD with her primary care doctor, end-of-life care negotiations could have begun with a broad ACP discussion well before the hospital admission. The doctor could have elicited Mrs. G.'s relevant life history, personal values, goals for care, and specific treatment wishes. The doctor could have also asked Mr. G. and the daughters for their opinions. All could have begun to think deliberately about the end-of-life care that would "fit" Mrs. G.'s life and values best. At the same time the doctor could have explained the unpredictability of life-threatening illness and the resulting need for on-the-spot care negotiations. Such an exchange could have informed future end-of-life care negotiations and prepared Mrs. G. and her family for them.

Even if broad ACP did not take place beforehand, "negotiations of death" *do* certainly occur several times during Mrs. G.'s hospitalization. Early on, the first cardiologist, the neurologist, and the pulmonologist all reach different conclusions about Mrs. G.'s presenting problems and their implications. The first cardiologist concludes Mrs. G. is imminently dying after her cardiac arrest and the family should assemble at once for their final good-byes. The neurologist, in contrast, concludes Mrs. G. will surely live but in a severely impaired, perhaps permanently vegetative, state. He might consider the situation not dire enough to justify assembling the family for final good-byes. And the pulmonologist concludes Mrs. G. cannot benefit from further mechanical ventilation. He convinces the family to sign a document preventing all future ventilator use with her. Discussing the situation through the medical record, the three specialists reach an initial compromise treatment plan: Mrs. G. will be weaned off the ventilator but otherwise receive full life support.

The later heart arrhythmia and the subsequent debate about reventilation prompt the most intense "negotiations of death." Mrs. G. becomes unresponsive just beforehand. That leaves the second cardiologist, the daughters, and Mr. G. as the main parties to the negotiations. The cardiologist wants reventilation, the daughters oppose it, and Mr. G. vacillates inbetween.

Disagreements such as this make me associate meanings for the situation with the people who hold them. To the second cardiologist the situation means fixing a physiological problem, the arrhythmia, and thereby saving Mrs. G.'s life. To the daughters the situation means honoring Mrs. G.'s previously expressed goals for care: regaining some mobility, some independent self-care, and enjoyment of her family. If none of those goals are possible, the situation means implementing her wish for comfort care only. To Mr. G. the situation means the fear Mrs. G. could die soon. Such a fear would ordinarily prompt all life support measures, including the ventilator, to give her "every chance to live." But the situation also means prolonged suffering for her especially if she goes back on the ventilator. Furthermore, to Mr. G. the situation means the burdens of shouldering the critical treatment decisions on Mrs. G.'s behalf and of caring for her himself if she survives. To the whole G. family the situation means placing Mrs. G.'s fate in God's hands and trusting Him with the outcome. The parties involved disagree over how to treat Mrs. G. precisely because they derive such different meanings from her predicament.

Progress on the decisional impasse begins only when an ethics consultant is called in to negotiate a treatment plan. He first meets with the family. As a sympathetic outside party with good communication skills, he listens as the family air their views. Mr. G. sees the second cardiologist's reventilation proposal as "a ray of hope" for Mrs. G.'s survival; the daughters, merely as prolonging her suffering and dying. The daughters also air their frustrations with what they see as the cardiologist's imperious manner. The ethics consultant then steers the conversation to considering the pros and cons of reventilating Mrs. G. All eventually agree to adjourn to think about the decision overnight. The family use that time to pray about it. The next morning they tell the ethics consultant their decision: not to reventilate Mrs. G. They are convinced that hospice suits her better.

The ethics consultant realizes he must win the second cardiologist's support for the decision. That process proves easier than the ethics consultant anticipates. He meets privately with the cardiologist and explains the family's decision and their reasoning based on Mrs. G.'s life history and values. Despite his earlier rigid insistence on the ventilator, the cardiologist readily agrees to the family's decision and signs the order transferring Mrs. G. to the hospital's palliative care service.

The "negotiations of death" for Mrs. G., however, do not end there. The palliative care service staff at first do not understand whether they are to admit Mrs. G. to the nonterminal care ward or the terminal care ward. The ethics consultant must intervene again to explain her terminal condition and to negotiate her admission to the terminal care ward. Then the on-call palliative care doctor balks at sedating Mrs. G. when she grows restless that night. He fears overdosing her. The family must explain her discomfort and assure the doctor about their motives for requesting the sedation before he acquiesces and orders the sedation.

In the final analysis the "negotiations of death"—not the instructional and proxy-designating AD—provide the most substantial guidance for Mrs. G.'s end-of-life care. The AD issues insufficiently specific treatment instructions for Mrs. G.'s situation, and the family makes collective decisions despite the designations of individual advocates. In contrast to the AD, the on-the-spot "negotiations of death"—a kind of ongoing broad ACP—succeed in allowing Mrs. G. to die quietly and comfortably under hospice care as she always wanted.

Summary Points

1. Advance care planning aims to apply a patient's treatment wishes from earlier times of decision-making capacity to treatment decisions in future times of decision-making *in*capacity.
2. Traditional advance directives, the most common form of such planning, either issue specific treatment instructions based on the patient's wishes, or designate proxies who know the patient's wishes well enough to make decisions on the patient's behalf. The instructional directives in particular have many drawbacks.
3. However, one kind of instructional directive, Physician Orders for Life-Sustaining Treatment (POLST), can help terminal patients remain at home and thereby avoid unwanted hospitalization and life-sustaining treatments.
4. By drawing broadly on a patient's life history, personal values, and goals for care, "broad" ACP discussions with patients can give proxies wide-ranging guidance for future end-of-life care decisions.

To Learn More ...

1. Tulsky JA. Beyond advance directives: importance of communication skills at the end of life. JAMA. 2005;294(3):359–64.

2. Slomka J. The negotiation of death: clinical decision making at the end of life. Soc Sci Med. 1992;35(3):251–9.
3. Quill TE. Initiating end-of-life discussions with seriously ill patients: addressing the "elephant in the room." JAMA. 2000;284(19):2502–7.
4. Teno JM, Stevens M, Spernak S, Lynn J. Role of written advance directives in decision making: insights from qualitative and quantitative data. J Gen Intern Med. 1998;13(7):439–46.
5. Sudore RL, Fried TR. Redefining the "planning" in advance care planning: preparing for end-of-life decision making. Ann Intern Med. 2010;153(4):256–61.

References

1. Davidson KW, Hackler C, Caradine DR, McCord RS. Physicians' attitudes on advance directives. JAMA. 1989;262(17):2415–9.
2. Emanuel LL, Emanuel EJ. Decisions at the end of life. Hast Cen Rep. 1993;23(5):6–14.
3. Ditto PH, Danks JH, Smucker WD, et al. Advance directives as acts of communication: a randomized controlled trial. Arch Intern Med. 2001;161(3):421–30.
4. Perkins HS. Controlling death: the false promise of advance directives. Ann Intern Med. 2007;147(1):51–7.
5. Brock DW. Good decision making for incompetent patients. Hast Cen Rep. 1994;24(6):S8–11.
6. Reilly BM, Magnusson CR, Ross J, et al. Can we talk? Inpatient discussions about advance directives in a community hospital: attending physicians' attitudes, their inpatients' wishes, and reported experience. Arch Intern Med. 1994;154(20):2299–308.
7. Emanuel LL. Advance directives. Annu Rev Med. 2008;59:187–98.
8. Teno JM, Hill TP, O'Connor MA. Advance care planning: priorities for ethical and empirical research. Hast Cen Rep. 1994;24(6):S1.
9. Weir RF, Gostin L. Decisions to abate life-sustaining treatment for nonautonomous patients: ethical standards and legal liability for physicians after Cruzan. JAMA. 1990;264(14):1846–53.
10. Stelter KL, Elliott BA, Bruno CA. Living will completion in older adults. Arch Intern Med. 1992;152(5):954–9.
11. Gillick MR. Reversing the code status of advance directives? N Engl J Med. 2010;362 (13):1239–40.
12. Paulker NB, Nettles-Carlson B. The prevalence of advance directives: lessons from a nursing home. Nurs Pract. 1995;20(2):7–8, 17–8, 21.
13. Tulsky JA. Beyond advance directives: importance of communication skills at the end of life. JAMA. 2005;294(3):359–64.
14. Roter DL, Larson S, Fischer GS, Arnold RM, Tulsky JA. Experts practice what they preach: a descriptive study of best and normative practices in end-of-life discussions. Arch Intern Med. 2000;160(22):3477–85.
15. Tolle SW, Tilden VP, Nelson CA, et al. A prospective study of the efficacy of the physician order form for life-sustaining treatment. J Am Geriatr Soc. 1998;46(9):1097–102.
16. Citko J, Moss AH, Carley M, Tolle S. The National POLST Paradigm Initiative, 2nd Edition #178. J Palliat Med. 2011;14(2):241–2.
17. Slomka J. The negotiation of death: clinical decision making at the end of life. Soc Sci Med. 1992;35(3):251–9.
18. Carveth S. Standards for cardiopulmonary resuscitation and emergency cardiac care. JAMA. 1974;22(7):796–7.

19. Moss AH. Informing the patient about cardiopulmonary resuscitation: when risks outweigh benefits. J Gen Intern Med. 1989;4(4):349–55.
20. Taffet GE, Teasdale TA, Luchi RJ. In-hospital cardiopulmonary resuscitation. JAMA. 1988;260(14):2069–72.
21. Vitgelli CE, Cooper K, Rogatko A, Brennan MF. Cardiopulmonary resuscitation and the patient with cancer. J Clin Oncol. 1991;9(1):111–5.
22. Watnick S, Dirkx T. Kidney disease. In: McPhee SJ, Papadakis MA, Rabow MW. 2012 current medical diagnosis & treatment. New York: McGraw Hill Medical, 2012. p. 892.
23. Orentlicher D. Advance medical directives. JAMA. 1990;263(17):2365–7.
24. McIntyre KM. Implementation of advance directives: for physicians, a legal dilemma becomes an ethical imperative. Arch Intern Med. 1992;152(5):925–9.
25. Dresser R. Advance directives: implications for policy. Hast Cen Rep. 1994;24(6):S2–5.
26. Ditto PH, Hawkins NA. Advance directives and cancer decision making near the end of life. Health Psych. 2005;24(4 suppl):S63–70.
27. Euthanasia Education Council. A living will. 1969.
28. Eisendrath SJ, Jonsen AR. The living will: help or hindrance? JAMA. 1983;249(15):2054–8.
29. Joos SK, Reuler JB, Powell JL, Hickam DH. Outpatients' attitudes and understanding regarding living wills. J Gen Intern Med. 1993;8(5):259–63.
30. Fairman RP. Withdrawing life-sustaining treatment: lessons from Nancy Cruzan. Arch Intern Med. 1992;152(1):25–7.
31. Emanuel LL. Advance directives: do they work? J Am Coll Cardiol. 1995;25(1):35–8.
32. Silviera MJ, Kim SYH, Lange KM. Advance directives and outcomes of surrogate decision making before death. N Engl J Med. 2010;362(13):1211–8.
33. Lo B, McLeod GA, Saika G. Patient attitudes to discussing life-sustaining treatment. Arch Intern Med. 1986;146(8):1613–5.
34. Finucane TE, Beamer BA, Roca RP, Kawas CH. Establishing advance medical directives with demented patients: a pilot study. J Clin Ethic. 1993;4(1):51–4.
35. Loewy EH. Talking advance directives and medical practice. Arch Intern Med. 1994;154 (20):2265–7.
36. Mack JW, Cronin A, Taback N, et al. End-of-life discussions among patients with advanced cancer: a cohort study. Ann Intern Med. 2012;156(3):204–10.
37. Song MK. Effects of end-of-life discussions on patients' affective outcomes. Nurs Outlook. 2004;52(3):118–25.
38. Keating NL, Landrum MB, Rogers SO, et al. Physician factors associated with discussions about end-of-life. Cancer. 2010;116(4):998–1006.
39. Piemonte NM, Hermer L. Avoiding a "death panel" redux. Hast Cen Rep. 2013;43(4):20–8.
40. Smith TJ, Bodurtha JN. The "good planning panel". Hast Cen Rep. 2013;43(4):30–2.
41. Hofman JC, Wenger NS, Davis RB, et al. Patient preferences for communication with physicians about end-of-life decisions. Ann Intern Med. 1997;127(1):1–12.
42. Pfeifer MP, Mitchell CK, Chamberlain L. The value of disease severity in predicting patient readiness to address end-of-life issues. Arch Intern Med. 2003;163(5):609–12.
43. Peppercorn JM, Smith TJ, Helft PR, et al. American Society of Clinical Oncology statement: toward individualized care for patients with advanced cancer. J Clin Oncol. 2011;29(6):755–60.
44. Cueing Berlinger N. The conversation. Hast Cen Rep. 2013;43(4):29–30.
45. Maciejewski PK, Prigerson HG. Emotional numbness modifies the effect of end-of-life discussions on end-of-life care. J Pain Symptom Manage. 2013;45(5):841–7.
46. Coate Johnston S, Pfeifer MP, McNutt R for the End of Life Study Group. The discussion about advance directives: patient and physician opinions regarding when and how it should be conducted. Arch Intern Med. 1995;155(10):1025–30.
47. Quill TE. Initiating end-of-life discussions with seriously ill patients: addressing the "elephant in the room". JAMA. 2000;284(19):2502–7.
48. Zhang B, Wright AA, Huskamp HA, et al. Health care costs in the last week of life: associations with end-of-life conversations. Arch Intern Med. 2009;169(5):480–8.

49. Gillick MR. A broader role for advance medical planning. Ann Intern Med. 1995;123 (8):621–4.
50. Prouty CD, Mazor KM, Greene SM, et al. Providers' perceptions of communication breakdowns in cancer care. J Gen Intern Med. 2014;29(8):1122–30.
51. Doukas DJ, McCullough LB. The values history: the evaluation of the patient's values and advance directives. J Fam Pract. 1991;32(2):145–53.
52. Duc de la Rochefoucauld F. Moral maxims and reflections. London: Methuen &Co, 1912. Variously labelled Maxim 26 or 27.
53. Byock I. The meaning and value of death. J Palliat Med. 2002;5(2):279–88.
54. Larson DG, Tobin DR. End-of-life conversations: evolving practice and theory. JAMA. 2000;284(12):1573–8.
55. Weeks JC, Catalano PJ, Cronin A, et al. Patients' expectations about effects of chemotherapy for advanced cancer. N Engl J Med. 2012;367(17):1616–25.
56. Smith TJ, Longo DL. Talking with patients about dying. N Engl J Med. 2012;367(17):1651–2.
57. Matsuyama R, Reddy S, Smith TJ. Why do patients choose chemotherapy near the end of life? A review of the perspective of those facing death from cancer. J Clin Onc. 2006;24 (21):3490–6.
58. Perkins HS. The varied understandings of "terminal cancer" and their implications for clinical research and patient care. Med Decis Making. 2014;34(6):696–8.
59. Happ MB, Capezuti E, Strumpf NE, et al. Advance care planning and end-of-life care for hospitalized nursing home residents. J Am Geriatr Soc. 2002;50(5):829–35.
60. Castillo LS, Williams BA, Hooper SM, et al. Lost in translation: the unintended consequences of advance directive law on clinical care. Ann Intern Med. 2011;154 (2):121–8.
61. Virmani J, Schneiderman LJ, Kaplan RM. Relationship of advance directives to physician-patient communication. Arch Intern Med. 1994;154(8):909–13.
62. Tulsky JA, Fischer GS, Rose MR, Arnold RM. Opening the black box: How do physicians communicate about advance directives? Ann Intern Med. 1998;129(6):441–9.
63. Perkins HS, Cortez JD, Hazuda HP. Cultural beliefs about a patient's right time to die: an exploratory study. J Gen Intern Med. 2009;24(11):1240–7.
64. Finucane TE, Shumway JM, Powers RL, D'Alessandri RM. Planning with elderly outpatients for contingencies of severe illness: a survey and clinical trial. J Gen Intern Med. 1988;3(4):322–5.
65. Kirchhoff KT, Hammes BJ, Kehl KA, Briggs LA, Brown RL. Effect of a disease-specific planning intervention on surrogate understanding of patient goals for future medical treatment. J Am Geriatr Soc. 2010;58(7):2133–40.
66. Tunzi M. Advance care directives: realities and challenges in central California. J Clin Ethic. 2011;22(3):239–48.
67. Fagerlin A, Schneider CE. Enough: the failure of the living will. Hast Cen Rep. 2004;34 (2):30–42.
68. Fischer GS, Tulsky JA, Rose MR, Siminoff LA, Arnold RM. Patient knowledge and physician predictions of treatment preferences after discussion of advance directives. J Gen Intern Med. 1998;13(7):447–54.
69. Danis M, Southerland LI, Garrett JM, et al. A prospective study of advance directives for life-sustaining care. N Engl J Med. 1991;324(13):882–8.
70. Emanuel LL. Advance directives: What have we learned so far? J Clin Ethic. 1993;4(1):8–15.
71. Jacobson JA, Kasworm E, Battin MP, et al. Advance directives in Utah: information from death certificates and informants. Arch Intern Med. 1996;156(16):1862–8.
72. Schneiderman LJ, Kronick R, Kaplan RM, Anderson JP, Langer RD. Effects of offering advance directives on medical treatments and costs. Ann Intern Med. 1992;117(7):599–606.
73. Teno JM, Stevens M, Spernak S, Lynn J. Role of written advance directives in decision making: insights from qualitative and quantitative data. J Gen Intern Med. 1998;13(7):439–46.
74. Lynn J, Goldstein NE. Advance care planning for fatal chronic illness: avoiding commonplace errors and unwarranted suffering. Ann Intern Med. 2003;138(10):812–8.

75. Rubin SM, Strull W, Fialkow MF, Weiss SJ, Lo B. Increasing the completion of the durable power of attorney for health care: a randomized controlled trial. JAMA. 1994;271(3):209–12.
76. Doukas DJ, Gorenflo DW. Analyzing the values history: an evaluation of patient medical values and advance directives. J Clin Ethic. 1993;4(1):41–5.
77. Hickman SE, Nelson CA, Moss AH, et al. Use of the physician orders for life-sustaining treatment (POLST) paradigm program in the hospice setting. J Palliat Med. 2009;12(2):133–41.
78. Dunn PM, Schmidt TA, Carley MM, et al. A method to communicate patient preferences about medically indicated life-sustaining treatment in the out-of-hospital setting. J Am Geriatr Soc. 1996;44(7):785–91.
79. Wenger NS, Citko J, O' Malley K et al. Implementation of physician orders for life-sustaining treatment in nursing homes in California: evaluation of a novel statewide dissemination mechanism. J Gen Intern Med. 2012;28(1):51–7.
80. Hickman SE, Sabatino CP, Moss AH, Nester JW. The POLST paradigm to improve end-of-life care: potential state legal barriers to implementation. J Law Med Ethics. 2008;36 (1):119–40.
81. Hickman SE, Nelson CA, Perrin NA, et al. A comparison of methods to communicate treatment preferences in nursing facilities: traditional practices versus the Physician Orders for Life-Sustaining Treatment Program. J Am Geriatr Soc. 2010;58(7):1241–8.
82. Schmidt TA, Hickman SE, Tolle SW, Brooks HS. The Physician Orders for Life-Sustaining Treatment Program: Oregon emergency medical technicians' practical experiences and attitudes. J Am Geriatr Soc. 2004;52(9):1430–4.
83. Menikoff JA, Sachs GA, Siegler M. Beyond advance directives—health care surrogate laws. N Engl J Med. 1992;327(16):1165–9.
84. High DM. Why are elderly people not using advance directives? J Aging Health. 1993;5 (4):497–515.
85. Miles SH, Koepp R, Weber EP. Advance end-of-life planning: a research review. Arch Intern Med. 1996;156(10):1062–8.
86. Lee MA, Brummel-Smith K, Meyer J, Drew N, London MR. Physician orders for life-sustaining treatment (POLST) outcomes in a PACE program. J Am Geriatr Soc. 2000;48(10):1219–25.
87. Morrison RS, Olson E, Mertz KR, Meier DE. The inaccessibility of advance directives on transfer from ambulatory to acute care settings. JAMA. 1995;274(6):478–82.
88. Schneiderman LJ, Pearlman RA, Kaplan RM, Anderson JP, Rosenberg EM. Relationships of general advance directive instructions to specific life-sustaining treatment preferences in patients with serious illness. Arch Intern Med. 1992;152(10):2114–22.
89. Brock DW. What is the moral basis of the authority of family members to act as surrogates for incompetent patients? J Clin Ethic. 1992;3(2):121–3.
90. High DM. Families' roles in advance directives. Hast Cen Rep. 1994;24(6):S16–9.
91. Cohen-Mansfield J, Rabinovich BA, Lipson S, et al. The decision to execute a durable power of attorney for health care and preferences regarding the utilization of life-sustaining treatments in nursing home residents. Arch Intern Med. 1991;151(2):289–94.
92. Wolf SM, Boyle P, Callahan D, et al. Sources of concern about the Patient Self-Determination Act. N Engl J Med. 1991;325(23):1666–71.
93. LaPuma J, Orentlicher D, Moss RJ. Advance directives on admission: clinical implications and analysis of the Patient Self-Determination Act of 1990. JAMA. 1991;266(3):402–5.
94. Hardwig J. What about family? Hast Cen Rep. 1990;20(2):5–10.
95. Kjellstrand C. Who should decide about your death? JAMA. 1992;267(1):103–4.
96. Houts RM, Smucker WD, Jacobson JA, Ditto PH, Danks JH. Predicting elderly outpatients' life-sustaining treatment preferences over time: the majority rules. Med Decis Making. 2002;22(1):39–52.
97. Sudore RL, Fried TR. Redefining the "planning" in advance care planning: preparing for end-of-life decision making. Ann Intern Med. 2010;153(4):256–61.

98. Lynn J, Teno J, Dresser R, et al. Dementia and advance-care planning: perspectives from three countries on ethics and epidemiology. J Clin Ethic. 1999;10(4):271–85.
99. Fischer GS, Alpert HR, Stoeckle JD, Emanuel LL. Can goals of care be used to predict intervention preferences in an advance directive? Arch Intern Med. 1997;157(7):801–7.
100. McMahan RD, Knight SJ, Fried TR, Sudore RL. Advance care planning beyond advance directives: perspectives from patients and surrogates. J Pain Symptom Manage. 2013;46 (3):355–65.
101. Rosenfeld KE, Wenger NS, Kagawa-Singer M. End-of-life decision making: a qualitative study of elderly individuals. J Gen Intern Med. 2000;15(9):620–5.
102. Martin DK, Thiel EC, Singer PA. A new model of advance care planning: observations from people with HIV. Arch Intern Med. 1999;159(1):86–92.
103. Singer PA, Martin DK, Lavery JV, et al. Reconceptualizing advance care planning from the patient's perspective. Arch Intern Med. 1998;158(8):879–84.
104. Teno JM, Gruneir A, Schwartz Z, Nanda A, Wetle T. Association between advance directives and quality of end-of-life care: a national study. J Am Geriatr Soc. 2007;55 (2):189–94.
105. Roe JM, Goldstein MK, Massey K, Pascoe D. Durable power of attorney for health care: a survey of senior center participants. Arch Intern Med. 1992;152(2):292–6.
106. Emanuel LL, Danis M, Pearlman RA, Singer PA. Advance care planning as a process: structuring the discussions in practice. J Am Geriatr Soc. 1995;43(4):440–6.
107. Back AL, Arnold RM, Baile WF, Tulsky JA, Fryer-Edwards K. Approaching difficult communication tasks in oncology. CA Cancer J Clin. 2005;55(3):164–77.
108. Seymour JE. Negotiating natural death in intensive care. Soc Sci Med. 2000;51(8):1241.
109. A Living Will. New York: Concern for Dying, 1978.
110. Emanuel LL, Emanuel EJ. The Medical Directive: a new comprehensive advance care document. JAMA. 1989;261(22):3288–93.
111. Pearlman RA, Starks H, Cain K et al. Your life, your choices. Washington, DC: United States Department of Veterans Affairs, 2002. Accessed at www.hsrd.research.va.gov/publications/intern/ylyc.htm on 6 March 2007.
112. Texas Advance Directive Act, S.B. 1260, 76˙ Legislature (Texas 1999). Includes Texas' Directive to Physicians and Family or Surrogates (revised, 1999), Medical Power of Attorney (revised, 1999), and Out-of-Hospital Do-Not-Resuscitate order (revised, 2004) Accessed at www.oag.state.tx.us/elder/elder.shtml#legal on 6 Mar 2007.
113. Five wishes. Tallahassee, Florida: Aging with Dignity, 2000. Accessed at www.agingwithdignity.org on 24 Nov 2006.
114. Advance health care directives. Sacramento, California: California Medical Association, 2006. Accessed at www.cmanet.org/publicdoc.cfm/7 on 6 Mar 2007.
115. Volicer L, Cantor MD, Derse AR, et al. for the National Ethics Committee of the Veterans Health Administration. Advance care planning by proxy for residents of long-term care facilities who lack decision-making capacity. J Am Geriatr Soc. 2002;59(4):761–7.
116. Veterans Health Administration Advance Directives. Accessed at www.va.gov/vhapublications/viewpublication.asp?Pub_ID=1450 on 6 Mar 2007.
117. Medical Orders for Life-Sustaining Treatment. Albany, New York: New York State Department of Health. Accessed at www.health.state.ny.us/professionals/patients/patient_rights/molst on 24 Nov 2006.

Chapter 7
Proxy Decision-Making at the End of Life

Instructional advance directives (ADs) were intended to give patients with current decision-making capacity a say in their future treatment when they might lack that capacity [1, 2]. Patients use these directives to specify their treatment wishes in advance. Yet, as Chap. 6 explains, many practical problems have arisen with instructional ADs. They are signed by relatively few Americans [3], may not be located in the medical crisis [4, 5], and often do not anticipate key medical aspects of a patient's illness [4, 6, 7]. Furthermore, most such ADs issue vague instructions [4–6, 8–10] that may not reflect interim changes in patients' wishes [6], and apply only to late terminal diseases when few treatment decisions arise [2, 11] Instructional ADs, therefore, have had little impact on end-of-life care [1, 6, 10].

Disappointment over the usefulness of instructional ADs, however, has not diminished medicine's commitment to honor the treatment wishes of dying patients. But many of those patients cannot make or express decisions for themselves [2]. One study indicates that nearly half of American elders who died from 2000 to 2006 required end-of-life treatment decisions, and three-quarters of those patients could not make the decisions for themselves [2]. Therefore, health professionals (HPs) must often turn to surrogates, or "proxies," to make decisions for the patients. These proxies make more than two-thirds of end-of-life treatment decisions for nursing home residents, hospitalized patients with life-threatening illnesses [12], and intensive care unit (ICU) patients [13].

Proxies gain their decision-making authority in any of three ways. One involves the durable power of attorney (DPOA) for health matters. A patient who still has decisional capacity signs this legal document naming one or more proxies to make later treatment decisions if the patient ever cannot make those decisions for him- or herself. Recent laws and regulations have promoted DPOAs, and they are increasingly common in clinical practice. Even so, most patients still do not sign them. The other two ways identify proxies for patients who do not have DPOAs. One of those ways involves a legal hierarchy of preferred proxies. Many states have already legislated such hierarchies. The typical hierarchy lists in order of decreasing preference the patient's spouse, adult children, parents, other relatives, and

H.S. Perkins, *A Guide to Psychosocial and Spiritual Care at the End of Life*,
DOI 10.1007/978-1-4939-6804-6_7

sometimes attending doctors. Another way relies on medicine's tradition of using readily available family members as informal proxies.

While other chapters in this book (Chaps. 8 and 9) address the *emotions* proxies may experience as family to the dying patient, this chapter addresses the *processes* proxies may use in making treatment decisions for the dying patient. This chapter assumes throughout that the patient already either cannot or does not make treatment decisions for him- or herself, necessitating proxy intervention. This chapter also assumes the patient's proxies are immediate family, but they need not be. Some patients have no family, prefer certain distant family members over closer ones, or prefer proxies (such as friends or professional advisors) other than family. HPs, of course, must exercise good discernment in recognizing when a patient cannot make decisions for him- or herself and in identifying who among possible proxies can best make decisions on the patient's behalf.

To help HPs do so, this chapter begins by describing the characteristics of patients who lack decisional capacity. The chapter then discusses two particular brain injuries—death by brain criteria and the persistent vegetative state—that often cause such incapacity. The chapter next describes three methods proxies often use to make their decisions—the "purely" subjective method, the substituted judgment method, and the best interests method. As I will explain, I favor a fourth method instead—the constructive judgment method—for its grounding in the patient's life story and its broad clinical applicability. This chapter concludes by describing specific factors that can help or hinder proxy decision-making. The chapter illustrates its points with Mrs. G., the same patient as in Chap. 6.

The Case

Mrs. G. awakes with labored breathing one night, and her husband immediately calls 911. Just as paramedics arrive, Mrs. G. suffers a full cardiorespiratory arrest. Vigorous resuscitation efforts soon restore adequate breathing and blood pressures but leave Mrs. G. unresponsive. The paramedics transport her to the nearest hospital. Doctors at the hospital place her on a cardiac monitor in the emergency room, insert a tracheal breathing tube, transfer her to the intensive care unit (ICU), and begin mechanical ventilation there. Mrs. G.'s closest family, consisting of her husband and two adult daughters, arrive at the hospital shortly afterward.

A neurologist, a cardiologist, and a pulmonologist all examine Mrs. G. within a few hours of her ICU admission. Each specialist communicates a different assessment to the family. The pulmonologist commits to no specific long-term prognosis but suggests weaning Mrs. G. off the ventilator in the next day or two. He admits she might or might not die in the process. The neurologist believes she will probably live but cannot rule out an eventual vegetative state. He suggests a wait-and-see approach. The cardiologist thinks Mrs. G. is definitely irreversibly dying, and urges the family to say their last good-byes. These conflicting assessments leave the family confused about what to think.

Then the daughters remember that Mrs. G. said before, "I don't want to live if I cannot 'participate in life.' I don't want to be a vegetable." She also said that quality life for her means being able to move about somewhat independently (despite a painful constricted spine condition) and to recognize and enjoy her family. Convinced Mrs. G. would not want to live permanently unresponsive or attached to the ventilator, the family agrees to weaning her. They even take the pulmonologist's advice and sign an order never to start mechanical ventilation again. They brace themselves for her death. Unexpectedly, however, Mrs. G. begins breathing on her own during the weaning process. Her neurologic condition improves over the next few days. She regains a dull consciousness and nods weakly to questions. The family thinks she gives a flicker of recognition when they visit.

The family begins round-the-clock bedside vigils to ensure proper care for Mrs. G. Mr. G. stays all day every day, sleeping sporadically on a nearby couch. The two daughters take turns staying all night to notify the nurses whenever Mrs. G. needs tracheal suctioning to clear airway secretions obstructing her breathing. Although Mrs. G.'s DPOA officially appoints Mr. G. and the elder daughter as her proxies, he and both daughters together make all care decisions for her.

A different cardiologist rotates onto Mrs. G.'s case one week into her hospitalization. He detects a new, abnormal heart rhythm and insists Mrs. G. go back on the ventilator. "I know what is wrong with her heart," he tells the family. "And I can fix it. But she must go back onto the ventilator to 'rest,' or she will die." The cardiologist's comment gives Mr. G. renewed hope that Mrs. G. can survive, but communicates a sense of urgency for reventilating her. The comment also angers the two daughters. They see the second cardiologist as "pompous and overly focused on the heart," not on Mrs. G. as a "whole" person. Furthermore, the daughters are convinced Mrs. G. is irreversibly dying and should simply receive comfort care. They believe reventilation will "just torture" Mrs. G. by increasing her suffering to no purpose and "torture" the family by raising false hopes for her survival. In addition, the family will have to retract the difficult decision they made to prohibit further mechanical ventilation. When all the other doctors on Mrs. G.'s case defer to the cardiologist on the point, the daughters request an ethics consultation.

The hospital ethicist responds by first reviewing Mrs. G.'s instructional AD, DPOA, and current medical record. He next calls a meeting of the family and the HPs attending her. Mr. G., his two daughters, the general internist assigned to the ICU, the ICU's head nurse, and the ethicist attend. At that meeting the ethicist summarizes Mrs. G.'s hospital course for the group and then reads the most pertinent treatment instruction in her instructional AD, "I do not want life-sustaining treatment if I have a condition that my doctor concludes is … irreversible and, without … life-sustaining treatment, is expected to result in (my) death in a short time." Everyone concludes the instruction does not exactly apply because the doctors do not yet agree Mrs. G.'s condition is "irreversible."

The ethicist then asks each family member individually, "What do you think Mrs. G. would decide for herself in this situation?" Mr. G. answers first. Because he has known Mrs. G. the longest and the best, he feels most responsible for the decisions made on her behalf. He leans toward reventilating her, citing Mrs. G.'s prior oral wishes to receive every chance to live. He believes she would accept more mechanical ventilation to get that chance. The daughters gently disagree, citing another statement in Mrs. G.'s instructional AD that, if life-sustaining treatment serves only "to prolong dying … (in) a terminal condition," she wants comfort care alone. Discussion follows, but all finally decide they can neither interpret percisely what Mrs. G.'s written or oral wishes mean nor guess confidently what Mrs. G. would decide for herself in the situation.

The ethicist next asks the family, "What decision do you think would best serve Mrs. G. interests?" The family considers again Mrs. G.'s previously expressed wishes about having every chance to live but also about having enough quality of life to be able to move somewhat independently and to enjoy the family. Eventually Mr. G. asks, "Is there a chance she will ever do those things again?" The ethicist admits no one can predict for sure. But he adds that Mrs. G.'s age, past medical illnesses, and serious current medical condition weigh against long-term survival or much return of function. The family still cannot decide which treatment plan will best serve Mrs. G.'s interests. The ethicist finally adjourns the meeting to give the family overnight to consider what to do.

That evening Mr. G. and his daughters talk privately among themselves, pray together, and discuss reventilation with their minister and a close doctor friend from the church. They finally decide to forego the ventilator and to request hospice care instead. They call the ethicist about their decision the next morning. He takes the decision to the cardiologist and persuades him to accept it. The cardiologist then orders Mrs. G.'s transfer to the hospital's hospice ward. Mrs. G. dies there early the next day while Mr. G. and the younger daughter watch at the bedside.

When Does a Patient Lack Decision-Making Capacity?

Proxies, of course, gain their authority to make decisions for patients only when patients lack sufficient decision-making capacity to do so themselves. *Capacity* here means the ability to perform a specific task; *decision-making capacity*, the ability to make decisions based on one's personal values and a sound grasp of circumstances, available choices, and likely consequences of those choices [14]. In medicine, decision-making capacity underlies any valid informed consent. The decision-maker—patient or proxy—ideally understands all key facts about the disease, the options for treating it, the risks and benefits of each option, and the relative worth of those options according to the patient's own values. The decision-maker then chooses the best treatment option for the patient and requests it clearly and consistently.

Of course, no patient as decision-maker fulfills this ideal of decision-making capacity completely. A patient in a particular situation fulfills some aspects of capacity better than other aspects. Capacity may also change with time and circumstance: Capacity is not all-or-none forever but has varying degrees of more or less. Complicating judgments of it further, no precise threshold exists for determining whether a patient has sufficient capacity to make a particular decision. Neither ethics nor law provides such a threshold. Still, many experts agree on one point: Whatever one considers the general threshold, it should increase as the momentousness of the decision increases. That is, the more momentous the potential consequences of the decision, the greater should be the evidence that the decision-maker commands sufficient capacity to make that decision. Yet even this simple, intuitive idea raises some controversy. I, therefore, believe judgments of a patient's decision-making capacity necessarily involve imprecise "judgment calls." Naturally, the patient always deserves the benefit of the doubt. But, once a reasonable judgment of patient decision-making incapacity emerges, a conscientious, qualified proxy must step in to make decisions for the patient.

The Case

No doubt exists that Mrs. G. lacks decision-making capacity as she lies barely conscious early during her hospitalization. She gives no sign of understanding facts about her disease; weighing treatment options; or making clear, consistent treatment requests. Therefore, when the pulmonologist recommends weaning her from the ventilator, Mr. G. and the daughters—the people who know Mrs. G. best and whom she trusts most—rightfully step in to make the decision for her.

Serious Brain Injuries

A Common Justification for Proxy Decision-Making

Many brain injuries can cause decisional incapacity and, when so, require proxies to make care decisions for patients. Two especially serious injuries, death by brain criteria (DBC) and the persistent vegetative state (PVS), always cause such incapacity. However, the lay public and some HPs confuse these two conditions. That confusion is understandable: Both conditions are commonly called "brain death" and share one prevalent cause, cardiac arrest interrupting blood and oxygen flow to the brain. But the two conditions differ in other ways—alternate causes, diagnostic criteria, management, and prognoses. Proxies and HPs must clearly differentiate the two conditions to make correct diagnoses and hence to institute the best management strategies. Fortunately, recent medical advances that localize brain damage and specify its pathologic signs help distinguish DBC and PVS [15, 16]. This section contrasts them based on that new information (Table 7.1).

Table 7.1 Characteristics distinguishing death by brain criteria and the persistent vegetative state

Characteristic	Death by brain criteria	Persistent vegetative state
Definition	Complete brain destruction including for the brain stem: no arousal, thinking, or spontaneous respirations or other brainstem functions	Extensive brain damage but a functioning brain stem: a profound dementia with normal arousal but severely impaired ability to think. Differs from coma by normal arousal: "awake but not aware."
Common causes	Long-term interrupted blood flow to the brain over 6–20 min due to cardiac arrest, severe bodily trauma (such as in auto accidents or assaults), or ruptured major arteries	• Short-term interrupted blood flow to the brain such as from cardiac arrest or severe bodily trauma • Also metabolic disorders (such as carbon monoxide poisoning or prolonged low blood glucoses), severe infections, or end-stage dementias
Degree and location of brain injury	Destruction of both the higher and lower brain structures (i.e., the cerebral hemispheres *and* the hypothalamus and brain stem). Ineffective cardiorespiratory contractions, unable to support life, may continue for minutes to hours.	Severe damage to the higher-brain structures (the cerebral hemispheres) but little or no damage to the lower brain structures (the hypothalamus and brain stem). Effective brainstem functions may support life for years.
Spontaneous motor movements	None	Closing and opening eyes in normal sleep–wake cycles; slow scanning eye movements without fixation; grunts, yawns, grimaces; and primitive limb movements often present.
Response to noxious stimuli	Absent except for primitive spinal cord reflexes	Present including startle reflexes to loud sounds and pain
Spontaneous respirations	Absent	Usually absent at first (when coma is present) but eventually resume.
Brainstem reflexes[a]	Absent	Present
Discretionary confirmatory tests	• Cerebral blood flow studies show no flow. • Electroencephalogram (a "brain wave test") shows no waves (i.e., flatness or "isoelectricity").	• Cerebral blood flow studies show some flow. • Positron emission tomography ("PET scan") shows greatly depressed brain tissue metabolism.
Added considerations	Sedation, low body temperature, medically induced paralysis, and persistently low blood pressure must not be present. They can mimic death by brain criteria.	Patients under 40 have a greater likelihood of some neurologic recovery than do older patients. Recovery never includes the

(continued)

Table 7.1 (continued)

Characteristic	Death by brain criteria	Persistent vegetative state
		ability to communicate or to live independently.
Time to determine permanence of the brain-injured state	1–3 days	3–12 months
Management	Patient is dead: stop all treatments.	Patient is not dead. Treatment is controversial, but life support is often continued indefinitely. Patient may live for years. Survival depends on whether life-threatening complications such as pneumonias are treated.

[a]Brainstem reflexes include pupillary, oculocephalic (“doll’s eyes”), and oculovestibular (ear “calorics”) reflexes.
Data from Refs. [15, 22]

Death by Brain Criteria (DBC)

Historically, cessation of all cardiovascular and respiratory functions defined death. Doctors determined it by simply listening for a heartbeat and looking for chest expansions. If doctors could detect neither, they declared patients dead. But new technologies began to complicate those determinations in the 1960s. Widespread introduction of machines to perform such vital functions as electrically stimulating the heart or mechanically expanding the lungs suddenly kept some patients alive who would have been declared dead before. Doctors wondered whether death in those patients could ever be determined and, if so, how. The new life-support technologies had upended the ages-old cardiorespiratory criteria for death.

Organ transplantation exacerbated the problem from the late 1960s on. As the first transplantations (with kidneys) succeeded, the demand for transplantable kidneys and eventually other transplantable organs quickly exceeded supply. Meanwhile the traditional cardiorespiratory criteria for death became a significant obstacle to harvesting cadaver organs for transplantation. Organs lose their viability quickly when circulation and oxygenation stop. Harvesting must, therefore, occur within minutes after cessation of effective cardiorespiratory function so the cadaveric organs remain viable. But agonal, ineffective reflexive contractions of the heart and chest wall muscles, independent of the brain and unable to sustain life, may continue for minutes to days after apparent bodily death. The early laws about transplantation required all such signs to stop before organs could be harvested from cadavers. Doctors had to wait until organs had typically lost their viability. Thus, the old medical traditions for determining death and the earliest transplantation laws based on them prevented many beneficial transplantations.

Responding early on to the transplantation problem, a special Harvard Medical School committee proposed in 1968 that permanent loss of all brain function, which the committee called "irreversible coma," supplement the traditional cardiorespiratory criteria as a parallel definition for death [17]. Either definition, the committee suggested, could determine death. The new parallel definition eventually became known as "death by brain criteria." Just as with the cardiorespiratory definition, DBC was diagnosed solely based on the physical signs. They consisted of unresponsivity to external stimuli (and presumed unawareness of them), absence of spontaneous respirations or other movements, and absence of reflexes above the spinal cord. The committee declared the flat electroencephalogram (a "brain wave test") merely optional confirmation for the diagnosis. An influential presidential commission later added its support for the committee's idea by declaring, a person with "irreversible cessation of circulatory and respiratory functions, or … of all functions of the entire brain, including the brainstem, is dead" [18]. That commission urged states to codify both definitions under a new law, called the Uniform Determination of Death Act. Most states now have.

DBC typically follows either cardiac arrests, severe bodily trauma, or ruptured major arteries, in which blood flow to the brain stops for 6–20 min. Both higher (cerebral hemispheres) and lower (hypothalamus and brain stem) brain structures suffer extensive destruction. The higher destruction eliminates all thinking including learning, remembering, and awareness of one's self and surroundings. The lower destruction eliminates all brain stem reflexes and life-sustaining functions. Physical examination shows no response to pain (such as by groaning, grimacing, or withdrawing limbs), no "higher" reflexes (including pupillary, corneal, and gag reflexes), and no spontaneous respirations (by a standard protocol called the apnea test). These signs must persist for 6–72 h, depending on the underlying cause of injury. Sedation, low body temperature, medically induced paralysis, and persistently low blood pressure ("shock") must not be present because some aspects of those conditions mimic DBC. Despite the judgment involved in recognizing the signs of DBC, they produce good agreement among neurosurgeons and other experienced observers [19]. Therefore, the physical examination for DBC is reproducible and accurate. The diagnosis requires neither brain wave nor blood flow studies. (Brain wave tests are, in fact, sometimes misleading. Blood flow studies, however, always show absent blood circulation to the brain.)

No neurologic recovery occurs with DBC. Therefore, once diagnosed, DBC has clear-cut management: The patient is dead, treatment provides no benefits, and the doctor should stop all of it. General professional standards and all current law concerning DBC support this strategy.

Persistent Vegetative State (PVS)

PVS like DBC involves extensive higher (cerebral) brain structure destruction or damage but unlike DBC has preserved lower (hypothalamus and brain stem) brain structure and function. PVS also like DBC may result from severe blows or interrupted blood flow to the head. But PVS may have other causes, too, including metabolic disorders (such as carbon monoxide poisoning or prolonged low blood glucoses), severe brain infections (such as meningitis, encephalitis, or abscesses), and end-stage degenerative diseases (such as the dementias) [20, 21]. Blood flow studies show some circulation, but positron emission tomography (PET) shows severely depressed brain tissue metabolism.

A cardiac arrest such as Mrs. G.'s can first cause a coma, a state in which the patient is neither awake nor aware of self or surroundings. The coma may eventually evolve into a vegetative state, in which the patient awakens but remains unaware. That patient shows normal cycles of sleep and wakefulness on brain wave testing but still does not sense self or surroundings. The patient opens his or her eyes spontaneously and makes with them slow, scanning motions interrupted only by unstimulated brief jerks or reactions to external sounds. The eyes never fixate. Some higher reflexes including startle and arousal reflexes and brainstem reflexes (such as pupillary and other eye reflexes) return. The vegetative patient yawns, groans, or grimaces and moves arms and legs in primitive ways. The hands, for example, may reflexly grasp objects placed into them. Untrained observers may misinterpret such movements as conscious. Because the neurologic signs may change as swelling in the brain subsides over weeks or months, the diagnosis of a vegetative state or persistent vegetative state requires considerable time. A vegetative state can be diagnosed only at one month; a *persistent* vegetative state, only at three months after a cardiac cause or 12 months after other causes.

A few percent of PVS patients may regain some neurologic function months or years after the initial injury. Patients under 40 have a higher likelihood of some recovery than do older patients. But only about 0.1 % of diagnosed PVS patients eventually regain any consciousness or cognition, and even those patients suffer severe, permanent neurologic handicaps [22]. The regained neurologic function almost never includes the ability to communicate or to live independently [23].

Because PVS patients are severely brain damaged but not actually dead, management for them remains controversial. Meticulous nursing care and aggressive treatment of occasional, potentially life-threatening complications such as pneumonias can keep these patients alive for years. Karen Quinlan, for example, lived nine years in her PVS. On the other hand, the Council on Ethical and Judicial Affairs of the American Medical Association acknowledges that such a patient has the right to refuse life-sustaining treatment through a proxy, and the doctor has no duty to maintain a patient's life without prospects for consciousness. Recognizing the arduousness of PVS care and the extreme unlikelihood of future conscious life, the Council concludes, "Even if death is not imminent but a patient is beyond doubt permanently unconscious, … it is not unethical to discontinue … life-prolonging medical treatment…. (as long as) the dignity of the (PVS) patient (is) maintained at all times" [24]. The Council urges doctors, however, to consult neurologists and other colleagues before discontinuing life support in such situations. While many scholars support the Council's position, others question the accuracy of PVS diagnoses; note the possibility, though small, for PVS patients to regain consciousness; suggest legal culpability for stopping a PVS patient's life support; and worry about increasingly unjustified stoppages of life support on patients in general [22]. For these reasons, withholding life-support treatment from PVS patients still unnerves many doctors, and most choose to continue it indefinitely.

Remaining Controversies About DBC and PVS

Ever since the presidential commission endorsed the two parallel definitions of death—cessation of all cardiorespiratory function and DBC—doctors have accepted and used both widely. In the process they have implicitly endorsed three assumptions behind the commission's endorsement of those definitions: that only the living can die; that the living and the dead can be distinguished; and that, however death is determined, total brain deterioration accompanies it.

Still, controversies remain about especially the second and third assumptions and their practical implications. One controversy concerns whether the whole brain (as in current DBC) or only the higher brain (as in PVS) must cease functioning to justify a determination of death by brain criteria. Proponents of the whole-brain criterion understand death as the functional disintegration of the whole living organism [25]. They see the whole-brain criterion as reliably heralding such disintegration as inevitable and imminent. They also point out that cessation of all brain functions quickly shuts down all other bodily organs, thereby aligning the whole-brain criterion for death with the traditional cardiorespiratory criterion.

Proponents of the partial, higher-brain criterion, however, understand death as cessation of the two uniquely cerebral functions, consciousness and thinking. As a result these commentators consider PVS as death. They argue against the whole-brain criterion in two ways. First, they point out that the whole-brain criterion requires testing only the brain stem, not the higher brain where the mind functions. The underlying assumption is that, if the brain stem does not function, the cortex does not either. These people argue that external devices such as ventilators and pacemakers already can perform many of the brainstem's life-sustaining functions and may someday perform them all [26]. These commentators worry that future patients who depend on external devices for such brainstem life-support functions but still maintain consciousness and thinking ability may prematurely be declared dead under the whole-brain criterion. Second, these commentators believe the whole-brain criterion falsely separates the concepts of mind and brain. These commentators consider the two inextricably bound together. The roots of this disagreement over the separability of mind and brain extend back at least as far as ancient Greek and Hebrew philosophers. The Greek philosophers believed the mind separates cleanly from the brain (and the rest of the body) at bodily death, whereas the Hebrew philosophers believed that mind and brain—forever intermeshed—live and die together. As a practical point, however, the proponents of this partial or higher-brain criterion for determining death admit that no easy, quick, or reliable tests yet exist for testing cerebral functions separately from brainstem functions.

Yet another commentator does not see life and death as discrete states separated by a distinct moment in time [27]. Instead, she sees life as consisting of multiple dimensions and declining unevenly as its various dimensions deteriorate at different rates. She concludes that life does not suddenly shut off but "tapers off" gradually into death. This commentator, therefore, favors a zone of discretionary declarations of death based on brain and other bodily functions. PVS would provide this zone's highest functional boundary; cessation of cardiorespiratory function, its lowest. Doctors, says this commentator, should be allowed to declare death anywhere within that zone. This view coincides with the opinions of many doctors as reported in a recent survey [28].

While acknowledging some merit in these various thought-provoking opinions, I continue to support the current parallel definitions of death—cessation of all cardiorespiratory functions and cessation of all brain functions—as a purely practical matter. These definitions, though conceptually still rough and imperfect, have shown themselves useful in clinical practice. They rely on simple tests in the physical examination; have proven compatible with diverse legal, cultural, and religious beliefs about death; and have withstood the test of time. These definitions have served their purpose well. Only the most compelling reasons should justify changing them. PVS should remain distinct from death.

The Case

Early in Mrs. G.'s hospitalization the neurologist rules out DBC by the current whole-brain criterion. But he continues to entertain the possibility of a vegetative state. He also knows that, if Mrs. G. has one, it might be either temporary or permanent. He monitors her accordingly for the physical signs that predict poor neurologic outcomes such as vegetative states or other severe brain disabilities after cardiac arrest. (Those signs include absent corneal reflexes, pupillary reflexes, motor responses, and responses to painful stimuli at 24 h and absent motor responses at 72 h.) Unfortunately, no known signs reliably predict *good* neurologic outcomes [19].

Thus, Mrs. G.'s family and doctors wait for her brain injury "to declare its long-term nature." Mrs. G. herself might have considered an eventual PVS a life not worthwhile by her own criteria (being able to move purposefully and independently and to enjoy her family). Still, a PVS is not death. And Mrs. G. obviously cannot make decisions for herself. The family must continue to shoulder the decision-making burden for her. In fact, Mrs. G.'s groggy consciousness, occasional eye opening, and weak nods do not convince me that she *ever* regains the ability to make medical decisions for herself. Her proxies must make them for her until she dies.

Although Mrs. G.'s DPOA names only Mr. G. and the elder daughter as her official legal proxies, I believe the younger daughter also plays a legitimate role as a proxy for Mrs. G. throughout her hospitalization. I explain in the next section my reasons based on medicine's tradition of engaging available family as informal, on-the-spot proxies.

Who Should Serve as a Proxy?

When patients have already legally designated their proxies in DPOAs, those proxies should have considerable, but not absolute, priority over any alternative proxies. The exceptions arise when the designated proxies prove unable or unwilling to serve. Yet only a minority of patients actually complete DPOAs. For most other patients who become incapable of making their own decisions, HPs must turn to readily available family members to serve as informal proxies.

I can accept this practice for several reasons. First, most patients *want* family members as their proxies [6, 29–31]. Seventy percent of hospitalized elders in one study [6], 78 % of critically ill patients in another study [32], and 90 % of dialysis patients in a third study spontaneously chose family members as their preferred proxies [29]. In addition, states' legislated hierarchies of proxies reflect this general preference for family proxies. Second, medicine, ethics, and law all assume family members know a patient's treatment wishes better than any other possible proxies [33]. The underlying logic is that frequent, intimate interactions over years give family members the most accurate knowledge of patients' wishes. In today's fragmented health care, for instance, only family members may know a patient's treatment wishes from similar medical situations in the past.

Third, as the basic unit of human social organization [33, 34], the family imposes a special obligation on its members to protect their old and sick [35]. Even people who shirk this obligation surely believe it exists. Furthermore, intuition suggests the stringency of the obligation peaks for the nuclear family and decreases further out the family tree, and nearly always outweighs any parallel care obligation among nonfamily. And, fourth, the family consistently presents the most readily accessible option for making proxy medical decisions [36]. Not only do family members usually know the patient best, but they are also most available to make decisions in the medical crisis.

These arguments convince me that using family as proxies, whether formal or informal, typically protects patients best. It can also nurture shared values, privacy, and solidarity within families [37]. Nonetheless, the interests of family (including proxies) occasionally conflict with the interests of patients. Such conflicts of interests may be difficult for proxies themselves to recognize, even more so for outsiders such as HPs [38]. However, if HPs suspect a proxy–patient conflict of interest that jeopardizes patient welfare, the HPs should always look into the potential conflict. If further examination confirms it, they should take whatever steps necessary to protect the patient.

Nonfamily proxies are needed only when no family members can serve as satisfactory proxies. The proxy options then include patients' attending doctors, a hospital's ethicist or ethics committee, and court-appointed guardians. All have significant drawbacks. The doctors, for example, may be unduely influenced by the technologic imperative, which encourages treatment no matter how small the chances of benefit or how burdensome the side effects are. That imperative may sometimes blind doctors to patients' need for comfort care over long-shot treatment for cure. A hospital's ethicist or ethics committee or a court's appointed guardian may have even greater drawbacks. All three typically know less about patients' life histories before the current hospitalization than do the doctors. Furthermore, all three typically take considerable time for deliberation [39] and may increase the stress on patients or others by introducing strangers late into the decision-making process. And courts usually appoint as patient guardians lawyers who volunteer their service, do not know the patients personally, lack medical training, and can devote little time to the decision-making task. Those guardians often simply agree automatically to the doctors' recommendations without exercising independent judgment. Guardianship appointments by courts also incur great expense.

For all these reasons many states list attending doctors as the final option after every family option in their legislated proxy hierarchies is exhausted. I agree that attending doctors are generally the best, most practical proxy option for patients who lack satisfactory family proxies. But attending doctors acting as patients' proxies must exercise restraint not to overreach their authority. Many states' legislated hierarchies, therefore, require doctor proxies to consult other doctors before making proxy decisions.

The Case

Mrs. G.'s DPOA officially names Mr. G. and the elder daughter as her legal proxies. But the younger daughter plays an equally important role as an informal proxy. Together, the three make every proxy decision for Mrs. G. during her hospitalization. They all know Mrs. G.'s wishes about life support and have discussed those wishes extensively with her on at least two prior occasions. On one, when a stroke had left a distant relative dependent on tube feedings, mechanical ventilation, and around-the-clock nurses, Mrs. G. told her family, "I don't want to be a vegetable like that. I want to be able to participate in life." But on another, when a friend had lived unexpectedly after being weaned from a ventilator, Mrs. G. said, "You just never know. I want to live if I have any chance at all." These apparently contradictory wishes suggest ambivalence about mechanical ventilation—the very treatment the family must decide about during this hospitalization. When the daughters had previously pressed Mrs. G. to clear up the confusion, she simply said, "You and your father can decide for me. You'll know what to do." Thus, Mrs. G.'s wishes for the immediate family to serve together as her proxies are clear, but her wishes about mechanical ventilation are not. Still, when Mrs. G. loses decision-making capacity, Mr. G and the daughters step up responsibly to make all care decisions for her, and they prove themselves model proxies.

Three Commonly Used Methods for Making Proxy Decisions

The patient autonomy principle guides all widely accepted methods of medical decision-making these days. The principle not only lets patients who have sufficient capacity decide about care for themselves but also requires proxies, acting for patients' benefit, to decide about care for patients who lack that capacity. The overall goal in either case is to match care decisions to patients' particular wishes.

Three proxy decision-making methods, addressing slightly different circumstances, have gained extensive use under this principle. The methods fall into a logical priority order depending on the knowledge about a patient's particular care wishes [37, 40]. In the descending order of priority, the "purely" subjective method uses the patient's previously expressed wishes for the exact circumstances at hand; the substituted judgment method, the wishes that proxies imagine the patient would have but never expressed; and the best interests method, wishes that proxies believe would generally protect the patient's interests best. I describe the three methods in detail here, critique them, and note problems with each. I conclude all are flawed. I, therefore, recommend in the next section still another decision-making method that conforms to the autonomy principle by taking broad account of a patient's life story including personal traits, values, relationships, and experiences.

The Purely Subjective Method

Of the three methods this one requires the most precise knowledge of the particular patient's care wishes. The "purely" subjective method asks the question, What has the patient already decided about the care he or she wants in this situation? Acting merely as the patient's mouthpiece [41], the proxy expresses the specific treatment instructions the patient, while having decision-making capacity, left in either written or oral ADs. Some proxies prefer this method (when it applies) because it makes them feel that the patient, not the proxy, has made the treatment decisions. Unfortunately, as the prior chapter and the introduction to this chapter indicate, treatment instructions in written ADs encounter numerous practical problems [10, 42]. They include patients' infrequent completion of written ADs, the ADs' vague descriptions of clinical situations, patients' potential misunderstandings about benefits and burdens of treatments, the difficulty of accurately foreseeing the consequences of many clinical decisions, the frequent ambiguity of a patient's instructions, the possible changes in a patient's wishes with time or circumstances [43], and the tendency of many HPs (especially doctors) to ignore a patient's prior instructions in planning treatment. Treatment instructions in oral ADs have all those problems and at least one more—potentially inaccurate recall by hearers. Thus, the "purely" subjective method, based as it is on ADs, has rarely fulfilled its intended purpose of specifying a patient's wishes clearly and guiding their implementation in clinical situations.

Some of these problems arise in Mrs. G.'s case, severely limiting the potential advantage of her prior instructions. While two real cases had prompted the family to ask Mrs. G.'s wishes about life support, she discussed those cases only as hypotheticals for herself. Her opinions might have been different had she actually faced those situations as the patient. In addition, her expressed wishes might have been based on inaccurate information or mere suppositions about the cases. She may not have understood, for example, the benefits and burdens of tube feedings, mechanical ventilation, CPR, and other treatments in life-threatening situations. Furthermore, Mrs. G.'s oral instructions—wanting "to participate in life" and not wanting "to be a vegetable"—provided no way to verify independently the family's recall of them, and little specific guidance for the treatment decisions the family faced.

The Substituted Judgment Method

This method depends not on knowing the particular patient's precise, expressed wishes about some possible future crisis but on intuiting the patient's presumed, but unexpressed wishes about an actual current crisis [40, 44]. The method asks the question, What do you think the patient would want for care in this situation? [45, 46]. Like the "purely" subjective method the substituted judgment method may convince the proxy he or she acts only as the patient's mouthpiece by merely reporting decisions the patient would have made [47, 48]. But unlike the "purely" subjective method the substituted judgment method requires the proxy to use considerable imagination [49]—a daunting task for some—and to cope with uncertainty about the decision.

Substituted judgment, of course, depends critically on proxies' ability to guess patients' treatment wishes accurately. If proxies do not have that ability, the rationale for the method—preserving patient autonomy—collapses [34, 50, 51]. Many studies, therefore, have tested that ability. Table 7.2 summarizes some of them. Most studies have involved elderly, chronically ill patients and allowed them to choose their own proxies for study purposes. The chosen proxies have typically been spouses or adult children. The studies have compared proxy guesses and patient wishes about CPR, mechanical ventilation, and other life-support treatments in common clinical scenarios requiring proxy decisions: severe brain damage (as in stroke or dementia), advanced lung disease, advanced cancer, and terminal illness. Regardless of the type of proxy, clinical scenario, or treatment, proxies' guesses have barely risen above chance agreement with patients' actual wishes [3]. Perhaps most disconcertingly, other studies have indicated that proxy guesses do not improve if the patients designate their proxies beforehand [52] or discuss their treatment wishes with those proxies [29, 52, 53]. Furthermore, proxies' mistaken guesses do not trend consistently toward overtreatment or undertreatment, allowing for conscious compensation during decision-making. And proxies, even long-time spouses, still have exaggerated confidence in the accuracy of their guesses [54].

So why do proxies predict so badly? Human psychology suggests some reasons. First, proxies may have difficulty sorting through the avalanche of information in a medical crisis to identify facts crucial to decisions [52]. Second, the emotional stresses in critical-care or end-of-life situations may hinder clear understanding and accurate reasoning. Third, proxies may fall victim to certain incorrect assumptions. For example, a proxy may incorrectly believe the patient thinks exactly as does the proxy (the similarity fallacy), that certain demographic characteristics indicate the patient surely thinks in certain ways (the social projection or stereotyping fallacy), or that everyone in a certain group to which the patient belongs thinks alike (the consensus fallacy) [52–54].

Perhaps more subtle problems than inaccurate proxy guesses may also plague substituted judgment. One of those problems may be intentional nonuse by proxies [54]. Specifically, many proxies who say they know patients' treatment wishes may intentionally not use substituted judgment in making proxy care decisions. For example, only 66 % of proxies having foreknowledge of the incapacitated patients' wishes in one study [3] and only 4 % of spouses caring for the demented patients in another study [55] planned to use any or only such knowledge in making decisions for the patients. Still another problem may be lack of buy-in by patients themselves. Some may just not consider substituted judgment the best or only valid method for making proxy decisions for them [3].

Before any empirical studies raised concerns about the substituted judgment method, the presidential commission of the 1980s endorsed it as a worthwhile first try at promoting autonomy for patients who lack decision-making capacity. Most doctors subsequently adopted the method to guide proxy decisions when such patients leave no clear, applicable treatment instructions. Mrs. G. certainly left none about reventilation. The hospital ethicist, therefore, reasonably follows established clinical practice by using the substituted judgment method when he asks the family, "What do you think Mrs. G. would decide for herself in this situation?"

Yet I see no solutions for most practical problems with substituted judgment. Neither do many other observers. Some, in fact, dismiss it as an inaccurate, even harmful "charade" or "fiction" [34, 42, 50, 55–60]. I reluctantly conclude that, however reasonable the substituted judgment method once seemed, there must be a better approach to proxy decisions.

The Best Interests Method

Some people consider this method better overall for making proxy decisions than either the "purely" subjective method or the substituted judgment method. Unlike those methods the best interests method assumes the particular patient's care wishes cannot be known [37]. It, therefore, focuses on the *usual*, not the *particular*,

Table 7.2 Proxies' accuracy at guessing patients' treatment wishes

	Studies (Year)						
	Uhlmann (1988)[ab] [56]	Uhlmann (1988)[ab] [56]	Seckler (1991)[b] [57]	Hare (1992)[b] [58]	Sulmasy (1994)[ab] [59][d]	Suhl (1994)[b] [42]	Pruchno (2005)[b] [54]
Patients	Outpatients over 65 with at least one chronic disease	Outpatients over 65 with at least one chronic disease	Outpatients over 65	Outpatients, mean age = 44	Outpatients	Inpatients, mean age = 55	Current hemodialysis patients over 55
Proxies	Spouses	Attending doctors	Chosen by patients: 59 % next-generation relatives (mostly daughters); 32 % same-generation choices (spouses or friends)	Chosen by patients: 72 % spouses	Chosen by state's legal hierarchy: 51 % spouses, 37 % adult children	Chosen by patients: 52 % spouses, 16 % children	Spouses
Clinical Scenarios	Current health, stroke, or chronic lung disease	Current health, stroke, or chronic lung disease	Current health or moderate dementia	Permanent coma, diabetes needing limb amputation, or advanced cancer	Persistent vegetative state, coma with a small chance of recovery, or fatal disease with progressive brain damage	Persistent coma, terminal cancer, chronic lung disease, or progressive paralysis	Stroke, dementia, permanent coma, or terminal illness with or without pain
Proxy accuracy by treatment[b]							
• CPR	poor/slight	poor/slight	poor	slight	slight/fair	none/poor	–
• Mechanical Ventilation	poor/slight/fair	none/poor	–	slight	slight	poor/slight	–

(continued)

Table 7.2 (continued)

	Studies (Year)						
	Uhlmann (1988)[ab] [56]	Uhlmann (1988)[ab] [56]	Seckler (1991)[b] [57]	Hare (1992)[b] [58]	Sulmasy (1994)[ab] [59][d]	Suhl (1994)[b] [42]	Pruchno (2005)[b] [54]
• Surgery	—[c]	–	–	poor	none/poor	none/poor	–
• Dialysis	–	–	–	–	poor/slight/fair	–	poor

[a]This study reports cardiopulmonary resuscitation (CPR) alone and with mechanical ventilation. I listed results for CPR with mechanical ventilation under "mechanical ventilation." The other studies report separate results for mechanical ventilation alone.

[b]These studies used the kappa statistic to measure proxy accuracies. Kappa expresses agreement beyond chance and ranges from −1.00 to +1.00. Negative kappas mean agreement worse than chance with −1.00 indicating complete disagreement. Positive kappas mean agreement better than chance with +1.00 indicating complete agreement. A zero kappa means agreement completely by chance. The table below provides common verbal interpretations of kappa statistics. I used it to make my interpretations of kappa statistics for Table 7.2.

Kappa Statistic	Agreement beyond Chance
0.00 or less	None
0.01-0.20	Poor
0.21-0.40	Slight
0.41-0.60	Fair
0.61-0.80	Good
0.81-1.00	Excellent

From Ref. [60], with permission

Three additional studies reported no kappa statistics, just percent agreements between proxy guesses and patient wishes. These percent agreements had no corrections for chance. All three studies involved community dwelling elderly or seriously ill patients. One study, surveying proxies chosen specifically by the patients, asked about patient wishes concerning life-support treatments under clinical scenarios similar to those in the table [51]. A second study, surveying proxies chosen according to the local state's legislated hierarchy, asked about patient wishes concerning similar treatments and clinical scenarios [11]. And the third study, surveying spouses of patients, asked about patient wishes concerning mechanical ventilation in a terminal illness with pain or impaired thinking [4]. The percent agreements in all three studies approximated the percent agreements comparable to "poor" to "slight" agreement beyond chance in the studies reporting kappa statistics.

[c]A dash means the data were not reported.

[d]I estimated agreements from a small bar graph in the article.

patient's wishes. It asks the question, Which care decision would serve the *usual* patient's interests best in this situation?

While commanding intuitive appeal, the best interests method does present problems. I can think of four. First, unlike the "purely" subjective and substituted judgment methods the best interests method may burden proxies with feeling responsible for life-or-death decisions [47]. Specifically, the psychic trauma involved with choosing someone's death can make proxies feel like "Dr. Kevorkian" or "the patient's executioner." Second, opinions differ over how to assess patient interests and treatment outcomes under this method. Some experts favor "inter-subjective" assessments that answer the question, What *do* most proxies decide for patients in this situation? The standard for good decisions under this approach is actual majority decisions as ideally determined by empirical surveys. Other experts, however, favor assessments that answer the question, What *should* proxies decide for patients in this situation? The standard for good decisions under that approach is views about the good life as determined from sources considered free of idiosyncratic opinion and, thus, universally valid or "objective." The sources referenced most often include *The Holy Bible* and certain secular philosophies-of-living. Those philosophies-of-living sources may promote such aims as avoiding suffering, maintaining independence, fostering clear thinking, leaving a significant legacy, or being true to a faith or a philosophical idea [9, 49], Despite the stark differences between the two approaches to assessment—the intersubjective and the "objective"—I suspect most proxies use some combination of both.

Third, available best interests research has limitations for informing end-of-life proxy decisions. For example, it typically addresses only a few major treatments (e.g., CPR, mechanical ventilation, and cancer chemotherapy) in a few extreme conditions (e.g., severe brain injury, advanced lung disease, and terminal cancer) at relatively few medical care sites (often university settings). Some researchers, therefore, urge collecting additional data on proxy decisions about a broad range of treatments and medical conditions at a wide variety of care sites [1, 51, 61, 62]. Such data, of course, would not be available for years. Even when they do become available, using them will pose the persistent problem of applying inferences from broad studies to individual cases [51]. Furthermore, even if directly applicable data already exist for a patient's particular situation, HPs may not have them immediately available to help proxies make decisions. HPs may then still have to advise proxies based on the HPs' recall of their own experience. Yet recall and personal experience both have their biases. And, fourth, the best interests method gives little substantive guidance for decisions about the always seriously mentally impaired. Those with the least capacity pose the most controversy [63, 64]. Some commentators claim such patients have no interests at all except for comfort. I do not agree. Rather, I believe with one commentator that even severely mentally impaired patients have *many* other interests in addition to comfort. Their interests surely include privacy; concerned, respectful care for themselves [65]; and supportive care for their families and close friends.

I think that despite the problems involved most proxies use the best interests method in making end-of-life decisions. Mrs. G.'s proxies probably use it in deciding about reventilation for her even though they may not realize they are using that method. Still, as I explain in a later section, there must be a better method for making proxy decisions.

The Case

The hospital ethicist in Mrs. G.'s case guides her family exactly according to the commonly accepted priority sequence for the standard proxy decision-making methods. Unfortunately, none of the three standard methods precisely fits Mrs. G.'s situation.

First, the ethicist tries the "purely" subjective method. Mrs. G.'s written AD directs him to do so. It instructs her proxies to use the wishes she has written into her AD to decide treatment if she cannot decide it for herself. In particular, the AD states that she refuses CPR always, and other life-support treatments if "(1) I have a terminal condition that my attending physician … concludes is … irreversible and … expected to result in (my) death within a … short time or (2) I am in a coma or persistent vegetative state that my attending physician … concludes (is) irreversible." Ventilators surely fall under this AD's definition of life-support treatments, that is, any treatments "which sustain, restore, or replace a spontaneous vital function and … only … prolong the dying process when applied to a (terminal) patient …" But Mrs. G. meets neither condition 1 nor condition 2 for withholding the ventilator. Because the second cardiologist believes he can successfully treat Mrs. G.'s heart arrhythmia, he does not consider her irreversibly terminally ill. And, although the neurologist initially considers a coma or incipient vegetative state for Mrs. G., he can never identify all the criteria for either diagnosis. Thus, the wishes Mrs. G. wrote into her AD do not exactly pertain to the reventilation question facing her family.

Next the ethicist tries substituted judgment. Mrs. G.'s AD states that, whenever Mrs. G.'s explicit prior treatment wishes do not apply to the situation at hand, her family can "consent (to or) refuse (on her behalf) … all types of … health care (including) hospice care, surgical procedures, … and … mechanical … procedures affect(ing) any bodily function." Thus, acting in accordance with the substituted judgment method, the ethicist tries to define Mrs. G.'s likely wishes for the situation. He does so by asking each family member, What do you think Mrs. G. herself would decide about the ventilator? Unfortunately, Mrs. G.'s prior comments about ventilators are confusing. On one occasion she said she does not want to "live like a vegetable" on a ventilator, and on another occasion she appeared to say she would tolerate a ventilator to have "every chance to live." The family understandably does not know what to think.

The ethicist has his own reservations about the substituted judgment method. He knows about the many studies showing that proxies, even long-time spouses like Mr. G., do little better than chance at discerning patients' unexpressed wishes about life support. As such, Mr. G. and his daughters are as likely to guess wrong as to guess right about Mrs. G.'s wishes in this situation. The ethicist also believes that, regardless of what the family decides about the ventilator this time, other life-support decisions will follow. The family's guesses about Mrs. G.'s wishes might not be any more accurate for other life-support treatments than for the ventilator.

Once resigned to the inaccuracies of typical substituted judgment, the ethicist might try a variation of it. Rather than asking What would Mrs. G. herself likely decide? he might ask, What would most patients in this community likely decide? Early research suggests some clinical usefulness of such community-wide substituted judgments. Three studies in particular suggest that, in general, the majority treatment wishes of patients in a community approximate the wishes of individual patients at least as well as do proxy guesses [1, 43, 61]. Furthermore, as the wishes of individual patients for life support decline overall with age, the majority wishes of patients in the community approximate those individual patient wishes ever better while proxy guesses approximate them ever worse [43]. This community-based variation on typical substituted judgment offers the potential advantage of reducing proxy or doctor biases in decision-making. But the variation also has the disadvantage of lacking data about people's wishes in most end-of-life scenarios. I do not know, for example, any community-based data pertaining precisely to Mrs. G.'s case. Therefore, like the "purely" subjective method both versions of the substituted judgment method do not help make the reventilation decision for Mrs. G.

Finally, the ethicist tries the best interests method as a last effort at making that decision. Mrs. G.'s AD instructs her proxies to do so, saying, "If my desires cannot be determined based on my written or oral statements, my Patient Advocate should do what he or she believes to be in my best interests." Unfortunately, Mrs. G.'s prior, potentially contradictory statements about end-of-life care create considerable doubt about what constitutes her best interests under the circumstances. She has said she wants "every chance to live" but not "as a vegetable." Furthermore, she has said, if her doctors ever diagnose a terminal condition, she wants "only the medical attention necessary to make me comfortable … (and the opportunity) to live out my last days at home or in a hospice rather than in a hospital." But she has also said she does not want to burden her family with physical care (as might happen with home hospice care).

Translating these wishes into a coherent vision of her best interests puzzles everyone. Nobody can confidently say what Mrs. G.'s best interests are and whether reventilating Mrs. G. would serve those interests. Data show a low survival rate to hospital discharge—less than 5 % in any neurologic state—after an out-of-hospital cardiorespiratory arrest like Mrs. G.'s. That rate surely drops for elders like Mrs. G. with multiple coexisting serious diseases. Would survival to discharge merely prolong her dying or give her a realistic chance to live as she wishes? Would functional limitations, physical discomforts, doctor appointments, and medical procedures consume her remaining days to no purpose? Or would she be able to "participate in life," move around sufficiently, and "enjoy her family"? All these questions plague the family. The disagreement between the two cardiologists only exacerbates the family's quandary: The first cardiologist believes that Mrs. G. is imminently and irreversibly dying; the second, that she can be successfully treated for her heart arrhythmia, survive, and function adequately in the future. Thus, like the "purely" subjective and substituted judgment methods the best interests method also proves unhelpful for deciding about reventilation.

The ethicist and the family conclude that each of the three standard proxy decision-making methods fails to some degree. They are all imperfect at capturing Mrs. G.'s essence as a "whole person," at accounting for important clinical details in her case, and at pointing to a clear decision [41]. The family are left adrift. I believe another decision-making method might serve them and Mrs G. better.

Constructed Judgment, an Alternative Proxy Decision-Making Method

When three proxy decision-making methods already exist and have gained widespread usage, why is another method needed? The answer, I believe, lies in two important flaws the three methods share. One flaw, which I mentioned in the last section, involves proxies' inaccurate knowledge of what patients have already decided, what patients would decide, and what care would best promote patients' interests. This flaw arises from the patient autonomy principle itself [66]. That principle focuses so thoroughly on the patient's specific expressed wishes and interests that it often overlooks the importance of the patient's life context [13]. A patient's wishes or interests never speak for themselves. They always require interpretation in light of the patient's traits, values, relationships, past experiences, and current circumstances. Those aspects of life context fill out the patient as a "whole person" and suggest insightful interpretations of his or her previously expressed wishes or interests [41, 63].

Of course, as life context changes, the most fitting interpretations of a patient's wishes or interests will usually have to change, too. Biblical scholarship offers two useful directions for making such context-specific interpretations. First, one should state as accurately as possible "the original text." Next, one should interpret that "text" in light of both the original context and the new context [49]. Here, Mrs. G.'s written AD and earlier oral statements provide "the original text" of her wishes about ventilators. Her family and doctors have the job of interpreting those statements in both their original context (two other patients—one ventilated for years, the other successfully weaned) and new context (Mrs. G.'s own out-of-hospital cardiorespiratory arrest with its high likelihood of only short-term survival and severe brain damage).

The other flaw of the three standard proxy decision-making methods involves their unrealistic expectations. All three methods expect proxies to make well-informed, well-reasoned, and accurate life-and-death decisions under the most difficult circumstances. Many proxies in the medical crisis are emotionally distraught over patients' precarious physical states, are mentally unprepared for the stresses of decision-making, or are unfamiliar with the esoteric medical issues at hand. Nonetheless, proxies are often expected to grasp highly technical clinical information quickly (often with just one explanation) and to apply it immediately. They are also expected to focus exclusively on patients' interests even when other people (sometimes even the proxies themselves) may have strong, countervailing interests. In Mrs. G.'s case the family approach their proxy decisions with remarkable understanding and courage. They educate themselves about Mrs. G.'s medical problems and focus their decision-making on her interests alone, not their own. Many other proxies, however, are not so capable. The decision-making task as defined by any of the three standard methods simply demands too much of them.

I, therefore, favor an alternate proxy decision-making method that, I believe, avoids the two serious flaws of the standard methods. This method, called "constructed judgment," involves first understanding the major defining and unifying elements of a patient's life [41] and then constructing care decisions that "fit" those elements best. "Fit" here means reflecting faithfully the patient and the context of his or her life story [9, 35, 41, 52]. One author describes constructed judgment as shaping care decisions to the patient's "narrative" [8]. The process obviously requires knowing the most important influences on the patient's life [67]. Those influences may be character traits [68], personal values [46], interpersonal relationships, formative experiences [13], or interests [69].

The key question of constructed judgment is, Which care decision in this situation "fits" the patient's life story best? Answering that question depends on mutual education among the patient's proxies, others close to the patient, and the HPs attending the patient. The proxies and others close to the patient must teach the HPs the important facts about the patient's life; the HPs must teach the proxies and others the important facts about the patient's illness. Once fleshed out and adequately understood, the patient's life story and the medical facts together often present certain decisional options that "ring true" for the patient [8, 35, 41, 52]. I believe those options usually serve the patient best.

However, unlike the other proxy decision-making methods the constructed judgment method sometimes justifiably allows the interests of others to outweigh the interests of patients in care decisions [9, 70]. Two such situations come to mind. One occurs when patients *want* the interests of others to outweigh their own apparent interests: patients may want to sacrifice for their survivors. For example, the gravely ill homeowner may refuse expensive life support in order to save the family home for his children, and the slowly dying cancer patient may refuse home hospice in order to protect her family from the exhausting in-home caregiving involved. In such cases the sacrifices reflect patients' values, wishes, interests, and life stories. Mrs. G. herself previously expressed a willingness to sacrifice for her family. Specifically, she told them never to sell the family home just to fund her medical care, and she wrote into her AD that she would not want terminal home hospice care if it would "impose an untenable burden on my family."

The second situation in which the constructed judgment method allows others' interests to outweigh patients' interests occurs when patients' impaired judgment makes their immediate wishes deviate from lifelong values and risks serious harm to others important to the patients. A case from my own practice illustrates.

Police picked up a huge bear of an elderly man walking in his pajamas along an interstate highway and brought him to the local hospital. The man immediately demanded to be taken home. The doctors, however, searched his past medical records and found a diagnosis of an aggressive, argumentative dementia.

When the man's wife arrived, the doctors perceived her as a good, caring proxy for the man. The wife explained that, because the man hated hospitals, she had cared for him herself at home throughout his dementia. By then, however, he was requiring considerable physical care and sometimes wandered away from home. He occasionally even woke her up at night brandishing a knife over her. She admitted to being physically and mentally exhausted and fearing for her life. The doctors recommended transferring the man directly to a dementia-oriented nursing home, but their recommendation troubled the wife greatly. She explained that the man had loved her deeply and been a good husband. He had always wanted to live out his life with her at home, and she had promised him several times never to send him to a nursing home. Though worried now for her safety, the wife wanted to keep her promise to him.

The doctors eventually decided the matter by the constructed judgment method. They asked themselves the question, Which care decision best "fits" this man's life story? They concluded the man's lifelong love for his wife was the strongest relevant personal value for him. It justified sparing the wife further exhaustion and fear over honoring the man's immediate wish to go home. The doctors, therefore, persuaded the wife to allow a nursing home transfer. She agreed reluctantly. The transfer occurred, and the man lived out his life at a nursing home near the couple's home.

As this last case illustrates, the constructed judgment method sometimes requires doctors to exercise respectful assertiveness when proxies hesitate to make decisions that protect patients or proxies from unreasonable, potentially harmful patient wishes. Doctors may even need occasionally to assert the interests of proxies and others against the interests of patients. The doctors and other attending HPs may then need explicitly to assuage the guilt of proxies who accept a course of action seemingly contrary to patients' interests. Those situations require professional comportment, courage, and sensitivity from HPs. Some critics, of course, claim such situations permit doctors to run roughshod over patients. I disagree. Those situations demand self-critical discernment about what is right. But, as long as doctors employ great restraint and sound justification in promoting others' interests over patients' apparent interests, I believe the risk of power abuse is small [46].

Using the Constructed Judgment Method in End-of-Life Care

The constructed judgment method (like the other three decision-making methods) relies largely on the patient's principal doctor to help proxies reach well-reasoned decisions. But, because constructed judgment is new to most doctors, I recommend the facilitating doctor plan a systematic approach to it ahead of time. I offer here some suggestions from my clinical experience, but I realize other doctors will develop their own approaches based on personal style and clinical experience.

I suggest the doctor's approach to the constructed judgment method, while primarily serving patients' interests, also take account of proxies' needs in the decision-making process. Proxies typically have three: building strong relationships with HPs, receiving important communications, and developing trust in the quality of the patient's care [71]. Unfortunately, the fragmentation of today's American health care often works against satisfying those needs. That fragmentation, characterized by many caregivers and discontinuities in the process of care, will undoubtedly continue into the near future. Still, doctors and other HPs can take simple steps now to address proxies' needs.

Proxies need strong relationships with HPs to help cope with the unfamiliar, highly technical, and seemingly impersonal healthcare environment. One step in fostering those relationships is for HPs to identify themselves and their roles in the patient's care [71]. HPs should wear name tags, introduce themselves directly to proxies, and explain their responsibilities in the patient's care. I, for one, usually say, "Hello. I am Henry Perkins, one of the senior internal medicine doctors here. I have overall responsibility for Mr. Smith's care. Much of what I do involves diagnosing the problem, coordinating care, and communicating with specialist doctors on his behalf."

Most proxies also need frequent, accurate communications to feel abreast of a patient's condition. Continuity and consistency of information favor having the same HP provide communications over time. (That HP should typically be the principal doctor. However, other HPs should be allowed to provide information as proxies need it, but those HPs should inform the the patient's principal doctor what they said and when.) Fragmentation of care, of course, often subverts clear communication. For that reason the HP who assumes primary responsibility for informing a patient's proxies should provide a reliable means of contact and invite proxies to call as necessary to ask questions or discuss care. Doctors can give out their business cards; nurses, the public-access telephone number at their nurses' station.

Proxies also need to be able to trust the quality of patient care [71]. Certain experiences either erode or promote proxies' trust. Getting conflicting information or witnessing uncoordinated services erodes that trust. Fortunately, meeting proxies' prior two needs—building strong relationships with HPs and communicating effectively—fosters trust in the quality of care. Of course, expressing heartfelt concern for proxies, exercising patience with them, and explaining medical matters without using technical jargon also help.

I consider family conferences the most efficient format for making decisions by constructed judgment. (Although I describe these conferences in detail in Chap. 8, I mention a few relevant points here.) Using the constructed judgment method effectively, of course, requires forethought. In particular, setting up the discussion properly is critical. One doctor, who should typically be the main information source for patient and proxies, should take responsibility for leading the conference. He or she should invite all patient-designated proxies, other key family members and friends, and HPs closely involved in the patient's care to attend. And he or she should assemble these people in a quiet, private place and allow for unhurried discussion.

The lead doctor might open the conference by telling his or her name and role in the patient's care, introducing the other HPs present, and asking the non-HPs to introduce themselves. He or she might say, "I know this situation is hard for you. It is hard for everybody. For that reason we, the staff, want to support you and hope you will support us as together we try to reach decisions that fit Mr. Smith best." Then, using lay terms, the doctor might briefly summarize Mr. Smith's medical condition and state the matter to be decided.

Constructed judgment can then take place. Time-urgent, goal-driven Americans usually expect to address the most pressing issue immediately, not the initial, seemingly indirect descriptions of constructed judgment. The lead doctor might, therefore, begin by explaining, "Most people expect me to tackle the most pressing issue right away. But to help decide that issue, I first need to learn more about Mr. Smith [8, 13, 40, 48]. You can help me. So please have patience. The information I get from you about him now can help us make decisions a little later."

Having secured the group's cooperation in the constructed judgment process, the doctor might ask general questions such as What is Mr. Smith like? or How would you describe him? If participants have difficulty answering, the doctor might ask more specific questions such as What was Mr. Smith like before this illness? What made life worth living for him back then? What is most important to him in life now? What does Mr. Smith still look forward to? Which core personal values or traits of his (such as strength of character or will to live) [68] would he want to influence our decisions? or How has he reacted to being ill or hospitalized before? At some point the doctor should also learn about others whom care decisions for Mr. Smith may affect. Family and friends rarely volunteer that information. The doctor may, therefore, need to ask explicitly, How will care decisions for Mr. Smith affect others? and Would Mr. Smith want us to consider the interests of those others when we make care decisions for him?

When sensing an adequate patient description to inform care decisions, the lead doctor should summarize aloud what he or she has heard, invite confirmation or correction, and suggest initial care options. Only then should the doctor ask the central question of constructed judgment, What decision about the issue at hand "fits" Mr. Smith best? The prior patient description should permit participants to critique the initial care options and suggest others according to Mr. Smith's life story and core traits, values, and interests as well as the interests of others important to Mr. Smith.

Constructed judgment is obviously never strictly deductive: It always involves imagination and intuition. As a result each issue may have multiple acceptable decisions. But the patient's character and life story limit the possibilities [9, 41, 52]. Whatever final decision emerges must pass muster by the best "fit" or "authenticity" to the patient [8, 46].

The Case

In the family conference the hospital ethicist, Mrs. G.'s family, and the clinicians in attendance consciously apply the three standard proxy decision-making methods to the reventilation question. But, perhaps without realizing it, the participants simultaneously gather much of the information key to making a constructed judgment decision. Constructed judgment needs much of the same information as do the standard proxy decision-making methods but assembles that information in a different way to reach a decision.

Sources for the information about Mrs. G. include her written AD, her orally expressed wishes about quality of life, the family's intimate knowledge of her character and life story, and their observations of her in the hospital. Mrs. G.'s written AD designates Mr. G. and the elder daughter as her official proxies and expresses full confidence in them to make decisions about "all types of medical or mental health care, … hospice care, surgical procedures, diagnostic procedures, medications, and … mechanical … procedures …" But prior conversations also make clear Mrs. G. does not limit proxy decision-making to Mr. G. and the elder daughter; she wants the younger daughter to participate in the decision-making, too. The AD also states Mrs. G.'s wishes for comfort care only and no life support in minimally functional states (such as irreversible coma or a persistent vegetative state) or in imminently terminal illness. Prior end-of-life conversations with the family provide important additional information, offering insights into what Mrs. G. considers acceptable qualities of life. Specifically, those conversations indicate that Mrs. G. insists on "participating in life" and having relatively high function: She wants to move independently and to interact with her family.

At the same time the family understands some of Mrs. G.'s traits and values important for interpreting these stated wishes. Mrs. G. loves her family deeply and is willing to endure even great back pain to spend time with them. In fact, she has recently endured a painful car trip halfway across the country just to hold her first great grandchild. She has also worried about how Mr. G. would bear the stress of her illnesses, and she wants to save him from the unrealistic intention to care singlehandedly for her at home. She, therefore, adamantly rejects any home care arrangement that would burden him severely. But she also fears hospitalizations for the loss of personal control and of bodily privacy they involve.

The family make important observations of Mrs. G. in the hospital, too. They see her frightened, "haunted" look when she has difficulty breathing due to retained airway secretions. And they see her discomfort when she gags violently as the nurses suction those secretions out. The family also notice that Mrs. G. communicates ever less as her hospitalization goes on until she finally does not appear to recognize them or respond to them.

I believe conscious use of the constructed judgment method can supplement such bare-bones insights with helpful additional details about Mrs. G.'s life story. Those details can help tailor proxy decisions further to her. My experience suggests that the richer the descriptive detail about patients and their life stories, the more personalized the care decisions through constructed judgment, the clearer and more flexible the treatment plans, and the fewer the regrets of proxies afterward.

Constructed judgment for Mrs. G. must address the question in the end, Which decision—reventilating or not—best "fits" Mrs. G. and her life story? Answering might consider several points from the information already known about her. Mrs. G. said a quality life for her meant avoiding unnecessary discomfort, being able to move about independently, and enjoying her family. Using the ventilator certainly would hamper all those goals at least short-term. It would require having an uncomfortable tracheal tube and periodic airway suctionings. The ventilator would also prohibit her moving far from the bed, would prevent her from talking with her family due to the tube, and would limit family visits to the prescribed ICU visiting hours. Furthermore, the ventilator would compromise Mrs. G.'s sense of both personal control and bodily privacy. The ventilator would even cause the family distress, which Mrs. G. wants to avoid. And reventilation cannot guarantee an eventual outcome acceptable to Mrs. G.: She, like the one distant relative with a stroke, might never wean off the ventilator again. All these considerations, therefore, convince me that constructed judgment would justify not reventilating Mrs. G.

Why I Recommend the Constructed Judgment Method Over the Other Proxy Decision-Making Methods

Readers may react skeptically to constructed judgment because the other three proxy decision-making methods are so widely accepted. Why, then, do I recommend it? I have six main reasons. First, unlike the other decision-making methods constructed judgment does not demand an impossibly high standard of proxy performance. The other methods expect proxies to match their decisions exactly to patients' specific wishes or interests. But the complexities and vagaries of clinical reality constantly frustrate that expectation. Patients do not usually formulate their treatment wishes beforehand. Even if patients do, the wishes are often inadequately informed, inadequately considered, or inconsistent; and many patients do not communicate them to others. Proxies must guess at them in the crisis. But guessing quickly and accurately is difficult in complex, unfamiliar, and rapidly changing clinical situations. The stress involved may actually leave some proxies mentally paralyzed, incapable of deciding anything at all. Furthermore, the other decision-making methods imply—surely incorrectly—that only one right decision exists for each situation.

In contrast, the constructed judgment method keys care decisions directly to the traits, values, and experiences that have shaped particular patients' lives most. Proxies typically know the patients well enough to understand those traits, values, and experiences and their likely influences on the patients. Proxies need not divine patients' exact wishes or interests on the spot [36, 45]. And constructed judgment gives proxies the realistic task of choosing *any* option that reasonably "fits" patients and their life stories. The reasoning process often creates multiple acceptable options, and proxies do right for patients by choosing any of those options. That inherent flexibility makes good sense for minimizing the legalistic burdens of, say, clear and convincing evidence and for easing proxies' qualms after the decision [72].

Second, as I noted earlier, patients' wishes and interests do not speak for themselves. They require interpretation within the contexts of patients' life stories. Relationships, of course, are a key element of those life stories. Significant events for one person such as illness and treatment affect others, too, especially intimates [38, 52, 67]. The interests of both deserve consideration in care decisions. Yet the three standard proxy decision-making methods consider only the interests of patients. The constructed judgment method, in contrast, considers the interests of patients and others alike.

Still, some commentators worry that factoring others' interests into care decisions compromises patients' interests. That risk certainly exists. But I have only rarely seen flagrantly abusive proxy decisions that harm patients solely to benefit others. Other authors likewise report no such "rampant disregard" of patients' interests [3]. I suspect the more common problem is the unrecognized compromise of *others'* interests in care decisions [70]. As with the wife caring for her demented, knife-wielding husband, intimates do not typically assert their own interests when they conflict with patients' interests. HPs must sometimes assert the interests of others for them.

The constructed judgment method, then, involves striking some defensible, proportionate balance between the interests of patients and the interests of others. Patients' interests should ordinarily predominate, and exceptions should be rare. But one, which I mentioned before, occurs when patients *want* the interests of others such as spouses or children to override their own. Other possible exceptions occur when treatment offers little benefit to patients and might harm others severely [67, 73]. A rule of thumb that guides me in such situations is that the more distant the relationship of others to the patient, the less likely any harm to them will be severe enough to justify overriding the patients' interests. Withholding an expensive chemotherapy with little benefit for the advanced cancer patient, for example, might be justified to prevent family bankruptcy but not to reduce premiums for all Medicare beneficiaries.

Striking a balance between the interests of patients and the interests of others admittedly involves intuition. No precise calculation is possible. Thus, any balance is open to criticism. However, if proxies strike a balance in good faith and that balance considers patients seriously, HPs should be willing to defend it and the proxies who made it.

Third, the constructed judgment method also promotes the shared decision-making [46] most proxies want [74–78]. That decision-making involves sharing information between proxies and doctors with each party's communicating a different kind of information to the other. Proxies communicate their personal knowledge of the patients to the doctors; the doctors communicate their detailed knowledge of medicine to the proxies. Both parties offer observations of the patients, and the doctors offer specific care recommendations. Proxies and doctors then negotiate care decisions. Constructed judgment thereby fosters mutual respect and joint responsibility. No one person—proxy or doctor—bears the whole burden of deciding on care. Proxies contribute meaningfully to decisions and feel supported by the HPs, and doctors and other HPs gain insights about patients that help individualize care.

Fourth, of all proxy decision-making methods, only constructed judgment provides rich, broadly applicable patient characterizations [10]. The three standard methods provide merely sketchy characterizations suited at best to decisions about isolated treatments. The "purely" subjective method, which depends on a patient's prior written or oral wishes, has an additional limitation: That method applies only to clinical situations the patient imagined beforehand—typically the persistent vegetative state or an imminently terminal illness. In contrast, the constructed judgment method produces far more detailed patient characterizations, which can serve a broad range of treatment decisions even in situations the patient did not anticipate [8, 68].

Fifth, constructed judgment promotes reasonable discretion in interpreting patient wishes and interests. Unlike the other decision-making methods the constructed judgment method relies more on the good will of proxies and doctors than on the accuracy of their treatment guesses. The character of the decision-makers and their relationship to the patient often matter more than their specific treatment decisions [36]. Constructed judgment thereby transfers the decision-making process from the realm of legalistic evidence and strict logic [72] to the realm of intimacy and affection.

Sixth, whether they realize it or not, many proxies already use the constructed judgment method in end-of-life care decisions [35]. The personal characterizations that fuel this method often surface spontaneously among family and friends at times of a life-threatening crisis. People remember, for example, the patient's idiosyncrasies, will to live, or beliefs about God and life. People also remember significant life experiences shared with the patient. Those considerations, largely irrelevant to the standard proxy decision-making methods, provide the groundwork for the constructed judgment method. They also go a long way to creating personalized care.

In summary, I believe the constructed judgment method provides a powerful, intuitive, and practical process for making proxy decisions. That process involves trusted proxies; relies on rich characterizations of patients' formative traits, values, and experiences; promotes transparent consideration of others' interests; permits discretionary judgments within a range of decision-making options [79]; aims for decisions reflecting patients' character and life stories; and fosters mutual respect and shared responsibility for decisions between proxies and HPs.

The Case

I believe that the constructed judgment method could benefit Mrs. G.'s case in many ways. By portraying Mrs. G. in the context of her life story, constructed judgment could provide a rich informational base for making not just the reventilation decision but other care decisions, too, large and small. Constructive judgment could refocus those decisions from narrow treatment goals to broader personal goals. It might even give a coherent, long-term direction to her care. The emotional, relational, and spiritual aspects might gain importance, too, relative to the physiologic ones.

Constructive judgment might also help her family throughout the ordeal. Constructive judgment might give them a sense of being heard and respected by the HPs. The family could derive satisfaction from contributing to the informational base for Mrs. G. and to the decisions about her treatment. They would feel like trusted intimates acting for her out of love. Constructed judgment might also ease the family's grieving process by encouraging them early on to tell each other their memories of Mrs. G. Similarly, constructive judgment might minimize the family's later regrets—what one author calls the distressing "residue" of proxy end-of-life decisions [41]. Such regrets trouble Mr. G. in particular. Feeling most responsible for her care, he relives many times afterward the eventual decision to decline reventilation. He even tells his daughters ruefully, "Your mother really wanted to live." Perhaps the part of constructed judgment that would explicitly link that decision to Mrs. G.'s own life and values might convince him about the rightness of declining reventilation.

Finally, constructive judgment might also help the attending HPs including the second cardiologist. It might heal the rift between him and the family. He would learn about the "whole person" of Mrs. G. and could rethink his ventilator proposal in light of what he learns. His demeanor toward the family might change from a bullying arrogance to a respectful cooperation as he and the family negotiate a decision "fitting" for Mrs. G. And learning the details of Mrs. G.'s life through constructive judgment might increase the appreciation that he and the other HPs have for Mrs. G. and thereby their satisfaction in taking care of her.

Factors that Help or Hinder Proxy Decision-Making

HPs who attend decisionally incapacitated dying patients want to know which factors help or hinder proxies' decision-making for those patients. This section reviews some of the recent research on the topic. The reader should note again that I address here only the *process* of proxies' decision-making. I address proxies' *emotional reaction* to that process in Chaps. 8 and 9.

A study by Vig and colleagues proved especially useful for purposes of this review: Much of the information I report here came originally from that study [12]. Vig and colleagues wanted to describe the experiences of proxies who make decisions for critically or chronically ill patients. The researchers knew beforehand that proxies participate in about 75 % of care decisions for hospitalized patients with life-threatening diseases and in about 50 % of care decisions for nursing home residents. The researchers, therefore, interviewed the designated proxies of 50 chronically ill, elderly patients who had recently been hospitalized, enrolled in hospice, or died, and asked those proxies which factors had helped or hindered them in making care decisions for the patients. The factors proxies named fell mostly into four general domains: proxy personal characteristics and life circumstances, proxy social networks, proxy–patient relations and communications, and proxy–health professional relations and communications (Table 7.3). I take up the individual factors by their domains and illustrate the factors with quotes from interviewees of Vig and colleagues.

Proxy Personal Characteristics and Life Circumstances

Helping factors in this domain include having participated in similar proxy care decisions before. "I had lost both parents of the same (disease)," one proxy explained. "I had been through (the disease) before. I knew how to talk to (such a patient) and (which points) to bring up (for discussion) …" Another helping factor is using good strategies to cope with the stress of patient illnesses and proxy decisions. One proxy, for example, coped by "losing" herself in her artwork. And other proxies coped by drawing strength from their spiritual beliefs or religious communities. Still another helping factor is proxies' anticipating the effect of various decisions on *their* lives. As a result, proxies try to make only those decisions they can live with later in good conscience. For instance, a woman designated as proxy for an elderly neighbor had to make the difficult decision to stop life support for her. The proxy explained afterward, "(T)o do something that someone has asked me (not to do, here to use life support) would be a hard thing to live with …"

Hindering factors in the same domain include juggling competing responsibilities such as caring for oneself or others, and coping with one's own illnesses. One proxy illustrated both factors. While making decisions for her husband sick from cardiac surgery, one woman had to cope with her own newly diagnosed breast cancer and to undergo surgery for it. Another hindering factor is having long distances to travel to visit the patient. The long distance between one proxy's home and the hospital where her husband had been admitted severely limited the frequency of her visits to see him. She worried that her long absences in between visits gave the doctors too much freedom "to do what(ever)" they wanted with him. And still another hindering factor is financial worries such as about affording patients' follow-up nursing home care.

Proxy Social Networks

Helping factors here include having others besides the attending HPs to talk to about care decisions. Proxies especially value medically trained family and friends who can explain clinical events and advise about decisions. Another helping factor

Table 7.3 Factors that help or hinder proxy decision-making

Domain	Helping factors	Hindering factors
Proxy personal characteristics and life circumstances	• Having participated in similar decisions before • Using good coping strategies • Making only decisions one can live with in good conscience later	• Juggling competing responsibilities (such as care for self or others) • Coping with one's own illnesses • Having to travel long distances to the patient's site of care • Having financial worries (such as about affording the patient's follow-up nursing home care)
Proxy social networks	• Talking with others • Receiving emotional support from others • Reaching consensus on decisions	• Having disagreements with others over the "the one right" decision
Proxy–patient relations and communications	• Knowing the patient's treatment wishes and being able to implement them (especially if having promised to do so) • Making decisions that promote the patient's comfort • Learning about the patient's condition and being invited to participate in treatment decisions • Having health professionals readily available to answer questions and to give support • Getting clear, honest medical information in lay terms • Getting a doctor's recommendation about what to do	• Encountering a situation that prevents implementing the patient's treatment wishes • Having problematic emotional attachments to the patient (such as difficulty in "letting the patient go") • Having too many health professionals involved in the patient's care • Receiving inadequately coordinated communications from health professionals
Proxy–health professional relations and communications	• Being shown respect by health professionals for proxies' contributions to care decisions • Getting health professionals' assurance afterward that care decisions were right for the patient	

Adapted from Ref. [12], with permission

is receiving emotional support from others. And a third helping factor is being able to reach consensus with HPs and other involved parties on decisions. The corresponding hindering factor is not being able to do so. Disagreements and the resulting arguments with others over the "the one right" care decision stress proxies terribly. As one proxy of Vig and colleagues said, "(F)amily is family. And when (someone in the family is) dying, they all want to have their say … It (i)s a hard time …"

Proxy–Patient Relations and Communications

Knowing a patient's treatment wishes and being able to implement them are a hugely important helping factor in this domain because implementing those wishes gives proxies enormous satisfaction. Doing so makes proxies feel they have met their duty to the patient. "I had made a promise to (the patient)," one proxy said. "(If) you make that kind of commitment, you've got to do what (is necessary) … to see that it's fulfilled …" Making decisions that promote patient comfort is yet another helping factor. Those decisions, too, give proxies great satisfaction. And having been asked by the patient to learn about the patient's condition and to participate in care decisions as proxies is yet another helping factor. Such a request surely gives proxies decision-making legitimacy in their own eyes and a sense of the gravity of their responsibility. Yet knowing a patient's wishes and being prevented by circumstance from implementing them is an important hindering factor for proxies. That factor can plague them with deep regrets after a decision. One proxy gave a rueful example about her husband's death. "(Not being able to fulfill his wishes made the situation) difficult," she said. "I *did* [italics added] know his wishes … to have his demise at home. (But) we couldn't do it for him. We had … to take him to the hospital so that he would be comfortable in his last hours." Deep or complicated emotional attachments are still another hindering factor especially when a long-time intimate must decide to stop a patient's life support.

Proxy–Health Professional Relations and Communications

Good communications with HPs, of course, provide helping factors in proxy decision-making. The ready availability of HPs—especially doctors—and their willingness to answer questions and to give emotional support are one especially important helping factor. Clear, honest, and timely HP communications about the patient's condition and its treatment, and a doctor's recommendation about what to do are others. And being shown respect by HPs for the contributions proxies make to care decisions is still another. Doubtless, all those helping factors foster proxy trust in HPs. "(W)hen a doctor treats the spouse with a lot of respect and answers questions like they're important," one proxy said, "(that doctor gives) you the feeling of competence. I think Dr. F. made me feel like a very important part of the (care) team." One other important helping factor is HPs' reassurances after a decision that it was right for the patient. Those reassurances are especially significant after a decision to let a patient die. They may help proxies remember the thought process behind the decision and the HPs' support for it. Notably, reassurances may require some repetition to ease proxies' "executioner's guilt." [80] As trusted and respected authorities, HPs must not underestimate the consoling power of their reassurances.

Of course, poor communications with HPs create factors that hinder proxy decision-making. Two such factors arise regularly in today's American healthcare system: having many—perhaps too many—HPs involved in a patient's care and having inadequate coordination of HP communications. The wife of a patient who died during emergency surgery complained about both factors, "There were just too many people … too many different stories. I was told one thing (by one team). And, when another team would come through, they'd tell me something else. I was so confused … I didn't know what was going on. (Finally,) I said, 'I want to speak to one person and one person only. I can't take in all this …'"

While the study by Vig and colleagues provides many rich examples of helping or hindering factors, Mrs. G.'s case provided some others. I describe them in the next section.

The Case

Factors That Could Help Proxy Decision-Making in Mrs. G.'s Case

At least two factors from each domain of Table 7.3 might help Mr. G. and his daughters in their decision-making for Mrs. G. The three family members certainly possess potentially helpful *personal characteristics and life circumstances*—having witnessed similar medical situations and decisions concerning Mrs. G. before, using good coping strategies, and getting strong spiritual support. Mr. G. and the daughters know well from their close relationship with Mrs. G. her personality and likely reactions to medical care. For example, they saw her impatience during a prior hospitalization for blood clots, and they could reasonably expect a similar reaction from her during this hospitalization. The family also know the importance of good supportive coping strategies and adopt some for themselves early in Mrs. G.'s illness. They share reminiscences about her as a kind of life review. They relive, for example, Mrs. G.'s joy at holding her first great grandchild. They also post family pictures at Mrs. G.'s bedside to remind her and them about the importance of family. Furthermore, the family make fond new memories of Mrs. G. even while she was hospitalized. Mr. G., for instance, offers early on to give Mrs. G. one of his "famous kisses." When he asks her permission to do so, she smiles out of her stupor and immediately puckers up. Mr. G. asks her afterward whether she liked the kiss. She nods yes. In addition, the family lean heavily on their faith throughout the ordeal. Their prayers and discussions with the church pastor give them important strength and guidance for making decisions about Mrs. G.'s care.

The family's *social network* also provides helping factors for proxy decision-making. The family finds knowledgeable outsiders (such as a doctor friend from their church) who explain Mrs. G.'s medical predicament. Furthermore, church members provide important emotional support with visits and cards. The hospital nurses provide such support, too. In fact, one daughter praises the nurses as "attentive and compassionate" for encouraging the family to talk about the ordeal and then listening to them. "Those nurses (even) stand by us," the daughter explains, to the point of helping the family oppose the second cardiologist when the family disagrees adamantly with his reventilation plan.

Family–patient relations and communications provide even more helping factors. For one, the family already knows Mrs. G.'s general wishes against life support and hospitalization as expressed in her written AD and oral advance care planning statements. The family knows she wants no life support if she is ever persistently vegetative or imminently dying. They also know she prefers to die at home, not in a hospital, unless home care would excessively burden the family. As for other helping factors, the family learns everything they can about Mrs. G.'s condition, participates actively and knowledgeably in care decisions, and assists with care. The family, for instance, asks the doctors many questions about Mrs. G.'s condition: the pulmonologist about Mrs. G.'s early respiratory problems, the neurologist about a possible vegetative state, and the second cardiologist about the later abnormal heart rhythm. Furthermore, Mrs. G. has already asked her immediate family to serve as her proxies. Abiding by those wishes, the doctors explicitly invite the family to participate in decisions about her care. The family feels they have a legitimate place in the care discussions. The elder daughter believes the family contributes knowledge of Mrs. G. as a "whole" person and of her treatment wishes, however spotty. Both kinds of knowledge, she says, give useful guidance —"something to hang on to"—when the family and the doctors negotiate care decisions for her. And the family knows Mrs. G.'s wish, as written into her AD, to receive only comfort care if she is irreversibly dying. Therefore, when the family become convinced Mrs. G. is dying, they stay at the bedside around the clock to ensure her comfort. They request analgesics for pain, oxygen for breathlessness, and sedation for restlessness.

Proxy–health professional communications provide even more helping factors. For one, the doctors make themselves readily available to the family to answer questions. When a close relative from out of town comes to visit Mrs. G. for the last time, for instance, the pulmonologist takes considerable time to meet with him and to answer all his questions. The doctors also avoid speaking in medical jargon around the family. The doctors use only clear, honest lay language. For another helping factor the doctors consistently make specific care recommendations along the way. The first cardiologist recommends final good-byes soon after admission; the pulmonologist recommends no further ventilation after Mrs. G. weans off the ventilator early on; and the second cardiologist, disagreeing with both the first cardiologist and the pulmonologist, recommends aggressive reventilation later on.

Such specific recommendations give the family a clear focus for their deliberations (even if the family does not always agree with the recommendations). For another helping factor, many of the doctors make the family feel valued for their input into care decisions. The hospital ethicist, for instance, convenes a conference specifically to hear the family's views about reventilation. And, once the family has decided against it, the ethicist personally defends that decision to the second cardiologist and wins his acceptance of it.

Unfortunately, Mrs. G.'s case falls short with two other potential helping factors —good HP communications and reassurances. Problems with those factors might eliminate some important benefits for the family. One problem is the poor communication styles of some doctors. The second cardiologist, for one, strikes the family as "pompous, imperious," overbearing, and unwilling to listen. The family believe he sees Mrs. G. only as a physiologic specimen, not as a "whole person;" is overconfident about his ability to fix her heart; and never seriously considers the family's view that she is irreversibly dying. And for another, the young general internist helping to care for Mrs. G. does not "connect" emotionally with the family perhaps due to the internist's discomfort with dying patients. The other problem is the lack of the doctors' reassurance for the family that the heart-wrenching decision to decline reventilation and to allow Mrs. G. to die is right for her. Such reassurance might save the family—especially Mr. G.—severe regrets about the decision afterward.

Factors That Could Hinder Proxy Decision-Making in Mrs. G.'s Case

Unfortunately, at least one factor from each domain in Table 7.3 might also hinder the family's decision-making on Mrs. G.'s behalf. Perhaps the most prominent hindering factor under *proxy personal characteristics and life circumstances* is the daughters' competing job and family duties. The elder daughter supervises a busy education program in the local public schools and attends to her own husband; the younger daughter works full-time as an accountant, teaches Sunday School, and cares for her young children. Yet Mrs. G.'s illness forces the daughters to take on the added duties of caring for the frail Mr. G. They prepare meals for him, wash his clothes, and pay his bills. Their regular duties, the new duties of caring for Mr. G, and the overnight stays at Mrs. G.'s bedside quickly exhaust both daughters. Thankfully, some personal characteristics and life circumstances that complicate other proxy decision-making situations (such as long distances to the hospital, poor proxy health, and precarious family finances) do not apply here. The hospital is only a few miles from where Mr. G. and his daughters live, all three of them enjoy good health, and a generous health insurance plan covers nearly all of Mrs. G.'s care expenses.

The one hindering factor from the *proxy social network* involves the family's initial internal disagreement over the "right" decision about reventilation. The second cardiologist's claim, "I know what is wrong with Mrs. G.'s heart, and I can fix it," raises Mr. G.'s hopes that she can survive. He leans at first toward reventilation. Then, when the cardiologist adds, "Either you put Mrs. G. back on the ventilator, or she will die," Mr. G. feels compelled to do it. He cannot bear the thought of allowing Mrs. G. to die when treatment might save her. The daughters, however, disagree mightily. They believe the doctors originally acted too hastily to ventilate Mrs. G. without consulting her AD. The daughters also remember the pulmonologist's advice not to ventilate Mrs. G. again after her early successful weaning. Both daughters and Mr. G. subsequently signed the order to that effect. By the time the reventilation question arises, the daughters firmly believe Mrs. G. is irreversibly dying. They favor hospice care for her. The daughters ask themselves at the time, "Why is the (second cardiologist) bothering us again about ventilating Mom? The medical record shows we've already made that decision. He can't just swoop in as The Big Hero and upset everything and everybody. His plan would surely add to Mom's suffering and not improve her outcome." The cardiologist's aggressive plan particularly angers the daughters because they think it raises false hopes for Mr. G.

To their credit Mr. G. and his daughters take a constructive approach to the disagreement. They ask advice from others (especially their minister and the doctor friend from the church), pray individually and together, and talk through their disagreement. They eventually agree that not reventilating but giving hospice care instead are best for Mrs. G.

Two hindering factors from *proxy–patient relations and communications* might have adversely affected the family's decision-making, too: circumstances preventing implementing the patient's wishes, and problematic emotional attachments. Circumstances in Mrs. G.'s case that appear to prevent implementing her treatment wishes include the quick ambulance transfer, which left no time to find Mrs. G.'s AD and to send it with the paramedics. By the time the family arrives separately at the hospital, the emergency room doctors have already transferred Mrs. G. to the ICU for mechanical ventilation, a treatment she probably would not want. The elder daughter is dismayed. She does not understand that a full cardiorespiratory arrest like Mrs. G.'s requires immediate ICU admission, mechanical ventilation, and cardiorespiratory monitoring to maximize survival and functional outcome. Deep emotional ties to Mrs. G. may also complicate the family's decision-making. Any early, emotionally based hope for her to survive, for example, may conflict with an equally emotionally based hope for her not to suffer needlessly from treatment.

Two hindering factors from *proxy–health professional relations and communications* exist in Mrs. G.'s case as they do in much of today's American healthcare system: too many HPs involved in a patient's care, and poor communication among them. The G. family encounters numerous specialist doctors in Mrs. G.'s case. They include the emergency room doctors, a pulmonologist, a neurologist, two cardiologists, a hospital internist, a doctor-ethicist, and a hospice specialist. While valuing the doctors' expertise, the G. family become confused and frustrated when the many doctors work at cross-purposes to each other. Fragmentation of care often lies at the root of the problem. The specialist doctors typically finish their particular care tasks and then leave the case. The resulting discontinuities of care cause spotty or flawed communications among the doctors so that any one doctor often does not know what other doctors think or do.

The problems of fragmentation and poor communication create two particular controversies in Mrs. G.'s case. Soon after her admission the pulmonologist, the neurologist, and the first cardiologist each interpret Mrs. G.'s condition quite differently to her family. The pulmonologist expresses a good prognosis (for being weaned from the ventilator); the neurologist, a guarded prognosis (for not being left persistently vegetative); and the first cardiologist, a poor prognosis (for not being able to survive the hospitalization). The conflicting prognoses confuse the family. They do not know what or whom to believe. Likewise, when Mrs. G. weans successfully from the ventilator, the pulmonologist convinces the family never to ventilate her again and to sign such an order. However, because the second cardiologist apparently does not know about the order, he later urges the family to authorize ventilation again. The cardiologist never speaks directly with the pulmonologist because the pulmonologist has already left the case. The family, therefore, do not know to whom to turn for advice. Proxy decision-making and continuity of care surely suffer.

I believe Mrs. G. is irreversibly dying by the time the reventilation question arises. I do not believe that ventilating her again and treating whatever heart arrhythmia she has can save her for long. If they would, she would be trapped in a quality of life too poor to satisfy her. I conclude, therefore, the family's decision for hospice care instead of reventilation suits her best. That decision by constructed judgment is right for her.

Steps Health Professionals Can Take to Ease Proxy Stress in Making Decisions

As Mrs. G.'s case illustrates, many factors may help or hinder proxy decision-making. But HPs can support proxies through that task by promoting the helping factors and minimizing the hindering factors as possible. Some hindering factors, of course, cannot be changed: HPs may only be able to acknowledge them to proxies and to encourage good coping strategies.

Nonetheless, Vig and colleagues suggest 10 steps HPs *can* take to help proxies through decision-making [12]. First, HPs should encourage ongoing self-care by proxies. Proxies should protect their quiet time and participate in regular, restorative leisure activities such as hobbies. Second, HPs should urge proxies to recognize their important duties at home or work, order them by priority, and attend to them accordingly. Proxies should not allow decision-making for the patient to disrupt completely their other essential activities. Third, HPs should be readily available to proxies to listen and to give emotional support. Fourth, HPs should minimize the number of HP communicators to patient, proxies, and family. One designated HP communicator is ideal but unrealistic in most cases. Two or three communicators are more practical, but they must coordinate information among themselves and communicate it to other HPs to avoid confusing proxies with inadvertent inconsistencies. The communicators should ordinarily include the patient's principal attending doctor and at least one nurse. Fifth, HPs should invite proxies into all important care deliberations, encourage proxies to express their views, and consider those views respectfully.

Sixth, HPs should encourage proxies to consult trusted outsiders about particularly difficult decisions. Medically trained family members, clergy, and other patients in similar circumstances may have useful insights to contribute. As appropriate, HPs might offer in-house referrals to medical specialists, chaplains, or social workers. HPs might also suggest proxies discuss any outside opinions with their HP communicators. Seventh, HPs should promote consensus among all parties who have either a formal or informal role in making decisions for the patient. Eighth, HPs should encourage proxies to bring up for consideration as early as possible any potential future problems such as financial shortfalls or intrafamilial rifts. Ninth, HPs should remind proxies as necessary that HPs attending the patient want to support the proxies and to share with them the responsibility for care decisions. Proxies should not feel they must shoulder all the decision-making responsibility by themselves. And tenth, HPs should always show commitment to providing incapacitated, dying patients the best care possible.

Summary Points

1. The three standard proxy decision-making methods now in common use aim to match care decisions to patients' exact wishes or interests. I believe that aim is virtually impossible to achieve due in part to proxies' consistently inaccurate guesses of those wishes and interests. I, therefore, recommend instead the constructed judgment method, which aims to find care decisions that best "fit" patients' traits, values, experiences, and life stories.
2. Health professionals can ease proxies' stress in decision-making by providing timely, understandable medical information; respecting proxies' opinions; sharing with proxies the responsibility for decisions; and providing reassurances about good decisions afterward.
3. Healthcare professionals must know the differences between the persistent vegetative state and death by brain criteria, and be able to explain those differences to proxies in lay language.

To Learn More …

1. Buchanan A, Brock DW. Deciding for others. Milbank Mem F Q. 1986;64 (suppl. 2):17–94.
2. Sulmasy DP, Snyder L. Substituted interests and best judgments: an integrated model of surrogate decision making. JAMA. 2010;304(17):1946–7.

3. Vig EK, Starks H, Taylor JS, Hopley EK, Fryer-Edwards K. Surviving surrogate decision-making: What helps and hampers the experience of making medical decisions for others. J Gen Intern Med. 2007;22(9):1274–9.
4. Yarborough M. Deciding for others at the end of life: storytelling and moral agency. J Clin Ethic. 2005;16(2):127–43.
5. Cranford RE, Smith HL. Some critical distinctions between brain death and the persistent vegetative state. Ethics Sci Med. 1979;6(nn):199–209.

References

1. Smucker WD, Houts DM, Danks JH, et al. Modal preferences predict elderly patients' life-sustaining treatment choices as well as patients' chosen surrogates do. Med Decis Making. 2000;20(3):271–80.
2. Silveira MJ, Kim SYH, Langa KM. Advance directives and outcomes of surrogate decision making before death. N Engl J Med. 2010;362(13):1211–8.
3. Vig EK, Taylor JS, Starks H, Hopley EK, Fryer-Edwards K. Beyond substituted judgment: how surrogates navigate end-of-life decision-making. J Am Geriatr Soc. 2006;54(11):1688–93.
4. Moorman SM, Carr D. Spouses' effectiveness as end-of-life health care surrogates: accuracy, uncertainty and errors of overtreatment or undertreatment. Gerontologist. 2008;48(6):811–9.
5. Tilden VP, Tolle SW, Nelson CA, Fields J. Family decision-making to withdraw life-sustatining treatments from hospitalized patients. Nurs Res. 2001;50(3):105–15.
6. Puchalski CM, Zhong Z, Jacobs MM, et al. Patients who want their family and physician to make resuscitation decisions for them: observations from SUPPORT and HELP. J Am Geriatr Soc. 2000;48(5 suppl.):S84–90.
7. Bloche MG. Managing conflict at the end of life. N Engl J Med. 2005;352(23):2371–3.
8. Berger JT, DeRenzo EG, Schwartz J. Surrogate decision-making: reconciling ethical theory and clinical practice. Ann Intern Med. 2008;149(1):48–53.
9. Brudney D. Choosing for another: beyond autonomy and best interests. Hast Cen Rep. 2009;39(2):31–7.
10. Perkins HS. Controlling death: the false promise of advance directives. Ann Intern Med. 2007;147(1):51–7.
11. Sulmasy DP, Terry PB, Weisman CS, et al. The accuracy of substituted judgments in patients with terminal diagnoses. Ann Intern Med. 1998;128(8):621–9.
12. Vig EK, Starks H, Taylor JS, Hopley EK, Fryer-Edwards K. Surviving surrogate decision-making: What helps and hampers the experience of making medical decisions for others. J Gen Intern Med. 2007;22(9):1274–9.
13. Swigart V, Lidz C, Butterworth V, Arnold R. Letting go: family willingness to forgo life support. Heart Lung. 1996;25(6):483–94.
14. Buchanan A, Brock DW. Deciding for others. Milbank Mem Fund Q. 1986;64(suppl. 2):17–94.
15. Cranford RE, Smith HL. Some critical distinctions between brain death and the persistent vegetative state. Ethics Sci Med. 1979;6(4):199–209.
16. The Task Force on Death and Dying of the Institute of Society, Ethics, and the Life Sciences. Refinements in criteria for the determination of death: an appraisal. JAMA. 1972;221(1):48–53.
17. A definition of irreversible coma: report of the Ad Hoc Committee of the Harvard Medical School to Examine the Definition of Brain Death. JAMA. 1968;205(6):337–40.

18. The President's Commission for the Study of Ethical Problems in Medicine and Biomedical and Behavioral Research. Defining death: medical, legal, and ethical issues in the determination of death. Washington, D.C.: U. S. Government Printing Office, 1981.
19. Booth CM, Boone RH, Tomlinson G, Detsky AS. Is this patient dead, vegetative or severely neurologically impaired? Assessing outcome for comatose survivors of cardiac arrest. JAMA. 2004;291(7):870–9.
20. Easton DJ. Coma and related disorders. In: Stein JH. Internal medicine. St. Louis: Mosby; 1994. P. 1015.
21. Zivin JA. Approach to cerebrovascular diseases. In: Goldman L, Ausiello D. Cecil textbook of medicine. New York: Saunders; 2004. P. 2285.
22. Council on Scientific Affairs and Council on Ethical and Judicial Affairs, American Medical Association. Persistent vegetative state and the decision to withdraw or withhold life support. JAMA. 1990;263(3):426–30.
23. Simon RP. Persistent vegetative states and brain death. In: Goldman L, Ausiello D, editors. Cecil textbook of medicine. New York: Saunders; 2004. p. 2276–7.
24. Council on Ethics and Judicial Affairs and Council on Scientific Affairs, American Medical Association. Current opinions, section 2.20. Chicago IL: American Medical Association; 1989.
25. Bernat JL, Culver CM, Gert B. On the definition and criterion of death. Ann Intern Med. 1981;94(3):389–94.
26. Younger SJ, Bartlett ET. Human death and high technology: the failure of the whole-brain formulations. Ann Intern Med. 1983;99(2):252–8.
27. Emanuel LL. Reexamining death: the asymptotic model and a bounded zone definition. Hast Cen Rep. 1995;25(4):27–35.
28. Payne K, Taylor RM, Stocking C, Sachs GA. Physicians' attitudes about the care of patients in the persistent vegetative state: a national survey. Ann Intern Med. 1996;125(2):104–10.
29. Hines SC, Glover JJ, Babrow AS, et al. Improving advance care planning by accommodating family preferences. J Palliat Care. 2001;4(4):481–9.
30. Hornung CA, Eleazer GP, Strothers HS, et al. Ethnicity and decision-makers in a group of frail older people. J Am Geriatr Soc. 1998;46(3):280–6.
31. Kelly B, Rid A, Wendler D. Systematic review: individuals' goals for surrogate decision-making. J Am Geriatr Soc. 2012;60(5):884–95.
32. The SUPPORT Principal Investigators. A controlled trial to improve care for seriously ill hospitalized patients. The study to understand prognoses and preferences for outcomes and risks of treatments (SUPPORT). JAMA. 1995;274(20):1591–8.
33. High DM. Families' roles in advance directives. Hast Cen Rep. 1994;24(6):S16–8.
34. Emanuel EJ, Emanuel LL. Proxy decision making for incompetent patients: an ethical and empirical analysis. JAMA. 1992;267(15):2067–71.
35. Elliott BA, Gessert CE, Peden-McAlpine C. Decision making by families of older adults with advanced cognitive impairment: spirituality and meaning. J Geront Nurs. 2007;33(8):49–55.
36. Lynn J. Why I don't have a living will. Law Med Health Care. 1991;19(1–2):101–4.
37. Hackler JC, Hiller FC. Family consent to orders not to resuscitate: reconsidering hospital policy. JAMA. 1990;264(10):1281–3.
38. Nelson JL. Taking families seriously. Hast Cen Rep. 1992;22(4):6–12.
39. Capron AM. Where is the sure interpreter? Hast Cen Rep. 1992;22(4):26–7.
40. Winzelberg GS, Hanson LC, Tulsky JA. Beyond autonomy: diversifying end-of-life decision-making approaches to serve patients and families. J Am Geriatr Soc. 2005;53 (6):1046–50.
41. Yarborough M. Deciding for others at the end of life: storytelling and moral agency. J Clin Ethic. 2005;16(2):127–43.
42. Suhl J, Simons P, Reedy T, Garrick T. Myth of substituted judgment. Arch Intern Med. 1994;154(1):90–6.

43. Houts RM, Smucker WD, Jacobson JA, Ditto PH, Danks JH. Predicting elderly outpatients' life-sustaining treatment preferences over time: the majority rules. Med Decis Making. 2002;22(1):39–52.
44. Shalowitz DI, Garrett-Mayer E, Wendler D. The accuracy of surrogate decision makers. Arch Intern Med. 2006;166(5):493–7.
45. Berger JT. Patients' interests in their family members' well-being: an overlooked, fundamental consideration within substituted judgments. J Clin Ethics. 2005;16(1):3–10.
46. Sulmasy DP, Snyder L. Substituted interests and best judgments: an integrated model of surrogate decision making. JAMA. 2010;304(17):1946–7.
47. Braun UK, Naik AD, McCullough LB. Reconceptualizing the experience of surrogate decision making: reports vs genuine decisions. Ann Fam Med. 2009;7(3):249–53.
48. Barnato AE. Arnold RM. The effect of emotion and physician communication behavors on surrogates' life-sustaining treatment decisions: a randomized simulation experiment. Crit Care Med. 2013;41(7):1686–91.
49. Potter RL. The hermeneutical process and clinical ethics. J Clin Ethics. 1995;6(1):88–91.
50. Emanuel EJ. What criteria should guide decision makers for incompetent patients? Lancet. 1988;1(8578, 23 Jan.); 170–1.
51. Tomlinson T, Howe K, Notman M, Rossmiller D. An empirical study of proxy consent for elderly persons. Gerontologist. 1990;30(1):54–64.
52. Rid A, Wendler D. Can we improve treatment decision-making for incapacitated patients? Hast Cen Rep. 2010;40(5):36–45.
53. Fagerlin A, Ditto PH, Danks JH, Houts RM, Smucker WD. Projection in surrogate decisions about life-sustaining medical treatments. Health Psych. 2001;20(3):166–75.
54. Pruchno RA, Lemay EP, Feild L, Levinsky NG. Spouse as health care proxy for dialysis patients: Whose preferences matter? Gerontologist. 2005;45(6):812–9.
55. Mezey M, Kluger M, Maislin G, Mittleman M. Life-sustaining treatment decisions by spouses of patients with Alzheimer's disease. J Am Geriatr Soc. 1996;44(2):144–50.
56. Uhlmann RF, Pearlman RA, Cain KC. Physicians' and spouses' predictions of elderly patients' resuscitation preferences. J Geront (Med Sci). 1988;43(5):M115–21.
57. Seckler AB, Meier DE, Mulvihill M, Paris BE. Substituted judgment: how accurate are predictions? Ann Intern Med. 1991;115(2):92–8.
58. Hare J, Pratt C, Nelson C. Agreement between patients and their self-selected surrogates on difficult medical decisions. Arch Intern Med. 1992;152(5):1049–54.
59. Sulmasy DP, Haller K, Terry PB. More talk, less paper: predicting the accuracy of substituted judgments. Am J Med. 1994;96(5):432–8.
60. Landis JR, Koch GG. The measurement of observer agreement for categorical data. Biometrics. 1977;33(1):159–74.
61. Shalowitz DI, Garrett-Mayer E, Wendler D. How should treatment decisions be made for incapacitated patients and why? PLoS Med. 2007;4(3):e35. Published 13 March 2007. doi:10.1371/journal.pmed.0040035.
62. Varm S, Wendler D. Medical decision making for patients without surrogates. Arch Intern Med. 2007;167(16):1711–5.
63. Wikler D. Patient interests: clinical implications of philosophical distinctions. J Am Geriatr Soc. 1988;36(10):951–8.
64. Venkat A. Surrogate medical decision making on behalf of a never-competent, profoundly intellectually disabled patient who is acutely ill. J Clin Ethics. 2012;23(1):71–8.
65. Spike JP. Care versus treatment at the end of life for profoundly disabled persons. J Clin Ethics. 2012;23(1):79–83.
66. Jecker NS. The role of intimate others in medical decision making. Gerontologist. 1990;30(1):65–71.
67. Yarborough M. Continued treatment of the fatally ill for the benefit of others. J Am Geriatr Soc. 1988;36(1):63–7.
68. Breslin JM. Autonomy and the role of the family in making decisions at the end of life. J Clin Ethics. 2005;16(1):11–9.

69. Hardwig J. The problem of proxies with interests of their own: toward a better theory of proxy decisions. J Clin Ethics. 1993;4(1):20–7.
70. Torke AM, Petronio S, Purnell CE, et al. Communicating with clinicians: the experiences of surrogate decision-makers for hospitalized older adults. J Am Geriatr Soc. 2012;60(8):1401–7.
71. Boyd EA, Lo B, Evans LR, et al. "It's not just what the doctor tells me": factors that influence surrogate decision-makers' perceptions of prognosis. Crit Care Med. 2010;38(5):1270–5.
72. Lo B, Rouse F, Dornbrand L. Family decision making on trial: Who decides for incompetent patients? N Engl J Med. 1990;332(17):1228–32.
73. Strong C. Patients should not always come first in treatment decisions. J Clin Ethics. 1993;4 (1):63–7.
74. Degner LF, Sloan JA. Decision making during serious illness: What role do patients really want to play? J Clin Epidemiol. 1992;45(9):941–50.
75. Deber RB, Kraetschmer N, Irvine J. What role do patients wish to play in treatment decision making? Arch Intern Med. 1996;156(13):1414–20.
76. Mazur DJ, Hickam DH. Patients' preferences for risk disclosure and role in decision making for invasive procedures. J Gen Intern Med. 1997;12(2):114–7.
77. Kraetschmer N, Sharpe N, Urowitz S, Deber RB. How does trust affect patient preferences for participation in decision-making? Health Expect. 2004;7(4):317–26.
78. Saba GW, Wong ST, Schillinger D, et al. Shared decision making and the experience of partnership in primary care. Ann Fam Med. 2006;4(1):54–62.
79. Lynn J. Procedures for making medical decisions for incompetent adults. JAMA. 1992;267 (15):2082–4.
80. Tilden VP, Tolle SW, Nelson CA, Thompson M, Eggman SC. Family decision making in foregoing life-extending treatments. J Fam Nurs. 1999;5(4):426–42.

Chapter 8
The Impact of Terminal Illness on the Family

When a patient becomes terminal, his or her care assumes a kind of regularity. Almost no new diagnostic questions arise. Additional history, physical examination, and laboratory tests contribute little. And changes in treatment are rare. Nursing care predominates over medical care. Yet this sad phase affects more than just the patient: it also affects close family members. They face their own emotional traumas prompted by the patient's decline toward death. As his or her care becomes largely set, the attention of health professionals (HPs) turns more and more to the family and the effect of the patient's impending death on them [1, 2].

Hospice philosophy rightfully supports attention to the family as a key part of quality end-of-life care. But that attention is not merely an indirect means to provide benefits for dying patients. The few studies about such benefits show them to be few in number and small in magnitude [3]. Instead, this attention aims to provide direct benefits for family survivors themselves. The rationale behind the effort lies in family systems theory, which sees a family as an interdependent social network [4]. A crisis for one in this network is a crisis for all [4–6]. Thus, family members automatically become "additional patients" in any end-of-life scenario.

Many studies confirm the impact of a patient's serious or terminal illness on close family members. One study of patients cared for at home and likely to die within six months documented that impact quantitatively. Over one-third of families shouldered heavy caregiving burdens, nearly one-third lost all or most of their savings, almost as many lost a main source of income, and one-sixth had to change important life plans because of the patient's illness. Overall, more than half of families suffered at least one of these major traumas [7]. Though not so clearly documented, other severe traumas undoubtedly occur in the long-term psychological, social, and spiritual realms as families try to adjust to a future without the patient. For these reasons the American College of Critical Care Medicine has concluded, "Support of the psychological and spiritual health of the family is … essential (to) … care for the critically ill (patient)." [8].

H.S. Perkins, *A Guide to Psychosocial and Spiritual Care at the End of Life*,
DOI 10.1007/978-1-4939-6804-6_8

I likewise believe that HPs attending a dying patient must supportively address the effects of the patient's illness on the family. As HPs work to give the patient a comfortable, respectful death, they must also try to ease its traumas for family survivors. Such attention surely helps the family achieve some acceptance of the death, find meaning in it, face it bravely, and begin to reconfigure their lives for afterward.

This chapter, therefore, addresses HP attention to the needs of the family. (Following prior precedents, I use the term "family" to include anyone having a significant emotional relationship with the patient and sharing mutual support with him or her. Under this definition some people considered here as "family" may not be officially related to the patient by blood or marriage [8].) Accordingly, both this chapter and the next one focus on attention to the dying patient's family but take up different aspects of that attention. This chapter focuses on general care for the family of patients dying in healthcare institutions (including hospitals, nursing homes, and inpatient hospices) where most Americans still die. The next chapter focuses instead on specific care for family caregivers of patients dying at home. The different focuses make for the different contents.

This chapter begins by outlining the three phases of a terminal cancer. Each phase forces the family to adapt to distinctive conditions. The chapter next describes emotional reactions, stressors, and stress relievers the family may experience during a patient's terminal illness. The chapter then discusses the distinctive intensive care unit environment where many dying patients still find themselves late in their illnesses. The chapter also suggests interventions HPs in any institutional healthcare setting might try to help family members endure a patient's terminal decline. Much of the emphasis here falls on strong, supportive HP-family communications. Finally, the chapter addresses the question, Do families expect too much from end-of-life care? Throughout this chapter the case of a woman dying of brain cancer illustrates many of the points.

The Case

Mrs. M. has been a devoted family woman, a capable homemaker, and an accomplished amateur athlete all her adult life. But at 79 she begins to have unusual fatigue, persistent leg weakness, and occasional confusion that interfere with her everyday activities. Within months she needs a walker to move around, loses weight, and can no longer balance the household checkbook. Mr. M., her husband, takes her to several neurologists, but none can determine a diagnosis. They all say it must "declare itself" with time.

Mrs. M.'s function continues to decline over the subsequent months. She loses the ability to walk short distances even with the walker. Mr. M. must transport her in a wheelchair. She often skips dinners at the retirement center where Mr. and Mrs. M. live. Going to the common dining room simply requires more energy than she can muster. Perhaps most alarming to her and the family, she begins to misidentify her children and grandchildren and to forget the major events in their lives.

One day Mr. M. returns from golf to find Mrs. M. paralyzed in her legs and too weak to sit up. He calls an ambulance to take her to the local hospital where she is admitted for yet another diagnostic workup. The doctors suspect a stroke at first, but the early tests are inconclusive. After two weeks a brain scan, a special X-ray of the head, finally reveals Mrs. M.'s diagnosis: a brain tumor. The attending hospitalist, a senior doctor employed full time on the hospital's wards, informs Mrs. M. and her family. The diagnosis plunges the whole family into a crisis, which lasts through Mrs. M.'s death six months later. Along the way Mr. M. and the five adult M. children experience many of the stressors and stress relievers common when a loved one is dying.

The Family Experience of the Phases of a Terminal Cancer

As Chap. 5 explains, few Americans die from sudden, unexpected causes; most die from previously diagnosed, chronic, and eventually fatal diseases [9]. Two kinds of these diseases, the organ system failures (such as heart failure and severe emphysema) and the progressive frailties (such as Alzheimer's dementia), vary widely in their courses. Yet the third kind, the cancers, follow remarkably consistent courses with a predictable sequence of phases. Mrs. M.'s cancer certainly does so. At each phase of such a fatal cancer, the family must adapt to a new reality concerning the patient's illness. This reality thrusts on them new decisions, new tasks, and new visions of the future [10]. Identifying these phases explicitly as they occur can help the family understand the course of a cancer, anticipate the future, and plan for it.

Table 8.1 displays side by side two systems for classifying the phases of a cancer's progression [6, 11]. I show both systems because both have benefits. The Rait/Lederberg system, the less detailed one, has three phases [11]. The diagnostic phase involves changes in the patient that lead to a cancer diagnosis. The ameliorative phase involves treatment to ease the cancer's symptoms or to stall its progression. And the terminal phase involves waning control over the cancer despite treatment, changes in the main treatment goals from prolonging survival to enhancing comfort and meaning, and facing the patient's approaching death. In contrast, the McCorkle/Pasacreta system, the more detailed one, uses seven phases [6]. It divides the first two Rait/Lederberg phases into two phases each and the last one into three phases. While sharing many of the same meanings, the McCorkle/Pasacreta system characterizes cancer progression more specifically than does the Rait/Lederberg system. Nonetheless, for simplicity's sake, I use the Rait/Lederberg system to structure the following discussion of cancer progression.

Table 8.1 Phases of a terminal cancer

Rait/Lederberg system[a]		McCorkle/Pasacreta system[b]	
Phases	Description	Phases	Description
Diagnostic	Abnormal symptoms and signs in the patient lead to a cancer diagnosis.	Initial	Abnormal symptoms and signs appear.
		Critical	A medical crisis arises, leading to a cancer diagnosis.
Ameliorative	Treatment improves the cancer or its effects.	Therapeutic	Treatment begins.
		Stable	Treatment eases the cancer's symptoms or stalls its progression.
Terminal	Treatment no longer controls the cancer. The main treatment goals change from prolonging survival to enhancing comfort and meaning in remaining life. Both patient and family face the approaching death.	Unstable	Treatment no longer stalls the cancer's progression.
		Downward	Steady physical and/or mental deterioration follows.
		Imminently dying	Death occurs within hours, days, or weeks.

[a]Data from the original construct of D Rait and M Lederberg as described in Ref. [11]
[b]Data from Ref. [6]

The Diagnostic Phase

This phase begins when significant, persistent changes in the patient prompt a medical evaluation and a doctor then diagnoses a cancer. Many families react to the diagnosis with shock or disbelief. But some may also feel relief. This relief stems from having identified a definite diagnosis, eliminated prior uncertainties, and been reassured that medicine has encountered the problem before. Just putting a diagnostic name to the problem suggests controlling, or at least containing, it. Establishing a diagnosis may also bring some conceptual order to the situation by making sense of the patient's prior symptoms.

Along with the family's other reactions anxiety is sure to accompany the new reality of the cancer. It signals a different, more menacing future than either patient or family previously anticipated. The patient's death suddenly feels closer and more real than before [12]. The family often react by reflexively pledging all available resources—physical, emotional, medical, and financial—to battling the cancer, but the family cannot fully appreciate the trials ahead. Patient and family alike may sometimes experience a mental "fog" that hinders clear thinking or decisive action at the time. Upheaval may also occur in roles, relationships, and responsibilities until a new family normality settles in. All these changes may cause family members bewilderment, fear, or even anger [12, 13].

Family members then search for effective strategies for coping with the crisis. Education and interpersonal contacts are usually two of them. Prompted by unfamiliar medical concepts and practices, family members may conduct their own research into the patient's disease. They may even seek second opinions, formal or informal. Family members may also seek emotional support from other people including other family members, friends, clergy, other patients with the disease, or knowledgeable acquaintances.

The Case

The outpatient workup yielding no diagnosis for months frustrates Mrs. M. and her family as her disability steadily increases. Her progressive leg weakness, for example, makes her go from needing a walker to walk, to needing a wheelchair to move about, to needing help just to sit up. As a former athlete, Mrs. M. becomes especially discouraged. After yet another unproductive doctor visit late on, Mrs. M. exclaims in frustration, "The doctors don't realize what (abilities) I've already lost!"

When finally unable to sit up on her own, Mrs. M. is admitted to the hospital for an intensive, thorough diagnostic workup. The attending hospitalist comes to perform her admission history and physical examination and asks Mrs. M. why she has come. Mrs. M. responds out of her discouragement, "I've come here to die. Everybody lives too long, and it's now past my time. I just want to die." The hospitalist assesses her medically and begins diagnostic tests. The tests, however, do not yield a diagnosis right away. But the hospitalist visits Mrs. M. daily and talks frequently with her family. He assures them the doctors are conscientiously pursuing a diagnosis, and the lack of one to date frustrates them as well as the family. After a week Mr. M. begins to wonder whether the doctors already know Mrs. M.'s diagnosis and are just not disclosing it. He says impatiently, "The doctors still have not told us what is wrong after all this time."

The brain scan finally reveals the diagnosis, a brain cancer, and the hospitalist tells Mrs. M. and her family. He uses many good techniques for breaking bad news. He gathers Mrs. M. and her family in the family waiting room when it is vacant, brings tissues, and sits down next to Mr. and Mrs. M. The hospitalist then discloses the diagnosis sensitively in clear, everyday language. Mrs. M. and some other family members begin to cry quietly. Mr. M. remains stoic as he comforts Mrs. M. The hospitalist waits patiently for family members to speak again. One daughter eventually says, "At least we know now. That is something (helpful)." Mr. M. then asks the doctor, "How long does she have to live?" The hospitalist says treatments do exist. They delay the cancer's progression but do not provide a cure. He estimates survival in terms of months to a year. He then adds empathetically, "I know this news is difficult for you to hear. You may need some time to take it in. You might research brain tumors at medical Web sites or get a second opinion if you wish. You might also want to discuss the situation with people whom you trust—say, other family members, your family doctor, or your priest. Those discussions may help you think clearly about what to do next."

The hospitalist then explains he is not a specialist in treating these tumors, but the hospital employs such specialists. He promises to consult several including a neurosurgeon, an oncologist, and a radiation therapist. He anticipates all three will examine Mrs. M. and make their treatment recommendations within a day or two. He promises to meet again with Mrs. M. and the family after that to help them make treatment choices.

Before leaving, the hospitalist gives the family his contact telephone number and invites them to call with questions at any time. (Although some doctors worry that patients or families will abuse a doctor's contact telephone number by calling too much, I find most people do not call as much as they should. They do not want to "bother" the doctor off-hours.) The hospitalist also offers to meet later with other key relatives (such as Mrs. M.'s other adult children who live far away and could not attend this family conference). And he continues to support Mrs. M. and her family with daily visits on subsequent days. His accessible, honest, dedicated approach quickly wins the family's admiration.

The diagnosis of Mrs. M.'s brain cancer stuns the family for its implications. The diagnosis means Mrs. M. will live much less time than they expected. The family also realize their usual functioning as a family must change. As the matriarch, Mrs. M. performed many functions that maintained family cohesion. She made critical family decisions with Mr. M. and saw reliably to his and others' needs. Mr. M. and the children never anticipated Mrs. M.'s being unable to perform those functions. But her increasing disability and approaching death will prevent her from doing so. She must relinquish her past caretaking role for a future care-receiving one. Others must take over her usual family duties.

The Ameliorative Phase

The word *ameliorative* means to make better or more tolerable. It refers here to reducing the "violence" of a cancer. The word covers treatments aimed at curing a cancer and treatments aimed only at easing its symptoms, shrinking its size, or slowing its progression. [Cure, of course, is more likely with some cancers (such as the skin- or blood-based cancers or the hormonally controlled cancers) than others (such as the cancers of the internal solid organs).] The ameliorative phase begins with the first treatments against the cancer.

Symptom relief is, of course, always a primary aim of treatment during this phase. Cancer pain usually yields well to treatment through meticulous titration of analgesics, but other common cancer symptoms such as fatigue, loss of appetite, or shortness of breath respond less well. Depression is common and can be serious. It deserves vigorous treatment. Other mental conditions, however, are rare [14, 15]. If depression is not present, a cancer patient's alertness, sociability, interest in the outside world, and hopefulness may extend throughout the illness [14].

If treatment achieves significant amelioration, the cancer may enter a relatively stable, arrested phase for weeks, months, or longer. Both patient and family settle into a routine. Monitoring and treatment continue. The family's distress subsides [16]. Still, the family may have to arrange for the patient's day-to-day care. Much of it takes place at home. One family member typically assumes major responsibility for that care, either performing it him- or herself or overseeing workers hired in to perform it. Other family members often assume only minor responsibility. Over time the principal family caregiver can become overburdened with all the patient's care requirements. They can easily exhaust the caregiver physically and mentally. They may eventually prompt the patient's permanent transfer to a nursing home or inpatient hospice [17, 18]. Curiously, such a transfer reduces only a little the burdens family caregivers feel. The reason may involve caregivers' continuing worry about the patient's condition, or the caregivers' ongoing tasks that healthcare institutions do not perform. Those tasks may include scheduling doctor appointments; arranging for supportive visits from family, friends, clergy, and others; renewing prescriptions; or paying medical bills. The next chapter discusses in detail typical caregiver tasks and reactions.

The Case

After disclosing Mrs. M.'s diagnosis, the hospitalist requests consultations from a neurosurgeon, an oncologist, and a radiation therapist. The hospitalist reassembles Mrs. M. and her immediate family after the consultants make their assessments. He mentions the three possible treatment options—surgery, chemotherapy, and radiation—and summarizes the consultants' opinions about them. The neurosurgeon believes the tumor is too strategically placed in Mrs. M.'s brain to permit surgery. He cannot resect all the tumor without doing severe, permanent neurologic harm to her. The oncologist concedes that chemotherapy offers great toxicity and no benefit. And the radiation therapist says radiation could shrink, but not eliminate, the tumor. Still, shrinkage would give Mrs. M. some comfort and functional improvement for up to four months.

The hospitalist acknowledges to Mrs. M. and her family the difficulty of the treatment decision. He reemphasizes that the tumor will eventually regrow with any of the three treatments, and that the side effects of some treatments (surgery and chemotherapy in particular) may be worse than the disease. He also repeats his prediction, now based on confirmatory opinions from the specialists, that Mrs. M. has only months to live regardless of her treatment. The hospitalist then invites questions from the family and says finally that, if Mrs. M. were his mother, he would prefer radiation for her. He then leaves Mrs. M. and her family time in private to discuss their options and to make a choice.

When the hospitalist returns about 30 min later, Mrs. M. has made her choice. "I don't want to prolong this (suffering)," she says. "I'll take the radiation." To make sure she understands the implications of her decision, the hospitalist repeats that the radiation provides only temporary improvement in symptoms and function. "I understand that," replies Mrs. M. soberly. "I have already lived too long. I am ready to die and only want to be kept comfortable now." The hospitalist reassures her that radiation and follow-up hospice are her best options in that case. The rest of the family agree. The hospitalist then commits to the overall goal of comfort care and promises the family he will not abandon them: He will visit regularly as long as Mrs. M. remains in the hospital. Impressed with his demeanor, the elder daughter says afterward, “He is really on our side.”

Mrs. M. begins her radiation treatments immediately. Her headaches recede, and her sense of humor returns. She also regains her ability to feed herself and to stand with assistance. She soon can be discharged from the hospital. She and her family decide that due to her ongoing care requirements she should go to the rehabilitation ward at the retirement center rather than to the M.s’ independent-living apartment. Her improvement continues after her transfer to the rehabilitation ward. She regains the ability to transfer into and out of bed and to walk a short distance. She feels so good at one point that she asks to return to the M.s’ independent-living apartment to catch up on her housework! Mr. M. wisely dissuades her. The improvement, however, plateaus after about two months. Some disabilities remain.

To feel some control over events, Mr. M. starts to monitor the details of Mrs. M.’s day-to-day condition. He takes notes about her physical and mental changes, interactions with the ward staff, and food intake. He reports his observations to Mrs. M.’s doctors. He also stays at her bedside long hours each day, coming early to help her get up for the day and staying late to talk with her. He often just sits there, holding her hand as she sleeps. He also takes her out occasionally to ride in the car or to visit the retirement center’s patio garden.

The M.s’ adult children also take on significant care responsibilities for Mrs. M. and the rest of the family. The elder daughter, who lives closer than the other children, assumes many of Mrs. M.’s former matriarchal duties. This daughter coordinates visits from out-of-town siblings and boards them at her home. She also organizes family celebrations at holidays. Perhaps most importantly, she provides much of the emotional support for Mr. M. by sitting with him at Mrs. M.’s bedside for several hours each afternoon. She also eats frequent dinners with him to ensure he gets good nutrition, and she takes him to local cultural events as a diversion. The other adult children also play their parts. Two sons who work in business, for example, help Mr. M. plan how to pay for Mrs. M.’s nursing care.

The Terminal Phase

This phase signifies the time when no treatment works in fighting against the patient’s cancer. Symptoms increase. Function decreases. The cancer progresses relentlessly onward to the patient’s death [19]. Hope for cure fades.

While easy in concept, pinpointing the start of this phase is difficult in practice. That time is often clear only in retrospect. But practical matters—medical, financial, and others—often require some prospective estimate of it. Different methods can make those estimates but produce different results. Cancer patients in one study said the terminal phase starts when a doctor declares the patient terminal [20]. By that definition the terminal phase has a median duration of 69 days. One expert, however, has said the terminal phase starts when doctors end treatment aimed at cure. By that definition the phase has a median of 45 days [19]. And hospice patients in yet another study, 80 % of whom had cancer, said the terminal phase starts when hospice care begins [21]. By that definition the phase has a median duration of 36 days.

The different estimates suggest that some people may consider a cancer patient already terminal while others may simultaneously consider the patient not yet terminal. People may, therefore, reach different conclusions about treatment. People who believe the patient already terminal will tend to favor treatment only for comfort while people who believe the patient not yet terminal may favor continued aggressive treatment for cure or control of the cancer. Unfortunately, each camp typically cannot understand why the other thinks as it does. As a result, treatment decisions for the patient can become contentious and difficult to resolve.

Other factors complicate the estimates further. The duration of the terminal phase varies according to initial patient function: The better the patient functions at the start of this phase, the longer its duration [19]. Therefore, the medians cited before may only roughly approximate the duration of a particular patient's terminal phase. Furthermore, changes in a patient's symptoms, signs, and functional abilities may be subtle and fluctuate with time, making irreversible cancer progression difficult to discern at any time.

However people time it, the terminal phase marks a definitive shift in treatment goals from combating tumor progression to controlling symptoms and supporting function [19]. The whole phase may last up to years but usually lasts only weeks to months. It includes certain downward functional milestones. Still, the downward progression is never steady or exactly predictable. Variations from average occur frequently [22]. In general, patient awareness and functional abilities begin to decline at about two to three months before death. Many patients insist they still have much to live for at this time [14]. Symptoms begin to increase at about two to four weeks before death [15]. Complete relief of symptoms may not be possible then even with the best medical care. Finally, noticeable bodily changes mark the minutes surrounding death. Heartbeat and breathing slow and grow irregular even though the eyes may remain open. The skin becomes pale and waxy in appearance, and the muscles relax. The bowels and bladder may discharge. Body temperature then drops slowly to room temperature over the next several hours, marking the end of physiologic dying [22].

After rising in the diagnostic phase and falling in the ameliorative phase, family stresses rise again in the terminal phase [12]. The changes in the patient's symptoms and physical abilities—often unexpected and sometimes sudden—fuel the stresses. The family must make frequent adjustments in how they react. They often feel "off balance."

The stresses for the family in this phase may arise from physical or emotional sources in the cancer patient or from emotional sources in the family members themselves. The increase in the patient's symptoms increases the stresses for the family. They suffer vicariously by seeing the patient suffer physically. The increased symptoms may also serve as a constant, unwelcome reminder of the patient's approaching death [16, 23]. Meanwhile, the patient's function decreases, manifesting eventually as the loss of basic self-care abilities. The patient may lose the abilities to bathe, dress, feed, and toilet him- or herself. The care burden and its accompanying stresses increase concomitantly for family caregivers [12, 16].

Emotional changes in the patient may also stress the family, testing their patience and flexibility. These changes may make family members wonder whether they still really know the patient. For example, the patient may incur a depression, causing him or her to lose interest in personal relationships or previously pleasurable activities. The patient may become withdrawn or apathetic. The patient may no longer recognize the important cues or use the reciprocal pleasantries that facilitate normal social discourse. The patient may withdraw so much as to lose the ability to anticipate others' feelings or to perceive their kindnesses. He or she may no longer express thanks for favors, for example. The patient may also level hurtful accusations or make excessive demands on the family, making them doubt the adequacy of their caregiving. The patient may also openly express the wish to die, causing a disheartening effect on the family. Such emotional changes in the patient make relationships one-sided. The patient absorbs what the family gives without giving much back. The family may feel emotionally drained, abused, or taken for granted as a result.

The family's own emotions surely cause stress, too. The family may struggle to reconceive the patient's formerly healthier self as a newly sick and dying self. They may even cling tenaciously to the old self, trying to preserve it by repeating memories of the patient's earlier appearance and accomplishments. (Some reminiscing, of course, is healthy, and I encourage it through life reviews.) The family may also sense that the time with the patient is running out. It seems increasingly precious. Conflicted feelings result: The family may simultaneously feel gratitude for the time left with the patient and resentment at foreseeing its end coming soon. The family may sometimes go to great lengths to prolong that time [13, 24] by demanding overly aggressive life support beyond any benefit to the patient.

The Case

After several months' improvement Mrs. M.'s symptoms and functional abilities deteriorate again. Brief fluctuations occur at first, whipsawing the family. The improvements buoy them; the declines depress them. The declines soon predominate. Many become permanent. Mrs. M.'s headaches turn severe. The rehabilitation ward's analgesic regimens cannot control them. She also loses the ability to stand or walk and becomes confined to bed and wheelchair. All these changes indicate that, as the doctors predicted, her brain tumor is growing again. Mrs. M. has already received the maximum doses of radiation. Only analgesic treatments remain now. The rehabilitation ward doctor declares her terminal and recommends hospice for pain control. Mr. M. resists the hospice recommendation at first. He wants to believe Mrs. M. can still be cured. But he finally accepts the grim reality and concedes, "Hospice is the only option left." By all definitions Mrs. M. has entered her terminal phase.

Her decline accelerates, following almost exactly the typical terminal decline of a cancer. Her mood changes first. Mrs. M.'s personality becomes affected. She turns highly critical of Mr. M. and complains about everything from his pushing her too fast in the wheelchair to his ordering too much food or the wrong kinds of food for her meals. She even complains that he "never" visits her. Though nearly always unfounded, the constant complaints distress Mr. M. He begins to think he will never please her no matter how hard he tries. Mrs. M. also occasionally demands physician-assisted suicide. Mr. M. finds the idea especially discouraging for its implication that she wants to leave him. Physician-assisted suicide, however, is not possible because it is illegal in the state where the M.s live.

Her mental acuity deteriorates next. Mrs. M. forgets family members and facts she knew only a few weeks before. She cannot remember family even from the pictures on her ward room wall. And she shows ever less interest in leaving the room and sleeps ever more. A rapid physical decline follows. Physical therapy no longer helps her. She eventually can no longer sit up and becomes confined to bed alone. She eats little and loses more weight. Even hospice's intensive analgesics do not completely control her headaches.

The family watch these changes with dismay. Whenever the family notice changes in Mrs. M., they anxiously ask the ward staff, "Do you see what we see in her?" or "What do these changes mean?" Untrained to respond to such questions, the staff avoid answering. The resulting uncertainty fuels the family's anxieties. Meanwhile family members develop varying views on the situation. One daughter describes Mr. M. as "the hopeful one" but herself as "Debbie Downer," the pessimistic one. "She used to be such a vibrant woman," this daughter says. "It's hard to see her becoming so disabled."

The family must simultaneously do the hard emotional work to adjust to Mrs. M.'s rapid decline. They must relinquish their hope for cure and accept her approaching death. One son concedes sadly, "I don't think she will live much longer." The family can hope only for her comfort at the end. The family must also adjust to losing half of their long-term emotional support system, consisting of Mr. and Mrs. M. Their life as a family will change dramatically after Mrs. M. dies.

Furthermore, the family must cope with their own immediate distresses. They learn not to rely on the HPs for psychosocial guidance. Because the hospice staff supervise terminal care from afar, they have little direct contact with Mrs. M. or her family. And the nurse's aides, who conduct most of the direct patient care on the ward, have little training in family psychosocial issues at the end of life. The aides, therefore, avoid these awkward issues. The aides eventually seem to disappear completely as Mrs. M.'s death turns imminent. The family feel abandoned to their own grief. "We never see the aides anymore," complains one daughter. "So we have to support each other."

The frequent visits from immediate family members help, though. Mr. M., of course, spends nearly all day at Mrs. M.'s bedside. The elder daughter, who lives close by, leaves her job early to join him in the late afternoons and early evenings.

And each child from out of town visits about every second month. In addition, the family organizes one last reunion for Mrs. M. on a holiday weekend late in her illness course. All the extended family attend, and Mrs. M. appears to recognize them. But she responds little and sleeps in her wheelchair through most of the event.

As with most deaths after prolonged, fatal illnesses, predicting Mrs. M.'s time of death proves difficult. Mrs. M slips into a coma early in her decline. The rehabilitation ward staff believe she is imminently dying and summon the family for a bedside death watch. But tests soon reveal a bladder infection. Mrs. M. wakes up with antibiotics, asks why she is still alive, and becomes upset that she has not already died. The family go home again, baffled at the false alarm. After such a mishap the ward staff hesitate to predict Mrs. M.'s time of death too soon again.

However, when Mrs. M. turns suddenly much worse several weeks later, the supervisory ward nurse tells Mr. M. and the elder daughter that Mrs. M. will not live through the evening. Mr. M. and the daughter keep a bedside vigil until midnight. Still Mrs. M. remains alive. The nurse then suggests Mr. M. and the daughter go home to sleep for a few hours. They do and return early the next morning. Mrs. M. is breathing but unresponsive. Mr. M. and the daughter spend another hour with her before she dies peacefully.

Family Reactions to a Fatal Illness

Families have many reactions to the diagnosis and management of a loved one's potentially fatal illness. The reactions are too numerous to describe them all here. But I do address two common reactions to the diagnosis—denial and uncertainty—and reactions to two basic management decisions—the site of care and the use of opioid analgesics.

Denial

Some family members react with denial to the diagnosis of a loved one's potentially fatal illness [25]. *Denial* is the rejection, conscious or not, of the true significance of an event in order to ease the fear, anxiety, or other distressing effects associated with the event [26]. Denial has three aspects, which occur virtually simultaneously: encountering the threatening event; substituting a less threatening, false significance for the true significance of the event; and experiencing decreased distress as a result. Denial can be either beneficial or harmful [25]. Beneficial denial allows the affected person time to assimilate distressing information and to muster strategies to cope with it. This kind of denial is short-term and allows one's mind "to catch up" with the troubling new reality. Harmful denial, in contrast, prevents the affected person's long-term adaptation to that reality. For example, harmful denial may prevent family members from engaging the patient in important end-of-life conversations such as to ask forgiveness, reconcile, or say good-byes.

One expert suggests (and I agree) that, when encountering a patient or family in denial, HPs adopt a watch, wait, and communicate approach. HPs should show respect by not directly challenging the denial. Instead, HPs should give emotional support and maintain open communications to encourage people to examine their denial voluntarily in time [26].

Uncertainty

Even if people do not deny the true significance of an event, they may encounter considerable uncertainty about it. Because modern life is highly predictable for Americans, uncertainty unnerves them. Yet a potentially fatal illness raises considerable uncertainty due partly to the arcane science and procedures of medicine and partly to the inherent unpredictability of the illness itself. That uncertainty often leaves family members confused and distressed [16].

Family members may sometimes have to make difficult medical decisions despite the emotional chaos caused by the uncertainty [1]. Those decisions may pit medicine's great life-saving power against the family's heartfelt wishes for patient comfort. In such situations the family might manage their uncertainty best by acknowledging it openly; identifying reasonable, concrete expectations for the patient's course of illness; setting goals related to those expectations; and taking logical steps to achieve the goals. HPs can help by providing timely information and advice. Family well-being may be at much at stake as the patient's: Family members may scrutinize their decisions afterward and bear severe, long-term regrets about some. HPs can ease the regrets by emphasizing to the family, as justified, that the decisions were the best under difficult circumstances and by explicitly sharing responsibility for the decisions.

Sites of Care

Many families emerge from terminal care situations with definite opinions about hospices, acute care hospitals, and nursing homes as sites for such care. Three studies have explored those opinions in detail. The studies first identified patients who had received terminal care in any of the three kinds of sites and then interviewed family members about that care. One study involved patients who had died usually from cancers or strokes in any of the kinds of sites [10]. Another study involved patients who had died from cancer in either home hospices or acute care hospitals [27]. And the third study involved patients with cancer who had been declared terminal (and may have already died) and had received care in either a combined inpatient/outpatient hospice or "conventional" settings [28]. The first study focused on the process of care [10]; the second, on its outcomes [27]; and the third, on both its process and outcome [28].

Opinions about the process of terminal care were both positive and negative (and sometimes self-contradictory) for all three kinds of sites (Table 8.2). Yet each kind of site received mostly positive comments. Home hospices had the highest percentage of positive comments (91 %); acute care hospitals, the next highest (69 %); and nursing homes, the lowest (51 %). Positive comments about home hospices mentioned the nurse's attentiveness, assessment skills, informativeness, emotional support, and involvement of the family in care [10, 28]. Positive comments about acute care hospitals mentioned the doctors' communication style (if open and collaborative) and the nurse's attentiveness and compassion, and positive comments about nursing homes mentioned prompt attention from staff HPs and the feeling of them as extended family [10].

Negative comments about home hospices mentioned poor staff training and excessive family expectations of what hospice does; negative comments about acute care hospitals mentioned insensitive communications, poor diagnostic or prognostic judgment, treatment delays, and inadequate pain control; and negative comments about nursing homes [10] mentioned poor staff training, staff inattentiveness (especially to dying patients), and doctor inaccessibility [10]. The "Patterns of Care" part of the next section, "Family Stressors in Terminal Illness," elaborates on some of these negative family opinions about terminal care sites.

Overall, families craved friendly, supportive relationships with HPs during terminal care. They especially wanted more time and attention from doctors. Families hated what they saw as "assembly line" terminal care in acute care hospitals. One observer claimed HPs there spend too little time with patients due to reimbursement restrictions. Furthermore, she said those HPs tend to view patients only as their diseases, and, therefore, "treating the 'whole individual' … (i)s not part of the

Table 8.2 Family members' opinions about the process of terminal care at various sites

Site	Positive comments	Negative comments
Home hospices	Nurses' attentiveness Nurses' assessment skills Nurses' informativeness Nurses' emotional support Nurses' involvement of family in care	Poor staff training Excssive family expectations of what hospice does
Acute care hospitals	Doctors' communication style good (if open and collaborative) Nurses' attentiveness Nurses' compassion	Insensitive communications Poor diagnostic or prognostic judgment Treatment delays Inadequate pain control
Nursing homes	Prompt attention from health professionals Feeling of staff as extended family	Poor staff training Staff inattentiveness (especially to dying patients) Doctor inaccessibility

Data from Refs. [10, 28]

equation." [29] Many family members simply feel abandoned when the patient's condition becomes terminal [10].

Perhaps for those reasons families expressed greater satisfaction with terminal care outcomes in hospices than in acute care hospitals [27, 28]. Supporting comments mentioned hospice's ability to meet patients' physical and psychosocial needs and families' psychosocial needs [27].

Opioid Analgesics

Opioid medications, called "narcotics" when used for illicit, nonmedical purposes [30], derive from poppies and play an important role in pain relief. They have gained justified, widespread use in medicine. As with all other medications the opioids can have both beneficial and harmful effects. But careful monitoring, assessment, and dosing can maximize pain relief benefits and minimize harms.

Nonetheless, the potential harms make many patients, family members, and HPs hesitate to treat pain aggressively with opioids. These people often have fears due to confusion over three terms—*tolerance, physical dependence, and addiction* [31]. The significant negative connotations of those terms cause many people to reject adequate opioid analgesia automatically. Tolerance refers to the body's adaptation to a medication such that maintaining the same bodily effect over time requires increasing doses of the medication. Physical dependence refers to the body's acclimation to a medication such that stopping or reducing its dosage precipitously can produce powerful withdrawal symptoms. And addiction refers to an actual disease associated with taking one or more drugs; with genetic, psychosocial, and environmental influences; and with drug craving and compulsive drug taking despite harmful personal consequences. Those consequences include severely impaired emotions, behaviors, and interpersonal relationships [31, 32].

Opioids can relieve nearly all pain safely and effectively in patients with cancer or other terminal illnesses. Still, such patients may get tolerance or physical dependence [33]. Sensible medical oversight can manage both conditions adequately. Notably, these patients do *not* get the disease of addiction: They do not have uncontrolled, self-destructive behaviors from their opioids. Therefore, patients having pain with their cancers or other terminal illnesses should not be denied adequate opioid analgesia out of misguided fears of addiction. Rather, those patients should receive whatever opioid doses they need to gain good relief [33]. Such patients in my experience request only the doses they need to be comfortable and no more.

The Case

Perhaps because Mrs. M. suffers such severe initial disabilities, her family never deny she has a serious illness. A likely cause at her age is cancer, and they know that. The family do, however, experience many uncertainties even after her cancer is confirmed. For example, they do not know for sure whether .the cancer will respond to treatment, how great her improvement will be and how long it will last, how they can help care for her, and when she will die. But the family take a remarkably constructive approach to their uncertainties: They ask questions and learn as much as they can about Mrs. M.'s situation. They use the hospitalist as an important source of information. They ask him questions about diagnosis and treatment, and he answers as best he can.

Opioids, begun in the hospital, control most of Mrs. M.'s pain until the end of her illness. Mr. M. and his adult children occasionally call the attending doctors for supplemental doses when the usual doses do not relieve an especially severe headache. Mr. M. with his public health education understands opioid use and does not worry about addiction with Mrs. M.

Soon after Mrs. M.'s hospital admission Mr. M. rightfully assumes primary decision-making responsibility for her whenever she cannot make decisions for herself. (He always consults the children, however.) It is he who makes the final decision about hospice transfer several months later. By then all other ameliorative treatments have been exhausted, and only treatments for comfort remain. Her death is close. He agrees that hospice is "the only option left."

The subsequent hospice care produces clear benefits. The hospice intake nurse provides a thorough, sensitive assessment of Mrs. M.'s situation and some on-the-spot information and emotional support for the family. She also optimizes the opioid regimen and orders a mechanized hospital bed for Mrs. M.'s comfort and ease of care. But the hospice care also disappoints the family in significant ways later on. Perhaps the family expect too much, but I do not think so. Some of their disappointments arise from the organizational relationship between the hospice staff and the rehabilitation ward staff. The hospice staff provides little direct, hands-on patient care. Rather, they advise the ward staff who give the care. The hospice staff track Mrs. M.'s course only through her medical record, and Mrs. M. and her family rarely see them as a result. Unfortunately, the nurse's aides who do give Mrs. M.'s direct care have little training or experience in treating a brain cancer patient. They can provide little information about the meaning of her symptoms or signs. The family ask, "Do you see what we see? What does it mean?" and get no answers. Furthermore, Mr. M. and his children receive no orientation to the working relationship between hospice and the rehabilitation ward. They do not know what to expect and must figure out that relationship on their own. An explanatory brochure would help.

The family also comes to think the ward staff overemphasize the medical and underemphasize the psychosocial even as Mrs. M. nears death. The family wish the staff would sometimes just forego the routine vital signs and talk with them about the trauma of watching a loved one die. Even when the family request specific advice about caring for Mrs. M., the ward staff is ill-prepared to give such advice. The family ask for ways to spur Mrs. M.'s memory when she begins to forget family members. The staff suggest only posting family pictures on her ward room wall. "We have already done that," exclaims one daughter in frustration. The staff offer no other suggestions. Compounding the family's disappointment, visits from the ward staff become ever less frequent as Mrs. M. nears death, making the family feel abandoned.

Family Stressors in Terminal Illness

Because the family is as much a part of terminal care as is the patient, HPs must be able to recognize stressors that affect families in terminal care situations. Table 8.3 lists some of those stressors, each with multiple references in the medical literature. The stressors cluster into five categories: features of the illness, the family's response to the illness, information about the illness, communications from HPs, and patterns of care. I briefly describe these stressors by category. The M. family experienced many of them during Mrs. M.'s illness.

Features of the Terminal Illness

Observing a terminal patient's illness often stresses the family even more than does their actual caregiving for the patient [6]. Symptoms, functional decline, and problem behaviors trouble families deeply, symptoms perhaps most of all [34]. Pain, confusion, fatigue, loss of appetite, and other symptoms prompt empathy in the family, and that empathy brings heartache. Symptoms that increase also remind family members starkly about the patient's approaching death [16, 23]. Functional decline also stresses the family. It forces them to revise downward their expectations of the patient and to increase their caregiving efforts for him or her [35]. And problem behaviors such as wandering, aggression, incontinence, and uninhibited actions likewise stress the family due to the embarrassment or danger involved [35, 36].

The Family's Response to the Illness

Powerful distresses ripple through a family when a loved one is critically ill or dying. Those distresses develop into full-blown psychiatric illnesses for some. In fact, 69 % of family members of critically ill intensive care patients incur depression; 35 %, anxiety [37–39]. Family members, of course, often become depressed or anxious just because the patient is seriously ill. But family members may also become depressed or anxious because the *patient* becomes depressed or anxious. Depression may prompt the patient to withdraw emotionally. Family members sense that withdrawal and can become depressed by it. As I mentioned before, depression may also prompt the patient to lash out in anger. He or she may excessively criticize family members or level hurtful accusations at them. The

Table 8.3 Stressors for families during a terminal illness

Category	Stressors	References
Features of the illness	Patient's symptoms	[18, 23, 34, 41, 43]
	Patient's functional decline	[6, 35]
	Patient's problem behaviors	[25, 35, 36]
Family's response to the illness	Depression or anxiety	[37–39]
	Grief over losses in the patient's appearance or abilities	[46, 70]
	Perception that patient has lost his or her identity in the healthcare system	[46, 47, 71]
	Unhealthy coping strategies: avoidance, hesitance to request needed help	[18, 23, 24, 41]
	Conflicting wishes about the patient's future: prolong life versus maintain comfort	[46, 51]
Information about the illness	Too much, not enough, or wrong kind of information about the patient	[24, 42, 46–49, 70]
	Inability to reconcile own observations of patient with professionals' explanation of what those observations might mean medically	[46, 70, 71]
	Uncertainty about patient's overall prognosis for symptoms, function, and survival	[1, 34, 46, 71, 72]
	Lack of information about what patient will likely experience when life support stopped or patient dies	[1, 8, 46, 49, 73, 74]
Communications from health professionals	Condescending or disrespectful address	[34, 46, 48, 49, 70]
	Jargony speech	[46, 70, 75]
	"Blocking" statements	[24, 29, 42, 76]
Patterns of care	Unfamiliar, chaotic nature of healthcare delivery	[8, 34, 46, 52, 70, 71, 73, 75]
	Institutional care settings	[10, 34, 45]

people targeted may blame themselves for somehow failing the patient. A subsequent depression for them may linger a long time after the patient dies.

In addition to the stresses of depression and anxiety, family members typically feel stress from the patient's progressive losses in appearance, ability, or identity [35]. One author calls these losses "the little deaths" that occur throughout dying [13]. As they accumulate, the losses in appearance or ability chip away at the patient's image. Family members may grieve each loss. Family members may also experience stress from the patient's loss of identity caused by having an impersonal disease in a seemingly impersonal healthcare system. The HPs are not necessarily to blame: They do not intend that loss and may not realize it happens. To them the patient has always been a patient. They usually do not know him or her from before the illness. Under the circumstances the patient easily seems to become his or her disease. Family members may, therefore, be dismayed at the resulting loss of the patient's personal history and feel compelled to tell others, especially the HPs, about the patient's prior life—his or her values, accomplishments, and relationships. Far from clinically useless, this information can prove valuable in personalizing the patient's terminal care.

Two unhealthy coping strategies may also cause stress. One is avoidance, the strategy that ignores the patient's terminal illness or tries to wish it away [18, 24]. One study suggests Euroamericans use avoidance to cope more than do African Americans [40]. The other strategy is the family's reluctance to ask HPs for important help when needed [23, 41]. Family members may hesitate to "bother" busy HPs, or may believe the time required from HPs for a reply robs sick patients of valuable care time. Still, if ongoing, either unhealthy coping strategy causes problems to fester and stresses to mount.

One last stressor here comes from a family's typical wishes for a patient: to live as long as possible and to live as comfortably as possible. The two wishes often conflict in treatment decisions about potentially fatal illnesses. Trying for longer life may sacrifice comfort; trying for comfort may sacrifice longer life. The conflict creates enormous stresses for family members who must decide.

Information About the Illness

Any terminal illness stresses families over information—whether they get too much, not enough, or the wrong kind of it. Families usually complain about not getting enough. They particularly want to know (among other things) what their own observations of the patient mean medically; what the patient's overall prognosis is for symptoms, function, and survival; and what the patient will likely experience at death. In giving this information, HPs should try to engage family members "where they are." Efforts to inform, therefore, should take account of family members' education and experience. HPs should tailor their explanations accordingly. A good way to do that involves encouraging family members' questions, listening carefully, and using family members' own words in responses. These techniques *can* help minimize the stresses of hearing bad news. While I do not elaborate any more here on informing techniques, I do so in other parts of this book. Chapter 4 focuses on this topic in particular.

Communications from Health Professionals

Health professionals sometimes communicate in ways that unintentionally cause family stress. Three stress-causing communications occur especially frequently: condescending or disrespectful address, jargony speech, and "blocking" statements. The first two communications can occur in any medical situations; the third typically occurs only in critical or end-of-life situations.

Modern medicine intimidates the medically untrained, often making them feel ignorant or stupid. Condescending or disrespectful address by HPs reinforces those feelings. HPs, therefore, must avoid it. In fact, I urge HPs to resist the current trend of calling patients and family members automatically by their first names. That practice, I believe, diminishes people. In contrast, using surnames ("Mr. Jones" or "Mrs. Smith") shows respect. HPs, therefore, should use surnames unless patients or family members invite them to do otherwise. With any illness, but especially a terminal one, that sign of respect acknowledges the gravity of the situation and the valued status of patients and family members.

Jargony speech also increases families' stress by impeding their grasp of medical matters. As a result, families may feel disadvantaged or marginalized in care discussions. HPs should refrain, then, from using technical terms whenever possible. If using technical terms is necessary, HPs should define them explicitly. But HPs should generally communicate with patients and families in colloquial terms, say, using the seventh-grade vocabulary of a typical local newspaper.

"Blocking" statements also stress families by blocking, or cutting off prematurely, potentially distressing discussions [42]. HPs use them when discussions may involve volatile issues, may threaten HPs emotionally, or may entrap them in conversations they cannot easily escape. Blocking statements avoid awkward, uncomfortable situations.

The statements may take several forms. One form minimizes or "normalizes" a person's distress ("The patient is not really all that bad, Mr. Jones." or "Of course, you feel anxious. It is only natural in these situations."). Another form abruptly substitutes a more comfortable subject for a less comfortable one ("Let's talk now about the patient's latest laboratory results. We can talk later about whether he is actually dying.") And still another form ignores the uncomfortable subject altogether. All three forms alienate people by trivializing their distress. Blocking statements suggest it does not warrant serious attention. They, therefore, make HPs seem distant and insensitive.

Patterns of Care

To families of patients, medical care can seem like an "alternate universe" [29] with unfamiliar, sometimes incomprehensible concepts and practices. Medical care delivery can also seem chaotic. Incomprehensible laboratory reports, differing diagnostic assessments and treatment suggestions [1], rotating staff assignments [24], patient transfers [8, 24], and many other features are unsettling. Family members feel adrift. Everyday intuitions seem to fail them. Quandaries become magnified in life-or-death situations. "You are not medically trained", says a family respondent in one study. "If the HPs ask, 'Should we do this or that?' I d(o)n't know what to tell them." Stress results.

The different characteristics of institutional care settings add to the stress. Families must acclimate to each new setting with its advantages and disadvantages.

The intensive care unit (ICU) provides orderly, competent high-tech care. Low patient-to-nurse ratios, usually one or two to one, make for ready access to HPs. Though families may understand little about esoteric ICU technology, they often consider the patient safest there. The hospital ward provides less intensive, less technical care with higher patient-to-nurse ratios and less access to HPs overall. Families correctly perceive less attention given to each patient on the hospital ward than in the ICU. The nursing home provides the least intensive, least technical patient care of all. Most is basic, long-term bodily maintenance. Nurse's aides, trained perhaps for only 6–18 months, do most of the direct care. One or two nurses supervise them on-site, but doctors visit only occasionally. Families may understand this care best but worry about its possible gaps. As a result, families may feel obliged to provide any care beyond basic maintenance.

The hospice movement of the 1960s and 1970s introduced a welcome, comfort-oriented care alternative to the other terminal care settings. Hospice care typically occurs in patients' homes, in nursing homes, or in special hospice inpatient units [10, 27]. Hospice consultants oversee that care while others (such as family members in patients' homes or staff in nursing homes) deliver it. The consultants provide caregivers one-on-one instruction [43], considerable say in care delivery, and some attention to the emotional and spiritual needs of patients and caregivers [28], Many families feel most supported with hospice. Its terminal care generally has shown better pain relief, patient satisfaction, family satisfaction, and cost effectiveness than has terminal care in acute care hospitals [21]. But hospice still has its problems. Delayed referrals [44], some inconvenient inpatient locations [45], exaggerated family expectations for complete symptom relief [10], ongoing related family burdens (including caregiving), lack of primary doctor involvement [45], and inadequate training or performance of hospice personnel [10, 45] can all cause families stress.

The Case

Stressors from each category bedevil the M.s during Mrs. M.'s terminal illness. All three stressors from the features-of-illness category certainly cause trouble. As the tumor regrows after radiation, Mrs. M.'s symptoms increase and her function decreases, alarming the family. She suffers frequent headaches, only partially relieved by opioids. Unfortunately, nothing relieves her other symptoms such as fatigue and poor appetite. Mrs. M. also experiences a steep, steady decline in function. In less than a month she goes from riding in the car and walking with Mr. M. in the retirement center's garden patio to being confined to bed and sleeping much of the time. She also begins to have intermittent flashes of anger. Many are directed at Mr. M. A stoic man, he endures these flashes of anger with love and patience. Mrs. M.'s deterioration, however, discourages the whole family. The elder daughter speaks for them all when she says, "Mom used to be such a vibrant woman. It's hard to see her so sick."

Of the family-response stressors, only two cause trouble. The family, of course, grieve the loss of the person Mrs. M. was. Mr. M. misses her repartee and golf outings with him. Likewise, the children miss the many ways Mrs. M. cared for them both as children and adults. These griefs cause stress in the family's decisions for Mrs. M. The hospice decision illustrates. The family want to preserve for as long as possible whatever traces of Mrs. M.'s former self-image remain. They even hope for a miraculous cure. But they also want to keep her comfortable. The decision about hospice transfer pits their hope for an eventual cure (and a return of Mrs. M.'s old self) against their wish for her comfort. Mr. M. in particular wavers over hospice at first but finally agrees to it. He explains dejectedly, "It's the only option left." To their credit the M. family do not perceive the healthcare system as a threat, do not adopt unhealthy coping strategies (such as avoidance), and do not hesitate to ask for the help they need.

All four information stressors also arise. While the M. family members never appear to get too much information, they sometimes get too little. Some responses from the rehabilitation ward staff to family questions offer an example. The family see changes in Mrs. M. and ask the staff, "Do you see the changes we see?" and "What do they mean?" But the staff never answer directly, frustrating the family. The family also have difficulty reconciling some of their observations with the doctors' assessments. After the radiation the family see a dramatic improvement in Mrs. M.'s function. Their hope for a prolonged survival increases. But the doctors discourage that hope. They warn that Mrs. M.'s function will deteriorate again soon. They still predict a survival of only a few months. The doctors explain that a quick improvement after radiation paradoxically portends a rapid tumor resurgence later. Confused by this explanation, the family cannot square their optimistic observations with the doctors' pessimistic predictions.

This mix of too little information and confusing information creates for the family considerable uncertainty about how and when Mrs. M. will die. The incident of the bladder infection offers an example. When Mrs. M. becomes unresponsive for the first time during her illness, the ward staff call the family in for a bedside death vigil. The staff say Mrs. M. is imminently dying. The family assemble from across the country only to witness Mrs. M. regain alertness with antibiotics. She is definitely not dying right away. The family return home not knowing how much to trust the ward staff's future death predictions.

While jargon almost never arises in the HPs' speech, the other two communication stressors—disrespectful address and blocking statements—do. Mr. M. and his family are not overly demanding, but the nurse's aides sometimes show them discourtesy, even disrespect. The aides, for example, tell the family, "Call us if you need us." But when the family do call, some aides criticize the family for calling. Furthermore, some aides use blocking strategies to respond to the family's questions. The aides either ignore the family's questions altogether or answer with a quick "I don't know." Both responses annoy the family.

Finally, from among the patterns-of-care stressors, overly restrictive visiting hours do not apply here. The rehabilitation ward allows the family to visit Mrs. M. at any time. Yet the other two patterns-of-care stressors—an institutional care site and the seeming chaos of health care—do apply. As Mrs. M. nears hospital discharge, she and her family must choose disposition between the M.s' independent-living apartment and the rehabilitation ward. They wisely choose the rehabilitation ward. While home offers comfortable, familiar surroundings, only the ward offers the nursing care Mrs. M. will eventually need. But the ward room is never quite home for Mrs. M. although the family try to make it feel like home. They bring plants from the M.s' apartment and post family pictures on the wall. They also bring everyday clothes for Mrs. M. to wear instead of the institutional gowns. Despite all these efforts the room does not completely shed its institutional feel. The family regret that result.

Furthermore, the care on the rehabilitation ward sometimes seems chaotic. The ward appears to lack important practice procedures and schedules. As a result the M. family have difficulty anticipating when the doctors will round, when meals will be delivered, how to request supplemental analgesic doses for Mrs. M.'s headaches, and how to get Mrs. M. cleaned promptly and her bedsheets changed when she becomes incontinent. Making the apparent chaos even worse, the nurses and nurse's aides visit the bedside ever less frequently as Mrs. M. nears death. Mr. M. and the elder daughter must assume much of her basic care. By the end Mr. M. rarely leaves Mrs. M.'s room, becomes visibly stressed and exhausted, and loses weight.

Family Stress Relievers in Terminal Illness

Certain factors can help counteract a family's stressors during a terminal illness. These stress relievers cluster into five categories, four of which match exactly the categories for stressors: the family's response to the patient's terminal illness, information about the illness, communications from HPs, and patterns of care (Table 8.4). The one different category is the family's response to the patient's death. I describe elements of each category in the following sections.

The Family's Response to the Patient's Terminal Illness

Many of the stress relievers in this category arise out of positive aspects of family members' relationships—with the patient, themselves, other family members or friends, and God or a spiritual community. When accessible, reliable, and genuine, such relationships support families even during the most trying parts of end-of-life experiences.

Positive aspects of a relationship with the patient show themselves partly in family members' "true portrayal" of the patient—strengths and weaknesses alike—at the critical time of a terminal illness. By accurately representing the patient's life history and values, family members can help make decisions for the patient consistent with his or her character and help HPs individualize care accordingly. In the

Table 8.4 Stress relievers for families during a terminal illness

Category	Stress relievers	References
Family's response to the illness	Accurate representation of patient's history and values to health professionals	[34, 51, 71]
	Use of healthy coping strategies	[18, 40, 73, 77, 78]
	Maintenance of social support	[40, 77]
	Maintenance of strong faith or spirituality anchors	[48, 51, 70–72]
Information about the illness	Information about care site, family's own observations of patient, and patient's likely experience at death	[1, 8, 48, 49, 73, 74]
Communications from health professionals	Inclusion and respect demonstrated	[34, 73]
	Compassion demonstrated	[10, 79]
Patterns of care	Care in an intensive care unit or inpatient hospice	[1, 8, 10, 28, 43, 79]
	Open, convenient visiting hours	[8, 48]
	Visits by patient's primary care doctor	[34, 49]
Family's response to the patient's death	Belief that patient's death was not blameworthy	[46, 48, 51, 54, 71, 75, 78]
	Grieving in emotionally, culturally, and spiritually meaningful ways	[8, 46, 48, 51, 53]

process family members can gain the satisfaction of serving the patient faithfully. As the common wisdom goes, the HPs have knowledge of the clinical medicine, but the family has knowledge of the patient. Good individualized care requires both [46, 47].

Positive aspects of a relationship with oneself show themselves through family members' adoption of healthy strategies to cope with a patient's terminal illness. Cognitive reappraisal is one such strategy. It involves reappraising the sad experience of the patient's approaching death as also an opportunity for the family member's personal growth. That growth might involve deepening one's appreciation for the human condition, one's compassion for others, or one's faith and spirituality [18]. African Americans tend to use cognitive reappraisal frequently [40]. Active problem-solving is another such strategy. It involves identifying problems caused by the patient's illness and actively trying to solve them [18]. Euroamericans tend to use that strategy frequently [40]. Regardless of ethnic or cultural background, of course, many people use both strategies simultaneously.

Positive aspects of relationships with other family members and friends provide another stress reliever, social interaction. It produces two important benefits: It reminds family members about the empathetic support of others and broadens family members' perspective beyond the immediate terminal illness. And positive aspects of relationships with God or a spiritual community provide for many people still another stress reliever, a spiritual anchor for weathering a patient's terminal illness and reconfiguring one's life healthfully afterward.

Information About the Illness and Communications from Health Professionals

As I say many times in this book, informing and communicating lay the foundation for all good end-of-life care. That is especially true in the care for families. Because a later section of this chapter, General Care of the Family during Terminal Illness, describes broad strategies for informing families, this section addresses only a few specific interventions aimed at relieving stress.

One of those interventions is orientation to the care site. Medical institutions often seem unfamiliar, foreboding places to families. "You are not medically trained," explains one family member in an ICU study. "You have to put your … trust in people (there) you have never seen (before) and you don't know." [1] Furthermore, those HPs have esoteric ideas, speak an esoteric language, and follow esoteric customs. All the unfamiliarity intimidates the medically untrained.

Some of the stress from the unfamiliarity, I believe, can dissipate with a simple orientation on admission to the care site. Such an orientation not only acclimates families to the environment but also boosts their trust in the staff. Some parts of the orientation might be delivered verbally by HPs; some, in a short written brochure given to families [39]. The verbal orientation should include the names, professions (doctor, nurse, etc.), and specific care duties of the HPs attending the patient. (My own experience suggests how often HPs forget to give this information.) The verbal orientation might also include some coaching about how to function at the site (such as where to find the public restrooms and the cafeteria, which telephone number to use to contact the nursing station, how to relate most effectively with the staff, and how to act at the bedside).

The written brochure might explain general care practices at the site [8]. That brochure should certainly give the visiting hours and explain the rationale for them: to allow the patient to rest and the staff to do their caregiving tasks. (But reasonable accommodations might be made in special cases [8, 48].) The brochure might also encourage families to ask the patient's primary care doctor to visit even if that doctor does not currently care directly for the patient. Those visits increase continuity of care for the patient and provide social support for the family [34, 49].

Too little information about a patient's illness, of course, can cause a family stress. HPs have a duty to disclose as much information as the patient wishes and should do so in ways that help the family understand the information properly [50]. Judgment is required about what, when, and how much to disclose. HPs must often decide these particulars by a best guess based on family members' apparent needs and tolerances. But the goal should always be conveying the essence of the patient's illness and its implications for the patient's future.

One conscious focus for information giving might be family members' own observations about the patient. Family always observe the patient and want to know what their observations mean. Without guidance family members may draw incorrect, needlessly stress-producing conclusions. HPs should, therefore, periodically ask family their observations and respond with the best possible medical interpretations. Another conscious focus for information giving might be family members' usual worry about whether their loved one will die uncomfortably. HPs should address that worry explicitly when the patient's death approaches. They might ask the family such questions as Are you concerned about what will happen to your loved one when we stop the ventilator? or Do you worry about what a cancer patient like your relative experiences in the last hours of life? HPs may then be able to reassure families that most foreseeable deaths occur peacefully.

Other parts of this book describe in detail communication strategies for end-of-life care. I make here only one general comment about stress-relieving communication strategies aimed at families. The best of those strategies have three characteristics: broad inclusion of family members (unless the patient requests otherwise), respectfulness for all people involved and their views, and compassion. Empathy, or sensitivity to the feelings of others, underlies all these characteristics.

Patterns of Care

Just as some patterns of care create stress for families, other patterns of care provide stress relief. Such relief sometimes comes with a patient's admission to an intensive care unit or an inpatient hospice. While those sites pursue opposite goals of care, both ensure close attention by HPs. That attention reassures families their relatives are receiving capable, conscientious care—whether it be aggressive, curative care in an ICU or comfort-oriented, noncurative care in an inpatient hospice. Some experts even say good hospice care is as intensive as good ICU care.

Patterns of good terminal care also reduce family stress by removing unnecessary obstacles to patient–family–HP interaction. Short or inconvenient visiting hours are one common obstacle; discontinuities of attending HPs, another. When a patient turns terminal, visiting hour limitations should be relaxed for patient and family convenience. At the same time the scheduling of attending HPs, including the nurses, should maximize continuity of care. That continuity allows patient and family to get to know the HPs and to build trust in them during an emotionally fragile time. Continuity of the principal doctor is especially important. The M. family attest to that fact when describing a past instance of their family doctor's care for Mrs. M. She had developed a pneumonia, and the doctor decided to treat her at home. But he conscientiously visited her there every day until she recovered. As the elder daughter explained, "The doctor really cared about Mom and was willing to go the extra mile to make sure she got well. We all felt good about him and trusted him because he continued to see her."

The Family's Reaction to the Patient's Death

Two reactions to the patient's death may serve as important stress relievers for the family. One is believing that no one, especially family members, failed the patient in the care process and thereby deserves blame for the death. Three arguments may be invoked to justify this belief. First, the patient received maximal medical treatment. All reasonable life-sustaining treatments were used, and none were withdrawn prematurely. They all simply failed. Second, no medical mistakes caused the death: The patient received the best care possible, free of fatal errors. And third, the patient was simply bound to die at that time [46, 51]. No human effort could have thwarted his or her predetermined fate.

The other reaction is expressing grief in emotionally, culturally, and spiritually meaningful ways. Reminiscing is one important way. For that reason I mention it throughout this book. Sharing stories preserves for survivors "remnants" of the patient after death. Other ways include saying final good-byes, keeping bedside vigils, witnessing the patient's death, and attending funerals or memorial services afterward [46, 51–53]. Parts of Chaps. 10–13 describe emotional, cultural, and spiritual expressions of grief further.

The Case

The M. family benefits from every category of stress reliever but especially from the response-to-the-terminal-illness category. The M.s represent Mrs. M.'s life history and personal values accurately and make decisions that reflect that history and those values. The decision for radiation illustrates. The family understand that family relationships and athletics are key to Mrs. M.'s idea of a worthwhile life. The pain and disability from her tumor disrupt both. She cannot remember her children and grandchildren or play golf or tennis. These disabilities frustrate her. She finally concludes, "Everybody lives too long. It's past my time (to die)." Therefore, when deciding on treatment, she chooses radiation—the most comfortable, least invasive ameliorative treatment. She justifies the choice by saying "I don't want to prolong this (illness). I am ready to die. I just want to be comfortable now." And the family support her choice. They know that, if Mrs. M. cannot recover enough to participate in the two most important aspects of her life, she does not want to live at all.

The M. family meanwhile use the healthy coping strategy of active problem-solving. For example, Mr. M. takes Mrs. M. for rides in the car and walks with her in the patio garden to stimulate her mind. The elder daughter posts family pictures on the ward room wall to remind Mrs. M. about the children and grandchildren. And the sons plan financing for Mrs. M.'s rehabilitation stay by reviewing the M.s' Medicare coverage and their savings. All the children help take care of

Mr. M., too. They spell him at bedside care whenever they come to town. And the elder daughter ensures Mr. M. gets adequate nutrition and rest. She even takes him to local theater productions as a diversion.

Furthermore, the family maintain their social supports. Frequent visits or telephone calls with each other prove most important. Those contacts bolster family solidarity. The family also welcome visits from Mrs. M.'s friends and get to know other patients and families on the rehabilitation ward. These outside contacts provide an added boost. In addition, the elder daughter consults a colleague at work about Mrs. M.'s situation. This colleague experienced recent cancer deaths of both parents and can explain some of what is happening to Mrs. M. She helps give the daughter an "emotional grounding" for adapting to Mrs. M.'s decline.

The M. family benefit from the other categories of stress relievers, too, but not so extensively. In the information-about-the-illness category the family get some notice about when Mrs. M. will die. The rehabilitation ward's nursing supervisor tells Mr. M. and the elder daughter on Mrs. M.'s last day that she is imminently dying. The nurse first predicts death before midnight. When that does not happen, the nurse urges Mr. M. and the daughter to go home, sleep, and return in the morning. They do and come back in time to spend about an hour with Mrs. M. before she dies.

In the communications-from-HPs category the family benefit especially from the hospitalist doctor. He shows inclusiveness and respectfulness by including all available family members in Mrs. M.'s care planning conferences, by patiently answering all the family's questions, and by listening carefully to their opinions. He also shows compassion. For instance, he discloses what he would favor as treatment for his own mother in such a situation, and he continues to visit the M. family for support even after Mrs. M. transfers off his service to begin radiation.

In the patterns-of-care category the family benefit from the open visiting hours on the rehabilitation ward. Because Mr. M. lives at the retirement center, he and his family may visit the rehabilitation ward at any time. The open hours allow him to run errands during normal working hours and to visit Mrs. M. off-hours.

In the reaction-to-the-death category the M. family benefit by grieving Mrs. M.'s death in emotionally and culturally meaningful ways. One involves hosting a reception in Mrs. M.'s memory after her death. The extended family and many friends attend. A review of Mrs. M.'s life forms the focus for the reception. A grandson shows a continuous loop slide show of Mrs. M.'s life. And both Mr. M. and the elder daughter give short reminiscences. Mr. M. tells how as a sweaty teenager riding his bicycle he first met Mrs. M. And the elder daughter tells about Mrs. M.'s care as a mother. The daughter closes her reminiscence by saying poignantly, "I realize now all the little things Mom did for us kids were not so little after all."

A Special Case: Family Stressors and Stress Relievers in the Intensive Care Unit (ICU)

Mrs. M. was never admitted to an intensive care unit (ICU). Nonetheless, recent statistics show an association—though surely not cause and effect—between ICU stays and deaths in America. About one-fifth of Americans die shortly after ICU stays [54], about half of Americans dying in hospitals spend some of their last three days in ICUs [55], and about one-tenth of patients in ICUs have decisions made to withdraw life support [37]. Patients dying after an ICU stay usually have a severe organ system failure such as heart, liver, or lung failure, but recent studies show that patients with advanced cancers increasingly undergo aggressive treatments sometimes involving the ICU [56, 57].

A patient's ICU stay traumatizes the family because they vicariously experience some of the patient's suffering and may have to make life-or-death decisions on the patient's behalf [54]. HPs working in an ICU, therefore, need to know the ICU's special stressors for families and ways to relieve those stressors.

One special stressor is the alien nature of the ICU. While the ICU is familiar to the HPs working there, it is not to families of patients. In fact, one author calls the ICU a "vortex," or whirlpool, for families [1]. The term invokes a rapid, uncontrolled, dizzying experience. The unstable, frequently changing medical conditions of patients; the inherent uncertainties and unfamiliar concepts involved with diagnoses; the esoteric, highly sophisticated treatments; their potentially life-or-death consequences; and the aggressive "rescue culture" make families feel lost or marginalized in the ICU [1, 55]. The multitudes of HPs attending each patient contribute to the bewilderment of families. And no guarantee exists that ICU care follows a patient's presumed wishes or provides overall benefit [1]. This alienness and other stressors, too, may sometimes make the ICU seem threatening to patients and hostile to families. Many family members understandably react with depression or anxiety [37].

HPs can take a few simple steps to ease families' stresses in the ICU. With the ICU environment seeming so alien and threatening, HPs might initially orient families to the ICU and how it works [4, 8]. A brief orienting brochure (like the one I suggest elsewhere for Mrs. M. on the rehabilitation ward) might help [39]. Furthermore, HPs in the ICU should wear name tags, introduce themselves explicitly, and explain their roles in the patient's care [8]. HPs should also take extra time to discuss the patient's case with the family at admission. Families hunger for information then. They want to know diagnosis, treatment, and prognosis and the reasoning behind each [4]. HPs should specify the important areas of uncertainty, too. All explanations should be plain and honest, avoiding medical jargon [4]. HPs should invite families to ask questions and should anticipate important questions families might not think to ask. Nurses should be empowered to contribute to the explanations: Families may need the nurses to interpret the doctors' comments [29]. After admission, families should receive updates at least daily and with any significant changes in the patient's condition [4]. And visiting hours should be made as liberal as practicable on a case-by-case basis [4].

HPs must also appreciate families' emotional turmoil during a patient's ICU stay. That turmoil may impair families' ability to understand medical information. HPs may, therefore, have to repeat information several times or in several ways for families to grasp it [37]. Families also need explicit emotional support to cope with the severe illnesses typical for an ICU. This support might come in several ways. One is helping families fashion realistic hopes for outcomes. Another way is promising good, conscientious care and then delivering on that promise. And still another way is demonstrating to families that the ICU staff genuinely care about the patient and the family [4]. That assurance creates the foundation for family trust in the staff.

General Care of the Family During a Terminal Illness

The prior discussion about family stressors and stress relievers during a terminal illness suggests ways HPs can help families beyond just performing a dying patient's physical care or coaching the family as they perform it. That help might include referring family caregivers for caregiving education, individual counseling, group support, or respite care. Sections in the next chapter discuss those specific interventions. This section, in contrast, discusses general approaches to care for *all* family members experiencing a loved one's terminal illness. I organize these approaches into four categories—overall communication, family conferences, special communication roles for nurses, and potentially helpful perspectives for families.

Overall Communication

Many experts believe good communication is "an essential component of end-of-life care." [42] Communication is especially important to families of dying patients. In fact, nearly half of family complaints about current terminal care involve communication [10]. Room for improvement surely exists. Unfortunately, the research on communication with families in terminal situations is "slim," [42] leaving large knowledge gaps about what might constitute improvement. Therefore, much of what I say here necessarily relies on expert opinion and my own experience. Ideas for improvement may change with additional research.

HPs should concentrate their communication efforts mostly, but not exclusively, on families of the sickest patients [58]. Regular, gentle efforts to engage these families are probably more important for meaningful communication than any one particular technique [10, 19, 54]. Still, quality end-of-life communications involve

HP accessibility and continuity; honest, timely, clear, and useful information; sensitive interaction marked by compassion; and sound guidance in decision-making [10]. The doctor serving as primary communicator for the patient's medical team should lead the effort. That doctor should put the family at ease and inspire trust [58]. He or she should give a reliable contact telephone number and encourage the family to call with questions or concerns [34, 45, 49, 59].

The doctor should surely strive for continuity in the doctor–family relationship. One threat to that continuity is transfers among doctors or care institutions while the patient is still alive. The doctor, therefore, should extend himself or herself after patient transfers to maintain those relationships. Only an occasional, brief telephone call to the family may be necessary to give them important emotional support. The doctor should also make regular, supportive visits to the patient while he or she is dying and a call or visit to the family afterward. Although some of those efforts may not get reimbursed (but all should be), I urge doctors to make them anyway. A doctor's giving personal condolences provides important consolation for grieving families. Empathy and a sense of professional duty should motivate the doctor in doing so.

Information is useful to families, of course, only if it is consistent and clear. Thus, HPs who give information—doctors, nurses, and others—should coordinate it with the rest of the medical team to minimize inconsistencies. The information delivery should also explain key concepts and repeat them as necessary for understanding [1, 10, 60]. Pictures help. In addition, good communication requires periodic informational updates (perhaps every day for rapidly changing, critically ill patients and every other day for less rapidly changing dying patients). Updates should occur even more frequently with serious downturns in the patient's condition.

The most momentous decisions families make in end-of-life situations involve either stopping life support or focusing on comfort-only care. Chapter 7 addresses those decisions in detail. I mention here only two points. First, most families want to make those decisions together with the patient's doctor. They welcome the doctor's input and consider it carefully. The doctor's opinion carries great weight. I urge the doctor to make a clear recommendation and to give its rationale. And second, many families struggle with the spiritual implications of those decisions and want advice from either the institution's chaplain or their own spiritual advisor. The doctor or other HPs attending the dying patient should offer the family timely spiritual referrals during momentous decisions according to the family's preferences [53].

Quality end-of-life communication must always demonstrate compassion, literally "suffering with." Family in a terminal situation need others to understand something of their distresses and to empathize. Therefore, HPs must avoid the conversational "blocking" maneuvers I discussed before. Those maneuvers stop awkward or emotionally charged conversations before they even start [42]. Instead, HPs should encourage family members to talk about their distresses. And HPs should explicitly acknowledge distresses when they come up. HPs should listen carefully and respond empathetically. HPs might say such things as "It must be hard to face the death of your brother." or "I can remember feeling confused and overwhelmed when my own mother was dying. I imagine you have some of the same feelings now." Compassion, of course, is not automatic and cannot be faked. It must be consciously cultivated. It develops only with effort. But grieving families will recognize—and appreciate—HPs' compassion when it is genuine.

Compassion for dying patients and their families involves special obligations. One relates to disclosing deaths to family. I believe that should occur in person whenever possible. If family members are nearby but not immediately present when a patient dies, the HP communicator might call and prepare them initially over the telephone. The HP might say something like "I'm afraid Mrs. Jones has taken a turn for the worse. I think you should come. Please meet me at the nurses' station so I can update you before we see her together." Then when the family arrive, the HP should be readily available, should lead the family to a private area, and should disclose the death there before the family and the HP see Mrs. Jones in her room.

Family Conferences

Current trends toward shared decision-making and family involvement in care make family conferences a cornerstone of ICU care. They ease family depression and anxiety during serious patient illnesses, and the trauma of family bereavement after patient deaths. Every leading critical care society now recommends these conferences [39, 58]. I believe they should become a central part of family care with *all* patients who have deteriorating, potentially fatal illnesses.

Family conferences should begin early and recur frequently in critical or terminal situations in the hospital [8, 58]. While these conferences may often be informal, HPs should prepare for them with the same thoughtfulness with which they prepare for other key clinical events [55]. HPs should plan ahead what information to give and how to present it. The conferences should be multidisciplinary (at least with doctors and nurses) and structured. They should not only give information but also encourage an exchange of views among patients, family members, and HPs [8]. Some experts recommend structuring the conferences after the VALUE acronym. VALUE stands for value what family members say, acknowledge the family's emotions, listen, ask questions to understand the patient and his or her life story, and elicit questions from the family [39, 58]. The conferences should aim overall to support families and to facilitate consensual decisions.

The little research that exists about family conferences suggests four points. First, family satisfaction with the conferences parallels the percentage of time family members have to speak [54]. Doctors, therefore, should speak less and listen more than they ordinarily do [54]. One study shows doctors talk, on average, for more than four minutes to begin such a conference. Overall they talk, on average, for 71 % of the time, leaving only 29 % of the time for family members to talk [54]. Those percentages should probably come closer to 50–50 %. Second, doctors should express more empathy than they do. They tend to respond more often to informational cues than to emotional ones from the family [42]. Perhaps for that reason doctors average only 1.6 empathetic comments per conference. Empathetic comments are not difficult or time-consuming. Doctors need only acknowledge how emotionally draining having a loved one so ill is or how burdensome making life-or-death decisions for another is [61]. Third, conferences should end with an agreement about an action plan. And, fourth, the family may benefit from receiving specific information, printed or verbal, on the patient's disease if the patient is critical or on bereavement if the patient is terminal [58].

Special Communication Roles for Nurses

Nurses often play important roles in end-of-life communications with families. In fact, nurses probably disclose patient deaths to families more often than do doctors [1]. Because they spend so much time at the bedside, nurses have many opportunities to get to know family members and to talk informally with them. Nurses can use these conversations as an avenue for exchanging important information. Nurses might ask family members about the patient; his or her life history and values; and the family's own needs, feelings, and expectations for care [55]. The nurses might also explain how the medical system works, who the various attending HPs are and what roles they play in the patient's care, and how families can help give care at the bedside [4, 22]. Furthermore, because family members often do not adequately understand the information doctors give [55], nurses might need to explain that information in terms the family can understand [1]. The doctors, of course, should encourage the nurses to offer those explanations [29]. The nurses' informal interactions with family members and the nurses' generally good instincts about reading people's moods can also yield valuable assessments of family stressors [8]. The entire medical team should encourage nurses to report those assessments and should consider them in care of the family.

Potentially Helpful Perspectives

HPs can help families cope with end-of-life ordeals through not only good communication strategies but also helpful perspectives. I find four perspectives especially helpful to families. First, people ordinarily live as though life will always be what it is now. But the death of a close loved one compels change. Therefore, whenever a patient nears death, HPs should encourage family members to prepare for change by imagining life after the patient dies. Second, people may paradoxically believe at different times they have full control or no control over death. Reality lies in between: Patients and families can control some things but not others. The healthiest approach to a death, then, is to identify the things one can control and to change them as needed, and to accept as is the things one cannot control. Third, different family members realize the inevitability of a patient's approaching death at different times. Some family members may, therefore, have to wait for others to "catch up" to that realization. Important treatment decisions may sometimes have to be postponed as a result. But important, inevitable end-of-life decisions such as stopping futile life support should never be postponed more than briefly (and certainly not indefinitely). And fourth, imminent death may trouble family members more than it troubles the dying person. As one insightful 102-year-old woman says, "When you're old, you get tired. You have done so much that you get tired of living. You just want to rest (from life)." For many patients, therefore, the closer death comes, the more it looks like a friend.

The Case

Communication effectiveness, the heart of these suggestions for family care, differs markedly between Mrs. M.'s hospitalist on the hospital ward and her nurse's aides on the rehabilitation ward.

The hospitalist uses an exemplary communication style. He assumes primary responsibility for informing the family about all medical aspects of Mrs. M.'s case. He convenes family conferences at critical points in her illness such as when the brain tumor is first diagnosed and when ameliorative treatment should be decided on. He delivers important information in a timely, clear way and encourages questions. He also listens carefully to what family members say. And he ensures continuity of communication by providing a contact telephone number and inviting the M. family to call him with questions. He visits Mrs. M. and her family regularly even after the diagnostic workup ends and she transfers to the radiation oncology service for treatment. The family particularly appreciate the hospitalist's empathetic guidance when he says he, too, would favor radiation if his mother were in a similar predicament. Above all, he never dismisses Mrs. M. as just another dying patient. And he frequently reminds the family to take basic care of themselves despite the hardships of Mrs. M.'s illness. He urges them to eat regularly, to take breaks from the hospital vigil, and to rest properly. This communication style earns him the family's full trust. As one daughter says, "We felt the doctor was looking out for us. He was on our side."

Perhaps only partly through any fault of their own, the nurse's aides on the rehabilitation ward use a sharply contrasting, deficient communication style. They are often preoccupied with the mechanics of care (such as measuring blood pressures, delivering meal trays, and changing bedsheets) and rarely think to ask the family how they are coping. The aides also respond with blocking maneuvers when the family volunteer their distresses. If a new symptom or sign appears, the family ask the aides anxiously, "Do you see what we see? What does it mean?" The aides either ignore those questions or block further discussion with a curt "I don't know." In addition, the aides offer no new suggestions when the family ask for ways to trigger Mrs. M.'s memory. And none of the ward staff including the nurses and the aides explain what the family can expect from hospice. Thus, when hospice is engaged, the family are severely disappointed by what they perceive as a lack of direct hospice involvement in Mrs. M.'s care.

The aides' regrettable communication style understandably puts the family off. They come to believe the aides are either too distracted or too disinterested to address family psychosocial issues. The family feel isolated in their grief. The whole rehabilitation ward staff, aides and all, obviously need education about communicating with families in end-of-life situations.

Can End-of-Life Care Satisfy Families?

Health professionals genuinely aspire to give compassionate, high-quality end-of-life care. As I say throughout this book, such care includes personal attention to patients and their families; honest, clear, and timely communications; effective symptom control; and empathetic responses to survivors' grief. When done well, end-of-life care eases the patient's dying and consoles the family. It also gives HPs enormous satisfaction.

Yet the landmark 1995 SUPPORT study exposed significant shortcomings in the end-of-life care actually delivered [62]. Doctors in the study rarely knew the wishes of their patients about such care. And, even when informed about those wishes, the doctors rarely followed them in treating patients. Many patients, therefore, died in ways they did not want.

Those disturbing results spurred powerful responses from prominent medical institutions. The Institute of Medicine conducted a comprehensive review of end-of-life care and recommended ways to improve it [63]. *The Journal of the American Medical Association* published a series of case commentaries giving practical lessons in quality end-of-life care [64]. And the Hastings Center, a medical ethics think tank, published a special report addressing the question, Why has improving end-of-life care been so difficult? [65] Many improvements have followed, and more will surely come. HPs must incorporate them into practice as they appear.

Behind all the reform efforts lies the assumption that, if the healthcare system makes enough of the right improvements, end-of-life care can eventually satisfy families. But the public's lingering dissatisfactions with it make me wonder: those dissatisfactions seem as persistent now as ever [66, 67]. Families want pain-free, "dignified" deaths for loved ones; more quality time with them near the end of life; continuity of care and clear communications from attending HPs; and timely changeovers to purely comfort care. Families claim these wishes are often not now realized [8, 44]. They express the most dissatisfaction with communications about diagnosis, its implications, and the cause of death [67] and with psychosocial support for patients and families [8].

I believe not only that HPs strive constantly to improve their end-of-life care but also that some problems in dying can never be completely solved. Every death, I think, will fall short of the ideal—that quick, clean, comfortable, and sometimes heroic death of Hollywood movies [8]. Families expecting that ideal will always be dissatisfied [55].

Which problems, then, persistently fuel families' dissatisfactions with end-of-life care? I can think of six. Three occur at the level of individual patient cases; three, at the level of healthcare institutions.

One of the three at the level of individual patient cases is the inherent uncertainty of diagnosis and prognosis [46]. The medically untrained often misperceive medicine as an exact, deductive science. They want—and expect—quick, accurate information in life-threatening crises. People simply do not understand that, like a jigsaw puzzle, diagnosis requires assembling pieces of clinical data into a coherent picture. Gathering and arranging those pieces take time. The wait involved upsets many families anxious for a quick definition of the crisis [48]. HPs should address that anxiety and try to assuage it while everyone awaits a firm diagnosis. They can reassure the family that the HPs have not forgotten them in their distress, and that the doctors are pursuing an expeditious, orderly workup. Of course, even when a firm diagnosis emerges, a firm prognosis may not. Complex life-threatening diseases often defy accurate, specific outcome predictions. In fact, accurate predictions about death may be possible only hours before a death occurs. Thus, the only realistic approach in most dying situations is a frustrating wait-and-see one.

Another problem at the level of individual patient cases is the arcane care technology used in many life-threatening situations. Because families do not understand that technology, they are wary of it and may even feel manipulated by it. Circumstances also force families to trust stranger specialists who control the technology and whose intentions and skills the families can assess only superficially. Common, everyday wisdom may help little. It may even mislead or confuse.

I believe American popular culture is partly responsible for families' confusions over the technology. Movies and television shows create unrealistic expectations of medical care by depicting it too simplistically. Compared to media depictions, real-life medical care seems slow, inept, chaotic, inefficient, impersonal, and self-serving. I strongly reject those impressions, but they do influence families. Discrepancies between what families see and what they expect to see in medical care often disturb them.

And a third problem at this level of individual patient cases is the harmful emotional baggage a family brings to life-threatening medical crises [68]. That baggage is unique to each family [22], illustrating Tolstoi's point that "every unhappy family is unhappy in its own way." [69] Unfortunately, emotional baggage dictates much of a family's response to terminal illness and is not immediately apparent to outsiders such as the attending HPs. Acrimonious feelings may lie dormant during ordinary circumstances only to erupt suddenly under the stresses of end-of-life situations. Old grievances, rivalries, codependencies, and guilts may surface with a vengeance. Even family members may not recognize those emotions or realize they have been simmering below the surface of life. HPs attending the dying patient, therefore, may have to decipher the family's unique dynamics and sensitivities on the spot. "Feeling their way along in the dark," HPs must guess at how to respond. Unintentional slights or unmet expectations may occur, leaving some family members hurt or angry and some HPs regretful.

One of the obvious institutional problems contributing to family dissatisfaction with end-of-life care is the legal, organizational, and economic limitations governing modern American medical care. Most nonHPs do not appreciate the huge constraints created by such limitations. Though sometimes nuisances in practice, all those limitations have a grounding in history. Past mistakes or abuses have prompted institutions to create laws, policies, or regulations to avoid the same mistakes or abuses in the future. Past mistakes of medical disclosure and abuses by forced treatment, for example, prompted laws to protect the privacy of personal medical records and to honor patient treatment refusals. Yet, if too many or too rigid, such laws, policies, and regulations may have the unintended effect of actually *obstructing* good care and frustrating patients, families, and HPs in the process.

Another institutional problem prompting family dissatisfaction with end-of-life care is fragmentation of care. That fragmentation can cause miscommunications, conflicts, and inconsistencies among institutions and HPs, leaving patients and families confused. The fragmentation can also blur lines of responsibility: If unable to identify any one doctor or other HP as ultimately responsible for a patient's care, a family may see no clear ultimate source of answers to their questions or concerns.

A third institutional problem causing families' dissatisfaction with end-of-life care is current cost-cutting measures. American healthcare institutions have had to cut patient care staff to make budget under reduced reimbursements. At the same time, health insurers require HPs to provide ever more documentation for the reimbursements they do get. These measures leave few staff and little time to care for patients. Furthermore, healthcare staffing has shifted to hiring ever less extensively trained (and cheaper) caregivers. Where fully trained nurses once provided much of the direct patient care, medical assistants and nurse's aides do so now. These basic-level HPs have too little training or experience to address adequately the nuanced psychosocial and spiritual needs of dying patients and their families. In short, powerful institutional forces work against meeting such needs even though doing so determines much of families' satisfaction with end-of-life care.

Readers should not misunderstand me here: I *do* believe family satisfaction with end-of-life care can be increased despite all the countervailing forces. Individual HPs can make improvements in their patient care practices and should try. But big institutional improvements, lying beyond the influence of individual HPs, are also necessary. Unfortunately, as the immediate representatives of healthcare institutions, individual HPs may face families' anger at the institutions. HPs should not take that anger personally. Rather, they should calmly hear families out, sympathize with their complaints, perhaps explain something of institutional constraints, and make whatever accommodations possible at the patient care level. Being professional at those times requires absorbing, not reflecting, people's anger and responding constructively.

The Case

The M. family have well-founded dissatisfactions with Mrs. M.'s end-of-life care. But almost none arise at the individual case level. Admittedly, Mrs. M.'s enigmatic presentation baffles her doctors at first, causing uncertainty and delay in making the diagnosis. Mr. and Mrs. M. become understandably impatient. The wait distresses them, creating momentary angry thoughts. At one point Mr. M. suspects the doctors of purposefully concealing the diagnosis. Mrs. M. meanwhile grows frustrated with her progressively disabling, undiagnosed illness. "The doctors don't realize what (abilities) I've already lost!" she exclaims with exasperation. The doctors never directly address these distresses. Yet, once a brain scan reveals the tumor, Mr. and Mrs. M. forget the prior wait. They are saddened at the diagnosis but grateful for the technology which made it clear painlessly and noninvasively.

In addition, the M.s are about as well-adjusted as any modern American family can be. They bring little harmful emotional baggage to Mrs. M.'s dying. They keep each other informed, visit Mrs. M. frequently, and share care duties for her and each other. The family also encounter no significant disagreements over the most momentous decisions involved with Mrs. M.'s illness: Which ameliorative treatment would serve her best? Should she return from the hospital to her independent-living apartment or to the retirement center's rehabilitation ward? and When should she transfer to hospice care? Sympathy and love enable the family to support each other throughout the ordeal. Their dissatisfactions with care, therefore, do not arise from individual-case-level problems.

Rather, their dissatisfactions arise from institutional-level problems. Laws, regulations, policies, and entrenched practices sometimes interfere with Mrs. M.'s care. For example, policy on the rehabilitation ward requires accepting at face value any patient's expressed refusals, however, nonsensical. When Mrs. M. becomes confused late in her illness, she refuses several baths in a row without giving a reason. She becomes soiled and cannot keep her bedsheets clean. Aware of the inevitable health risks and Mrs. M.'s penchant for cleanliness throughout her life, Mr. M. finally insists the staff bathe her. An argument follows, but the staff eventually relent. They bathe her against her expressed wishes.

Furthermore, cost cutting on the rehabilitation ward causes understaffing, and care suffers. The M. family say the staff invite them "to call us any time you need something." But, when the family do call for help, the overworked staff sometimes come very late (only after an hour or more) or not at all. The family soon decide to rely as much as possible on themselves to give Mrs. M.'s basic personal care. Even that strategy causes problems. The understaffing disrupts ward routine, making it unpredictable. Because Mr. M. helps Mrs. M. eat, he must wait for her meals to arrive at her room. But the deliveries are erratic. Mr. M. sometimes misses his own meals at the retirement center's dining room because he is waiting for Mrs. M.'s meals to arrive at the rehabilitation ward. Despite such frustrations Mr. M. thanks the ward staff for whatever help they do give Mrs. M. He knows she depends on them for some care, and he wants to avoid alienating them for fear they will retaliate by withholding care from her.

Probably the M.s' most serious dissatisfaction arises from the fragmentation of Mrs. M.'s care. That fragmentation becomes especially problematic after she transfers to hospice care on the rehabilitation ward. The organizational complexity of care increases as a result. The new division of labor requires the hospice staff to monitor her care and to give instructions, and the rehabilitation ward staff to implement the instructions. Pain control improves, but most other problems do not. A prominent cause for the spotty care is the lack of coordination between the two staffs. The ward staff sometimes do not understand the hospice staff instructions but do not ask for clarification. The frequent changes of ward staff exacerbate the follow-through problem. Good communication between caregivers or shifts occurs only intermittently. Furthermore, the hospice staff appear unaware that miscommunications may exacerbate some unaddressed problems including Mrs. M.'s insomnia, intermittent angry outbursts, and loss of appetite and the family's emotional reactions to her dying. The elder daughter says ruefully, "We rarely see the hospice people. And the ward staff change the bedsheets but don't ask us how Mom or we are doing. We feel abandoned."

The M.s' experience shows how much impact such institutional-level problems have on patients and families during end-of-life care. The problems are huge, but I believe we can fix many. Improving family satisfaction with end-of-life care surely depends on doing so.

Summary Points

1. The typical fatal cancer has three phases: the diagnostic, the ameliorative, and the terminal. Each phase presents families with distinct adjustment challenges.
2. Health professionals' communications with families during a terminal illness should be timely, honest, informative, supportive, and compassionate.
3. Regular family conferences should be a part of patient care in all deteriorating, potentially fatal illnesses.
4. Important factors—some case-specific, others institutional—can erode family satisfaction with end-of-life care.

To Learn More ...

1. Davidson JE, Powers K, Hedayat KM, et al. Clinical practice guidelines for support of the family in the patient-centered intensive care unit: American College of Critical Care Medicine Task Force 2004–2005. Crit Care Med. 2007;35(2):605–22.
2. Hanson LC, Danis M, Garrett J. What is wrong with end-of-life care? Opinions of bereaved family members. J Am Geriatr Soc. 1997;45(11):1339–47.

3. Kirchhoff KT, Walker L, Hutton A, et al. The vortex: families' experiences with death in the intensive care unit. Am J Crit Care. 2002;11(3):200–9.
4. Scheuneman LP, McDevitt M, Carson SS, Hanson LC. Randomized, controlled trials of interventions to improve communication in intensive care: a systematic review. Chest. 2011;139(3):543–5.
5. Tulsky JA. Interventions to enhance communication among patients, providers, and families. J Palliat Med. 2005;8(suppl 1):S95–S102.

References

1. Kirchhoff KT, Walker L, Hutton A, et al. The vortex: families' experiences with death in the intensive care unit. Am J Crit Care. 2002;11(3):200–9.
2. Heyland DK, Rocker GM, O'Callaghan CJ, Dodek PM, Cook DJ. Dying in the ICU: perspectives of family members. Chest. 2003;124(1):392–7.
3. Griffin JM, Meis LA, MacDonald R, et al. Effectiveness of family and caregiver interventions on patient outcomes in adults with cancer: a systematic review. J Gen Intern Med. 2014;29 (9):1274–82.
4. Kleinpiel RM, Powers MJ. Needs of family members of intensive care unit patients. Appl Nurs Res. 1992;5(1):2–8.
5. Brody EM. "Women in the middle" and family help to older people. Gerontologist. 1981;21 (5):471–80.
6. McCorkle R, Pasacreta JV. Enhancing caregiver outcomes in palliative care. Cancer Control. 2001;8(1):36–45.
7. Covinsky KE, Goldman L, Cook EF, et al. The impact of serious illness on patients' families. JAMA. 1994;272(23):1839–44.
8. Davidson JE, Powers K, Hedayat KM, et al. Clinical practice guidelines for support of the family in the patient-centered intensive care unit: American college of critical care medicine task force 2004–2005. Crit Care Med. 2007;35(2):605–22.
9. Schulz R, Newsom JT, Fleissner K, DeCamp AR, Nieboer AP. The effects of bereavement after family caregiving. Aging & Mental Health. 1997;1(3):269–82.
10. Hanson LC, Danis M, Garrett J. What is wrong with end-of-life care? Opinions of bereaved family members. J Am Geriatr Soc. 1997;45(11):1339–47.
11. Nijboer C, Triemstra M, Tempelaar R, et al. Patterns of caregiver experiences among partners of cancer patients. Gerontologist. 2000;40(6):738–46.
12. Clipp EC, George LK. Dementia and cancer: a comparison of spouse caregivers. Gerontologist. 1993;33(4):534–41.
13. Hurley AC, Volicer L. Alzheimer disease: "It's okay, Mama, if you want to go. It's okay". JAMA. 2002;288(18):2324–31.
14. Lawton MP, Moss M, Glicksman A. The quality of the last year of life of older persons. Milbank Q. 1990;68(1):1–28.
15. Morris JN, Suissa S, Sherwood S, Wright SM, Greer D. Last days: a study of the quality of life of terminally ill cancer patients. J Chron Dis. 1986;39(1):47–62.
16. Hebert RS, Schultz R. Caregiving at the end of life. J Palliat Med. 2006;9(5):1174–87.
17. Grande GE, Todd CJ, Barclay SIG. Support needs in the last year of life: patient and carer dilemmas. Palliat Med. 1997;11(3):202–8.
18. Weitzner MA. The family caregiver of the older cancer patient. Hematol Oncol Clin North Am. 2000;14(1):269–81.
19. McCusker J. The terminal period of cancer: definition and descriptive epidemiology. J Chron Dis. 1984;37(5):377–85.

20. Young HY, Myung KL, Seon YK. Impact of awareness of terminal illness and use of palliative care on the survival of terminally ill patients with cancer: a prospective cohort study. J Clin Onc. 2011;29(18):2474–80.
21. Christakis NA, Escarce JJ. Survival of Medicare patients after enrollment in hospice programs. N Engl J Med. 1996;355(3):172.
22. von Gunten CF, Ferris FD, Emanuel LL. The patient-physician relationship. Ensuring competency in end-of-life care: communication and relational skills. JAMA. 2000;284 (23):3051–7.
23. Ferrell BR, Borneman T. Pain and suffering at the end of life for older patients and their families. Generations. 1999;23(1):12–7.
24. Schulz DS, Carnevale FA. Engagement and suffering in responsible caregiving: on overcoming maleficence in health care. Theor Med. 1996;17(3):189–207.
25. Vos MS, Putter H, van Houwelingen HC, de Haes HCJM. Denial and social and emotional outcomes in lung cancer patients: the protective effect of denial. Lung Cancer. 2011;72 (1):119–24.
26. Rabinowitz T, Peirson R. "Nothing is wrong, doctor": understanding and managing denial in patients with cancer. Cancer Invest. 2006;24(1):68–76.
27. Dawson NJ. Need satisfaction in terminal care settings. Soc Sci Med. 1991;2(1):83–7.
28. Kane RL, Klein SJ, Bernstein L, Rothenberg R, Wales J. Hospice role in alleviating the emotional stress of terminal patients and their families. Med Care. 1985;23(3):189–97.
29. Tomasic V. The outsider. Hast Cen Rep. 2012;42(1):24–6.
30. Williams N, DeBattista C. Psychiatric disorders. In: Papadakis MA, McPhee SJ, Rabow MW. Current medical diagnosis & treatment 2016. New York: McGraw-Hill Education; 2016. p. 1078.
31. Rose M. Prescription opioids: risk management and strategies for safe use. Continuing Med Educ Calif Phys Phys Assistants. 2015;140(3):56–60.
32. Pace CA, Samet JH. In the clinic: substance abuse disorders. Ann Intern Med. 2016;164(7): ITC50-1.
33. Rabow MW, Pantilat SZ. Palliative care & pain management. In: Papadakis MA, McPhee SJ, Rabow MW. Current medical diagnosis & treatment 2016. New York: McGraw-Hill Education; 2016. p. 72.
34. Tilden VP, Tolle SW, Garland MJ, Nelson CA. Decisions about life-sustaining treatment: impact of physicians' behavior on the family. Arch Intern Med. 1995;155(6):633–8.
35. Diwan S, Houghham GW, Sachs GA. Strain experienced by caregivers of dementia patients receiving palliative care: findings from the Palliative Excellence in Alzheimer Care Efforts (PEACE) program. J Palliat Med. 2004;7(6):797–807.
36. Quayhagen MP, Quayhagen M. Alzheimer's stress: coping with the caregiver role. Gerontologist. 1988;28(3):391–6.
37. Pochard F, Azoulay E, Chevret S, et al. Symptoms of anxiety and depression in family members of intensive care unit patients: ethical hypothesis regarding decision-making capacity. Crit Care Med. 2001;29(10):1893–7.
38. Azoulay E, Kentish-Barnes N, Chevret S, et al. Risk of post-traumatic stress symptoms in family members of intensive care unit patients. Am J Respir Crit Care Med. 2005;171 (nn):987–94.
39. Lautrette A, Darmon M, Megarbane B, et al. A communication strategy and brochure for relatives of patients dying in the ICU. N Engl J Med. 2007;356(5):469–78.
40. Haley WE, Roth DL, Coleton MI, et al. Appraisal, coping, and social support as mediators of well-being in Black and White family caregivers of patients with Alzheimer's disease. J Consult Clin Psychol. 1996;64(1):121–9.
41. Hinds C. The needs of families who care for patients with cancer at home: are we meeting them? J Adv Nurs. 1985;10(nn):575–81.
42. Tulsky JA. Interventions to enhance communication among patients, providers, and families. J Palliat Med. 2005;8(suppl 1):S95–102.

43. Newton M, Bell D, Lambert S, Fearing A. Concerns of hospice patient caregivers. ABNF J. 2002;13(6):140–4.
44. Teno JM, Casarett D, Spence C, Connor S. It is "too late" or is it? Bereaved family member perceptions of hospice referral when their family member was on hospice for seven days or less. J Pain Symptom Manage. 2012;43(4):732–8.
45. Carline JD, Curtis R, Wenrich MD, et al. Physicians' interactions with health care teams and systems in the care of dying patients: perspectives of dying patients, family members, and health care professionals. J Pain Symptom Manage. 2003;25(1):19–28.
46. Chambers-Evans J, Carnevale FA. Dawning of awareness: the experience of surrogate decision-making at the end of life. J Clin Ethics. 2005;16(1):28–45.
47. Meeker MA. Family surrogate decision making at the end of life: seeing them through with care and respect. Qual Health Res. 2004;14(2):204–25.
48. Abbott KH, Sago JG, Breen CM, Abernathy AP, Tulsky JA. Families looking back: one year after discussion of withdrawal or withholding of life-sustaining support. Crit Care Med. 2001;29(1):197–201.
49. Radwany S, Albanese T, Clough L, et al. End-of-life decision making and emotional burden: placing family meetings in context. Am J Hospice Palliat Med. 2009;26(5):376–83.
50. Drane JF. Becoming a good doctor: the place of virtue and character in medical ethics. Kansas City MO: Sheed & Ward; 1995.
51. Schenker Y, Crowley-Matoka M, Dohan D, et al. I don't want to be the one saying, "We should just let him die": intrapersonal tensions experienced by surrogate decision makers in the ICU. J Gen Intern Med. 2012;27(12):1657–65.
52. Torke AM, Alexander GC, Lantos J, Siegler M. The physician-surrogate relationship. Arch Intern Med. 2007;167(11):1117–21.
53. Perkins HS. Treating both dimensions of the body. Commentary on "providing guidance to patients: physicians' views about the relative responsibility of doctors and religious communities". South Med J. 2013;106(7):407–8.
54. Wendler D, Rid A. Systematic review: the effect on surrogates of making treatment decisions for others. Ann Intern Med. 2011;154(5):1336–46.
55. McDonagh JR, Elliott TB, Engelberg RA, et al. Family satisfaction with family conferences about end-of-life care in the intensive care unit: increased proportion of family speech is associated with increased satisfaction. Crit Care Med. 2004;32(7):1484–8.
56. Curtis JR, Patrick DL, Shannon SE, et al. The family conference as a focus to improve communication about end-of-life care in the intensive care unit: opportunities for improvement. Crit Care Med. 2001;29(2):N26–33.
57. Azoulay E, Soares M, Darmon M, et al. Intensive care of the cancer patient. Ann Intensive Care. 2011;1(nn):5.
58. Andrejak C, Terzi N, Thielen S, et al. Admission of advanced lung cancer patients to intensive care unit: a retrospective study of 76 patients. BMC Cancer. 2011;11(1):159.
59. Scheuneman LP, McDevitt M, Carson SS, Hanson LC. Randomized, controlled trials of interventions to improve communication in intensive care: a systematic review. Chest. 2011;139(3):543–5.
60. Emanuel EJ, Fairclough DL, Slutsman J, Emanuel LL. Understanding economic and other burdens of terminal illness: the experience of patients and their caregivers. Ann Intern Med. 2000;132(6):451–9.
61. Rothchild E. Family dynamics in end-of-life treatment decisions. Gen Hosp Psychiatry. 1994;16(4):251–8.
62. Selph RB, Shiang J, Engelberg R, Curtis JR, White DB. Empathy and life support decisions in intensive care units. J Gen Intern Med. 2008;23(9):1311–7.
63. The SUPPORT investigators. A controlled trial to improve care for seriously ill hospitalized patients: The study to understand prognoses and preferences for outcomes and risks of treatment. JAMA. 1995;274(20):1591–8.

64. Field MJ, Cassel CK, eds., for the Committee on Care at the End of Life, Institute of Medicine. Approaching death: improving care at the end of life. Washington, D.C.: National Academy Press; 1997.
65. McPhee SJ, Rabow MW, Pantilat SZ. Perspectives on care at the close of life. JAMA, various issues, 2000–8.
66. Improving end of life care: Why has it been so difficult? Special Report. Hast Cen Rep. 2005;35(6).
67. Baker R, Wu AW, Teno JM, et al. Family satisfaction with end-of-life care in seriously ill hospitalized adults. J Am Geriatr Soc. 2000;48(5):S61–9.
68. Malcrida R, Bettelini CM, Degrate A, et al. Reasons for dissatisfaction: a survey of relatives of intensive care patients who died. Crit Care Med. 1998;26(7):1187–93.
69. White DB, Malvar G, Karr J, Lo B, Curtis JR. Expanding the paradigm of the physician's role in surrogate decision-making: an empirically derived framework. Crit Care Med. 2010;38 (3):743–50.
70. Braun UK, Beyth RJ, Ford ME, McCullough LB. Voices of African American, Caucasian, and Hispanic surrogates on the burdens of end-of-life decision making. J Gen Intern Med. 2008;23(3):267–74.
71. Hayes CM. Surrogate decision-making to end life-sustaining treatments for incapacitated adults. J Hospice Palliat Nurs. 2003;5(2):91–102.
72. Apatira L, Boyd EA, Malvar G, et al. Hope, truth, and preparing for death: perspectives of surrogate decision makers. Ann Intern Med. 2008;149(12):861–8.
73. Vig EK, Starks H, Taylor JS, Hopley EK, Fryer-Edwards K. Surviving surrogate decision-making: What helps and hampers the experience of making medical decisions for others. J Gen Intern Med. 2007;22(9):1274–9.
74. Limerick MH. The process used by surrogate decision makers to withhold or withdraw life-sustaining measures in an intensive care environment. Oncol Nurs Forum. 2007;34 (2):331–9.
75. Maguire P, Pitceathly C. Improving the psychological care of cancer patients and their relatives: the role of specialist nurses. J Psychosom Res. 2003;55(5):469–74.
76. Haley WE, LaMonde LA, Han B, Burtan AM, Schonwetter R. Predictors of depression and life satisfaction among spousal caregivers in hospice: application of a stress process model. J Palliat Med. 2003;6(2):215–24.
77. Wiegand D. In their own time: the family experience during the process of withdrawal of life-sustaining therapy. J Palliat Med. 2008;11(8):1115–21.
78. Heyland DK, Rocker GM, O'Callaghan CJ, Dodek PM, Cook DJ. Dying in the ICU: perspectives of family members. Chest. 2003;124(1):392–7.
79. Tolstoi L. Anna Karenina. (Constance Garnett, trans.) New York: The Modern Library Classics; 2000. p. 3.

Chapter 9
Care for Family Caregivers

An Overview of Family Caregiving

Between 15 and 44 million Americans [1–3]—approximately one person in every four households [4]—provide long-term, unpaid care for chronically ill or dying relatives. These caregivers constitute "a vast, silent volunteer army," [5] America's "stealth weapon against chronic illness." [6] They give desperately needed care for historically underserved patients while saving the American healthcare system billions of dollars annually [7].

One author describes the typical family caregiver as a 46-year-old woman with a high school education; a full-time job; an annual household income of $35,000; and no formal medical training [6]. She is the patient's wife if the patient is terminally ill, or the patient's daughter or daughter-in-law if the patient is chronically but not terminally ill. And she devotes between 18 and 120 h per week to the patient's care [5, 8]. She views this caregiving as a second, full-time job [4, 9]. If she is a daughter or daughter-in-law of the care recipient, she often also has a husband and growing children whom she must care for at the same time [10]. "The American system of long-term care," one author concludes, "would collapse without" this selfless, dedicated family caregiver [6].

This chapter addresses the plight of those people who care for chronically ill or dying relatives. The chapter begins by reviewing reasons why families have a duty to care for their ill or dying. The chapter then describes typical aspects of family caregiving including care tasks, stressors, potential stress relievers, and impact on caregivers. Next the chapter describes family caregiving with the dementias, the fatal diseases for which the most data on the topic exist. (I believe caregiving with the dementias often resembles caregiving with other fatal diseases such as the cancers and the organ system failures. Many lessons about caregiving with the dementias should apply equally well to those other diseases.) The chapter then discusses interventions aimed at supporting family caregivers of dying patients and closes with suggestions

H.S. Perkins, *A Guide to Psychosocial and Spiritual Care at the End of Life*,
DOI 10.1007/978-1-4939-6804-6_9

about how health professionals (HPs) in particular can help. To illustrate many of its points, the chapter uses an elderly man who suffers a hip fracture and then requires family care at home for many months. He develops a dementia in the meantime and finally dies after a severely debilitating stroke.

The Case

Mr. L is an 88-year-old retired utilities executive with a long-standing cardiac arrhythmia, for which he takes a blood thinner. One day he trips on the sash of his bathrobe, falls, and breaks a hip. He undergoes a successful hip replacement but requires institutional rehabilitation for several months afterward due in part to pressure sores that develop on his feet and back. Mr. L. is finally allowed to return to his independent-living apartment in the retirement center where he and his wife live. She and the L.s' daughter take over most of his care there. They clean and dress his pressure sores; manage his complicated medication regimen; oversee his diet and rest; and coordinate his many general medicine, orthopedics, and physical therapy appointments. A visiting nurse comes twice a week while his pressure sores heal.

This in-home care for Mr. L. extends from months into years. Mrs. L. and the daughter notice distinct mental and physical changes in him over that time. He begins to lose his way in the once familiar hallways just outside the apartment. Someone must walk with him to keep him from getting lost. His stamina decreases gradually, and bowel incontinence occurs occasionally. Embarrassed at the bowel incontinence, Mr. L. stops going to the retirement center's common dining room for dinners. He eats them instead alone with Mrs. L. in the apartment. Mrs. L. must either cook them or bring them in. Still, his appetite decreases, and his weight drops. In addition, he begins to need help from Mrs. L. or the daughter to shower and dress. And despite being an avid lifelong reader and sports fan, Mr. L. loses interest in both books and sports. At first he skims books he used to devour eagerly; soon he no longer even opens books. He sometimes still turns on televised ball games but does not watch them. He cannot remember scores even while the games are being played.

One evening Mrs. L. finds Mr. L. drooling and unresponsive in his favorite easy chair. She calls 911. Paramedics respond and transport Mr. L. to the closest hospital. Brain scans over the next few days show a large but stable bleed. Mr. L. has had a hemorrhagic stroke likely due in part to the blood thinner. He regains minimal responsiveness after a few days but cannot make decisions for himself. He chokes regularly on his diet of thickened liquids. His doctors press Mrs. L., the daughter, and the son (who has just arrived from out of town) about whether they want a permanent feeding tube for him. Deciding Mr. L. would never want one, they say no. Because Mr. L. will no longer have adequate nutrition or hydration, the doctors conclude he is dying. They discharge him back to his independent-living apartment with in-home hospice care. The family attends him there until he dies quietly several days later.

The Familial Duty to Care for the Ill and the Dying

"Family caregiving," insists one expert, is key to "sustaining patients at the end of life." [4] Agreeing with that idea, many Americans endorse a family's duty to care for their ill or dying. But that duty has more than just widespread public support; it also has strong conceptual support, more so in American culture and ethics than in law.

Various authorities attest to the American cultural grounds for the duty. One prominent author, for example, sees the family's duty to give care as "traditionally" American and says that American families provide more care for their chronically ill or dying elders than do governmental or charitable service organizations. But this author also observes that fulfilling that duty falls disproportionately to women, and it is a "deeply internalized social stricture and (an) ingrained behavior" for them [11]. The author then backs up that view with results from her own research. She interviewed three generations of American women—grandmothers, mothers, and granddaughters—and found all three "resoundingly endorse(d their) filial care (duties) for elders." [11] Another American expert, who cared for her quadriplegic husband for years, adds that many American family caregivers—men and women alike—willingly undertake that caregiving duty for its intrinsic psychological, social, and spiritual benefits [12]. Even one prominent American medical professional organization acknowledges the importance of family caregiving. The American College of Critical Care Medicine states that, as ever more restrictive spending constraints take hold, the American healthcare system relies increasingly on families for long-term care of their ill or dying [13].

Governments in Great Britain and the USA have tried to legislate a family's duty to give care. While many people would agree with the logic behind them, the resulting laws have proven largely ineffective at promoting family caregiving. In 1601, Great Britain passed the Poor Law, which shifted the costs of caring for the ailing poor from the general public to the families of those people. The government justified the law by saying, "(T)hey who protected (us in) the weakness of our infancy are entitled to our protection in the infirmity of their (old) age." Similarly, in 1983 the US federal government declared that states could pass laws requiring adult children to pay for the care of their elderly parents. The government argued, first, that the capable adult child's *implicit* duties to care for an ill or aging parent mirror the parent's earlier *explicit* duties to care for the young child and, second, that the reciprocal child-to-parent caregiving duties [11] already have nearly the force of legal contracts. The government's declaration sparked considerable controversy. One opponent of the declaration argued that government can require a parent to care for a child but not for a child to care for a parent because only the parent has a choice in creating the parent–child relationship [14]. This author concluded that care by a child for a parent must always be voluntary. The author also warned that trying to legislate good behavior by families risks unintended bad consequences due to legal inflexibility and heavy-handedness. Surely for those reasons few, if any, American states subsequently adopted compulsory family caregiving laws.

Nevertheless, the family duty to care for the ill and the dying does have a solid foundation in ethics. One ethicist, for instance, recognizes a duty of adult children to care for their ill or dying parents but not based on some "debt of gratitude" to parents as asserted in past governmental declarations [15]. This ethicist analyzes the issue instead by contrasting favors (which incur debts) with willing sacrifices (which do not). Favors involve agreements, explicit or implicit, between two equally astute people that one will bear some burden for the other. The beneficiary thereby incurs some comparable debt or obligation to be repaid the benefactor in the future. Expected mutual gain motivates both parties. (Africans provide an interesting example of favors. An African who borrows a basket from a neighbor repays the implicit debt of the favor by returning the basket full of fruit from the family garden. Westerners, of course, follow similar customs about borrowing but in typically Western ways.) The ethicist goes on to reason that parents do not raise their children on favors, expecting the children to repay debts from past childcare with future elder care. Parents and young children obviously make no agreements, explicit or implicit, as equally astute parties expecting mutual gain. This ethicist believes that parents instead raise their children by voluntary, loving sacrifices. I agree.

How, then, does this ethicist justify a genuine duty of adult children to care for their ill or dying parents? She suggests that, ideally, parental sacrifices in child-rearing foster mutual affection. A lasting, loving parent–child bond results. Reciprocal caregiving duties grow naturally from that loving, intimate bond. Thus, affection, rather than indebtedness, motivates care in both directions. Of course, the burdens and benefits of caregiving sometimes get out of balance in one direction or the other: Parents need to provide more care when children are young and developing, and adult children need to provide more care when parents are old or dying. The loving parent–child relationship, however, never keeps a strict account.

I believe this parent–child relationship benefits more than just parents and children. It benefits those in other familial relationships, too—grandparents and grandchildren; brothers and sisters; aunts, uncles, nieces, and nephews. The parent–child relationship in its best form serves all as an inspiring model for caring and caregiving throughout the family.

I conclude that culture and ethics mostly, and law less so, establish a family's duty to care for its ill or dying members. Mutual affection, intimate bonds, and shared personal histories underlie that duty. Many families recognize it and respond accordingly with loving, selfless caregiving [16].

But not all families do. Some either do not acknowledge their caregiving duties or do not fulfill them conscientiously. When so, a patient's principal doctor, supported by the patient's other HPs, should remind family members about their duty to provide care. Reasonable limits to such care, of course, apply and may vary with families' circumstances. The doctor may need to explore those circumstances with each family. Still, many doctors hesitate to confront derelict families about this care duty. Doctors may think their moral authority inadequate to prevail in such situations. But doctors actually have more moral authority than they think. I, therefore, believe they should use it to ensure families do their part in providing care for ill or dying family members. Some patients may depend on doctors to advocate for them in this way.

The Case

Mrs. L. and her daughter certainly recognize their duty to care for Mr. L., and they act on it admirably during his long decline. One of the two women nearly always stays with him in the apartment. And they perform many care tasks for him from helping him shower to dressing his wounds. They also monitor his condition, engage the healthcare system as needed on his behalf, and provide him companionship. Unfortunately, the L.s' retired son does not help much. In fact, when he visits, he often irritates the daughter by suggesting additional tasks for her such as dusting the furniture or cleaning the shower without offering to help her. Furthermore, whenever Mr. L. asks for personal care in the son's presence, the son automatically calls his mother or sister to give it. The daughter begins to resent her brother. And, as Mr. L. becomes sicker, the son makes ever fewer visits home. Still, Mrs. L. and the daughter faithfully give Mr. L. care throughout.

The daughter tries to understand the difference in attitudes about Mr. L.'s care between the son and the two women. The daughter cites an illustrative incident. Mrs. L. plans a trip to visit the son in California early in Mr. L.'s illness. The daughter agrees to stay home with Mr. L. in the apartment while Mrs. L. is gone. But Mrs. L. cancels her trip at the last moment because she cannot emotionally bear to leave Mr. L. The daughter says of the incident, "I think giving care is more important to women than to men. We (women) just cannot allow ourselves to back off from it."

A Model of Family Caregiving

Having concluded that families do have a duty to provide care for their ill and dying, I devote the rest of this chapter to describing that care. I use a simple model, depicted in Fig. 9.1, to show four main components of family caregiving. On the far left the figure shows typical caregiving tasks. In the middle the figure shows stressors and stress relievers that may affect caregivers as they perform those tasks. I define here *stressors* as hardships that challenge a person's adaptive capacity, and *stress relievers*, or "resources" as other authors call them, as factors that ease the impact of

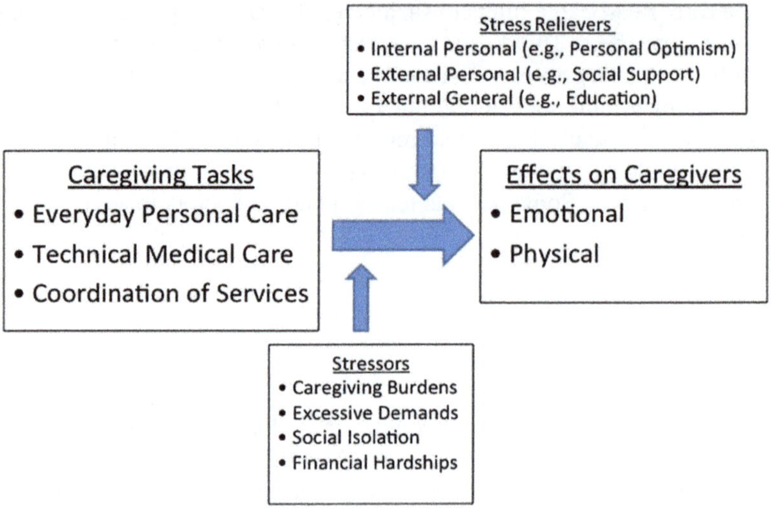

Fig. 9.1 A model of family caregiving

the stressors [17]. On the far right the figure shows some effects of caregiving on family caregivers once relevant stressors and stress relievers have had their impact [17]. The figure uses arrows to show the interrelationship among the main components. The figure also lists some subcomponents under each main component. In the next few sections, I discuss separately each main component and its subcomponents.

Typical Family Caregiving Tasks

Family caregivers have traditionally given an ill or dying relative basic physical and spiritual comfort [18]. They have administered treatments to ease a patient's physical symptoms. They have also tried to bolster the patient's spirits to encourage hope and the will to live. Family caregivers still give such care today. But, as the American healthcare system has evolved recently, family caregivers in the USA have been required to take on many additional care duties. The system offers ever more sophisticated state-of-the-art treatments but must also control expenses. One way the system controls expenses is by discharging patients from hospitals and specialty clinics earlier in their illnesses than ever before. Family caregivers must, therefore, assume complex care tasks at home that were formerly performed only by HPs in hospitals or clinics. Of course, as the woman who cared for her quadriplegic husband explains, home care for the severely ill or dying still has its many mundane "nasty chores." [12].

One reason family caregiving is such hard work is that caregivers must attend simultaneously to various domains of care. There are at least three: everyday personal care, technical medical care, and coordination of outside services. Caregivers play multiple roles within each domain, and the priority of the roles may change suddenly and frequently as the patient's medical condition and functional abilities change. For example, within "everyday personal care" caregivers' roles may include personal aide, homemaker, and companion; within "technical medical care" the roles may include nurse, care manager, and medical decision-maker; and within "coordination of outside services" the roles may include scheduler, advocate, and bill payer [19]. Caregivers must be able to move quickly and deftly among the various domains and their associated roles. The challenges for family caregivers, therefore, are daunting: The care they must provide imposes huge changes in personal relationships while exacting tremendous physical and emotional tolls. The woman who cared for her quadriplegic husband says with a mix of weariness, frustration, anger, and resentment that her husband's sudden, severe, and permanent disability forced her immediately to make the unwanted transition from wife to caregiver. That transition, she notes, required courage, "grit, and persistence." [12].

Everyday Personal Care

An interview study of nearly 1000 terminal patients and their family caregivers describes the patients' typical care needs in this domain. Many of the patients in the study had cancer, some were cared for in inpatient settings but others in outpatient settings, half had at least moderate pain, and some were bedridden. Nearly all the patients had at least some care needs, and more than a third had moderate-to-high care needs. Patient predictors of high care needs included elderly age, poor physical function, fecal or urinary incontinence, and poverty [20].

Many of the patients' care needs were mundane but time- and energy-consuming for caregivers. Sixty-two percent of patients needed help with transportation (including riding to and from doctor or hospital visits); 55 %, with homemaking (including house cleaning, shopping, and preparing meals [10]); and 26 %, with bodily care (including bathing, dressing, and eating) [20, 21]. A second, complementary study that compared the care needs of dementia and nondementia patients provided the insight that dementia patients have more bodily care needs than do nondementia patients [22]. (In a later section of this chapter, I contrast in detail the care needs of dementia and nondementia patients.)

Technical Medical Care

The study of nearly 1000 patients I cite above also found that 29 % of its patients had needs for basic technical care such as medication administration or dressing changes. Some patients had needs for even more complex technical care such as intravenous infusions or breathing treatments. The healthcare system automatically expects family caregivers to assume these care tasks willingly and to learn largely on their own the skills involved. But those skills are neither easy nor intuitive for the medically untrained. Family caregivers may fear learning medical procedures initially on loved ones. And, even after learning the procedures, caregivers may still lack confidence in performing the procedures. Thus, the process of learning and performing procedures often intimidates and frustrates family caregivers. Yet the healthcare system is often insensitive to such feelings. It may even treat as lazy or incompetent the caregivers who balk at learning or performing medical procedures on their own [19].

Coordination of Outside Services

Along with giving everyday personal care and technical medical care, family caregivers must often find and coordinate the outside services a patient needs. In fact, among the eight advanced, or "instrumental," activities of daily living (which include shopping, preparing meals, and administering medications), coordinating outside services is the one that elders most consistently need help with [22]. To give that help, caregivers may need to serve in multiple ways as liaisons between the patient and the healthcare system. Caregivers often must attend the patient's medical appointments to explain the patient's current condition and special needs [19], to maintain continuity of care [5], and to guard against medically related misunderstandings [19]. Caregivers must also implement the care plan [5], assess its results [23], recognize problems when they arise, and decide when those problems present a medical emergency requiring immediate professional attention. In addition, caregivers may have to arrange payment for medical services. That task often involves negotiating with health insurance companies about bills, coverages, and due dates [24]. Caregivers may occasionally even have to work extra jobs or use their own money to pay the bills [12]. By serving as patient liaisons in these ways, family caregivers understandably can feel flattened by the healthcare bureaucracy.

The Case

Mrs. L. and her daughter perform many care tasks for Mr. L. in all three domains. Though Mr. L. initially manages some basic personal care himself (such as feeding, toileting, and transferring between bed and chair), he consistently needs help with showering, dressing, and walking farther than a few steps. He has particular difficulty putting on his socks and stepping over the small ledge into the shower. Mrs. L. and her daughter must help with both those tasks. They also shop for his food, clothing, and any special equipment he needs. The son provides little help except to drive Mr. L. to his haircuts. All along Mrs. L. and her daughter insist their most important everyday personal care for Mr. L. involves keeping him company at home to reassure him that he is not abandoned.

Mrs. L. and her daughter also perform several technical medical care tasks for Mr. L. such as dressing and cleaning his pressure sores and managing his complicated blood-thinning regimen. The home care nurse provides brief instruction about both tasks, but Mrs. L. and her daughter must learn the necessary skills mostly on their own. They eventually perform the tasks adequately. Mrs. L. and the daughter dress and clean the established pressure sores twice daily and examine Mr. L. regularly for new pressure sores. The women also administer Mr. L.'s blood thinner, record his doses and outside blood-thinning test results, and contact his doctor for any dosing changes if the test results fall outside the therapeutic range.

In addition, the daughter coordinates Mr. L.'s outside care services. She makes his many medical appointments and drives him to them because bad eyesight prevents Mrs. L. from doing either. The daughter must always juggle Mr. L.'s medical appointments against her own personal schedule. "The medical system makes its appointments for Dad as though I have nothing else to do," she says. "It is up to me to schedule my work and the rest of my life around those appointments. They take a lot of time, and no one except me sees that. I may need to take an hour to drive Dad to a doctor's appointment and an hour to drive him home. We may wait for an hour in the doctor's waiting room. Then the appointment may last only fifteen minutes. It's a big time commitment for me." The daughter also schedules the renewals of Mr. L.'s medications. The new diversion restrictions on opioid prescriptions are especially burdensome to her. She must pick up each opioid prescription new from the doctor every month, take the prescription to the pharmacy, wait as the pharmacy fills the prescription, and then bring it home. As if such care tasks were not enough, long-term family caregivers such as Mrs. L. and her daughter face other stressors, too. I describe some of the most common caregiver stressors in the next section.

Caregiver Stressors

Caregiver stressors, which I defined earlier as hardships challenging a caregiver's adaptive capacities, can be divided into two kinds, primary and secondary [16, 23]. I address each kind separately.

Primary Caregiver Stressors

Primary, or direct, stressors arise directly from patient–caregiver interactions during caregiving. Primary stressors may arise from the patient's distressing symptoms, mobility problems, or other diminished capacities or from the caregiver's burdens, inadequate care skills, or fatigue. I use caregiver burdens—both care tasks and other burdens—to illustrate primary caregiver stressors.

Caregiving for chronically or terminally ill relatives burdens family caregivers both physically and emotionally. The caregiving usually involves mundane tasks such as bathing, dressing, and feeding patients. These tasks require from caregivers both physical strength and patience. The caregiving may also involve technical tasks such as performing specialized manual procedures or operating sophisticated medical equipment. Those tasks often require quick learning and precise execution. They also cause caregivers great anxiety. Unfortunately, both types of tasks increase steadily in number, frequency, and difficulty as patients become sicker and approach death. The associated stressors for caregivers increase simultaneously.

One obvious way to gauge the burdens of caregiving is to measure the time commitment involved. Many family caregivers intuitively consider their caregiving a full-time job [4, 9], and data from two studies support that intuition. One study provides the data for calculating the average intensity of family caregiving for hospice patients: Principal family caregivers average about 14 h of caregiving per day and other family members, about 3.5 h per day. At the same time, hospice professionals average about 3.5 h of caregiving per day per patient [25]. The other study reports the average duration of family caregiving for dying patients: about 2 years for cancer patients and 5.5 years for dementia patients [7, 26]. Thus, the time commitment required from family caregivers to care for chronically or terminally ill patients is huge—surely far more than what noncaregivers imagine.

Some caregiving burdens result directly from that huge time commitment. For instance, caregiving may interfere with caregivers' own life plans. Schooling, vocational advancement, child-bearing, and medical care may all have to wait until the caregiving ends. In addition, the caregiving often exhausts caregivers physically and mentally. They may look to hospice for relief but be disappointed when that relief is not as extensive as they expect. People commonly assume hospice takes over all of a dying patient's physical care, freeing the family to focus on life reviews, good-byes, funeral plans, and other emotional work. But the data in the prior paragraph suggest a different story: Hospice personnel mostly supervise the care, leaving much of its actual performance to the family. The reason for this odd practice is the Medicare requirement that volunteer caregivers, usually family, perform most of the physical care. Thus, while offering valuable direction and moral support, hospice provides family caregivers little relief from direct, everyday end-of-life caregiving.

Caregiving burdens certainly include more than just physical care tasks. The burdens also include the emotional turmoil of witnessing patients' distresses and decline. Family caregivers, for instance, see the regrettable effects of symptoms on patients. Though pain can often be adequately controlled, other symptoms such as fatigue, nausea, shortness of breath, and terminal stupor cannot. The accompanying distresses for patients may disturb family caregivers deeply. Caregivers may struggle to understand why modern medicine does not provide adequate relief. Caregivers may rail at doctors, the healthcare system, fate, or even God for patients' persistent distresses. Meanwhile, caregivers also face the sad prospect of seeing loved ones decline, knowing relief for patients and caregivers alike may come only at patients' deaths.

Two more caregiving burdens deserve special mention. One is patient misbehavior. Examples include abusive remarks, threats of harm, thrashing about, and lack of expressed thanks for caregivers' labors. Such misbehavior disheartens caregivers and erodes their satisfaction with their care. Living full-time with the patients only worsens the misbehavior's effects by making caregivers feel trapped in the abuse. Caregivers may eventually harbor secret wishes for patients to die and simultaneously feel guilty for having those wishes. The severe emotional distresses in caregiving make such wishes common.

The other caregiving burden deserving special mention is the obstacles to accessing HPs for advice. Family caregivers need timely, dependable access to HPs for direction when problems arise unexpectedly outside normal appointment times. Some obstacles may come from caregivers themselves due to their general reluctance to contact HPs: Caregivers hesitate to "bother" HPs with their caregiving problems or to appear incapable of solving the problems independently. Other obstacles may come from the contact procedures. Caregivers may not know whom or how to call after hours. And still other obstacles may come from the HPs who field after-hours calls. HP responses may be excessively delayed, unwelcoming, or unhelpful (sometimes because responders do not know the patients well). As the obstacles to access increase, caregivers' willingness to contact HP off-hours understandably decreases. Even if fearful for the patients, caregivers may often forego contact and try to manage the problems as best they can on their own.

Secondary Caregiver Stressors

Unlike the primary, or direct, caregiver stressors, the secondary, or indirect, caregiver stressors arise as "spillover" effects from caregiving into other parts of a caregiver's life. These secondary stressors may include changes, conflicts, or overload in life roles; social isolation; financial hardships; deterioration in self-image; and sacrifice of life dreams. I use life-role overload, social isolation, and financial hardships to illustrate secondary caregiver stressors.

Family caregivers face many demands in their roles inside and outside the caregiving relationship. Those demands sometimes conflict. If the conflicts become too burdensome, they create life-role overload [11]—the feeling of caregivers that they must fulfill more roles than possible. The many roles may include spouse, parent, employee, breadwinner, or homeowner as well as caregiver. Life-role overload hits most often the middle-aged woman who tries simultaneously to hold down a job, raise her own family, and maintain a household while she cares for the sick relative. Some caregivers who suffer life-role overload initially try to ease the strain by cutting down on sleep, recreation, or personal solitude. But self-renewal suffers, and caregivers eventually become exhausted and depressed. Other caregivers try easing the strain by quitting their paid jobs, but then face yet another problem, loss of income. The strain from life-role overload predictably intensifies as patients become sicker and their care needs increase. Caregivers may eventually lose all hope of ever fulfilling their various role responsibilities adequately.

Family caregivers may also face social isolation from the patient and other people. Social isolation from the patient occurs as caregivers lose intimate and affectionate engagement with him or her [12, 23]. When getting sicker, for example, a patient may turn increasingly inward as he or she concentrates all physical and emotional strength on staying alive, loses interest in the outside world, or experiences receding consciousness. One woman described such an experience with her dying husband: He would stare trancelike for long periods as though she were not there, making her feel dreadfully alone. Such isolating experiences obviously undermine meaningful companionship with the patient.

Social isolation may also occur from other people including HPs, family, and friends. Fragmentation in current health care causes some of the isolation from HPs. Transfers from one healthcare institution to another offer a prime example. Personnel obviously differ among healthcare institutions. Patients and their family caregivers, therefore, must adapt to different HPs with every transfer, and continuity of relationships suffers. Unfortunately, the overall busyness of health care and the lack of reimbursement for so-called social visits make maintaining former patient–family–HP relationships nearly impossible. Furthermore, the emotional stresses at the end of life exacerbate this problem with transfers from acute care to hospice care. Such a transfer usually involves a complete change of doctors, nurses, and social workers. Family caregivers may suddenly feel cut off from the HPs who gave the patient prior acute care and became trusted friends to patient and family in the process [10]. Caregivers may actually grieve for those lost relationships while also having to develop new working relationships with the hospice professionals [27]. In addition, many transfers to hospice care occur only days before patients die, leaving little time for close relationships with hospice personnel to develop.

Isolation from family and friends may occur, too, with time. These people often rally to support patient and caregiver immediately after a catastrophic diagnosis or event, but that support fades as the initial shock of the bad news subsides [19]. Other priorities in life reassert themselves, and these people resume their own busy schedules. Visits or offers to help decrease as a result. Of course, caregiving itself can contribute to social isolation. The long duration, arduousness, and mundaneness of the work [10] can leave caregivers too tired or preoccupied to pursue social contacts. Some caregivers may even wish *not* to have such contacts to avoid other people's inadvertently insensitive comments, annoying criticisms, or unwanted advice. Those caregivers may intentionally isolate themselves [18]. And, as patients become sicker and their care needs increase, caregivers' time available for socializing diminishes [4, 22]. Still, I have observed that church contacts can be some of the most supportive, fulfilling, and enduring for caregivers even late into patients' illnesses [28].

While a few families face no financial hardships from caregiving [10], many families face severe ones. According to one study, two-thirds of families caring for terminal patients experience significant financial hardships directly from the caregiving [29]. Those hardships may be a loss of wages, pension credits, or job advancements due to care-related absences [6]; increased health insurance premiums due to care expenses; or unreimbursed, out-of-pocket expenses [29] due, say, to hiring care helpers for the home or buying personal care supplies at a nursing home [30]. Over one-quarter of family caregivers must miss work, decrease their regular job hours, or quit their jobs altogether to provide care [29]. And approximately one-sixth of family caregivers for ambulatory cancer patients and one-third for terminally ill patients must spend down family assets, ask another family member for money, take out a bank loan, or apply for government welfare benefits just to pay the medical bills [21, 29, 31]. Some caregivers who already work one job must take a second.

I have noted over the years that families willingly make financial sacrifices to ensure loved ones get the care they need. Unfortunately, many families do not realize all the expenses of health care until the bills arrive months later. I, therefore, advise families to check regularly any pending medical bills. I also urge families to plan ahead for payment. Then, if financial hardships loom, I can direct family members early to available financial help.

The Case

Mrs. L. and her daughter certainly face many common primary and secondary stressors of caregiving. The primary stressor, caregiving burdens, takes a heavy toll on both women. Mrs. L. must spend several hours and considerable physical effort getting Mr. L. showered and dressed each morning. She also has anxiety about cleaning and dressing his pressure sores correctly. When the pressure sores are slow to heal, for example, she worries that her care technique is faulty. And, when Mr. L.'s appetite diminishes and his eating habits turn irregular, Mrs. L. frets about dosing his blood thinner safely. Mr. L.'s neediness also makes Mrs. L. sleep fitfully. She worries at night about what would happen and whom to call if something goes wrong with him. She sleeps even worse than usual when Mr. L. moves from the bed in the bedroom to the recliner in the living room where she cannot watch him. Her constant worry and accumulating sleep deficit leave Mrs. L. ever exhausted.

Caregiving burdens affect the daughter, too. Though busy in her own life, the daughter feels guilty about the caregiving tasks for Mr. L. she does not get to. But she also experiences frustrations with some of the tasks she does get to. "I feel I should give Dad the best care possible for comfort," she explains. "If he says he needs something like a space heater or a bathroom rug, I rush out to buy one right away. But then he often does not use what I get for him. That frustrates me." At the same time the daughter resents shouldering so much of Mr. L.'s care when she sees her brother as doing so little.

Secondary stressors also have an impact on both women. Mrs. L., for example, experiences a disheartening social isolation. Although her daughter visits almost daily and her son visits occasionally, Mrs. L. misses contact with her friends. Mr. L. refuses to go to the common dining room for dinner, the retirement center's main social event of the day. As a result Mrs. L. eats dinner alone with Mr. L. in their apartment. She tries lunching out with friends but feels guilty about leaving Mr. L. for even a short time. She stops the lunches when Mr. L. repeatedly says afterward, "Lunch must have been good. You were away an awfully long time." Mrs. L. would like to have friends visit at the apartment and they offer to, but Mr. L. refuses visitors most days. Mrs. L. responds ruefully, "Since (Mr. L.) broke his hip, I don't feel a part of the outside world. I am a prisoner in my own apartment. I have no freedom. I have lost my life." Observing her mother's social isolation, the daughter adds, Family members "are the forgotten people in these (caregiving) situations. We're invisible to others so long as we keep the patients alive and prop them up. Someone has to give them care, but we miss (our other family and friends) in the meantime."

The secondary stressors, life-role overload and social isolation, affect the daughter, too. She experiences severe life-role overload with working her full-time job, raising teenagers as a single mother, and helping to care for Mr. L. That overload causes her to neglect herself. She skimps on sleep, stops exercising, and gives up nearly all of her social life. Still, she thinks she never has enough time to care for everyone in her life—her own children, Mr. and Mrs. L., and herself. Like her mother she sacrifices a lot of herself to care for others.

Fortunately, Mrs. L. experiences few financial hardships from the caregiving for Mr. L. Mr. and Mrs. L. have good health insurance, which covers most of his medical expenses, and Mr. L.'s pension from the utility company covers the other expenses. The daughter, in contrast, experiences significant financial hardships. The caregiving for Mr. L. causes her so much absenteeism from work that her boss cuts her hours. She loses wages and benefits. The persistent exhaustion resulting from juggling her job, her family, and Mr. L.'s care eventually forces her to retire sooner than she planned. She must then live more frugally than before. She can no longer afford two of her favorite activities, going to the movies and eating out with girlfriends. She says sadly, "This is not the retirement I imagined for myself."

Family Caregiving in Dementia: A Special Case

Mr. L.'s decline begins with a hip fracture and ends with a fatal stroke. But I believe he acquires a dementia in between even though he does not die from it. Nonetheless, many Americans *do* die from dementias these days.

Most people think of the dementias as chronic diseases, not fatal ones. But they are both. Over five million Americans had the most common kind of dementia, Alzheimer's, and nearly 85 thousand died from it in 2013 [32]. That number accounted for 3.3 % of American deaths, making Alzheimer's dementia the sixth leading cause of death in the USA that year [33].

A dementia is any progressive intellectual decline characterized by impairments of short-term memory and at least one other cognitive function such as word-finding, navigation in familiar surroundings, thinking and planning, or previously mastered motor skills. The impairments must be severe enough to compromise a person's social or vocational function. Alzheimer's is one of several dementias, which differ among themselves in duration but not outcome. Alzheimer's dementia, for example, may last up to 20 years [9, 30] while the other dementias last only a few years [17]. Still, all the dementias eventually cause severe physical and mental disability and death [34, 35]. They should, therefore, be considered fatal diseases.

As they run their courses, the dementias create enormous monetary and nonmonetary costs for the USA. The nation's estimated monetary costs for Alzheimer's dementia run about $400 billion per year [32]. The other dementias add to that sum. All Americans bear some of these monetary costs even if only indirectly. Yet the nonmonetary human costs—physical, social, and emotional—may be even greater than the monetary costs, and family caregivers of dementia patients bear most of those costs directly.

The enormity of these costs likely accounts for why the dementias, especially Alzheimer's, have prompted so much research on family caregiving with them. Cancer has also prompted some research on family caregiving, but the organ-failure diseases such as heart, lung, or liver failure have prompted almost none.

Though limited, the research specifically contrasting the caregiving burdens of various chronic or terminal illnesses consistently documents similar care tasks but higher care burdens for the dementias than for the other illnesses. Table 9.1 shows the differences reported in one nationwide survey of family or friend caregivers to chronically ill American seniors [22]. Caregiving with demented patients averaged several more hours per week and more often required administering medications, helping with various basic activities of daily living, and managing incontinence than did caregiving with nondemented patients. Caregiving with demented patients also involved arranging or overseeing outside patient care services (such as those related to personal care or home modifications) more often than did caregiving with nondemented patients.

In addition, caregiving with dementia patients created more strains on caregivers than did caregiving with nondementia patients. Caregivers of dementia patients suffered emotional or physical problems more often and more severely than did caregivers of nondementia patients. Some dementia patients' misbehaviors such as wandering, screaming, or being violent may be partly to blame [9, 36, 37]. Caregiving with dementia patients may also rob caregivers of more restorative activities and time than does caregiving with nondementia patients. More caregivers of demented patients in this study sacrificed leisure activities (such as vacations, hobbies, or other recreational activities) and time with other family than did caregivers of nondementia patients. Furthermore, fewer caregivers of dementia patients than caregivers of nondementia patients reported that others did their fair share of caregiving. The insufficient restorative activities and time, and the consistent overwork in caregiving may explain the widespread depression and anxiety among dementia caregivers.

Table 9.1 Contrasts in caregiving for chronic dementia and nondementia illnesses[a]

Contrasting characteristic	Dementia[b] caregiving	Nondementia[b] caregiving
Average hours of caregiving per week	17	13
Basic care tasks required[c]		
• Administering medications (%)	55	37
• Dressing (%)	47	28
• Getting out of bed or chair (%)	46	34
• Bathing (%)	39	23
• Toileting (including getting back and forth to the bathroom) (%)	38	23
• Feeding (%)	33	16
• Managing incontinence (%)	26	11
Arranging or overseeing outside patient care services[c] (%)	64	53
Caregiver mental or physical problems due to caregiving[c] (%)	22	13
Strain on caregiver[d]		
• Emotional	3.0	2.2
• Physical	2.4	1.8
Personal sacrifices of caregiver[c]		
• Forgoing vacations, hobbies, or other leisure activities (%)	55	41
• Spending less time with family than before (%)	52	38
Others do their fair share of the caregiving[c] (%)	59	74

From Ref. [22], with permission

[a]Caregiving in this study meant giving unpaid care during the past year to someone, usually a relative, who was more than 50 years old.

[b]*Dementia* was defined as having an "Alzheimer's (diagnosis), confusion, dementia, or forgetfulness"; *nondementia*, as having any other chronic medical illness.

[c]Percentages of caregivers reporting the particular item.

[d]Ratings for strain ranged along a scale of one to five, least to most.

How, then, might HPs help family caregivers of demented patients? I offer a few care suggestions here, but the reader should understand them as merely supplementing or reinforcing the general care suggestions for chronically and terminally ill patients at the end of this chapter. Furthermore, the reader should keep in mind that most suggestions here are based only on expert opinion. They make sense but do not yet have adequate validation in data.

I suggest strongly that the principal doctor of either the dementia patient or the caregiver consciously track the caregiver's physical and emotional conditions. To do so, the doctor may simply need to ask periodically how the caregiver is coping with the strains of caregiving. The doctor might then use the caregiver's reply as a springboard to further help as needed. For example, as the reply indicates, the doctor might teach the caregiver about dementia and its course, help solve particular caregiving problems, or identify and encourage healthy coping skills [38]. And *every* HP attending the dementia patient, not just the principal doctor, should stay alert for signs of caregiver depression or anxiety. If those signs appear, the HP should take steps to ensure the caregiver receives adequate treatment [36]. And, of course, the HP should urge every caregiver to limit life-role responsibilities sensibly, to maintain social contacts, and to safeguard restorative activities [37]. Such tactics are key to enduring the many years of caregiving for a dementia patient.

If the patient lives long enough, dementia care will eventually exhaust the caregiver. The doctor may then need to step in, urging, perhaps insisting, the family transfer the patient to a nursing home or inpatient hospice [9, 30]. The aim is to gain expert help from experienced HPs and to distribute caregiving burdens among many people. At such times both nursing homes and inpatient hospices are good options for satisfactorily relieving the caregiver of most of the physical caregiving burden. But the transfer may not relieve him or her of the emotional burden: The caregiver may feel guilty about placing the patient in an institution or may worry about turning over the patient's care to strangers. Hospice transfers for dementia patients may, moreover, involve special obstacles not involved with nursing home transfers. For example, many families do not consider the hospice option (perhaps because they do not think of dementia as a fatal illness), doctors may have qualms about certifying a dementia patient terminal (that is, likely to die within six months) [17, 30], and the relatively scarce inpatient hospices may be inconveniently located for the family's access.

The Case

Mr. L. was never officially diagnosed with a dementia, and Mrs. L. adamantly insists he never had one. But I think he did. Mr. L.'s frequent repetition of stories or comments during the same conversation, his inability to remember scores of ball games even as they were being played, and his getting lost in familiar areas such as the hallway outside the apartment all suggest memory problems. In addition, his loss of interest in reading, a lifelong passion, suggests mental apathy. Furthermore, Mr. L.'s steadfast refusal to go outside the apartment for dinners, his insistence on Mrs. L.'s staying with him in the apartment, and his refusal to have visitors there suggest significant social impairment during those last years.

Furthermore, Mr. L.'s fundamental care needs and the care Mrs. L. and the daughter must give him fit better the pattern of a dementia than a nondementia illness. Mrs. L. and her daughter devote many hours to his care, administering his medications, helping him bathe and dress, and managing his occasional incontinence. They must also schedule outside services such as appointments with the home health nurse for wound care and with various doctors for follow-up medical care. Caregiving for Mr. L. also creates severe emotional strains for both women. Mrs. L. says she has lost her freedom and social life due to the caregiving for Mr. L. She feels imprisoned in the apartment. The daughter likewise has to sacrifice leisure activities and resents her brother for not doing his fair share of the care. The daughter must eventually retire from her job and incurs new financial hardships as a result.

All these factors—Mr. L.'s medical presentation, the nature of his care needs, and the effects of his care on Mrs. L. and the daughter—convince me he had a chronic dementia such as Alzheimer's disease during those final few years. His illness course until his rapidly fatal hemorrhagic stroke follows a slowly progressive frailty pattern typical for a chronic dementia. (See Chap. 5 for a graphic depiction of such a course.) While diagnosing a dementia early would not have changed the care tasks Mrs. L. and the daughter had to perform, it might have helped them to anticipate future problems with Mr. L.'s care, to understand their reactions, and to receive assistance tailored specifically to long-term caregivers of dementia patients.

The Impact of End-of-Life Caregiving on Family Caregivers

End-of-life caregiving obviously has great impact on caregivers. That impact is especially great on family caregivers due to their close personal ties to the patients. It may have social, financial, and legal dimensions, but its emotional and physical dimensions are usually the most powerful [39]. I focus my discussion here on those two dimensions. In doing so, I follow the literature's usual approach of discussing the negative and positive effects of the emotional dimension separately but all the effects of the physical dimension together.

The Negative Effects of the Emotional Dimension

Extensive research has documented the negative emotional effects of caregiving on family caregivers. The most consistent [1, 3, 21, 40] and widespread [8, 41] of those effects is depression. Mrs. L. and her daughter surely had it with their chronically low moods, poor sleep, fatigue, and other depressive symptoms. Their experience is typical. One study found over two-thirds of family caregivers suffer diagnosable depression [41], sometimes with anxiety. The depression may begin during the chronic nonterminal phase of care and increase later during the terminal phase when patients' symptoms worsen, their care needs increase, and caregivers must face the sad reality of patients' impending deaths.

What predisposes family members to depression during caregiving? Associations exist with both patient and caregiver factors. The patient factors include dementia [36], distressful symptoms, mobility problems [42], and intense care needs [23, 29]. The association between patient dementias and caregiver depressions is especially strong: Nearly, all studies on the topic show it [36]. It must, therefore, be cause and effect. Patient dementias likely cause caregiver depressions due to patients' progressive loss of function; their occasional problem behaviors; and caregivers' frequent discouragement and decreased life satisfaction under the circumstances [17, 36, 38].

The caregiver factors associated with caregiver depression include young age [29], female gender [8, 43], Euro-American ethnicity [44], and low socioeconomic status [22]; and poor physical health, inadequate coping skills, and minimal social support [23, 45]. Thus, young caregiving wives of chronically or terminally ill husbands are especially prone to this depression [26]. Caregiving often robs these women of intimate affection, vocational opportunities, economic security, and social contacts. Such losses drastically change the women's images of themselves and their expectations for the future. These women may come to resent shouldering the heavy caregiving duties for their husbands ill so early in life when serious illnesses and heavy care burdens are not supposed to occur. The young woman who attended her quadriplegic husband for years certainly resented her sudden, unexpected, and permanent change from wife to caregiver as required by his injury [12]. Such women often have to bear the full burden of supporting and raising children while caring for their ill husbands. Furthermore, in cases of disease, they may worry that either genetics or contagiousness may make them or their children susceptible to their husbands' diseases. And due to their young ages these women may lack adequate coping skills, some of which develop only with time and life experience. For all these reasons young caregiving wives tend to suffer depression more commonly than do other caregivers. HPs have a particularly important duty to monitor these women for depression and to guide them to psychiatric treatment when necessary.

Also of note, caregiver depressions from terminal situations may not resolve quickly after patients die. In fact, an early study of intimate caregivers of AIDS patients revealed that caregivers' depressions lasted up to three years after patient deaths [46]. Friends and family, therefore, should exercise patience even if caregivers' depressions require some months to resolve. During that time other people should avoid giving premature, gratuitous advice that caregivers "get over" their depressions and move rapidly on with life.

Other negative emotions in addition to depression may also accompany family caregiving. One such emotion is resentment. As a family caregiver said in one study, "Even if you love (the care recipient), you can find yourself resenting" that person for the caregiving burden he or she causes [47]. The resentment often comes with depression but need not [40]. Guilt may also accompany caregiving. Family caregivers may feel guilty, for example, about not giving "good enough" care or "complete" care [27, 38]. They may also feel guilty about harboring the secret wish, however understandable, that the patient would just die to relieve the caregiver's own burden [47]. And grief, of course, may accompany caregiving, too. Family caregivers may grieve a patient's many functional losses as the patient declines [9]. Unlike in early life when all functional milestones are gains, all functional milestones near death are losses [27]. Each loss may cause family caregivers grief. They may also grieve a patient's transfer to a nursing home or similar institution when the patient can no longer receive adequate care at home [30]. That transfer may symbolize the reality of the patient's approaching death. And, finally, family caregivers grieve the approaching loss of the patient's companionship at death.

The Positive Effects of the Emotional Dimension

Because people's distresses always command the most attention, researchers have usually focused on the negative emotional effects of families' end-of-life caregiving. But that caregiving may also have positive emotional effects. Much of the research about these effects has involved caregiver ratings of life satisfaction. Those ratings are typically elicited using such contrasting statements as "These are the best years of my life" and "Compared to other people, I get down in the dumps too often." [45] Life satisfaction ratings run on average lower for family caregivers than for noncaregivers [8] but vary widely. Some ratings are remarkably positive. A few studies, for example, have shown the curious "dual function" phenomenon: The greater the burden of care, the more rewarding some family caregivers rate the experience [8, 41, 44].

In that light, one study found that nearly all the caregivers of dying AIDS patients remembered at least one positive emotional event from the experience [46]. The authors of the study offered several possible explanations for this result. First, caregivers may see terminal caregiving as serving important purposes. Caregivers in the study, for example, may have seen their caregiving as a way to show patients love and to ensure them dignity in dying [46]. Second, caregivers may have refocused from global, unachievable medical goals to smaller, achievable non-medical ones. Caregivers may have then derived satisfaction from achieving those smaller goals. For example, when the deaths of their patients became imminent, caregivers may have refocused from the unachievable goals of curing patients' diseases and saving their lives to the smaller, achievable goals of nursing the patients conscientiously and enjoying intimate moments with them. Third, caregivers may have changed their perspectives to see important positive meanings in ordinary events. Just a laugh or a compliment may supply those meanings. Caregivers may then draw on those positive, meaningful memories as a buffer against depression during later difficulties [46].

Some research has also sought associations between life satisfaction in end-of-life caregiving and caregiver characteristics. Some positive associations include the caregiver characteristics of male gender, good physical health, higher education, and comfortable family income. Another association involves upbeat subjective caregiver appraisals—but not necessarily objective measures—of patient distress, caregiving burdens, and social support [41, 45]. Thus, though still unproven as cause and effect, caregiver characteristics—especially a positive attitude—may play an important role in providing satisfaction during end-of-life caregiving.

Physical Effects

While much research addresses the emotional effects of long-term caregiving, remarkably little addresses its possible physical effects. The sparseness of the research on that topic creates difficulties for drawing conclusions. Adding to the difficulties is the lack of sharp distinction in some studies between the effects of caregiving and of bereavement. Some experts, therefore, see much of the current research about long-term caregiving and its physical effects on caregivers as sketchy, weak, or inconsistent [28, 36]. Those experts hesitate to draw even tentative conclusions.

But other experts see the research as suggesting links between long-term caregiving and increased blood pressure, abnormal cardiovascular tests [36], decreased immune system effectiveness (and hence increased infections) [45], poor wound healing, and injuries related to frailty [4]. Most importantly, the research may suggest a link to increased overall mortality. One expert even proposes that long-term caregiving be considered an independent risk factor for overall mortality [2]. Nonetheless, direct links between long-term caregiving and physical effects on caregivers still require considerable research validation.

But an indirect link may also may exist: Family caregivers may neglect their own physical health due to devoting themselves to their sick relatives. Caregivers' preoccupation with the patients' care needs, fatigue from long hours of caregiving and loss of sleep, and tendency to ignore their own health problems rather than to seek help may all contribute to this neglect. I intuitively think this link is real.

In summary, current research indicates that long-term end-of-life caregiving often harms family caregivers' emotional health and may harm their physical health. Additional clarifying research is needed especially for physical health. But meanwhile HPs face the challenge of helping family caregivers such as Mrs. L. and her daughter manage the emotional and physical problems that accompany end-of-life caregiving. In the next section I describe interventions that can help.

Aids to Help Family Caregivers Provide Terminal Care

Family caregivers must often learn terminal care by trial and error. This "learn as you go" process involves various caregiver needs and many uncertainties. It also creates great anxiety for caregivers. HPs' overseeing the terminal care of patients can help family caregivers by suggesting proven aids to the learning process. I discuss in this section the knowledge and caregiving skills *that family caregivers themselves say they need to learn* and aids shown to help in doing so.

Knowledge and Skills Caregivers Must Learn

What knowledge and skills, then, do family caregivers themselves say they need to learn to give care during terminal illnesses? One early study asked this question of family caregivers of cancer patients. (Although the patients here were not necessarily terminal at the time, I believe much of the knowledge and many of the skills these caregivers mentioned pertain directly to terminal patients.) Overall, nearly one-third of these caregivers felt that they performed inadequately with the physical care of patients; over half, that they performed inadequately with the emotional care of patients and of close family and friends [31]. Table 9.2 lists the knowledge and skill needs in those two areas and two important additional skill needs (related to finances and communications) that caregivers often said they lacked. Caregiver knowledge and skill needs mentioned under the physical care of patients included understanding the patient's disease and its course, knowing the specific tasks of home care, learning physical care procedures (such as administering injections or providing feedings), and managing treatment side effects. Caregiver knowledge and skill needs mentioned

Table 9.2 Knowledge and skill needs of family caregivers of cancer patients

General area of need	Specific needs[a]	Percent of family caregivers citing the need[a]	Possible intervention(s) to help in learning[b]
Physical care of the patient			
	Understanding the disease	25 %	Education
	Knowing specific tasks of home care	20 %	Education
	Learning physical care procedures	15 %	Education
	Managing treatment side effects	15 %	Education
Emotional care of the patient and others			
	Addressing the patient's fear of death	23 %	Counseling
	Coping with seeing the patient's distress	22 %	Counseling
	Reassuring the patient	20%	Counseling
	Addressing other family members' fear about susceptibility to the patient's disease	19%	Counseling
Finances			
	Solving financial problems due to caregiving	16%	Financial counseling
Communication in the health care system			
	Communicating effectively and efficiently with health professionals	16%	Education, counseling

From Ref. [31], with permission

[a]The table includes only those skill needs cited by at least 15 % of family caregivers in the study.

[b]Interventions refer to those listed in Table 9.3.

under the emotional care of patients and other people included addressing the patient's fear of death, coping with the patient's distress, reassuring the patient, and addressing the fears of others about susceptibility to the patient's disease. The specific additional skill needs cited by caregivers were solving serious financial problems due to the caregiving, and communicating effectively and efficiently with HPs.

Aids to Terminal Caregivers' Learning and Well-Being

The knowledge, skills, and well-being of family caregivers surely affect the well-being of terminal patients in their care. Therefore, HPs have *a duty to the patients* to address the relevant needs of family caregivers. HPs must be able to suggest to caregivers aids for learning what they need to learn to care adequately for the patients and themselves.

I believe such aids fall into three broad categories: the internal individual, the external individual, and the communal. The first two categories of aids are *individual* because they vary by individual caregivers. I add the label *internal* to those individual aids originating inside the caregiver (say, in his or her character) and *external* to those originating outside the caregiver (say, in life circumstances particular to him or her). Relatively little research has studied the individual aids, but both the internal and the external categories appear to lie largely beyond HPs' control. I use caregiver optimism to illustrate internal individual aids and social support to illustrate external individual aids. Considerable research *has* studied the communal aids, and HPs *can* influence those aids by helping to develop them and then using them in practice. I use caregiver education, support groups, and several other interventions to illustrate communal aids.

Caregiver optimism illustrates the internal individual aids. By *optimism* I mean a person's general disposition toward positive expectations [42]. I believe one's optimism is mostly set early in life but can still vary somewhat according to later circumstances. Optimism can also vary according to its focus. In terminal care, for example, a caregiver's optimism may vary depending on whether it focuses on mastering skills needed for the patient's immediate care (such as giving injections or examining for pressure sores) [1, 31] or on longer term care goals (such as providing the patient a comfortable death or achieving emotional intimacy with the patient) [23]. Overall, optimism helps caregivers endure care ordeals in key ways: It counterbalances the sadness that inevitably accompanies a patient's physical decline and death. Optimism also promotes good physical health for caregivers [3] and protects them against depression [42].

Social support illustrates the external individual aids. Visits by family or friends certainly can meet important caregiver emotional needs. But those visits can also provide opportunities for sharing problems and insights about caregiving. If the visitors are not immediate family or close friends, I recommend limiting visits to about 15 min. Such brief visits convey support but do not excessively interrupt caregiving duties or exhaust caregivers with socializing. Still, the visits can allow time to talk about the situation if caregivers wish and to communicate understanding, reassurance, affection, and comfort [48]. Overall, though, visitors should allow caregivers to set the direction for these visits. Visitors should not force the conversation. In particular, they should resist the temptation to fill silences with insignificant chatter.

As important as "talking with" and "being present for" are, social support must occasionally also involve "doing for." [27, 48] A caregiver may sometimes be unable to accomplish the ordinary tasks of life due to physical exhaustion, mental paralysis, or simply lack of time. Family or friends might then offer to shop, prepare meals, or do housework for the caregiver. Family or friends might also offer to watch the patient while the caregiver rests or does other tasks. I realize that a caregiver may sometimes reflexively and unwisely decline offers of help. Such a caregiver may require persistent offers of help or explicit assurances that accepting help does not mean the caregiver has "failed" at the patient's care [30]. Family or friends may rarely even have to disregard politely but steadfastly a caregiver's refusal of badly needed help. But, however well-intentioned, helpers must use great caution about giving help despite a caregiver's explicit refusal. Just knowing that family and friends care and are willing to help can sometimes provide the emotional lift a caregiver needs [48].

As with caregiver optimism, limited research suggests social support improves caregiver well-being and alleviates depression [41, 45]. HPs of dying patients, therefore, should ask caregivers about social support and encourage family and friends to give it in ways caregivers appreciate. Such encouragement should especially focus on the caregivers most likely to need the support: seniors, men, spouses giving care alone at home [48], and long-term caregivers.

Five kinds of interventions illustrate the communal aids for caregivers: caregiver education, psychological counseling, support groups, respite care, and patient training. Unlike caregiver optimism and social support, all these communal aids have undergone extensive research for their impact on family caregivers giving terminal care. One article summarizes 78 of the relevant studies [49]. Each study compared one or more interventions against no intervention. Most patients in the studies had dementias, but some had cancers or strokes. The caregivers were nearly all spouses or adult children of the patients, and the caregiving was intense and long-term: It averaged 30 h per week for four years.

The particular interventions within each kind of communal aid had similar methods and purposes. Caregiver education provided information and resources to help caregivers address disease-related problems. Psychological counseling involved ongoing one-on-one meetings between counselors and caregivers. The counseling typically used cognitive–behavioral techniques to help caregivers solve problems, manage time and emotions, or protect leisure-time activities. Support groups gathered caregivers to meet each other, to discuss common problems and their solutions, and to gain moral support. Trained experts led some of the groups. Respite care (including adult day care) relieved caregivers briefly from caregiving duties and looked after patients in the meantime. And patient training aimed to improve patients' affect and living skills. Demented patients, for example, might attend anger management or memory improvement classes.

Table 9.3 The impact of communal aids[a] on family caregivers of dying patients

Kind of communal aid[b]	The magnitude of impact on family caregivers			
	Knowledge and skills to cope with caregiving[c]	Feelings of burden[c]	Self-rated depression[c]	Self-rated well-being[c]
Caregiver education	Moderate[d]	None	Moderate[d]	Moderate[e]
Psychological counseling	Moderate[d]	Small	Small	Small
Support groups	Small	Small	None	None
Respite care	Small	Small	Small	Small
Patient training	None	None	Small	Large[e]

Adapted from Tables 3 and 7 in Sorensen et al. [49], with permission

[a] *Communal aids* are those interventions that work through collective, often institutional, practices, policies, or resources to serve family caregivers.

[b] Caregiver education provided information and resources about disease-related problems. Psychological counseling typically used cognitive–behavioral techniques to help caregivers solve problems, manage time and emotions, and protect leisure-time activities. Support groups gathered caregivers to discuss common problems and gain moral support. Respite care (including adult day care) relieved caregivers briefly from caregiving duties while seeing to patients. And patient training aimed to improve patients' affect and living skills.

[c] The word descriptions designate mean impact sizes as follows: "none," less than 0.20; "small," 0.20–0.39; "moderate," 0.40–0.59; and "large," greater than 0.59.

[d] Only these interventions had impacts at least moderately better in magnitude and statistically different (at $p < 0.01$) than no interventions.

[e] For caregiver self-rated well-being, caregiver education had a moderate impact and patient training had a large one. Neither was statistically different from no interventions perhaps due to wide dispersion of the data.

Table 9.3 summarizes the impact of each kind of aid intervention on four caregiver outcomes—knowledge and skills, feelings of burden, self-rated depression, and self-rated well-being. Two kinds of interventions, caregiver education and psychological counseling, had statistically significant, though only moderate impacts on caregiver knowledge and skills. One kind—caregiver education—had a similar impact on self-rated depression. These impacts typically persisted for six months or longer even after the interventions stopped. Unfortunately, none of the kinds of interventions had statistically significant impacts on feelings of burden or self-rated well-being. (The moderate impact of caregiver education and the large one of patient training on self-rated well-being failed to reach statistical significance perhaps due to wide dispersion of the data.) Some studies, however, tested combinations of aid interventions, and the combinations often produced statistically significant, large impacts over all outcomes except self-rated depression [3, 50].

Notably, the communal aid interventions benefited some caregivers more than others: caregivers of nondemented patients more than caregivers of demented patients (for all outcomes) [3]; adult children caregivers more than spouse caregivers (for all outcomes); older caregivers more than younger ones (for all outcomes except self-rated depression); and women caregivers more than men caregivers (for knowledge and skills and feelings of burden) [49].

Table 9.3 should prove useful in guiding HPs' recommendations about such interventions. Because education about patients' diseases provides a broad impact across all caregiver outcomes, all caregivers should receive it. If education alone does not adequately meet caregiver needs, HPs might recommend adding other interventions. Those interventions might target specific outcomes that caregivers need. For example, psychological counseling would likely improve a caregiver's knowledge and skills; patient training would likely improve a caregiver's sense of well-being. HPs might thereafter recommend additional interventions based in part on caregiver demographics. For example, if a young daughter (such as Mr. L.'s) cares for her moderately demented father and education does not adequately improve her sense of well-being, the HP might anticipate further improvement by adding, say, patient training and recommend that.

The Case

Internal individual aids are hard to assess from the outside. That is certainly true about the personal optimism (or lack of it) with which Mrs. L. and her daughter view Mr. L.'s illness because both women guard their emotions so closely. The only way to assess optimism and other internal individual aids in such cases is to ask. With Mrs. L. and her daughter, their personal doctors or pastors are probably best situated to do so.

In comparison, the external individual aids are easier to assess. Both Mrs. L. and her daughter, for example, have obvious social supports. Mrs. L. has friends at the retirement community who offer to visit her, to run errands for her, and to take her to lunch. While these friends do frequently run errands for Mrs. L., they can arrange only rare social visits at the apartment or lunch dates out with her because Mr. L.'s frequent "bad days" often scuttle social plans at the last minute. And the daughter does have a few close friends she vents to confidentially about the frustrations of caregiving. Those friends listen patiently and allow her "pity parties" to run their course. The friends wisely offer only occasional unsolicited advice about how to handle the problems.

Notably, the communal aid interventions can likely help both Mrs. L. and her daughter. Their demographics are favorable for such interventions: Both are women, Mrs. L. is a senior, and the daughter is, of course, an adult child of the patient. Though Mrs. L. and her daughter insist they have had no communal aid interventions, they actually have. Concerning caregiver education, the home health nurse teaches them about care for pressure sores. The nurse demonstrates good wound cleaning, checks Mrs. L. and the daughter as they practice it, and measures the sores faithfully over time to document progress. The measurements prove crucial in convincing Mrs. L. and her daughter that the sores are healing.

Concerning psychological counseling, the daughter consults her internist informally about the stresses of long-term caregiving for her father. She complains that her father is "only playing helpless" when bathing and dressing. She thinks he is merely trying to gain attention. The internist tries a simple psychological technique in response. Suspecting a dementia with Mr. L., the internist explains to the daughter that Mr. L. may not be willfully refusing, say, to put on his socks. He may have just lost the ability to do so. The daughter thinks about the idea between appointments and tells the internist at the next one that she accepts his idea. Furthermore, she says that it has dramatically increased her patience for helping her father.

Concerning respite care, the internist urges the daughter and her mother to implement their plans for a vacation from caregiving. The internist provides them contact information for several reliable respite care agencies which can take care of Mr. L. during the vacation. Then with her mother's approval the daughter arranges a vacation to Florida for both of them and schedules care coverage with one of those agencies. Unfortunately, shortly before the vacation begins, Mrs. L. cancels it again because she still "cannot bear to leave (Mr. L.) with a stranger."

Concerning patient training, the home health nurse perceives the frustration of Mrs. L. and her daughter with Mr. L.'s arduous, time-consuming care. The nurse knows she can do little to change Mr. L.'s care needs. In fact, they will probably increase with time. The nurse decides, however, to encourage Mr. L. to express thanks to Mrs. L. and their daughter while he can. The nurse begins to prompt Mr. L. to say thank you after each wound cleaning and other care tasks. Though Mr. L. often forgets without specific cuing, Mrs. L. and the daughter say his thank-yous (whenever they do come) help dispel their feeling of being taken for granted.

And concerning a support group, meetings with other long-term caregivers might help Mrs. L. and her daughter cope with their caregiving burdens. A caregivers' group could support the women emotionally and allow them to learn from other caregivers' experiences. A retirement community such as Mrs. L.'s would be an ideal place for such a group, but I do not believe one exists there now.

Specific Steps in Helping Family Caregivers

Helping family caregivers requires meeting them where they are. Unfortunately, many family caregivers believe the American healthcare system falls far short of doing that. They feel frustrated and undervalued by the system. Many caregivers think it lacks clear goals for care; explicit, transparent decision-making procedures; and long-term anticipatory planning beyond just the next follow-up appointment [12]. Family caregivers also feel the system takes them for granted, expecting them to provide whatever outside care patients need whenever patients need it regardless of the inconvenience, expense, or hardships involved. Caregivers feel "invisible" to the system. They also believe it trains them poorly for their care tasks, rarely solicits their opinions, and almost never asks about their needs [6]. Caregivers desperately want the system—especially the doctors—to respect them, to educate them, and to recognize their hard work [5, 12, 18]. The American healthcare system obviously needs to improve its dealings with family caregivers.

One key objective in meeting family caregivers where they are and in improving the manner of dealing with them should be making caregiver needs a major focus in the care of dying patients [6, 24]. Doctors must lead that effort, but how? Table 9.4 suggests a general approach and some specific steps the doctor might take. The general approach involves paying attention to the caregiver as well as to the patient. The method in doing so uses frequent, open, and bidirectional communication between doctor and caregiver [19]. Both sides must feel free to speak and obliged to listen. Naturally, addressing the caregiver's problems may sometimes be done better in the patient's absence than presence. The doctor must use good judgment on that point.

The specific steps helpful to the caregiver separate into those related to care of the patient and those related to care of the caregiver. The steps related to care of the patient involve giving the caregiver the information and skills necessary to care properly for the patient outside the hospital or clinic. The doctor should certainly give the caregiver important information about the patient's condition at critical

Table 9.4 Steps for helping the family caregiver of a dying patient

General approach 1. Pay attention to the caregiver as well as to the patient. 2. Encourage frequent, open, and bidirectional communication between doctor and caregiver.
Specific steps for helping the caregiver care for the patient 1. Give the caregiver important information about the patient's illness especially at critical times. 2. Discuss the implications of the medical facts for the patient and the caregiver. 3. Ask the caregiver's opinion about the patient's condition and care. 4. Assign one nurse or medical assistant to teach the caregiver the skills necessary for the patient's care at home. Review those skills with the caregiver periodically thereafter. 5. Provide a contact telephone number for the caregiver to call with questions.
Specific steps for helping the caregiver him- or herself 1. Observe the caregiver for signs of physical or emotional problems. 2. Ask the caregiver occasionally, "How are you holding up?" 3. Visit the patient and the caregiver at home to detect patient problems and caregiver stresses that might not be apparent during office visits. 4. Always urge self-care for the caregiver. 5. If caregiver problems develop, address them promptly. 6. Give the caregiver contact information or referrals to organizations or individual professionals who can help with problems. 7. If the caregiver is nearing exhaustion or unable to sleep, urge either temporary respite care for the caregiver or permanent institutional placement for the patient.

Data from Refs. [2, 6, 19, 24, 31]

times in the illness. Those times occur at least at the diagnosis of a potentially fatal disease, at the start of new treatments (including any surgeries), at disease recurrence after a remission, and at the patient's imminent death [24]. The doctor, of course, should never just state medical facts but rather should help the caregiver understand the implications of those facts for both patient and caregiver. The doctor should also ask regularly for the caregiver's opinion of the patient's condition and care. The caregiver's reply may indicate important misunderstandings, which the doctor might then try to correct. Importantly, if the doctor's explanation of the patient's condition does not satisfy the caregiver, the caregiver should ask questions until getting adequate answers. The doctor should encourage such dialogue.

The doctor must also ensure the family caregiver learns the care skills necessary to provide the patient's care at home. To do so, the doctor might assign a particular nurse or medical assistant to teach those skills to the caregiver. The training often must address how to give medications on a regular dosing schedule, how to change dressings, how to turn and clean the patient in bed, how to perform physical therapy exercises, and how to operate medical equipment [6]. After teaching those skills, the assigned staffer should review them periodically thereafter with the caregiver. The staffer should also give the caregiver a reliable contact telephone number to call if questions about care procedures arise [24].

The other special steps to help, those involved with the care of the caregiver, should begin with the doctors observing the caregiver for signs of physical or emotional problems [31]. One simple technique is to ask the caregiver occasionally, "How are *you* holding up?" Another valuable but little-used and somewhat time-intensive technique is to visit the patient and the caregiver at home. Such home visits sometimes reveal patient problems and caregiver stressors not apparent during office visits [6]. If specific caregiver problems appear, the doctor should naturally address them promptly. The doctor need not assume primary responsibility for solving the problems if the caregiver is not already the doctor's patient. But the doctor should provide information or offer referrals to organizations or individual professionals who can help [24, 31].

One problem, caregiver self-neglect, occurs frequently. Often so thoroughly preoccupied with the patient's health, the caregiver neglects his or her own. The doctor should, therefore, always emphasize proper self-care. The caregiver's maintaining his or her best possible health is key to giving the patient the best possible care. Yet the caregiver may sometimes need the doctor's permission, even insistence, to take time to care for him- or herself [6]. Finally, if the caregiver is nearing exhaustion or unable to sleep [23], the doctor should urge temporary respite for the caregiver [31] or even permanent institutional placement for the patient [2]. Unfortunately, both options may present emotional or financial obstacles for some caregivers and may not fully relieve caregiver emotional distress [8].

A Comment About Caregiver Refusals of Help

Family caregivers sometimes refuse services HPs believe the caregivers need. One British study, for example, reported that 40 % of family caregivers giving terminal care at home need some help. It often involves housework. Yet nearly half of caregivers refuse the help when offered [51]. They give several reasons for refusing. Some say they do not want to relinquish their independence, "dignity," or familiar routines to get the help. Others consider the status quo "satisfactory" despite a decreased quality of life without the help. And still others deny themselves the help because they believe that other caregiving families need it more [51] or that it increases the workload of already overworked HPs [52]. Of course, some caregivers refuse the help because they cannot afford it or they feel duty-bound to give all the care their loved ones need.

If a family caregiver refuses sorely needed help, I suggest the HP not automatically accept the refusal at face value. Follow-up exploration of reasons makes sense. The HP might begin that exploration by saying, "This service is available, and I think it would help you. Please help me understand your reasons for not taking it." Once clear about the caregiver's reasons for refusing, the HP may be able to remedy some of the reasons or persuade the caregiver that others are unfounded. But, if good listening, empathy, and careful reasoning do not change the caregiver's mind, the HP may have to accept the refusal for the time being, monitor the caregiver's need, and perhaps suggest the service again later if the caregiver still needs it.

Summary Points

1. Families have a cultural and ethical duty—if not strictly a legal one—to care for their chronically ill or dying.
2. Family caregivers are especially vulnerable to depression. Therefore, health professionals attending chronically ill or dying patients should observe family caregivers for signs of depression. If it occurs, the health professionals should ensure caregivers receive adequate treatment for it.
3. Education and psychological counseling best help families cope with burdensome caregiving duties. But combinations of interventions often work better than does any intervention alone.

To Learn More …

1. English J. What do grown children owe their parents? In: O'Neill O, Ruddick W, eds. Having children: philosophical & legal reflections on parenthood. NY: Oxford University Press, 1979. P. 351–6.

2. Ory MG, Hoffman RR, Yee JL, Tennstedt S, Schulz TR. Prevalence and impact of care giving: a detailed comparison between dementia and nondementia caregivers. Gerontologist. 1999;39(2):177–85.
3. Sorensen S, Pinquart M, Duberstein P. How effective are interventions with caregivers? an updated meta-analysis. Gerontologist. 2002;42(3):356–72.

References

1. Schulz R, Quittner AL. Caregiving for children and adults with chronic conditions: introduction to the special issue. Health Psychol. 1998;17(2):107–11.
2. Schulz R, Beach SR. Caregiving as a risk factor for mortality: The caregiver health effects study. JAMA. 1999;282(23):2215–9.
3. Hebert RS, Schulz R. Caregiving at the end of life. J Palliat Med. 2006;9(5):1174–87.
4. Rabow MW, Hauser JM, Adams J. Supporting family caregivers at the end of life: "They don't know what they don't know." JAMA. 2004;291(4):483–91.
5. Zuger A. Arming the unsung heroes of health care. NY Times. 26 Oct 1999; p. F8.
6. O'Hara D. How can you care for the invisible caregiver? Am Med News. 17 September 2001; 26–7.
7. Schulz R, Mendelsohn AB, Haley WE, et al. End-of-life care and the effects of bereavement on family caregivers of persons with dementia. N Engl J Med. 2003;349(20):1936–42.
8. Haley WE, LaMonde LA, Han B, Narramore S, Schonwetter R. Family caregiving in hospice: effects on psychological and health functioning among spousal caregivers of hospice patients with lung cancer or dementia. Hospice J. 2001;15(4):1–18.
9. Prigerson HG. Costs to society of family caregiving for patients with end-stage Alzheimer's disease. N Engl J Med. 2003;349(20):1891–2.
10. Cantor MH. Strain among caregivers: a study of experience in the United States. Gerontologist. 1983;23(6):597–604.
11. Brody EM. "Women in the middle" and family help to older people. Gerontologist. 1981;21 (5):471–80.
12. Levine C. The loneliness of the long-term care giver. N Engl J Med. 1999;340(20):1587–90.
13. Davidson JE, Powers K, Hedayat KM, et al. Clinical practice guidelines for support of the family in the patient-centered intensive care unit: American college of critical care medicine task force 2004–2005. Crit Care Med. 2007;35(2):605–22.
14. Callahan D. What do children owe elderly parents? Hast Cen Rep. 1985;15(2):32–7.
15. English J. What do grown children owe their parents? In: O'Neill O, Ruddick W, editors. Having children: philosophical & legal reflections on parenthood. NY: Oxford University Press; 1979. p. 351–6.
16. Pearlin LI, Mullan JT, Semple SJ, Skaff MM. Caregiving and the stress process: an overview of concepts and their measures. Gerontologist. 1990;30(5):583–91.
17. Diwan S, Houghham GW, Sachs GA. Strain experienced by caregivers of dementia patients receiving palliative care: findings from the palliative excellence in Alzheimer care efforts (PEACE) program. J Palliat Med. 2004;7(6):797–807.
18. Fisher I. Families providing complex medical care, tubes and all. NY Times. June 7, 1998; 1, 21.
19. Levine C, Zuckerman C. The trouble with families: toward an ethics of accommodation. Ann Intern Med. 1999;130(2):148–52.
20. Emanuel EJ, Fairclough DL, Slutsman J, et al. Assistance from family members, paid care givers, and volunteers in the care of terminally ill patients. N Engl J Med. 1999;341(13):956–63.

21. Emanuel EJ, Fairclough DL, Slutsman J, Emanuel LL. Understanding economic and other burdens of terminal illness: the experience of patients and their caregivers. Ann Intern Med. 2000;132(6):451–9.
22. Ory MG, Hoffman RR, Yee JL, Tennstedt S, Schulz TR. Prevalence and impact of caregiving: a detailed comparison between dementia and nondementia caregivers. Gerontologist. 1999;39 (2):177–85.
23. Weitzner MA. The family caregiver of the older cancer patient. Hematol Oncol Clin North Am. 2000;14(1):269–81.
24. McCorkle R, Pasacreta JV. Enhancing caregiver outcomes in palliative care. Cancer Control. 2001;8(1):36–45.
25. Newton M, Bell D, Lambert S, Fearing A. Concerns of hospice patient caregivers. ABNF J. 2002;13(6):140–4.
26. Clipp EC, George LK. Patients with cancer and their spouse caregivers. Cancer. 1992;69 (4):1074–9.
27. Almberg BE, Grafstrom M, Winblad B. Caregivers of relatives with dementia: experiences encompassing social support and bereavement. Aging Mental Health. 2000;4(1):82–9.
28. George LK, Gwyther LP. Caregiver well-being: a multidimensional examination of family caregivers of demented adults. Gerontologist. 1986;26(3):253–9.
29. Tilden VP, Tolle SW, Drach LL, Perrin NA. Out-of-hospital death: advance care planning, decedent symptoms, and caregiver burden. J Am Geriatr Soc. 2004;52(4):532–9.
30. Hurley AC, Volicer L. Alzheimer disease: "It's okay, Mama, if you want to go. It's okay." JAMA. 2002;288(18):2324–31.
31. Hinds C. The needs of families who care for patients with cancer at home: are we meeting them? J Adv Nurs. 1985;10(6):575–81.
32. www.cdc.gov/aginginfo/alzheimers.htm; Accessed 05 March 2016.
33. www.cdc.gov/nchs/faststats/leading-causes-of-death.htm, Citing data from Health United States 2014, Table 20; Accessed 21 Apr 2016.
34. Harper GM, Johnston CB, Landefeld CS. Geriatric disorders. In: Papadakis MA, McPhee SJ, Rabow MW, editors. Current diagnosis & treatment 2016. New York: McGraw-Hill Education; 2016. p. 57–60.
35. Aminoff MJ, Kerchner GA. Nervous system disorders. In: Papadakis MA, McPhee SJ, Rabow MW, editors. Current diagnosis & treatment 2016. New York: McGraw-Hill Education; 2016. p. 1002–7.
36. Schulz R, O'Brien AT, Bookwala J, Fleissner K. Psychiatric and physical morbidity effects of dementia caregiving: prevalence, correlates, and causes. Gerontologist. 1995;35(6):771–91.
37. Zarit SH, Todd PA, Zarit JM. Subjective burden of husbands and wives as caregivers: a longitudinal study. Gerontologist. 1986;26(3):260–6.
38. Haley WE, Levine EG, Brown SL, Bartolucci AA. Stress, appraisal, coping, and social support as predictors of adaptational outcome among dementia caregivers. Psycholog Aging. 1987;2(4):323–30.
39. Schulz R, Newsom J, Mittelmark M, et al. Health effects of caregiving: the caregiver health effects study: an ancillary study of the cardiovascular health study. Ann Behav Med. 1997;19 (2):110–6.
40. Williamson GM, Shaffer DR, Schulz R. Activity restriction and prior relationship history as contributors to mental health outcomes among middle-aged and older spousal caregivers. Health Psychol. 1998;17(2):152–62.
41. Cheng WC, Schuckers PL, Hauser G, et al. Psychosocial needs of family caregivers of terminally ill patients. Psychol Rep. 1994;75(3):1243–50.
42. Kurtz ME, Kurtz JC, Given CW, Given B. Relationship of caregiver reactions and depression to cancer patients' symptoms, functional states and depression—a longitudinal view. Soc Sci Med. 1995;40(6):837–46.
43. Quayhagen MP, Quayhagen M. Alzheimer's stress: coping with the caregiver role. Gerontologist. 1988;28(3):391–6.

44. Pruchno R, Patrick JH, Burant CJ. African American and white mothers of adults with chronic disabilities: caregiving burden and satisfaction. Fam Relat. 1997;46(4):335–46.
45. Haley WE, LaMonde LA, Han B, Burtan AM, Schonwetter R. Predictors of depression and life satisfaction among spousal caregivers in hospice: application of a stress process model. J Palliat Med. 2003;6(2):215–24.
46. Folkman S, Moskowitz JT. Positive affect and the other side of coping. Am Psychologist. 2000;55(6):647–54.
47. Gavzer B. How can we help? Parade Mag. July 16, 2000; 4–5.
48. Clipp EC, George LK. Caregiver needs and patterns of social support. J Gerontol Soc Sci. 1990;45(3):S102–11.
49. Sorensen S, Pinquart M, Duberstein P. How effective are interventions with caregivers? an updated meta-analysis. Gerontologist. 2002;42(3):356–72.
50. Lorenz KA, Lynn J, Dy SM, et al. Evidence for improving palliative care at the end of life: a systemic review. Ann Intern Med. 2008;148(2):147–59.
51. Grande GE, Todd CJ, Barclay SIG. Support needs in the last year of life: patient and carer dilemmas. Palliat Med. 1997;11(3):202–8.
52. Ferrell BR, Borneman T. Pain and suffering at the end of life for older patients and their families. Generations. 1999;23(1):12–7.

Chapter 10
Culture and End-of-Life Care

The ascendancy of medical science has changed clinical medicine dramatically over the last century and a half. That science has emphasized objectivity over subjectivity. As a result, clinicians' observations of disease have gained ever more prominence over patients' descriptions of illness. Scientifically based clinical knowledge almost alone has come to be perceived as "real," [1, 2] and ideal medical practice as objective and value-free, undistorted by subjective feelings or biases [1, 3]. This view has produced powerful consequences including the contemporary "evidence-based" medicine movement, which requires scientific data to justify patient care practices.

Patients, however, have recently rebelled against such "objective, scientific" medical care. They insist once more that health professionals treat them as persons with feelings, relationships, and unique life contexts, and not merely as bodies with diseases. Patients rightly perceive their illnesses as having both a physical aspect and important nonphysical psychological, social, and spiritual aspects. Patients want their health professionals to address all those aspects. This patients' rebellion has coalesced into movements variously called holistic care, humanistic care, hospice, and death with dignity. Health professionals have had to reawaken to the subjectivity of illness.

Culture influences powerfully those subjective, nonphysical aspects of illness [4–6]. Culture determines when the community considers a person ill, how that person communicates his or her distress [3], what benefits and burdens an illness confers, which treatments are considered worthwhile, and what the illness means for the patient and the community at large [6–10]. Culture assumes particular prominence during terminal illness when dramatic, life-saving medical rescue becomes impossible and all can only wait for death to come. At that point the terminal patient and his or her survivors look to culture to provide ways for coping with the mystery and fearsomeness of dying [2, 11, 12]. Attending health professionals must be able to recognize and respect culturally based perspectives at that time.

This chapter addresses such perspectives on the end of life. The chapter defines culture; discusses key concepts derived from it; points out some of the practical

H.S. Perkins, *A Guide to Psychosocial and Spiritual Care at the End of Life*,
DOI 10.1007/978-1-4939-6804-6_10

implications for patient care in general and end-of-life care in particular; suggests ways health professionals might initially anticipate the cultural views of dying patients and their survivors; and describes LEARN, a method for eliciting people's culture-based views about illness and for negotiating treatment plans accordingly. The case of a seriously ill Mexican man, a former patient of mine, illustrates many of the points.

I distinguish the terms *ethnic group* and *race* from the outset. Ethnic group refers to an actual group of people who share values and beliefs and are, therefore, identifiable by themselves and others as "an ethnic group." [13, 14] Race, in contrast, refers to a grouping based on some vague perception of skin color [14, 15]. That grouping is merely social convention [16] and has no objective biological basis, not even a genetic one [3, 15, 17]. Some authors claim the concept of race arose in the eighteenth century to perpetuate biases about a human hierarchy grounded in European prejudices [5, 15, 16]. But the concept of race, I believe, has proven specious over time and should now be avoided [18]. For that reason I attribute cultures throughout this chapter to ethnic and other groups but not to races.

The Case

Senor B., an obese 63-year-old Mexican had lived 23 years as an undocumented alien in the USA. He presented one day as a new patient to my public clinic in South Texas. He spoke little English but told me through the clinic's Spanish interpreter that severe shoulder pain prevented him from working his usual jobs as a car mechanic and gardener. As a result he could no longer support his family. He also mentioned long-term heartburn, frequent thirst and urinations (and his suspicions of diabetes), and his erectile dysfunction. When I asked about herbal medications, Senor B. admitted to taking a friend's pills called *El Pajaro*, "The Bird," without relief. Senor B. smoked and drank "a little;" lived with his Mexican wife, also an undocumented alien; and still had a home in Mexico. He had no health insurance.

Over the next year I diagnosed severe shoulder tendonitis and uncontrolled diabetes, and treated Senor B. for both problems. Despite my repeated explanations otherwise, Senor B. steadfastly believed his oral diabetes medications allowed him to eat whatever he wanted. He gained weight. When the oral medications no longer controlled his diabetes, I recommended insulin. He refused, insisting that some of his diabetic friends became "addicted to insulin," went blind, and then died from it. All my efforts to persuade him to take insulin failed. So he continued the oral medications while his diabetic symptoms and blood glucoses worsened. He began to miss clinic appointments but always returned whenever I specifically called him to do so.

Senor B. then began to experience chest pressure and shortness of breath that limited his activity. The new symptoms suggested imminent risk for heart muscle damage, "a heart attack." The electrocardiogram showed prior damage, and the urine and blood tests showed his kidney function was deteriorating. I recommended immediate hospitalization, a cardiology referral, and cardiac catheterization if necessary. I also explained to Senor B. that he would likely need dialysis soon, and that the kidney specialists could begin preparations for it during hospitalization for his heart problems. Senor B. refused all my recommendations including hospitalization. He said he trusted me "because you take an oath when you go to school," but he did not trust "the government hospital." He believed he would die during "the heart surgery." He also feared the hospital would report his undocumented status to the police, and they would deport him.

I suggested Senor B. discuss his medical problems with his wife and return to the clinic as soon as possible with her afterwards. They returned a few days later. Senor B. refused all recommendations again. He insisted his fate was set, and no treatment could delay his death. He and his wife wanted to move back to the rural Mexican village where they had grown up and had their house. He wanted to die there.

I tried desperately to persuade him to take treatment at the public hospital instead. I pointed out that he could still enjoy some good life ahead but only if he got urgent treatment for his heart. That treatment was available in our hospital but surely not in any hospital in rural Mexico. I also promised that our hospital would not report him to the police or ask him to pay more than what he could easily afford. In fact, the hospital might forgive his entire bill. I finished by saying how sad I would be to see him forego the life-saving treatments we offered.

Senor B. thought for a moment and then replied, "I have to die from something, doctor. This is it. I want to live out my life back home with my family and friends."

I tried to negotiate a compromise, asking Senor B. to talk at least with the heart and kidney specialists before making a final decision. He reluctantly agreed. I made quick appointments for him with the specialists and a follow-up appointment with me afterwards. Senor B. kept none of those appointments, and I never saw him again. I suppose he and his wife returned to their home village in Mexico, and he died there.

I do not know all the ways Senor B. and I differed over the care I offered. Our differing demographics surely suggested that cultural differences made us see it differently. Senor B. was a Spanish-speaking, undocumented immigrant Mexican, probably Catholic, and educated no more than through primary school. In contrast, I am an English-speaking American citizen, Protestant, and educated through medical school. Even more strongly than the demographic differences, our conflicting general views about medical care also suggested cultural differences. Senor B. distrusted medical science and refused many of the treatments I recommended because he apparently considered them burdensome, ineffective, or even harmful. And he feared "the government hospital" as being evil and serving itself before its patients. In contrast, I trusted medical science and the treatments I recommended, and I believed they offered Senor B. clear benefits. I also viewed the public hospital as beneficent and serving patients well.

Years later I still wonder whether a more knowledgeable, more skilled approach to the cultural differences in this life-or-death situation might have served Senor B.

better than the approach I took. If so, he might have accepted the life-saving treatments I offered, and gone on to live a longer, fulfilling life.

Key Concepts About Culture

Culture is the values, beliefs, and behaviors a distinct group of people shares, uses to interpret their life experiences, and transmits to others [2, 6, 9, 18]. Culture exists in all such human groups. No person or institution avoids it: none is culture-free [2]. In fact, culture comprises a vital part of the conceptual perspectives, or frameworks, we all use to find meaning in our lives [14]. Much of that meaning comes to us through culturally based symbols and rituals [16].

This definition of culture has an important implication: A culture is identified only by its characteristic core values, beliefs, and behaviors—not by demographic labels such as race, nationality, or ethnicity. Still, many researchers including me use such labels to refer to culture because they are convenient and colloquial (though imperfect) markers for it. While I use demographic labels to refer to specific cultures in this chapter, the reader must keep in mind that the true basis for identifying cultures is values, beliefs, and behaviors, not demographic labels.

As universal human experiences, illness and death are core cultural topics [5, 18]. So important are they that medical anthropology, a specialized field within general anthropology, has emerged to focus on them [10]. Work in that field has already identified four common misconceptions about culture, illness, death, and medicine.

Misconception Number 1: Mainstream, Scientifically Based Medicine Offers the Only Valid Forms of Healing

This view, a kind of "cultural imperialism," rejects any culturally based approaches to healing that differ from mainstream, scientifically based medicine. Those alternative approaches range from acupuncture and chiropractic to folk medicines and spiritual healing rituals. Staunch cultural imperialists in medicine regard those approaches as ineffective at best and dangerous at worst [1]. Yet acupuncture, certain folk medicines, and other alternative healing techniques have recently proven beneficial in some situations. Cultural imperialism has unfortunately deprived mainstream medicine of the benefit of those approaches for many years.

Cultural imperialism contrasts starkly with the cultural humility Dr. Joseph A. Carrese, a doctor, anthropologist, and ethicist, models in his medical practice. His humility involves considering one's own cultural views as provisional truths subject to change with new experiences. It also involves fostering an interest in learning about patients' alternative cultural views and taking those views seriously when making treatment plans for patients. I return to the cultural humility idea later in this chapter when I discuss the LEARN method for resolving cross-cultural conflicts.

Misconception Number 2: Culture Is Homogeneous

This view claims that all members of a culture always think the same way. The logic here is simplistic and incorrect [3]: Because members of a culture share the same core cultural values, beliefs, and behaviors, they must always reach the same conclusions on specific issues. Education about cultural competence may unwittingly promote this view [5]. Lectures and journal articles list the supposedly distinctive views of specific ethnic cultures. And some medical anthropologists even provide clinical consultations focused on general ethnic cultural characteristics that may influence a particular patient's views [19]. Learners may incorrectly conclude they can deduce a patient's views from knowing just, say, his or her ethnic background. Thus, every Euro-American (EA) will insist on making his or her own medical decisions, and every African American (AfA) will distrust the healthcare system [20]. Such well-intentioned efforts at cultural sensitization may simply teach overly rigid stereotypes. Individual variations can get lost, and an inflexible "cookbook" approach to cultural differences can develop [1, 7, 9]. In reality, though members of a culture share its specific core values, beliefs, and behaviors, they may differ on many other values, beliefs, and behaviors that can prove determinative on certain issues [14]. Many commentators thoroughly decry stereotypes, ethnic and others, for that reason [2, 5, 11, 21].

I, however, have a more balanced opinion about stereotypes. I believe they have benefits as well as hazards. I see two benefits in particular. Cultural stereotypes can alert health professionals to views other than their own. (Teaching of cultural competence must always instill respect for those other views within broad limits.) Stereotypes can also help health professionals initially approximate a particular patient's views based on ethnic group, gender, or other characteristics linked to culture. While avoiding overgeneralizing, health professionals can use stereotypes as a place to start, but never to end, explorations of an individual patient's views. Health professionals must always resist the temptation to pigeonhole a patient too quickly or too rigidly and be willing to entertain his or her variations on a general cultural pattern [3, 6, 9, 22].

Misconception Number 3: Culture Is Static

According to this view, whatever core values, beliefs, and behaviors have characterized a culture before will always characterize that culture. It will never change. One form of this attitude is essentialism, the idea that a culture's core values, beliefs, and behaviors are innate in its members and hence do not change [16, 23]. But cultures obviously *do* change in response to changing circumstances. The Industrial Revolution, the American Civil War, urbanization, growing affluence, and the Internet all changed American core cultural views about self-worth, education, communication, freedom, personal rights, and communal obligations. And in medicine the tragic AIDS epidemic produced similarly dramatic cultural changes in attitudes about professional obligations and patient rights. Such powerful external events may force cultures to change in response [16].

Misconception Number 4: All Differences Among People from Different Cultures Are Cultural Differences

This view attributes interpersonal differences to culture first and foremost. We often assume that, when cultures differ, they cause our differences. But different life experiences, education, logic, and communications can also cause differences among people from different cultures [24].

An actual case illustrates differences that initially seemed cultural but were not. EA doctors hospitalized a stuporous, elderly AfA woman with an acute pneumonia [25]. Believing she was imminently dying, the doctors asked her family for permission not to resuscitate, intubate, or ventilate her. The family refused to give that permission. The doctors initially suspected cultural differences between EA tendencies to stop life support early and AfA tendencies to continue it late into a terminal illness. Extensive negotiations eventually revealed another basis for the disagreement. The woman had been treated for a similar pneumonia in the past. Her doctors at the time had told the family she would die. Yet she had surprised everyone by surviving. When facing a similar situation again, the family reasoned that, if the doctors had predicted incorrectly before, they could be predicting incorrectly again. The family refused permission to withhold life support for the second pneumonia because they thought the patient might still survive and they wanted to give her every chance to do so. The patient received full treatment (including mechanical ventilation) and survived to return home again. Thus, not culture but logic based on a similar prior experience prompted that family's second refusal to withhold life-support treatment all along.

Sources for Diversity Within Cultures

I believe diversity within cultures comes from three main sources—idiosyncrasies, subcultures, and variations in acculturation. The reader should note these sources are not mutually exclusive: Some causes for diversity may be categorized as originating in more than one source.

Idiosyncrasies are ideas a person shares with only a few, if any, other people. A patient treated on hemodialysis for long-standing kidney failure, for example, suddenly refuses further dialyses when he learns the dialysate baths contain potassium. He claims to have a "potassium allergy," a biological impossibility. The patient says exposure to potassium in the baths will cause a severe reaction and he will die. This belief may have arisen from the patient's peculiar misunderstanding about high blood potassium concentrations in kidney failure. A consulting psychiatrist identifies the man's idiosyncratic belief about a "potassium allergy" as an isolated fixed delusion. The man's doctors eventually convince him the baths are safe, and he resumes dialysis.

Subcultures are small groups defined by narrowly shared views within a larger culture defined by other, more widely shared views. Gender-specific views, for example, make the genders distinct subcultures within any larger ethnic culture. In this light one study suggests that men and women view the American healthcare system differently: Unlike men, many women in each of three ethnic groups—EAs, AfAs, and Mexican Americans (MAs)—believe the system empowers patients, benefits them, and respects their wishes [26]. Other narrowly shared views create additional, often partially overlapping subcultures, say, roughly according to age, education, vocation, or location of residence. Thus, every person belongs to many cultures simultaneously [7, 14].

Acculturation is the process by which a person from one culture assumes the values, beliefs, and behaviors of another culture over time [14, 27]. Acculturation usually refers to immigrants who settle in a new dominant ethnic culture. But the term can apply to *any* socialization from one culture to another. Medical school education illustrates acculturation without immigration: Medical students acculturate over time from the general nonmedical lay culture to the professional doctor culture [18]. A contrasting term, *assimilation*, refers to maintaining one's original cultural views while living in a new dominant culture.

Acculturation proceeds along multiple cognitive and behavioral dimensions at once, prompting anthropologists to track it variously through self-identification with either the former or new dominant culture; language usage; and preservation of former traditions [27, 28], food preferences, and social roles and duties [29]. Despite these multiple dimensions I imagine acculturation as moving along one composite continuum between the opposite poles of mere assimilation and full acculturation [9]. An immigrant starts life in a new dominant ethnic culture at the assimilation pole and moves some distance with time toward the full acculturation pole. How far he or she moves depends on how much of the dominant ethnic culture he or she adopts. The typical immigrant probably never achieves full acculturation.

A recent study supports such an acculturation process. Researchers studying popular views about breast and cervical cancer interviewed three groups of women in Los Angeles: Latinas, women who were born in Mexico or El Salvador; Chicanas, women who had Mexican ancestry but were born in the USA; and EAs, who had no Latin American ancestry and were born in the USA. Responses followed a distinct progression from an "Immigrant Model," reflecting the assimilation pole, to the "Biomedical Model," reflecting the full acculturation pole. The Immigrant Model, embraced wholly by the Latinas and partly by the Chicanas, attributed the cancers to direct physical trauma and generally unhealthy behaviors (such as smoking, promiscuity, or poor hygiene). In contrast, the Biomedical Model, embraced partly by the Chicanas and wholly by the EAs, attributed the cancers largely to recognized epidemiologic risk factors (such as genetics, family history, and exposure to certain sexually transmitted viruses) [30].

Health professionals who attend dying patients from homelands overseas may sometimes find an acculturation assessment helpful. One study provides a template for such an assessment. Researchers interviewed first, second, and third generations of MAs and Japanese Americans [28]. The first generations were the original immigrants, born in their homelands, and mostly elderly. The second and third generations were born in the USA and mostly middle-aged or young, respectively. The researchers assessed acculturation along three dimensions: the balance of knowledge between the homeland popular culture and American popular culture ("cultural knowledge"); the cultural orientation of everyday activities through language, customs, family structure, and gender roles ("cultural orientation in everyday activities); and "ethnic group self-identification."

Table 10.1 shows how the three generations differed. The first generations knew their homeland popular cultures better than American popular culture; had general homeland orientations in everyday activities; and claimed both homeland *and* American identifications. The second generations knew both their homeland and American popular cultures; had mixed homeland and American orientations in everyday activities; and split between either homeland *or* American identification. The third generations knew American popular culture better than their homeland popular cultures; split between homeland and American orientations in everyday activities; and split between either homeland *or* American identification. In summary, knowledge and orientation in everyday activities became progressively more acculturated to American culture with successive generations. But ethnic self-identification began as both homeland *and* American in the first generations and split into primarily homeland *or* American in the second and third generations [28].

How might health professionals easily and quickly assess acculturation in patients and family members? Using the same dimensions as in Table 10.1, Table 10.2 lists specific variables by which other researchers have assessed acculturation in these same ethnic groups [29, 31]. The variables suggest three simple, practical, and valid questions health professionals (HPs) might ask: [3] What language do you prefer to use (for immigrants from non-English-speaking homelands)? In which country were you, your parents, and your grandparents born? And which ethnic group do you consider yourself a member of? Though conceptually important, the psychological aspects of nonacculturation (such as keeping homeland values, maintaining connections to the homeland community, and feeling supported by that community) add little *discriminating* power to the three basic questions [29].

Table 10.1 Changes in acculturation over three generations of new immigrants to the United States

Dimensions of acculturation	Generation		
	First	Second	Third
Cultural knowledge	Homeland	Homeland *and* American	American
Cultural orientation in everyday activities	Homeland[a]	Homeland *and* American	American
Ethnic-group self-identification	Homeland *and* American	Homeland *or* American[b]	Homeland *or* American[b]

Data from Clark et al. [28]

[a]For cultural orientation in everyday activities, the first generation divides into two distinct populations: one with a high dominant homeland orientation and another with a lower but still dominant homeland orientation.

[b]For ethnic-group self-identification the second and third generations divide into two distinct populations each: one with a dominant homeland self-identification and the other with a dominant American self-identification.

The Case

The obvious demographic differences between Senor B. and me immediately revealed our different ethnic origins. In addition, though I did not realize it at the time, my initial clinical history largely answered the questions for assessing the more subtle issue of Senor B.'s acculturation to American society. That history revealed he had been born in Mexico, spoke only Spanish, and identified himself as Mexican. A full assessment lacked only confirmation that his parents and grandparents had also been born in Mexico. My history supplied other supporting information on the point, too. Senor B.'s speaking only Spanish and his care to avoid detection as an undocumented alien surely made him restrict his close contacts to other Mexican immigrants. And his distress about not being able to support his family and his apparently not having disclosed his health problems to his wife until very late suggested he followed the Mexican family patriarch's traditional bread-winning and protective roles. I should have realized early on that, despite having lived many years in the USA, Senor B. had embraced few of its mores and had not moved far from the assimilation pole along the assimilation-acculturation continuum. He had little acculturation to American culture. Only with more time caring for him could I assess his acculturation to mainstream medicine.

Culture and Medical Care in General

Having laid out basic concepts of culture, I now discuss in the next two sections some perspectives on culture and medicine. This first section addresses general perspectives related to illness and treatment, clinician–patient power differences; and family roles and responsibilities. The second section addresses more specific

Table 10.2 Assessing acculturation in research studies

Dimensions of acculturation [28][a]	Definitions [28][a]	Mexican-American examples [27, 31]	Japanese-American examples [29]
Cultural knowledge	Relative familiarity with homeland's popular culture over US popular culture. Reflects the balance of ethnic cultural influences.	• Birthplaces of parents and grandparents: Mexico versus the USA[b] • Knowledge of Mexico and US history, customs, and holidays • Most friends, neighbors, and close coworkers come from Mexico versus the USA.	• Birthplaces of self, parents, and grandparents: Japan versus the USA • Place where raised: Japan versus the USA
Cultural orientation in everyday activities	Kinds and degrees of activities that reflect homeland or US culture	• Language preference: Spanish or English • Adherence to traditional Mexican or American family structure and gender roles	• Language preference: Japanese or English
Ethnic self-identification	Identification as homeland or US ethnicity, that is, sense of belonging to homeland or US society	• Self-identification as Mexican, Mexican American, or American	• Self-identification as Japanese, Japanese American, or American

[a]Clark et al. originally considered "ethnic identity" the same as acculturation, but I consider them different. I believe "ethnic identity" is only a part of acculturation: Acculturation comprises a subjective component—"ethnic identity," "ethnicity," or ethnic self-identification—and an objective component—ethnic group membership based on observed behaviors. The last row of the table addresses the subjective component; the first two rows, the objective one. I use *acculturation* here in its sense covering both components. In addition, Clark et al. labeled what I call *cultural knowledge*, *cultural orientation in everyday activities*, and *ethnic self-identification* as "acculturative balance scale," "traditional orientation," and "face," respectively. I changed the Clark et al. labels to simplify them and to reflect the definition of acculturation in the text.

[b]Hazuda et al. also used the father's surname and the mother's maiden surname in their algorithm for differentiating Mexican Americans from other US ethnic groups.

perspectives related to end-of-life care including communication styles, advance directives (ADs), and informed consent in terminal illness. This second section presents several contrasts among major American ethnic cultures.

Illness and Treatment

Before modern medicine many cultures attributed illness to disharmonies inside and outside the body [8, 32]. Those disharmonies arose supposedly from imbalances of internal "humors" or from bodily attacks by external forces such as evil spirits. Illnesses occurred so frequently that they seemed like the body's natural state, and suffering seemed an inherent part of living. Some cultures still hold that belief today.

Twentieth-century medical science, however, radically changed many people's ideas about illness and treatment. Westerners in particular rethought their ideas about disease causation, treatment methods, and expectations for medical outcomes. The new medical science discovered germs, toxins, abnormal genes, anatomic defects, over- or underactive immune system responses, and other physical causes for disease. Westerners soon dropped many old ideas about nonphysical causes of illness including as a punishment from God or as a hexing by evil people or spirits.

People's attitudes about treatment changed, too. The old treatments addressing disharmonies had achieved little success [10]. Though the ill had sought treatment and hoped for benefits, they had rarely gotten them. People reacted by developing a pessimistic, fatalistic attitude toward diseases and their treatments. Some diseases such as cancer gave a quick "death sentence," [24, 33] which the patient had no hope of reversing [33]. Cancer relentlessly eroded the patient's physical powers, personality, and spirit until he or she died a typical cancer death [34]. Other diseases such as rheumatoid arthritis caused a progressive, crippling illness that defied treatment and lasted years. For both kinds of diseases, families administered at home the few, largely ineffective treatments available.

Then the new medical science introduced treatments that actually affected the course of diseases beneficially, sometimes dramatically. The revolutionary treatments included antibiotics; cancer chemotherapies; mechanized life support and intensive care; organ replacement therapies; and image-guided, minimally invasive surgical procedures. So many cures followed that people began to expect them. The treatment successes reinforced many Western cultural beliefs about the power of medical science, the duty to fight disease actively, the need to exercise self-discipline in pursuing disease prevention and treatment, and the importance of maintaining hopefulness in the face of disease.

The treatment successes also suggested to Westerners that God wishes good health for His people, that it is the body's natural state, and that good health can be effected by contemporary mainstream medicine. Diseases became problems to be solved.

Accordingly, Westerners abandoned their old fatalistic views. People began to see illnesses as temporary and expected the associated suffering to last only the short time needed for medicine's cures to take effect.

Clinician–Patient Power Differences

All clinician–patient relationships involve power differences [16] arising from the inherent differences in technical knowledge and from the frequent differences in personal health. The clinician has greater power based on greater technical knowledge and better health; the patient, less power based on less technical knowledge and poorer health. Ethnic and other cultural differences between clinician and patient only magnify those power differences.

Yet clinicians usually do not see those power differences. Two reasons may explain why. First, clinicians do not question the typical clinician–patient relationship as it is because medical practice largely mirrors the dominant, surrounding Western culture. Clinicians simply accept that relationship with its intrinsic power differences as the best or only way the relationship can be [8]. Second, clinicians like others with advantage often do not perceive their advantage. Clinicians, for example, usually underestimate their power to influence people and events inside and outside of medicine [3]. Vulnerable patients, of course, sense clinicians' power and patients' own relative powerlessness. As a result patients—especially minority patients—may not fully trust the healthcare system to serve their best interests. Because the clinician–patient power differences are real and important, clinicians must recognize them and manage them responsibly.

Family Roles and Responsibilities

Unlike EA culture, which champions the individual's autonomous decisions for him- or herself, non-EA cultures typically champion the family's collective decisions on behalf of the individual family member. Such family involvement provides the family member with reassuring security, dependability, and solidarity in unfamiliar or threatening situations [8]. In medical situations, for example, non-EA families take quite seriously their duties to keep vigil at the hospital, to monitor the patients' care, and to serve as liaisons with the clinicians [8, 10]. These families also provide some of the patients' direct, hands-on care in the hospital and most of that care at home [6]. For that reason many ethnic minority patients entrust their welfare completely to their families out of the belief, as one patient said, "My family will watch out for me when I need them to." [35].

Culture and End-of-Life Care Communication

Clinicians' bedside communications generally leave many American patients—particularly ethnic minorities—dissatisfied. For example, Hispanic Americans (HAs) and Asian Americans (AsAs) believe doctors do not spend enough time with them, listen to them, or consult them adequately in decisions about care [36]. These groups think they would receive better communications and care if they came from another ethnic group, undoubtedly EAs [36, 37]. HAs, AsAs, and AfAs also believe that either their poor English fluency or different lifestyles cause doctors to disrespect them [36]. These views suggest American clinicians need to improve their overall communications skills with minorities.

Those skills play an especially important role in end-of-life care. Many dying patients and their survivors want to tell health professionals their feelings about the dying experience, and culture fundamentally influences when and how people express those feelings. Health professionals must be prepared to listen and respond in culturally appropriate ways. This next section aims to help health professionals do so. It focuses on three important end-of-life communication topics that culture affects: communication styles, advance directives, and informed consent in terminal illness.

Communication Styles

Culturally determined communication styles range from the high-verbal, low-context to the low-verbal, high-context [2]. The high-verbal, low-context styles communicate most of their meaning by what is actually said or written and relatively little by the context of facial expressions, body postures, symbols, and rituals. Western cultures typically use these communication styles. In contrast, the low-verbal, high-context styles communicate relatively little meaning by what is said or written and most by context. Eastern cultures typically use those communication styles [6].

Matching communication styles between health professionals and patients or survivors assumes considerable importance as death nears. I suggest health professionals identify the communication styles dying patients and their survivors prefer, and try to use roughly those styles with them. Overall, the low-verbal, high-context styles fit end-of-life situations best [2] because they give people conceptual and emotional "space" to interpret the death in their own ways, to develop coping skills, and to reformulate hopes for the future [6]. Health professionals might, therefore, consciously shade their communication styles toward those low-verbal, high-context styles at such times.

Of course, all end-of-life situations involve ambiguities of meaning, which often emerge in the silences. Westerners used to high-verbal, low-context communication typically tolerate poorly the silences and their accompanying ambiguities.

Westerners prefer definitive diagnoses, clear disclosures, explicit treatment plans, and verbal interaction [37]. Nonetheless, health professionals should respect the silences and their ambiguities by resisting the impulse to fill the silences with talk. People need to make their own sense out of the ambiguities. Health professionals should concentrate instead on nonverbal gestures of caring. A hand on the shoulder or just a quiet, supportive presence may communicate the greatest comfort to dying patients and their survivors.

Advance Directives

As Chap. 6 explains, advance directives are documents that permit people to prepare for future mental incapacity by choosing beforehand the treatments and proxy decision-makers they want under those circumstances. Advance directives are grounded firmly in EA values (e.g., personal freedom, self-reliance, and orientation to the future) and beliefs (e.g., the future is not predetermined; one can make choices affecting it; and one should take responsibility for acting on those choices). In fact, one author calls advance directives an "artifact" of EA culture [22]. Advance directives may actually seem foreign, even threatening, to people who do not share those underlying EA values and beliefs. Perhaps for that reason AfAs, AsAs, and HAs are less likely than EAs to know about advance directives or to sign them [8, 12, 38, 39]. Furthermore, many from these minorities reject advance directives out of fear they may prompt discriminatory, adverse practices such as withdrawing life support prematurely.

Informed Consent in Terminal Illness

Medicine's informed consent standard may be another "artifact" of EA culture and its emphasis on personal autonomy [8]. This standard encourages disclosure of terminal diagnoses to patients and involvement of patients in planning end-of-life care (to the extent they wish). But many non-EA cultures consider such disclosure inhumane for causing patients distress, extinguishing hope, and even causing death to occur too soon [12]. These cultures may obligate families rather than patients to hear terminal diagnoses and to make subsequent care plans. One study in particular showed that more Korean Americans and MAs than either EAs or even AfAs want *not* to know about future metastatic cancer diagnoses or terminal prognoses and want to allow their families to make life support decisions for them under such circumstances [24]. Although I ordinarily try to allow great cultural leeway in end-of-life matters, I actively discourage withholding terminal diagnoses from *any* patients. I believe patients learn their diagnoses from context anyway, and most patients and survivors benefit from talking openly about the death to come.

Specific Ethnic Group Cultures and End-of-Life Care

I imagine American society as a tiered hierarchy of cultures. At the base sits American core culture. Its values include the preciousness of human life, the equal worth of human beings, and respect for the dead; [40] its beliefs include the existence of an all-powerful God or some other supreme focus for life; the inevitability of human death; and the existence of an afterlife [41]. Virtually all Americans hold those values and beliefs. Atop that core base sit the specific American ethnic group subcultures. They contain not only all the American core values and beliefs but also their own distinctive ones. Atop each ethnic group subculture sit even finer subcultures corresponding roughly to gender, age, or other characteristics.

This section describes the most distinctive end-of-life values and beliefs of that middle tier of ethnic group subcultures. I focus on four major ones—Euro-Americans (EAs), African Americans (AfAs), Asian Americans (AsAs), and Mexican Americans (MAs) (Table 10.3). An additional section of text but not of

Table 10.3 Contrasting common end-of-life values and beliefs among four major American ethnic groups

Value or belief	Euro-Americans (EAs)	African Americans (AfAs)	Asian Americans (AsAs)	Mexican Americans (MAs)
Person's control over the future	High [5, 42][a]	Moderate [42, 44, 59]	NR[b]	Low[34, 59, 95]
Desire for medical information	High [24]	High [24]	Low [24]	Low [24]
Main medical decision-maker (s)	Individual patient [24, 42, 43]	God and individual patient [5, 24, 44, 47]	Family [20, 24]	God, clinician, and family [24, 45, 59]
Spirituality	Low-moderate [44, 54]	High [44–46, 54]	NR	NR
Purpose of suffering	To regain health and function [20, 45]	To test faith and grow spiritually [45, 55]	To promote family harmony and maximize longevity [8, 20, 22]	To test faith [45]
Preference for life support	Low [20, 24, 38, 42, 45]	High [20, 38, 42–44, 54, 55]	High [8, 20, 22, 54]	Moderate to high [20]
Tolerance for disability	Low [45, 56]	High [45, 56]	NR	NR
Preference for hospice	High [5]	Low [5, 55, 56]	Low [3]	Low [3]

[a]Specific references for data occur in brackets.
[b]NR = Not reported.

the table addresses Native Americans (NAs). A subsequent section describes distinctive end-of-life values and beliefs of men and women and of young and old across the ethnic groups. I mention here only those values and beliefs addressed in the professional or academic literatures.

Although the text and table make unqualified statements (e.g., "EAs value medical information." or "MAs believe suffering tests one's faith."), the reader must not understand them as applying rigidly to everyone in a particular subculture under all circumstances. Many values or beliefs permit varied interpretations depending on circumstances. The reader, therefore, must avoid overgeneralizations and rigid stereotypes. The values and beliefs I mention here provide merely a starting place for health professionals to negotiate individualized care for patients.

I discuss EA values and beliefs first because EAs dominate contemporary American life by their numbers and influence. I then go on to discuss the comparable values and beliefs of the other three American ethnic groups, using the EAs' values and beliefs as an implicit foil.

Euro-Americans

Core Euro-American (EA) culture contains at least four beliefs relevant to end-of-life care. One belief, which I mentioned earlier, is that people *can* usually control their lives [16, 42] and should do so. (But one exception exists: An all-powerful God intervenes occasionally to override people's life plans for themselves.) Another belief is that people should be allowed considerable leeway, short of excessive interference with others, to make and implement their life plans [5, 24]. Yet another belief is that people's fundamental identities derive more from themselves as individuals than from their membership in families, ethnic groups, social classes, or other demographic groups. And still another belief, undergirded by the values of self-reliance and personal productivity, is that individuals earn self-worth through their deeds.

How might these beliefs affect EA wishes about end-of-life care? As patients, many EAs want as much information as possible about their illnesses even when that information is bad (Table 10.3). Furthermore, many EAs want to make as many of their own end-of-life decisions as possible. Accordingly, they more than other American ethnic groups try to control the end of their lives by signing ADs [42] to ensure the dying experiences they want and to relieve their families of difficult end-of-life decisions [20]. EAs are also most likely to discuss their personal treatment wishes beforehand with their clinicians [42] and families [43] and to expect those wishes to be followed in the end-of-life crisis.

In addition, among these four ethnic groups EAs take the most scientific, least spiritual perspective on end-of-life care [44, 45]. They tolerate suffering only for a reasonable chance to recover health and function, especially the ability to do self-care [20, 45]. They also hold the least favorable attitudes about life support and want to stop it sooner than do the other ethnic groups [20, 42, 45]. Thus, when treatment offers little chance for recovery to high function, EAs often exert their personal decision-making authority to refuse life support for themselves. Many want hospice care instead.

African Americans

Core African American (AfA) culture contains at least five distinct religious or spiritual beliefs relevant to end-of-life care [45]. One belief is that an all-powerful God actively controls human events [5, 44, 46, 47], has a plan for each person, and intends it for good [46]. Another belief is that life always presents hardships: illness is one [44]. The individual may never consciously reconcile the apparent inconsistency between these two beliefs: an all-powerful God's good will towards people on the one hand and their hardships (including illnesses) on the other. The individual may simply have to accept God's plan—the good and the bad together—without ever fully understanding it [46].

The next two beliefs, however, may provide some reconciliation by asserting unexpected benefits from illness or death. One of these beliefs is that, while sorely testing one's faith [48], the suffering that accompanies illness provides an opportunity for spiritual growth, [45], a source of hope, and important lessons in coping [44, 46, 49]. The other belief is that dying and death are a part of God's plan for everyone's good. The dying process ennobles a person as much as it disempowers him or her [33]. And death itself can sometimes appear to the dying person like a "welcome friend," a "going home," [44, 48] or the start of a better existence in the afterlife [46]. Still another belief relevant to end-of-life care is that clinicians serve as God's instruments of that care [44], but God alone ultimately controls the outcome of illness. [45]. He sometimes even saves lives unexpectedly through miracles [44, 47]. As an AfA woman explained in one study, "God has priority over living … (He can always) heal you." [5].

Evidence conflicts over whether core AfA culture also includes distrust for the American healthcare system [42, 50]. One recent study suggests AfAs believe the system generally "controls treatment" and "does not serve patients well." [39] In addition, some commentaries cite the many mistreatments of AfAs in the past as reasons why they *should* distrust the system: the involuntary medical experimentation on AfA slaves in the 1800s, the denial of curative antibiotics to AfA men during the Tuskegee Syphilis Study of the 1920s to the 1970s, the coerced sterilization of Southern AfA women in the mid-twentieth century, the federal government's sickle cell screenings and subsequent discrimination against test-positive AfAs in the 1970s, the suspicions about AIDS as a plot to control AfA reproduction in the 1980s [51, 52], and the stigmatization of AfA men in the 1990s due to a purported aggression gene. Present-day poverty, disproportionate incarcerations, and other ongoing social disadvantages of American minorities perpetuate the idea that America cares little about its AfAs and other minorities [52]. Such considerations convince many AfAs, so some commentators claim, that "racism (still) pervades American institutions" including the American healthcare system [51–53].

Nonetheless, other evidence contradicts that distressing view about AfA distrust of the American healthcare system. Some studies suggest AfAs *do actually trust* their clinicians based on perceptions of the clinicians' competence [45] and compassion [25, 49]. And, of course, AfAs still seek out clinicians' care. AfAs may, therefore, distrust the American healthcare system as a whole but trust their own particular doctors and nurses.

Gender differences complicate the issue further. One study suggests AfA men feel disempowered by the American healthcare system, but AfA women feel empowered by it. And, overall, AfA men fear harm from it, but AfA women expect benefit [26]. Conclusions about distrust of the American healthcare system, therefore, may really apply only to AfA men, not to AfA women.

In light of this confusing state of research and commentary, one expert's conclusion appears reasonable: Suspicions of AfA distrust are speculative and "largely unsubstantiated," but American medicine must still guard against even isolated instances of "insensitivity, neglect, injustice, or racism." [48].

Whatever distrust actually exists, AfAs appear to want considerable medical information from their attending clinicians (Table 10.3). AfAs also appear to use their generally deep spirituality to cope with serious illness. They interpret the suffering associated with illness as testing their faith and providing opportunities for spiritual growth. Furthermore, AfAs believe that both God and they have the power to make decisions affecting the course of illness. In addition, AfAs have strong wishes favoring life-saving treatment. Several reasons may explain those wishes. Religiously based sanctity-of-life values [48] may make life especially precious to AfAs. They may also believe they can (and should) live longer than other people [49]. And suspicions about ongoing discriminatory treatment practices may make AfAs especially wary about withdrawing life support prematurely [5, 44, 50, 54].

AfAs' disposition favoring life-sustaining treatments powerfully influences their views about many end-of-life matters. AfAs often reject advance care planning as fostering hopelessness [44], as decreasing a person's decision-making authority, and as diminishing his or her subsequent quality of care [42]. They also accept far greater disability and suffering to stay alive than do, say, EAs [55]. Furthermore, AfAs usually insist on aggressive life support and reject do-not-resuscitate (DNR) orders until late in a terminal illness [38, 42, 56]. And they consider talk about futile life support, hospice care, or costs of terminal care during the life-threatening crisis as devaluing life [54]. They completely reject suicide [49].

Asian Americans

Of the four American ethnic groups highlighted here, Asian Americans (AsAs) have the smallest professional medical literature on culture, health care, and end of life. The term *AsAs*, of course, encompasses many different nationalities, ethnic groups, and tribes, but I focus on the largest and most researched—ethnic Koreans, Japanese, and Chinese.

All these AsA core cultures emphasize the subordination of the individual's welfare to the group's [22]. The most important group is always the family [8], and all members have a responsibility for promoting the family's cohesion and harmony [56]. For instance, filial piety—faithfulness in duty to one's parents and other family elders—requires the young and healthy to care for the old and infirm [11]. Therefore, the dying process with its care rituals becomes more a relational experience for the whole family than just a physiologic one for the individual patient [8].

Other strong AsA traditions govern the dying process, too (Table 10.3) [12]. AsA clinicians and family members, for example, spare the patient from hearing a terminal prognosis due to the belief that merely mentioning death extinguishes the patient's hope for the future and precipitates death [5, 8, 22, 57]. Clinicians and family members sometimes even try to bolster the dying patient's spirits by making intentionally overly optimistic statements about prognosis. At the same time filial piety requires the adult children—particularly the eldest son [56]—to hear the true terminal prognosis and to make key life-support decisions on the elderly parent's behalf [24, 57]. The children often opt to keep the parent alive as long as possible. Ironically, terminal AsA parents may not want life support but accept it in deference to the obligation the children feel to give it [8, 12, 20, 56]. The power of this obligation and the perceived unpredictability of death discourage AsAs from performing advance care planning beforehand [57]. They may also reject home hospice care out of the belief that a dead person's spirit haunts the place where that person dies [5, 8].

Differences between the Western, typically high-verbal, low-context communication style, and the Eastern, typically low-verbal, high-context communication style, [2] complicate Western clinicians' end-of-life care for AsA patients. In such situations Western clinicians must consciously observe patients and families for meaningful nonverbal interactions. Trained interpreters can help. Still, misunderstandings will occur and may account for some of AsAs' dissatisfaction with their Western clinicians [58].

Mexican Americans

Hispanic or Latin Americans include people from typically Spanish- or Portuguese-speaking American countries. While each Latin ethnic group has its distinguishing cultural characteristics, I focus here on Mexican-Americans (MAs) because they are the largest Latin-American ethnic group in the USA and because Senor B. was one.

Core MA culture contains several key beliefs that influence views about end-of-life care. One belief is *familismo*, the idea that family plays the central role in one's sense of self and duty. Family forms the foundation for an MA's personal identity [59]. Therefore, as with AsA culture, MA culture expects each person to fulfill his or her duties within the family so that it can function as a cohesive, integrated unit. The resulting interdependence means the incapacitation of one family member, say, through serious illness, affects the whole family. Perhaps for that reason MA families together—not patients alone—often make major treatment decisions [24]. The senior men take the formal lead in those decisions, but the senior women wield great influence behind the scenes [56, 59].

Another belief asserts that powerful external forces control much of one's life, leaving relatively little control to the individual. Those forces include God, human authorities (such as governments, healthcare institutions, doctors, and other health professionals), and fate [34, 59]. As with AfAs and perhaps other ethnic groups, MAs believe a benevolent, all-powerful God can (and sometimes does) effect miraculous cures. This belief obviously inspires hope in the darkest life-threatening crises. MAs also believe that three specific values should govern relationships with powerful others. *Simpatia* involves maintaining harmony by avoiding unpleasant confrontations with others in general [59], *personalismo* involves keeping up a formal friendliness with perceived authorities, and *respeto* involves showing due respect to such authorities [56, 59]. Individual professional authorities (such as doctors and other health professionals) earn MA respect for their education; competence; [45] and selfless, trustworthy service to people in need. That respect prompts MA patients to defer many treatment decisions automatically to their attending clinicians [20]. In contrast, impersonal institutional authorities such as governments or hospitals command MA respect out of fear. MAs may fear institutional discrimination in health care due to their poor English, uninsured status, or difficulty in navigating the delivery system. Undocumented MA aliens, of course, always fear discovery and deportation by the government. MAs naturally respond to such fears with a general wariness towards institutions. Another possible external force in one's life is fate. Many MAs believe in *fatalismo*—the idea that fate has sealed their personal futures and fighting to change them is useless. In addition, one's personal future invariably involves suffering, and MAs must accept it when it comes. Yet another key MA belief mirrors a key AfA faith-based belief: that serious, life-threatening illness not only tests one's faith but also offers opportunities for spiritual growth [45].

These beliefs may directly affect MA preferences about end-of-life care in several ways (Table 10.3). Fatalismo may make expressing end-of-life care wishes beforehand seem pointless. MAs may expect little control over the end of life because they have sensed so little control in prior life. Advance care planning and informed consent may, therefore, bewilder these patients by trying to empower them at what seems to them an especially powerless moment [39]. MAs may also see those procedures as inviting institutional discrimination through intentional misapplication of patients' wishes. In addition, some MAs may *not want* to know their terminal diagnoses. These patients may believe knowing such diagnoses is the responsibility of other family members alone. And some family members may agree, asking clinicians to disclose serious diagnoses only to them and to allow them to make all important treatment decisions without patients' knowledge of diagnoses [24]. Clinicians, therefore, should ask MA patients before the crises the exact role they want for themselves and their families in knowing serious diagnoses and making treatment decisions. The high cohesiveness of MA families suggests that, if patients prefer collective family decisions, clinicians try to involve as many immediate family members as possible in making them [45]. Collective decisions themselves, of course, can cause problems. Disagreements may arise over the "right" decisions for the patient, and family members may regret their decisions later or blame each other if the illnesses go badly.

These key cultural beliefs may also explain MAs' tendency to continue life support until late in a terminal illness and to avoid hospice [20]. MAs' acceptance of life with inherent suffering [45] may promote this tendency. The suffering of earlier life may simply appear to transform naturally into the suffering of terminal illness. In addition, MAs may sometimes misinterpret a clinician's offer of life support as a recommendation for it. Such an offer may, therefore, unintentionally communicate false hopefulness to MA patients and families. MAs may also feel obligated to accept the offer just to try to maintain harmony with the clinician and to show him or her respect.

Two gender-specific cultural beliefs may also influence end-of-life care for MAs. One is *machismo*, or MA male identity, that requires a man to care for his family in two ways. The MA man must earn a living for his family. This breadwinning duty forms a core part of the man's self-respect. Thus, whenever the MA man becomes too ill to earn a living, his self-respect suffers. The MA man must also protect his family from unnecessary distress and difficult decisions. He does so in part by not discussing his medical problems with the family and by deciding about his end-of-life care on his own. The one other influential gender-specific belief is the women's strong sense of duty to care directly for dying family members. This belief may cause MA women to refuse hospice because it makes them feel they have not fulfilled their care-giving duty to the dying and thereby failed the whole family [6].

The Case

I realize some years later now that Senor B. took a typical MA man's view of his illnesses. He confronted them seriously only when he could no longer perform his primary cultural role as the family's breadwinner. That inability to support his family must have shamed him into coming to "the government hospital," an institution he deeply distrusted for the possibility it might initiate deportation proceedings against him. In addition, even when coming to the clinic, Senor B. always took a fatalistic attitude toward his illnesses. They would run their course, he thought, regardless of treatment. Furthermore, Senor B. appeared to shield his wife from the seriousness of his illnesses by not discussing them with her until that last clinic visit. Only then did Senora B. learn the gravity of his condition. At that point we three discussed the potentially life-threatening nature of Senor B.'s illnesses and the possible treatments for them. Senora B. was surprised at what she heard but appeared to have little say in what to do: Senor B. just declared that he did not want the treatments, that he and his wife would move back to Mexico, and that he would die there.

Unlike Senor B. I took a typical EA view toward his illnesses. I believed he could control much about them and about the rest of his future. I urged him to take that control by at least trying my treatments. All along I also urged him to bring Senora B. to clinic visits and to keep her fully informed about his illnesses and treatment.

Despite our different culturally based views about his illnesses, Senor B. and I did build a solid respect for each other. Senor B. respected me for my medical education and listened carefully to my advice though he rarely followed it. He also respected me enough to return (except for his last appointment) whenever I asked him to. And he trusted me. His trust arose from knowing, as he explained, that “you (doctors) take an oath when you go to (medical) school.” I, in turn, respected him for having lived the hard life of an undocumented alien. I also admired him for being definite about what he wanted out of life and for being fiercely independent in his views.

Still, I could not understand Senor B.’s refusal of seemingly ordinary treatments (insulin, dialysis, and perhaps coronary bypass grafts) that might save his life. *Fatalismo* surely influenced his attitude. He appeared to believe the diabetes and the atherosclerotic heart disease would run their courses no matter what. He simply concluded, “I have to die from something, doctor. It might as well be these things.”

Though I reluctantly accepted Senor B.’s refusing those life-sustaining treatments, I could not understand his refusing palliative end-of-life care, too. Senor B. steadfastly refused the advance care planning I offered, and he again took an attitude of *fatalismo*. “My wife already knows what I want. I don’t need to tell her,” he explained. “When that time comes, she will do whatever she needs to do.”

Perhaps *familismo* played a critical role, too, in Senor B.’s decision to move home to Mexico. He may have realized that, as he died, he would need the care and companionship of his whole family back in his native village. In addition, his family would have to meet a need of their own to provide him that care and companionship. Senora B. and the other women of the family, for instance, would expect to take care of him as he died and would feel cheated if they could not. And family members would need each other’s emotional support as they watched him die. I see now that, though medically questionable, Senor B.’s decision to go home may have been culturally well grounded.

Native Americans

In addition to these four large groups, other smaller but still important ethnic groups also live in the USA. Native Americans (NAs) are one. Over 5.2 million NAs, divided among 535 officially recognized tribes and perhaps another 100 unofficial tribe-like groups, reside here [www.census.gov/prod/cen2010/brief/c2010-10.pdf, accessed November 22, 2013]. Yet only a tiny professional medical literature describes NA culture [60], and an even tinier one describes NA end-of-life care.

Hepburn and Reed have written one of the few articles on general NA end-of-life care. These authors describe it in terms of beliefs about the natural life cycle, the individual's place in the community and his or her decision-making authority in medical matters, and NA customs of communication [60]. The natural life cycle, of course, entails birth, growth, and death. NAs believe people must never disrupt this cycle, and, in particular, must accept death when it comes. They must not fight it.

NAs also seek a proper allocation of decision-making authority among the individual, the family, and the tribe. With proper allocation the individual makes decisions (including end-of-life care decisions) for him- or herself and receives support for those decisions from family and tribe. Adult children never challenge the decisions of parents for themselves. But NAs may initially react warily to outside medical authorities: Non-NA health professionals must prove their trustworthiness before NAs participate in sensitive end-of-life discussions with them.

NA thinking also affects end-of-life communications in an important way: NAs believe language not only describes reality but creates it. Mere mention of death, therefore, causes it to occur, making NAs communicate about death in nuanced, indirect ways. Such communications may reference a hypothetical dying patient or another actual dying patient but not the immediate dying patient [61]. The resulting discussion may strike time-urgent, action-oriented, direct-speaking Western clinicians as meandering and oblique, but they must have patience with it. NAs may make decisions so subtly that non-NA clinicians may not realize the decisions have been made. I, therefore, urge non-NA health professionals to consult informants knowledgeable in NA culture before discussing death and other sensitive medical topics with NA patients or their families.

Despite many NA tribes the professional medical literature addresses only one in particular: the Navajos, the second largest tribe after the Cherokees. Carrese and Rhodes describe end-of-life care with the Navajos based on Dr. Carrese's service as an Indian Health Service doctor [61, 62]. The authors emphasize the importance of culturally appropriate communications. "With the tongue we have," one traditional Navajo healer tells Dr. Carrese, "we must be very careful what we say ... (when) treating the patient." Specifically, Navajos insist on always speaking positively and thereby communicating *hozho*, the "Beauty Way." The term encompasses the Navajo ideas of beauty, goodness, order, and harmony. Contrasting with *hozho*, direct negative speech (including talk about death) must be avoided. This prohibition obviously complicates end-of-life care planning with Navajos. Carrese and Rhodes point out that the Patient Self-Determination Act of 1991, the law requiring mention of advance directives to all newly hospitalized patients, encountered a severe unforeseen obstacle in Navajo hospitals. Compliance with the law mandated explicit discussion of death, alarming the Navajos because they believed such discussion would actually cause patients' deaths.

Another author describes Navajos' view of dying as a sacred journey that ends with the dead person's living eternally in the Great Spirit's presence [63]. Rituals, connecting the dying person to the rest of the universe in harmony, play a key role in sending him or her on this journey. I think health professionals who attend Navajos may be able to cite this belief as a comfort to dying patients and their survivors.

Gender and Age as Additional Cultures

Although ethnic cultures most easily come to mind, they constitute only one kind of culture. Other demographic characteristics suggest other kinds of cultures. I, therefore, believe every person belongs to multiple cultures at once, and most are nonethnic [7]. Each culture influences the person's total cultural make-up: The more closely a person embraces a particular culture's core values, beliefs, and behaviors, the more that culture shapes the person's views.

To illustrate nonethnic cultures, I briefly consider here gender- and age-associated cultures as they relate to patients. Because the literature on those cultures is vast, I cannot cover them comprehensively. I, therefore, limit my discussion to just a few points.

Patient Gender

Recent articles suggest that women and men patients differ dramatically in their descriptions of illness, the amount and kind of information they want from clinicians, their expectations for treatment outcomes, and their preferences for life support. Women often describe in their histories not only the physical symptoms but also the emotional and relational effects of their illnesses. Women also want considerable information about their illnesses and ask their attending clinicians relatively many questions. For example, women recovering from heart attacks often want detailed explanations of atherosclerosis and high blood pressure. Yet the amount and comprehensibility of the medical information they get leave some women dissatisfied [64]. Furthermore, women often want to share decision-making with their clinicians, prefer treatment plans based on clinician–patient consensus, and do not always expect treatment to cure their illnesses [65]. And women tend to prefer "a dignified death" to aggressive life support near the end of life [66]. Perhaps for those reasons terminally ill women often "hear" and accept clinicians' recommendations for comfort care only [67]. Overall, women feel empowered in the American healthcare system [26].

Men, in contrast, typically describe in their histories only the physical symptoms and not the emotional or relational effects of their illnesses. Men prefer relatively little information about their illnesses and ask relatively few questions of their attending clinicians [65]. The questions men do ask differ from women's. For example, men recovering from heart attacks ask mostly about sexual activity. Men tend to be satisfied with the information they get from their attending clinicians. Men also tend to leave medical decisions to the clinicians but *do* expect treatment to cure their illnesses eventually. And they tend to prefer aggressive life support even near the end of life. Perhaps for that reason terminally ill men do not always "hear" clinicians' recommendations for comfort care only. Overall, men do *not* feel empowered in the American care system [26].

Patient Age

Studies also suggest age differences in the preferences critically ill patients have about decision-making style and outcomes for care. Although many patients of all ages prefer to share decision-making with their clinicians, more elderly patients than younger ones prefer leaving decision-making completely to their clinicians [67]. In addition, fewer elderly patients than younger ones want to exchange quantity of life for quality of life (which may include symptom relief, control over one's surroundings, or financial and emotional support for the family) [68]. Yet among the terminally ill, more elderly patients than younger ones "hear" and accept clinicians' recommendations for comfort care only [67].

Characteristics such as education, income, military service, and vocation represent other cultures to which people may belong. But one pair of cultures plays a huge role in medical care especially near the end of life: the clinician culture and the patient culture. Both of those cultures appeared to have had important effects in Senor B.'s case. I contrast the two cultures in the next section.

The Contrasting Cultures of Clinician and Patient

Clinicians commonly make two assumptions about clinical medicine and culture: First, clinical medicine is scientifically based and, thus, culture- or value-free. And, second, cultural differences in clinical medicine arise only from *ethnic cultural* differences between clinicians and patients [25]. Both assumptions are incorrect and require explicit rebuttals before I can offer a better way for clinicians to think about culture in patient care. (Here, as elsewhere in this book, I use the term *clinicians* to mean all health professionals trained in the basic clinical sciences such as anatomy, biochemistry, and physiology. Although medical doctors may be the quintessential "clinicians," the word also applies to nurses, pharmacists, and other scientifically trained health professionals).

The First Incorrect Assumption: Medicine Is Culture- or Value-Free

Many claim that value-free, objective science characterizes all sound medicine. As such, clinical medicine, the applied branch of medical science, treats patients' bodily problems by supposedly objective scientific principles. Clinicians often see themselves as scientists pursuing benefits defined by objective scientific truth, not subjective views [69, 70]. According to these clinicians, any other clinical approach represents an interesting but misguided curiosity [1]. As culture-in-medicine expert Dr. Carrese observes wryly, scientifically trained clinicians believe they "have 'truth' but others merely have 'culture.'" [70].

Still, in treating the body, clinical medicine is always promoting some values over others. It may, for example, promote extending life over easing suffering, or treating disease over preventing it. Furthermore, clinical medicine cannot avoid the existence of multiple subjective medical realities. No single reality dictates the same presentation, response, and course of a particular disease in all patients. To the contrary, the variability of a disease strongly suggests the variability of subjective patient experiences with it. In that way disease—the objective abnormalities of structure or function in cells, tissues, organs, or organ systems—[19, 71, 72] differs from illness—the patient's subjective experience of feeling unwell. Renewed appreciation for that subjective reality has begun to refocus patient care back from disease to illness. Therefore, I am convinced the first assumption is wrong: Clinical medicine is neither value-free nor completely objective.

The Second Incorrect Assumption: Cultural Differences in Clinical Medicine Arise Only from Ethnic Cultural Differences Between Clinicians and Patients

Restated, this incorrect assumption claims that clinicians and patients share all the same *medically relevant* values, beliefs, and behaviors though sometimes differing over other, *ethnic cultural* ones. If so, clinicians and patients from the same ethnic culture would agree on every medical issue. But I have cared for many EA patients who differ from me and other EA doctors over diagnostic tests, vaccines, smoking cessation, exercise, diet, weight control, megavitamins and other alternative treatments, and adherence to mainstream treatments. These firsthand observations suggest to me that clinicians and patients often view the same medical circumstances in vastly different ways [71]—some consistent with mainstream medicine and some not [73].

A recent study supports this impression [74]. Researchers contrasted patients' stated beliefs and their doctors' perceptions of those beliefs about whether the patients' medical problems had a biological cause, the patients bore responsibility for causing those problems, the patients could control the problems, nonmainstream interventions could alleviate them, the problems had significant existential meanings, and the patients wanted a therapeutic partnership with their doctors. The doctors misperceived all six patient beliefs, often mistaking them for the doctors' own beliefs. Specifically, the doctors *over*estimated patients' belief that their medical problems had a biological cause and *under*estimated patients' beliefs that the patients bore some responsibility for causing the problems and could control them, that nonmainstream interventions could alleviate them, that the problems had some significant existential meanings, and that the patients wanted a therapeutic partnership with their doctors. The findings did not vary by patient ethnic group or doctor–patient ethnic concordance.

I believe clinical education creates much of the difference between clinician and patient perspectives. Such education, a form of socialization, instills a special, noninnate worldview in clinical trainees. That worldview, called "the clinical perspective," [75] grounds itself in specific values, beliefs, and behaviors often not held by the general population. The clinical perspective values hard work, initiative, and self-sacrifice in the service of others; encyclopedic knowledge of a specialty field; future orientation (especially concerning preventative measures to avoid future disease-related problems); scientifically tested, "evidence-based" interventions; clinical research with big sample sizes and quantitative results; strict deductive logic in diagnostic workup and treatment; and optimism about treatment outcomes. This perspective also employs specific concrete symbols (such as the white coat symbolizing authority and cleanliness) and ritual behaviors (such as daily inpatient rounds) [75].

These considerations convince me the second assumption is also wrong. In fact, I now believe its opposite: Even within the same ethnic culture, clinicians and patients belong to separate cultures: [69] A specific cultural gap *always* divides the two groups between a "clinician culture" and a "patient culture." [19, 71] The gap may be great or small, depending on the particular clinicians and patients, but it exists in every clinical encounter [5, 16, 19, 21, 75]. Bridging that gap requires recognizing it, understanding it, and planning creative ways to bridge it [74, 76]. To help do so in practice, Table 10.4 and the next subsections detail differences between the two cultures.

The Clinician Culture

So-called objective science, rationality, and a highly technical language characterize this culture [18, 19, 69, 71]. A typical perspective derived from these characteristics tries to reduce seemingly complex biological processes to simple concepts or laws [19, 75] and forms this culture's overall approach to patient problems. This "reductionistic" perspective appears in at least three ways. First, the science behind clinical medicine seeks the simplest explanations for complicated phenomena

Table 10.4 Clinician–patient cultural contrasts

Characteristic	Clinicians[a]	Patients[a]	Reference(s)
Overall approach to patient's problem	"Objective," reductionistic[b]	Subjective, holistic	[1, 19, 25, 70, 73, 74, 76]
Basis for approach	Medical science	Personal experience	[1, 19, 25, 70, 74, 76]
Focus of concern	Disease	Illness	[25, 71, 74, 76, 77, 83]
Language used	Technical	Colloquial	[19, 25, 71, 77]
Meaning(s) of patient's problem	Few: a puzzle to be solved, a defect to be repaired, or a foe to be defeated	Many: same as clinicians' meanings but other meanings, too (e.g., divine punishment, a way to command attention, a release from life's duties, or a reminder of mortality)	[73, 74, 76, 77]
Aim of medical care	Diagnosis and treatment	Symptom relief, functional improvement, and reassurance	[19, 25]
Weapons to be used	"Evidence-based," scientifically tested treatments	Established medical treatments and alternative "folk" treatments	[1, 4, 10, 70, 76]
Views of medical care	Egalitarian, personal, compassionate, sensitive, benevolent, altruistic. Time-efficient, objectively accurate. One objective truth, one reality	Technically competent, hope-inspiring. Sometimes elitist, power-wielding, callous, malevolent, bureaucratic, or self-serving. Many subjective truths, many realities	[1, 19, 70, 72, 76, 77, 79, 80, 83]
Control over medical events	Doctors only	Combination of God, doctors, and patient	[73, 83]
Grounds for medical decisions	Technical knowledge	Patients' values, experiences, and wishes	[19, 25]

[a]The word *clinicians* refers to all scientifically trained professional caregivers, quintessentially medical doctors; the word *patients* refers to all medically untrained recipients of medical care.

[b]*Reductionistic* refers to the view that all complex natural phenomena can be distilled into simple concepts or laws.

related to the human body [1, 77]. Second, in focusing narrowly on physical disease, clinical medicine often overlooks problems of the mind and social environment. And, third, practicing clinicians use mental "filters," which they learn and refine over years, to reduce patients' often bewildering constellations of symptoms and signs to recognizable patterns of disease [19, 75].

Decisiveness and focused action also characterize the clinician culture [21]. The urgency of some situations, of course, demands quick decisions, effective action, and time efficiency [69]. But in all clinical situations, diagnosing and treating disease are the main objectives, and "evidence-based" interventions the main tools. Clinicians see their work as solving pathophysiologic puzzles, repairing bodily defects, and defeating the disease foe. Once having diagnosed a disease, clinicians want to treat it, and they aim to cure it. Their professional stature depends on keen observational skills, an encyclopedic knowledge of diseases, sound deductive logic, insightful judgment, and extensive technical know-how about treatments.

Patient care here often resembles car repair: Any significant deviation from normal functioning requires fixing. The patient presents his or her malfunctioning "body vehicle" to the clinician-mechanic, who diagnoses the problem and makes the necessary repairs [4, 78]. (Doctors actually call this process "tuning up" the patient.) Unfortunately, the clinician-mechanic model sometimes focuses so intently on the patient's physical body problems that it overlooks pressing non-physical problems [75]. Still, clinicians see their patient care as genuinely humanistic. They consider it beneficent, sensitive, kind, compassionate [79], and patient-centered. They also consider it rigorously egalitarian and nondiscriminatory [79, 80] such that patients with similar clinical needs receive similar attention regardless of income or other demographics.

Clinicians, of course, play the lead role within the clinician culture [75]. Supposedly they alone have the expertise to control clinical outcomes [75]. They thereby command great authority in that culture while patients and families often feel powerless and vulnerable [19].

But like ethnic group cultures the clinician culture is not uniform. It contains subcultures of its own created by particular clinical professions and specialties. Doctors and nurses, for example, belong to separate subcultures due to their different professional goals and methods. Doctors focus on diagnosing diseases, improving physical symptoms and function, and studying their care scientifically. They rely heavily on medical technology to achieve those goals. Nurses, in contrast, focus on helping patients live independent and fulfilling lives. Nurses rely heavily on building positive relationships to achieve that goal [81]. Even among doctors, different specialties and therapeutic goals create different subcultures. Several doctor subcultures, for example, may coexist in intensive care units (ICUs). The attending surgeons may aim primarily to keep patients alive while the attending "intensivists," usually anesthesiologists or internal medicine-trained pulmonologists, may aim primarily to maximize patients' quality of life [82].

The Patient Culture

This culture differs dramatically from the clinician culture. Patients, not clinicians, play the lead here [75]. Patients understand medical problems as illnesses, that is, their own personal experiences of feeling unwell, [19, 71] and are the experts in them. Patients also use colloquial language to describe their illnesses: [19] A myocardial infarction is "a heart attack;" diabetes mellitus, "sugar;" and syncope, "a dizzy spell." That language helps shape patients' understanding of their illnesses.

In addition, illnesses are holistic, that is, affecting the whole patient. As such, they have multiple dimensions, including not just the physical but also the emotional, social, vocational, financial, existential, and spiritual [19, 73]. Different perspectives on those dimensions permit different interpretations of what an illness means. Like clinicians facing diseases, patients facing illnesses may see them as problems to be solved or alien attacks to be avoided, warded off, or surrendered to [75]. But patients may also see illnesses as moral weaknesses, divine punishments, ways to command attention, convenient releases from life's demands, reminders of bodily deterioration and mortality, or catalysts for spiritual growth [75, 77].

In the patient culture medical care exists to provide symptom relief, functional improvement, reassurance, and hope. Diagnosis and treatment are not ends in themselves but merely means to these other ends in patients' lives. Because patients determine those ends, they often want to participate in medical decisions. Patients want to make sure that decisions about their care accurately reflect their core values, experiences, and wishes [19].

This patient culture also suggests that, because patients' worldviews differ considerably from clinicians', patients' specific views about modern medicine also differ from clinicians'. Mixed feelings lie at the heart of patients' views. On the one hand, patients are awestruck at modern medicine's power over some diseases. Pneumonia, kidney failure, atherosclerosis of heart vessels, and other formerly deadly diseases now often yield to treatment. Modern medicine appears to work miracles in these cases. Its successes feed patients' hope for recovery in their own life-threatening crises [83]. On the other hand, patients often regard warily this highly specialized, highly technical care because they do not understand it [72, 79]. Modern medicine can seem like a dangerous monster [72], serving itself or clinicians better than the needy patients. Its care can sometimes seem callous or bureaucratic. Patients often feel modern medicine focuses too much on the physical and too little on the nonphysical or humanistic (especially during terminal illness). They crave meaningful, personalized relationships with their clinicians. Unfortunately, clinicians must often guess at such wishes because most patients, intimidated by clinicians' authority, do not ask explicitly for what they need [69]. Patients fear retaliation if they ask too much of their attending clinicians and thereby earn a reputation as being "difficult." Such mixed feelings, then, create for patients a curious combination of awe, trust, and wariness about modern medicine, and clinicians may have difficulty deciphering the exact combination for particular patients.

The patient culture also contains one other important existential element: the perception of limited control for patients over disease, illness, care, life, and death. Many Americans see God as having ultimate control over medical events despite the power of modern medicine [83]. They also see God as relinquishing some of His control to them. Some patients wish to exercise their control by making all important medical decisions themselves; others wish to share it with their clinicians.

The Case

The ethnic cultural differences in Senor B.'s case were obvious from the start, but the clinician–patient cultural differences were subtle. They occurred to me only later. As a typical doctor, I am disease- and action-oriented. I viewed Senor B.'s shoulder pain and chest pressure as disease puzzles to be solved. I felt obligated to diagnose the causes for his pains and looked to the medical sciences, mostly anatomy and physiology, for answers. The shoulder pain was the less serious problem: Strain or possibly a tear of the shoulder tendons was likely causing it. The chest pressure was potentially the more serious problem: Compromised blood flow to the heart muscle was likely causing it. I believed Senor B. could suffer a full "heart attack" at any time, thereby permanently damaging his heart muscle and perhaps dying. I wanted to take immediate action to treat the heart by admitting him to the cardiac ICU. I automatically assumed "the government hospital" would give Senor B. competent, compassionate, and nondiscriminatory care despite his status as a poor, undocumented alien. And I believed the cardiac ICU doctors would treat him with proper monitoring and interventions and save his life in the process. I tried to reassure him the hospital would give him whatever life-saving treatment he needed at a cost he could afford and would not initiate deportation proceedings against him.

As a patient, Senor B. took quite a different view of his case. He had illnesses, not merely diseases. First, the shoulder pain (among other symptoms) and then the chest pressure made him feel unwell. These illnesses had multiple dimensions for him including physical, social, vocational, financial, and emotional ones. Senor B.'s resulting inability to support his family jeopardized his position as patriarchal head of the family and compelled him to come to the public clinic, which he dreaded. He cared little about getting exact diagnoses but cared a lot about getting enough pain relief to work again. In fact, Senor B. was so desperate for relief that he sought treatment wherever he could find it: He used both mainstream treatments and alternative folk treatments. In addition, where I saw only clinical puzzles, Senor B. may have seen moral weakness (say, for his excessive eating, smoking, and drinking), divine punishment, or reminders of bodily deterioration and mortality.

He also felt a deep ambivalence about the public healthcare system he had landed in. He respected my clinical knowledge and wishes to help but also feared mistreatment at "the government hospital." Despite my optimism about treatment, Senor B. was always fatalistic. He was sure he would die from either his heart disease or "that heart surgery." Refusing my pleas to work him up for an impending heart attack, he said repeatedly, "God wants me to die from something. This must be it." Senor B.'s prior nonadherence to his diabetes regimen should have alerted me that he and I were often working at therapeutic cross-purposes.

As I recall his case now, I see that relatively few of our clinical management disagreements arose from ethnic cultural differences. Most of those disagreements actually arose from differences between my clinician culture and his patient culture.

Resolving Cultural Conflicts at the End of Life

One expert says cultures shape patients' perceptions of clinical reality [4]. Those perceptions differ from culture to culture, patient to patient, situation to situation, and time to time. Thus, clinician and patient are constantly renegotiating their different, culturally based factual interpretations, values, and expectations to reach agreement on the clinical reality they face together. I certainly tried mightily to negotiate with Senor B. a shared perception about the reality he and I faced. But I did not succeed much. A huge cultural gulf always separated us, and I needed a much broader perspective than I had to bridge it.

In searching for such a perspective, I have concluded recently that some of every person's cultures address death. We all derive so much identity, purpose, and meaning from our cultures [78] that we surely find in them insights about that momentous event. Cultures explain what death means to us and provide ways for us to cope with it. Cultures also shape our expressions of grief and can give us spiritual guidance [2, 5].

Unfortunately, cultures are so deeply ingrained and so rarely discussed that many of us do not know or cannot articulate our culturally based beliefs about death [11]. "Culture hides itself from its own," one writer observes [2]. Defining such beliefs for the first time at the end of life may be especially difficult just when they matter most. Yet patients and family members may find themselves doing that in front of health professionals as the end-of-life drama plays out. Health professionals must respond with insightful, practical, and compassionate help.

In doing so, health professionals must overcome the typically human discomfort at discussing death. They must also resist their usual time urgency and exercise the patience necessary to listen and to consider cultural differences thoughtfully [84]. Health professionals must always respect beliefs different from their own and never dismiss them as mere obstacles to doing what the health professionals want to do anyway [85]. Of course, respect for opposing beliefs does not mean automatically accepting them. But it does mean trying to understand those beliefs and to negotiate workable compromises within reasonable professional and personal bounds [86]. Health professionals should have a sensitive, systematic approach to do so. One is LEARN, the acronym for *L*isten, *E*xplain, *A*cknowledge, *R*ecommend, and *N*egotiate [87]. The next few paragraphs describe it.

Listen means eliciting the patient's perspective on the situation. I suggest the health professional take cues from the patient's ethnicity, gender, and age and on that basis make a preliminary mental list of culturally based views the patient might hold. This chapter's general descriptions of common American cultures, ethnic and otherwise, may help. The health professional can then ask specifically about those views if the patient draws a blank at first inquiry. (A reminder: Because this process does not set out rigid, preconceived ideas about what the patient *must* think, it is not cultural stereotyping. The process merely provides a place to start a focused inquiry [61]. Of course, the health professional must always keep an open mind and be prepared for deviations from prior culture-oriented expectations.) The health professional should use good listening techniques throughout. In general, he or she should respond to the patient with attentiveness and sympathy. The health professional should use the patient's own words whenever possible in replying, asking clarifying questions, or summarizing what the patient has said [87].

Explain means understanding one's own relevant cultural views and being able to explain them to others. A person's greatest cultural influences surely come from ethnic group, religion, gender, and education [88]. Those influences may emerge most powerfully in the critical-care or end-of-life crisis. The health professional, therefore, must be able to identify the possible sources of his or her cultural views, to explain those views clearly, to hear opposing views, and to anticipate the implications of all views for care.

Good, clear expression is a key to effective explanations. Both the verbal and nonverbal components of expression deserve forethought. Even when sharing the same native language with patient and family, the health professional must speak slowly, distinctly, and deliberately. He or she must explain key ideas, minimize the use of technical words, and clearly define the technical words whose use is necessary.

Not sharing the same native language naturally increases the difficulties of communication and the risks of miscommunication. When lacking easy fluency in the language preferred by the patient or family, the health professional *must* use a trained, skilled professional interpreter [89, 90], never untrained staff or family members of patients, to translate. Accurate medical interpretation requires specialized language use, exact transmission of meanings, and keen observation of interpersonal interactions (including nonverbal communications)—all of which a trained professional interpreter can provide. American law now requires hospitals and clinics to make available in person or by telephone the professional interpreters for many languages. The health professional should use these interpreters liberally. I prefer using in-person interpreters (whenever available) specifically to get their observations about nonverbal communications.

Even when using interpreters, the health professional should speak slowly and deliberately. He or she should look directly at the patient or family member; use short, simple sentences; and stop frequently, perhaps every sentence or two, to allow the interpreter to keep pace. The health professional should also never divert his or her attention, say, to writing progress notes or speaking to others, during interpreted conversations. To ensure accurate understanding in both directions, he or she should repeat back for confirmation any important information coming from the patient or family member and ask the patient or family member to repeat back any important information coming from the health professional.

As I said before, communication is nonverbal as well as verbal, and the health professional should appreciate both kinds. Good listening and good explaining, therefore, require good observation. With the emphasis in many non-Western cultures on the nonverbal, and the taboo in some about talking explicitly about death, the professional interpreter may be most valuable in helping the health professional understand indirect, nonverbal communications. Any face-to-face interpreted conversation, whether between Western and non-Western cultures or among strictly Western cultures, deserves a debriefing afterwards in which the health professional asks the interpreter in private for insights about the nonverbal communications.

Acknowledge means to identify explicitly a situation's cultural conflicts. This acknowledgment does not require a comprehensive knowledge of all cultures represented. Such knowledge is impossible even for professional anthropologists [91]. But the acknowledgment should reflect an appreciation for the main conflicts, a willingness to hear all perspectives, and a respect for each one. Otherwise, some parties may feel their views are not taken seriously.

I find that stating cultural conflicts as action questions, that is, questions about which actions to take in response to the conflicts at hand, helps start the resolution process [92]. Action questions in Senor B.'s case might have included the following: Should I insist he stop the *El Parajo* and all other nonmainstream medicines before I treat him? Should I even propose expensive life-prolonging procedures such as cardiac catheterization and hemodialysis when I know he is a poor, uninsured, and undocumented alien and worries about not being able to pay for them? [93] And should I require him to sign an outpatient do-not-resuscitate order if he steadfastly refuses my recommended life-saving treatments?

Recommend means offering preliminary proposals for resolving the cultural conflicts. Once those conflicts and relevant action questions get articulated, proposals for resolution often follow quickly [78]. Those proposals should acknowledge all perspectives represented, reflect whatever consensuses exist, and suggest initial actions to take. The proposals should permit any necessary adjustments later on.

Negotiate means performing a give-and-take about the proposals until the parties reach a generally agreed-upon, workable solution. The negotiation process sometimes requires persistence, patience, and flexibility [78]. And the success of any solution should be assessed in terms of accommodating as possible the basic cultural beliefs of patients first [85], families next, and health professionals and institutions last.

Some people advise negotiating until reaching a solution fully acceptable to all. That goal is sometimes impossible: Harmonizing seemingly conflicting cultural beliefs may never fully occur. Beliefs may be logically incompatible; parties may harbor hidden agendas, perhaps unconscious, that leave them unable to compromise; and life-threatening crises may limit the time available to formulate the perfect plan. Actual solutions, therefore, are often compromises, which may satisfy nobody fully. Neither ethics nor law provides clear guidance about when to accommodate and when not to in such highly charged crises [76]. I believe dying patients, through their proxies if necessary, should generally have the dominant say in end-of-life questions. Health professionals and institutions may have to extend themselves to accommodate patients' beliefs. Those beliefs may be overridden only when they will likely do serious, irreversible harm to others. Surely few patient beliefs meet those conditions.

The Case

Senor B.'s case still baffles me years later. Adhering to my professional training, I focused on his diseases. To me, they constituted his total medical reality. They also had established, "scientifically proven" treatments, which I was eager to use: insulin for the diabetes, aspirin and possibly arterial bypass grafts for the atherosclerotic heart disease, and eventual hemodialysis for the progressive kidney failure. I considered such care for him state-of-the-art, personal, and compassionate. Still, my approach rarely persuaded Senor B. to accept the treatments I recommended.

I realize now my perspective as a clinician blinded me to Senor B.'s quite different perspective as a patient. I had assumed we both saw his medical reality in the same highly clinical way. But his increasingly missed appointments and his steadfast refusals of nearly all my treatment recommendations should have made me suspect important differences between us. I should have explicitly asked his perspective, using the questions in Table 10.5 [4, 85]. The clinical history I had learned in medical school did not include them.

Though not explicitly asking these questions, my original clinical history with Senor B. answered some of them anyway. I only needed to see the answers in it. *How long do you think your illness will last?* His fatalistic attitude ("I have to die from something. This must be it.") strongly suggested Senor B. believed these medical problems would last unabated until he died and, in fact, would cause his death. *What problems does your illness cause you in everyday living?* Senor B. explained at the outset that his shoulder pain bothered him most because it prevented him from earning a living. He was thereby failing his family in his most important cultural duty. *How does this illness affect your body? How does your illness "work"?* Senor B. initially described not only his shoulder pain but also his thirst, frequent urinations, and erectile dysfunction. He had friends with diabetes and surely compared symptoms with them. He probably knew before his appointment with me that uncontrolled diabetes was causing these symptoms. *How serious are these problems to you?* Senor B. must have considered his symptoms serious enough to take the risk, as he perceived it, of coming to the public clinic. Of course, his inability to earn a living or to have intercourse with his wife surely undermined his sense of "macho" virility and, thus, played an influential role in

Table 10.5 Questions for eliciting a patient's perspectives on his or her illness

1. What do you think caused your illness?
2. Why did it start when it did?
3. How long do you think it will last?
4. How does your illness affect your body? That is, how does your illness "work"?
5. What problems does this illness cause you in everyday living?
6. How serious are these problems to you?
7. How have you already tried to treat your illness? How well did your treatments work?
8. How can I help you? What specific treatments do you think you need from me?
9. What results do you hope for from treatment?
10. What do you think is a realistic outcome?
11. What does this illness mean to you in thinking about yourself and your life?
12. What do you fear most about your illness?
13. Who else should be told about your illness?
14. Who helps you make decisions about your illnesses?

Data from Kleinman et al. [4]; Carter and Klugman [85]

prompting him to come. Furthermore, the frequent urinations from the uncontrolled diabetes must have caused him embarrassment at work and inconvenience at home. The later chest pains and shortness of breath must have also made him wonder about heart attacks and sudden death. *How have you already tried to treat your illness?* Because Senor B. was a Mexican immigrant, I did ask in my initial history whether he had tried Mexican herbals for his symptoms. He reluctantly admitted to using the traditional folk treatment, *El Pajaro. Who else should be told about your illness?* and *Who helps you make decisions about your illnesses?* Senor B. attended alone all his appointments except the last one. Although he occasionally mentioned his wife in our conversations, I did not know for a long time how much Senora B. asked about Senor B.'s medical problems, how much he told her, and whether he wanted to make medical decisions himself or to share them with her. However, I *did* know that MA women generally feel responsible for medical matters in the family, and that many MA husbands conceal their medical problems to keep their wives from worrying. I did not know Senor B. well enough to determine whether he and Senora B. fit these cultural patterns.

Unfortunately, my initial history did not answer most questions concerning Senor B.'s understanding of his illness: What do you think caused your illness? Why did it start when it did? What specific treatments do you think you need? What results do you hope for? What do you think is a realistic outcome? What does this illness mean to you? and What do you fear most about it? I surely could have gained a better understanding of his perspective by asking those questions. The answers might have helped me understand why Senor B. steadfastly rejected my life-saving treatments. The answers might have also enabled me to use the LEARN approach to reach with him a compromise treatment plan, incorporating many of his wishes and some of my recommendations.

Three points in Senor B.'s perspective continue to puzzle me today. I do not understand the fatalism that caused him to surrender to his disease without a fight. He apparently believed that mainstream medicine could not help him, and foresaw little use in even trying much of it. I also do not understand his belief that powerful external forces—individual human authorities (like me), powerful human institutions (like the hospital, the local police, and the federal immigration department), God, and fate—thoroughly controlled his life.

And I do not understand why Senor B. preferred to risk his life to return to his home in Mexico [93]. One expert says many dying immigrants have this wish. She gives several explanations for it [94]. First, many immigrants like Senor B. come to the USA to improve life materially for themselves and their families but never intend to stay permanently. Second, as these immigrants age, they become increasingly homesick, wanting more and more to return to familiar customs, familiar surroundings, and family back in the homeland. Third, when these immigrants become ill, they may need considerable care that only family will give. The many relatives back home can give it; the few, overly busy relatives in the USA cannot. And fourth, returning home may allow these immigrants to reestablish continuity among their past, present, and future; to resolve long-term, unfinished business back home; to gain a sense of closure for their lives; and to find comfort in the familiar when facing the vast unknown of death. This expert claims that returning home is a basic human need and an "organizing principle" for many immigrants at the end of life.

In sum, I do not know exactly why Senor B. left the clinic. The many cultural differences between us—ethnic, age-related, clinician–patient, and perhaps others—likely played a role. I also do not know what happened to him. He may have indeed returned home to Mexico and died there. But other cross-cultural cases since Senor B.'s have awakened me to the importance of cultural influences in end-of-life care. I have tried to change my practice habits as a result.

Summary Points

1. *Culture* is the values, beliefs, and behaviors that a distinct group of people shares, uses to understand life events, and transmits from generation to generation. Culture is not a race, nationality, or language group but a perspective that provides group members identity, meaning, and purpose for their lives.
2. Because health professionals and patients embrace markedly different health-related values, beliefs, and behaviors, every medical encounter involves at least two different cultures—the one of health professionals and the other of patients.
3. Health professionals might use the LEARN approach to identify and help resolve cross-cultural issues in patient care. LEARN stands for Learn, Explain, Acknowledge, Recommend, and Negotiate.
4. Cross-cultural issues are especially prevalent at the end of life.

To Learn More …

1. Crawley LM, Marshall PA, Lo B, Koenig BA. Strategies for culturally effective end-of-life care. Ann Intern Med. 2002;136(9):673–9.
2. Crawley LM. Racial, cultural, and ethnic factors influencing end-of-life care. J Palliat Care. 2005;8 Suppl 1:S58–69.
3. Kagawa-Singer M, Kassim-Lakha S. A strategy to reduce cross-cultural miscommunication and increase the likelihood of improving health outcomes. Acad Med. 2003;78(6):577–87.
4. Searight HR, Gafford J. Cultural diversity at the end of life: issues and guidelines for family physicians. Am Fam Physician. 2005;71(3):515–22.
5. Smith AK, Sudore R, Perez-Stable EJ. Palliative care for Latino patients and their families: whenever we prayed, she wept. JAMA. 2009;301(10):1047–57.

References

1. Beagan BL. Teaching social and cultural awareness to medical students: "It's all very nice to talk about it in theory, but ultimately it makes no difference." Acad Med. 2003;78:605–14.
2. Hallenbeck JL. Intercultural differences and communication at the end of life. Prim Care Clin Off Pract. 2001;28(2):401–13.
3. Barker JC. Cultural diversity—changing the context of medical practice. West J Med. 1992;157(3):248–54.
4. Kleinman A, Eisenberg L, Good B. Culture, illness, and care: clinical lessons from anthropologic and cross-cultural research. Ann Intern Med. 1978;88(2):251–8.
5. Koenig BA, Gates-Williams J. Understanding cultural difference in caring for dying patients. West J Med. 1995;163(3):244–9.
6. Kagawa-Singer M, Blackhall LJ. Negotiating cross-cultural issues at the end of life: "You got to go where he lives." JAMA. 2001;286(23):2993–3001.
7. Betancourt JR. Cultural competence—marginal or mainstream movement? N Engl J Med. 2004;351(10):953–5.
8. Berger JT. Culture and ethnicity in clinical care. Arch Intern Med. 1998;158(19):2085–90.
9. Carrillo JE, Green AR, Betancourt JR. Cross-cultural primary care: a patient-based approach. Ann Intern Med. 1999;130(10):829–34.
10. Clark MM. Cultural context of medical practice. West J Med. 1983;139(6):806–10.
11. Hallenbeck J, Goldstein MK, Mebane EW. Cultural considerations of death and dying in the United States. Clin Geriatr Med. 1996;12(2):393–406.
12. Searight HR, Gafford J. Cultural diversity at the end of life: issues and guidelines for family physicians. Am Fam Physician. 2005;71(3):515–22.
13. Vincent J. The structuring of ethnicity. Hum Organ. 1974;33(4):375–9.
14. Crawley LM. Racial, cultural, and ethnic factors influencing end-of-life care. J Palliat Care. 2005;8 Suppl 1:S-58–69 (p. 205).
15. Witzig R. The medicalization of race: scientific legitimization of a flawed social construct. Ann Intern Med. 1996;125(8):675–9.
16. Koenig BA. Cultural diversity in decision-making about care at the end of life. Presented to the Institute of Medicine Workshop: Dying, Decision-Making, and Appropriate Care. 02–3 December 1993, Washington, D.C.
17. Burchard EG, Ziv E, Coyle N, et al. The importance of race and ethnic background in biomedical research and clinical practice. N Engl J Med. 2003;348(12):1170–5.

18. Krakauer EL, Crenner C, Fox K. Barriers to optimum end-of-life care for minority patients. J Am Geriatr Soc. 2002;50(1):182–90.
19. Katon W, Kleinman A. Doctor-patient negotiation and other social science strategies in patient care. In: Eisenberg L, Kleinman A, editors. The relevance of social science for medicine. Dordrecht: Reidel Publishing Company, 1980. p. 253–79.
20. Blackhall LJ, Frank G, Murphy ST, et al. Ethnicity and attitudes towards life sustaining technology. Soc Sci Med. 1999;48(12):1779–89.
21. Brody H, Hunt LM. Moving beyond cultural stereotypes in end-of-life decision making. Am Fam Physician. 2005;71(3):429–30.
22. Berger JT. Cultural discrimination in mechanisms for health decisions: a view from New York. J Clin Ethics. 1998;9(2):127–31.
23. Fuller K. Eradicating essentialism from cultural competency education. Acad Med. 2002;77 (3):198–201.
24. Blackhall LJ, Murphy ST, Frank G, Michel V, Azen S. Ethnicity and attitudes toward patient autonomy. JAMA. 1995;274(10):820–5.
25. Perkins HS. Culture as a useful conceptual tool in clinical ethics consultation. Camb Q Healthc Ethics. 2008;17(2):164–72.
26. Perkins HS, Cortez JD, Hazuda HP. Advance care planning: does gender make a difference? Am J Med Sci. 2004;327(1):25–32.
27. Hazuda HP, Haffner SM, Stern MP, Eifler CW. Effects of acculturation and socioeconomic status on obesity and diabetes in Mexican Americans: the San Antonio heart study. Am J Epidemiol. 1988;128(6):1289–301.
28. Clark M, Kaufman S, Pierce RC. Explorations of acculturation: toward a model of ethnic identity. Hum Organ. 1976;35(3):231–8.
29. Meredith LS, Wenger N, Liu H, Harada N, Kahn K. Development of a brief scale to measure acculturation among Japanese Americans. J Community Psychol. 2000;28(1):103–13.
30. Chavez LR, Hubbell FA, McMullin JM, Martinez RG, Mishra SI. Structure and meaning in models of breast and cervical cancer risk factors: a comparison of perceptions among Latinas, Anglo women, and physicians. Med Anthropol Q. 1995;9(1):40–74.
31. Hazuda HP, Comeaux PJ, Stern MP, et al. A comparison of three indicators for identifying Mexican Americans in epidemiologic research. Am J Epidemiol. 1986;123(1):96–112.
32. Kraut AM. Healers and strangers: immigrant attitudes toward the physician in America—a relationship in historical perspective. JAMA. 1990;263(13):1807–11.
33. Powe BD. Cancer fatalism among African-Americans: a review of the literature. Nurs Outlook. 1996;44(1):18–21.
34. Sugarek NJ, Deyo RA, Holmes BC. Locus of control and beliefs about cancer in a multi-ethnic clinic population. Oncol Nurs Forum. 1988;15(4):481–6.
35. Johnson RL, Saha S, Arbelaez JJ, Beach MC, Cooper LA. Racial and ethnic differences in patient perceptions of bias and cultural competence in health care. J Gen Intern Med. 2004;19 (2):101–10.
36. Tucker CM, Herman KC, Pedersen TR, et al. Cultural sensitivity in physician-patient relationships: perspectives of an ethnically diverse sample of low-income primary care patients. Med Care. 2003;41(7):859–70.
37. Sullivan MC. Notes from the field—an adventure in multiculturalism. Bioeth Forum. 2001;17 (3–4):18–23.
38. Caralis PV, Davis B, Wright K, Marcial E. The influence of ethnicity and race on attitudes toward advance directives, life-prolonging treatments and euthanasia. J Clin Ethics. 1993;4 (2):155–65.
39. Perkins HS, Geppert CMA, Gonzales A, Cortez JD, Hazuda HP. Cross-cultural similarities and differences in attitudes about advance care planning. J Gen Intern Med. 2002;17(1):48–57.
40. Perkins HS, Shepherd KJ, Cortez JD, Hazuda HP. Exploring chronically ill seniors' attitudes about discussing death and postmortem medical procedures. J Am Geriatr Soc. 2005;53 (5):895–900.

41. Perkins HS, Cortez JD, Hazuda HP. Diversity of patients' beliefs about the soul after death and their importance in end-of-life care. South Med J. 2012;105(5):266–72.
42. Tulsky JA, Cassileth BR, Bennett CL. The effect of ethnicity on ICU use and DNR orders in hospitalized AIDS patients. J Clin Ethics. 1997;8(2):150–7.
43. McKinley ED, Garrett JM, Evans AT, Danis M. Differences in end-of-life decision making among Black and White ambulatory cancer patients. J Gen Intern Med. 1996;11(11):651–6.
44. Johnson KS, Elbert-Avila KI, Tulsky JA. The influence of spiritual beliefs and practices on the treatment preferences of African Americans: a review of the literature. J Am Geriatr Soc. 2005;53(4):711–9.
45. Braun UK, Beyth RJ, Ford ME, McCullough LB. Voices of African American, Caucasian, and Hispanic surrogates on the burdens of EOL decision making. J Gen Intern Med. 2008;23 (3):267–74.
46. Branch WT, Torke A, Brown-Haithco RC. The importance of spirituality in African-Americans' end-of-life experience. J Gen Intern Med. 2006;21(11):1203–5.
47. Barclay JS, Blackhall LJ, Tulsky JA. Communication strategies and cultural issues in the delivery of bad news. J Palliat Med. 2007;10(4):958–77.
48. Crawley LV, Payne R, Bolden J, et al. Palliative and end-of-life care in the African American community. JAMA. 2000;284(19):2518–21.
49. Mouton CP, Johnson MS, Cole DR. Ethical considerations with African-American elders. Clin Geriatr Med. 1995;11(1):113–29.
50. Shepardson LB, Gordon HS, Ibrahim SA, Harper DL, Rosenthal GE. Racial variation in the use of do-not resuscitate orders. J Gen Intern Med. 1999;14(1):15–20.
51. Gamble VN. A legacy of distrust: African Americans and medical research. Am J Prev Med. 1993;9(6 suppl):35–8.
52. Dula A. African American suspicion of the healthcare system is justified: what do we do about it? Camb Q Healthc Ethics. 1994;3(3):347–57.
53. Krakauer EL. Mistrust, racism, and end-of-life treatment. Hastings Cent Rep. 1997;37(3): 23–4.
54. Klessig J. The effect of values and culture on life-support decisions. West J Med. 1992;157 (3):316–22.
55. Crawley LM, Marshall PA, Lo B, Koenig BA. Strategies for culturally effective end-of-life care. Ann Intern Med. 2002;136(9):673–9.
56. Cykert S, Joines JD, Kissling G, Hansen CJ. Racial differences in patients' perceptions of debilitated health states. J Gen Intern Med. 1999;14(4):217–22.
57. Kwak J, Haley WE. Current research findings on end-of-life decision making among racially or ethnically diverse groups. Gerontologist. 2005;45(5):634–41.
58. Cooper-Patrick L, Gallo JJ, Gonzales JJ, et al. Race, gender, and partnership in the patient-physician relationship. JAMA. 1999;282(6):583–9.
59. Smith AK, Sudore R, Perez-Stable EJ. Palliative care for Latino patients and their families: whenever we prayed, she wept. JAMA. 2009;301(10):1047–57.
60. Hepburn K, Reed R. Ethical and clinical issues with Native-American elders: end-of-life decision making. Clin Geriatr Med. 1995;11(1):97–111.
61. Carrese JA, Rhodes LA. Bridging cultural differences in medical practice: the case of discussing negative information with Navajo patients. J Gen Intern Med. 2000;15(2):92–6.
62. Carrese JA, Rhodes LA. Western bioethics on the Navajo reservation: benefit or harm? JAMA. 1995;274(10):826–9.
63. White K. Living and dying the Navajo way. Generations. 1987;44–7.
64. Stewart DE, Abbey SE, Shnek ZM, Irvine J, Grace SL. Gender differences in health information needs and decisional preferences in patients recovering from an acute ischemic coronary event. Psychosom Med. 2004;66(1):42–8.
65. Elderkin-Thompson V, Waitzkin H. Differences in clinical communication by gender. J Gen Intern Med. 1999;14(2):112–21.
66. Bookwala J, Coppola KM, Fagerlin A, et al. Gender differences in older adults' preferences for life-sustaining medical treatments and end-of-life values. Death Stud. 2001;25(2):127–49.

67. Johnson MF, Lin M, Mangalik S, Murphy DJ, Kramer AM. Patients' perceptions of physicians' recommendations for comfort care differ by patient age and gender. J Gen Intern Med. 2000;15(4):248–55.
68. Bryce CL, Loewenstein G, Arnold RM, et al. Quality of death: assessing the importance placed on end-of-life treatment in the intensive-care unit. Med Care. 2004;42(5):423–31.
69. McIntyre N, Popper K. The critical attitude in medicine: the need for a new ethics. BMJ. 1983;287(6409):1919–23.
70. Taylor JS. Confronting "culture" in medicine's "culture of no culture." Acad Med. 2003;78:755–9.
71. Tirrell SE. The cultural divide between medical providers and their patients—aligning two world views. Bioeth Forum. 2001;17(3/4):24–30.
72. Ventres W, Nichter M, Reed R, Frankel R. Limitation of medical care: an ethnographic analysis. J Clin Ethics. 1993;4(2):134–45.
73. Street RL, Haidet P. How well do doctors know their patients? Factors affecting physician understanding of patients' health beliefs. J Gen Intern Med. 2010;26(1):21–7.
74. Pfifferling JH. A cultural prescription for medicocentrism. In: Eisenberg L, Kleinman A, editors. The relevance of social science for medicine. Dordrecht: D. Reidel Publishing Company; 1980. p. 197–222.
75. Feister A. What "patient-centered care" requires in serious cultural conflict. Acad Med. 2012;87(1):20–4.
76. Gilligan T, Raffin T. Whose death is it, anyway? Ann Intern Med. 1996;125(2):137–41.
77. Zola IK. Structural constraints in the doctor-patient relationship: the case of noncompliance. In: Eisenberg L, Kleinman A, editors. The relevance of social science for medicine. Dordrecht: D. Reidel Publishing Company; 1980. p. 241–52.
78. Kagawa-Singer M, Kassim-Lakha S. A strategy to reduce cross-cultural miscommunication and increase the likelihood of improving health outcomes. Acad Med. 2003;78(6):577–87.
79. Baruch JM. Dr. Douchebag: a tale of the emergency department. Hastings Cent Rep. 2012;42 (1):9–10.
80. Sharp LA. Denying culture in the transplant arena: technocratic medicine's myth of democratization. Camb Q Healthc Ethics. 2002;11(2):142–50.
81. Robertson DW. Ethical theory, ethnography, and differences between doctors and nurses in approaches to patient care. J Med Ethics. 1996;22(5):292–9.
82. Cassell J, Buchman TG, Streat S, Stewart RM. Surgeons, intensivists, and the covenant of care: administrative models and values affecting care at the end of life—updated. Crit Care Med. 2003;31(5):1551–9.
83. Jacobs LM, Burns K, Bennett Jacobs B. Trauma death: views of the public and trauma professionals on death and dying from injuries. Arch Surg. 2008;143(8):730–5.
84. Putsch RW. Cross-cultural communication: the special case of interpreters in health care. JAMA. 1985;254(23):3344–8.
85. Carter MA, Klugman CM. Cultural engagement in clinical ethics: a model for ethics consultation. Camb Q Healthc Ethics. 2001;10(1):16–33.
86. Pacquiao DF. Ethics and cultural diversity—a framework for decision-making. Bioeth Forum. 2001;17(3–4):12–7.
87. Berlin EA, Fowkes WC. A teaching framework for cross-cultural health care. West J Med. 1983;139(6):935–8.
88. Perkins HS, Supik JD, Hazuda HP. Cultural differences among health professionals: a case illustration. J Clin Ethics. 1998;9(2):108–17.
89. Orr RD, Marshall PA, Osborn J. Cross-cultural considerations in clinical ethics consultation. Arch Fam Med. 1995;4(2):159–64.
90. Jecker NS, Carrese JA, Pearlman RA. Caring for patients in cross-cultural settings. Hastings Cent Rep. 1995;25(1):6–14.
91. Mull JD. Cross-cultural communication in the physician's office. West J Med. 1993;159 (5):609–13.

92. Carrese JA, Perkins HS. Ethics consultation in a culturally diverse society. Public Aff Q. 2003;17(2):97–120.
93. Klugman CM, Bard JS. Futility on the border: commentaries. Hastings Cent Rep. 2013;43 (4):11–2.
94. Becker G. Dying away from home: quandaries of migration for elders in two ethnic groups. J Gerontol. 2002;57(2):S79–95.
95. Perez-Stable EJ, Sabogal F, Otero-Sabogal R, Hiatt RA, McPhee SJ. Misconceptions about cancer among Latinos and Anglos. JAMA. 1992;268(22):3219–23.

Chapter 11
The "Right" Time and Way to Die

Until about the mid-twentieth century Westerners lived constantly under the threat of death. It stalked them always and could pounce suddenly and unexpectedly, claiming even the healthiest. People had to be ready to fight it at any time. Not only immediate bodily survival hung in the balance. So did eventual postmortem disposition of the soul between heaven and hell. People, therefore, worried about fighting death as much in the "right" way as at the "right" time. *Ars moriendi*, the art of dying well, was a preoccupation for centuries [1].

With their eternal salvation at stake, Westerners naturally looked to religion for guidance in when and how to die. Thus, the Judeo-Christian faith provided much of that guidance. The faith's traditions contrasted God's great power and man's meager power over life-and-death matters. Those traditions also emphasized the importance of demonstrating in dying as well as in living such key Judeo-Christian virtues as faith, hope, charity, patience, obedience, and humility [1].

The Bible provided ancients many examples of dying at the "right" time and in the "right" way. In the Old Testament Abraham fathers both the Hebrew and Arabic races (**Genesis 15:5** and **Genesis 25:1, 6b**) before he dies "full of years, and … gathered to his people" (**Genesis 25:8**). The Jewish hero Samson cries, "O Lord God, remember me, … this once, … that I may be avenged upon the Philistines" as he pulls down his Philistine captors' house, killing himself and all inside it. And, "when (King) David's time to die (draws) near," he gives his son and successor, Solomon, ethical wisdom from a lifetime's reign. "Be strong," David says, And "keep the charge of the Lord your God, walking in His ways … that you may prosper" (**1 Kings 2:1a-3**). Similarly, in the New Testament Jesus himself prays at Gethsemane just before his death, "Father, if it be possible, let this cup pass from me. Nevertheless, not as I will, but as you will" (**Matthew 26:39**). He then voluntarily submits to arrest, interrogation, torture, and crucifixion. And Stephen, the first Christian martyr, looks to heaven and sees God's glory as he is stoned to death.

H.S. Perkins, *A Guide to Psychosocial and Spiritual Care at the End of Life*,
DOI 10.1007/978-1-4939-6804-6_11

Stephen finally cries, "Lord Jesus, receive my spirit. (And) … do not hold this sin against them" (**Acts 8:55, 59-60a**). Such Biblical stories showed believers when and how to die.

From these early Judeo-Christian traditions, detailed Western rituals about dying at the "right" time and in the "right" way evolved over time. By the Middle Ages dying people were expected to recognize the signs of imminent death. These people had to assemble their friends, ask the friends' forgiveness for wrongdoings the dying had done to them, likewise forgive any wrongdoings the friends had done to the dying, and publicly commend their own souls to God. By the eighteenth and nineteenth centuries dying people aspired to a "beautiful," romantic death. They tried to accept death with indifference and to plan for properly self-glorifying postmortem rites. Those rites might include a prescribed draping of the body, a dignified funeral procession through the streets, a grandly memorializing church service, and a burial with a suitably elaborate headstone.

Then during the twentieth century Westerners encountered the greatest changes ever in death rites. Two powerful forces prompted those changes. First, scientific medicine won out over other forms of care by developing reliable cures for many previously fatal diseases, especially the infections. And, second, patient autonomy took hold, causing patients to demand ever more information and more say about their treatment. Those forces simultaneously created new confusions about dying: The long-standing belief that God alone determines life or death suddenly had to accommodate a new reality in which medical science (through life-sustaining treatments) and patients (through accepting or refusing those treatments) also appear able to determine life or death. As a result, people now struggle with life-or-death decisions that never occurred before. People wonder, What is the "right" time and "right" way to die? [2]

This chapter explores those perplexing topics. It begins by describing Westerners' beliefs about the "right" time to die. Unfortunately, only a few articles address this topic. Nonetheless, I summarize the articles that do exist, and present tentative conclusions based on them. The chapter then addresses Westerners' beliefs about the "right" way to die. More articles address this topic than the first. Though mostly published since 2000, these articles provide a broad, consistent enough knowledge base to permit some firm conclusions. I offer a few here. The chapter also addresses three additional, focused topics—states considered worse than death, exaggerated expectations for end-of-life care, and physician-assisted suicide (PAS). These three focused topics represent important dimensions of current controversies about when and how to die. The chapter uses two cases to illustrate its points: The first, the case of a widower dying from heart failure, illustrates points about the "right" time to die; the second, the case of a man dying from prostate cancer, illustrates points about the "right" way to die. The reader should note that other chapters in this book also address some points about the "right" time and way to die. The reader may wish to consult those chapters while reading this one.

Case 1

Mr. D., an elderly Czech-American widower, lives at a retirement community. He retired many years ago from a long, successful career as an engineering supervisor at an airplane parts factory. He has appeared especially sad and withdrawn since his wife died two years ago. One daughter lives nearby, visits him weekly, and helps him with such tasks as cleaning and grocery shopping. This daughter lost her waitressing job several years ago and now struggles to support herself. Another daughter lives in a town close by and supports herself as a successful accountant. She visits Mr. D. only occasionally.

Mr. D. has multiple damaged heart valves due to childhood rheumatic fever. Various repairs over the years have been only modestly successful. Mr. D. now suffers frequent bouts of congestive heart failure. Each time he swells up, has difficulty breathing, and needs hospitalization "to drain off the salt and water." He has also had two other recent hospitalizations for pneumonias, for which he was admitted to the intensive care unit, intubated, and ventilated on a mechanical ventilator.

Mr. D. tells his daughter after the last pneumonia that he never wants to return to the hospital even if that means he will die. The daughter asks Mr. D.'s attending nurse practitioner at the retirement community about the wisdom of that request. The nurse practitioner urges against it before both Mr. D. and his daughter. The nurse practitioner notes the hospital doctors' opinion written into the medical chart that Mr. D. may live some more years with only periodic hospitalizations. With the nurse practitioner's approval the daughter refuses to let Mr. D. press his request for no further hospitalizations. She is inwardly relieved because she does not want Mr. D. to die.

Mr. D. spends the next summer at the farm where he grew up. His nephew, who now owns and runs the farm, occupies Mr. D. with minor repair projects. Mr. D. enjoys the stay. But, when he returns to the retirement community, he becomes sadder and more withdrawn than before. He loses interest in his hobbies of repairing household items and recycling computer parts. He attends a "last" local Czech folklife festival and bequeaths his car to his unemployed daughter. He even comments morosely to some elderly friends at the retirement community, "We are all just waiting around to die."

Then one day the retirement community's dining room staff find Mr. D. slumped over his dinner and unresponsive. They call the emergency medical service (EMS). By the time the EMS team arrives, Mr. D. has regained some consciousness. The team prepares him for transfer to the nearest hospital, but Mr. D. begs them not to take him. He says tearfully, "Please don't take me back to the hospital." The EMS team supervisor tries to persuade Mr. D. to go, but Mr. D. remains firm. The supervisor then telephones Mr. D.'s attending nurse practitioner to authenticate his expressed wish. She reluctantly does. The EMS team returns Mr. D. to his apartment where the daughter cares for him until he dies about 10 days later.

Mr. D.'s case poignantly raises the question, When is the "right" time to die? The first part of this chapter addresses that question.

When Is the "Right" Time to Die?

Noted bioethicist John Hardwig says, "When it comes to (dying), timing is everything." Hardwig goes on to specify the most difficult task each dying person faces as recognizing the "right" time to die and resisting the temptation to linger on. Modern medicine, however, has complicated the task. Disease over the centuries tended to cut many lives too short, but modern medicine with its powerful life-saving techniques can now support many lives perhaps for too long. Therefore, patients today, Hardwig believes, must determine when their life stories should end. "We need the wisdom," he insists, "to know when (to die)" [3].

Virtually all Americans share the fundamental belief that physical death is certain, but its timing is not [4]. People are left to discern, then, the "right" timing for physical death as best they can. They typically base such discernment on four fundamental beliefs: Each person has a "right" time to die, only God knows that time for sure, God has ultimate control over it (but people also have some influence over it), and people must recognize the signs for it. I take up these beliefs individually in the following sections. I use scripture to illustrate each of the first three beliefs because the Bible records so much of pertinent Judeo-Christian wisdom. I also use results from a recent multicultural study by my colleagues and me to illustrate all four beliefs, especially the last one. Health professionals giving end-of-life care must know about these beliefs to understand Americans' thinking about the "right" time to die.

Each person has a "right" time to die. Most Americans (and other Westerners, too) surely believe in one "right" time for each person to die. Many people, for instance, react to a person's death by saying, "It must have been his time (to die)." In contrast, they may react to a patient's unexpected recovery from a life-threatening disease by saying, "It must *not* have been her time (to die)." (As I was writing the earliest chapters of this book, doctors diagnosed me with many long, asymptomatic heart pauses. Some lasted up to nine seconds. Any one of those pauses could have continued on indefinitely, causing cardiac arrest and death. I was lucky: None obviously did. Many people might explain my good fortune in surviving as its just not being the "right" time for me to die.) Expressing this idea, one respondent in the study by my colleagues and me said, "(W)e're (all) going to have to die … but at the 'right' time." Likewise, another respondent said, "If (the doctor) had done all he could (to save me), then I would tell him, 'I'm ready to die because Jesus is waiting for me.'" Those circumstances would indicate to this respondent the "right" time for him to die [4].

Unfortunately, this belief may burden dying patients with the task of letting life go at just the one "right" or best moment. Patients may feel trapped by this standard between the opposing fears of dying too soon or too late. Dying too soon risks wasting additional rewarding life; dying too late risks enduring meaningless suffering after life has lost its "zest." [3] To this traditional view of a single "right" or best time to die, the new life-support technologies add another possibility: A person may have *more than one* "right" or good time to die. One good time to die may pass only to allow for a later, equally good or even better one. The "right" time may not even be an exact moment but a circumscribed block of time. The Psalms suggest such a possibility when urging people to number their days because "the years of life are threescore and ten or by reason of strength fourscore" (**Psalm 90:10, 12a**).

Only God knows for sure a person's time(s) to die. Many Americans believe exact knowledge of one's time to die rests with God alone, not humans. Two biblical statements suggest a Judeo-Christian grounding for this belief. Referring to God, The Psalms say, "(I)n your book were written, … the days that were formed for me when as yet there was none of them" (**Psalm 139:16**). The Book of Job says similarly, "Man's days are determined and (his) number of months is with you, (O God). You have appointed the bounds that (he) cannot pass, …" (**Job 14:5**). Responses in the study by my colleagues and me illustrate the belief further. One Euro-American (EA) told us, "How do you know when you're dying? (You can't know.) God is the only one (who knows) … when you're born and when you're staying." And an African American (AfA) said, People are "not God. (They) don't know how long (someone's) going to live," while a Mexican American (MA) said, Even doctors "can't be sure that (a person's) going to die." [4] Such responses suggest this belief occurs widely among Americans regardless of ethnic culture.

God has ultimate control over a person's time to die, but people have some influence over it, too. Many Americans believe God alone has ultimate control over life or death. This belief, which comes down from ancient times, appears in at least two passages of the Bible. The Psalms say, "My times are in (God's) hands" (**Psalm 31:15a**). Similarly, the Book of James says, Man does "not know about tomorrow … (He) ought to say, 'If the Lord wills, we shall live …'" (**James 4:14-5**). Eighty-three percent of respondents in our study also expressed this belief. As one respondent said, "It's God (who's in control) because … many times people should have been dead, but they're not. And (other) people should have lived (but) have not" [4].

Nonetheless, medicine's effective new preventative and life-saving technologies have prompted some people to modify that belief. They assert that, though God retains ultimate control over life or death, He sometimes allows humans limited influence over the time of death. In our study approximately one-fifth of respondents said that patients, doctors, and family members each have such influence. One respondent, for example, said about a patient's influence, "We have some (influence over) when we die by our conduct … (Some patients) have willed themselves to die … (Also,) if you smoke all your life, … that influences your (time to die)." [4] And another respondent said about the influence of the doctor and the family over his father's time of death, My father "was dead (from a cardiac arrest) for 25 min before the doctor got his heart going again. That's too long … I told the doctor, 'If his heart stops beating (again), take the machines off, and let him alone.'" [4] But another respondent warned against assisting in suicides. "People should (not help death along) like that Doctor (Kevorkian)," she said. "(I)t's God who should decide when you die." [4] God, of course, may use doctors as instruments of His plan. One respondent described his resuscitations from two cardiac arrests. "(T)he good Lord," he said, "was bringing me back, along with … the good doctors" [4].

People must recognize signs for the "right" time(s) to die. Hardwig believes that clinging to life late on is natural, but that we should not do so out of regret that some aspects of our lives could have been better [3]. Most causes for regret cannot be changed by then, and the regret must not blind us to the signs that our lives are ending. Hardwig says we must acknowledge that most regrettable things in our pasts cannot be changed. We should forgive ourselves for those things, he continues, and we should accept our lives—the bad with the good—for what they have been. He urges us to allow ourselves to die serenely [3].

Unfortunately, little wisdom about the signs of impending death comes down through the medical professional literature. But two studies do exist and can give helpful guidance.

One small study focuses on people's sense, as Hardwig says, that "their biographies are completed." [3] This study describes three elderly Britons and their thoughts about dying [5]. One man had led an interesting and fulfilling life, which had included achieving considerable professional success and heading a large family. But his wife had died recently, and the man sensed the time had come for him to die, too. "I have (already) lived my life," he explained, "and have much to be grateful for." Likewise, a woman had raised eight children on her own during World War II, outlived her husband by nearly 20 years, and seen great grandchildren born. She recognized shortcomings in her life but was generally satisfied with it. She, too, felt her time to die had come because life for her had become "rounded off." "There is a better life in store (for me)," she said. "If only God would take me." The third person, an angry man, felt he "had done his stint in life" and believed his "right" time to die had already passed. He described himself as "a victim of resuscitation": He had suffered a heart attack, received cardiopulmonary resuscitation, and survived. He believed the "heart attack was sent to take me off," and he had earnestly not wanted to survive it. Thus, all three of these people, according to the author of the study, clearly sensed that they had accomplished their purposes for living and that their lives were complete. They wanted their deaths to come soon.

Such insightfulness about the completeness of one's own life is rare. Most of us do not possess that insightfulness and must depend on more mundane signs for the "right" time to die. Therefore, the second study, one conducted by my colleagues and me, asked a multiethnic sample of Americans to identify the signs indicating people's "right" times to die [4]. Table 11.1 lists those signs and illustrates them [4, 6, 7]. The signs fall into three groups by source and frequency of mention in the sample. Most respondents mentioned *circumstances* as a source for such signs. Specifically, 88 % mentioned serious disease or poor bodily condition as signs for the "right" time to die, and 34 % mentioned "unnatural" causes such as accidents, murders, or medical errors as signs for *not* that time. Many respondents also mentioned *patients* as a source of such signs. About 50 % of respondents

Table 11.1 Commonly mentioned signs for the right time(s) to die

Source	Sign	Examples [references]	Percent of sample (%)[a]
Circumstances	Serious disease or poor bodily condition exists.	• "A good death (means) fading away when you're sick enough to die …" [6] • My father "had Parkinson's disease … and then he fell down and broke his hip … It was time for him to (die) … He was suffering." [4] • "(If) I couldn't function properly … (If) I didn't have … the ability to take care of myself, … (to) function mentally, I wouldn't want to live." [6] • "(T)he body gets so tired of struggling. (You) have worked all (your) life. (You've) had enough, and (you're) ready to go. (Y)ou're expiring your soul." [4]	88
	Only "natural" events—*not* accidents, murders, or medical errors—cause the death.	• "(G)etting injured on the freeway, becoming quadriplegic before you (die) and being totally dependent … That's a bad death." [6] • Mama "poured her glass of beer … When she drank the first drink, she started foaming from the mouth … Then she was dead. Took her just like that. We don't know how (her murderers poisoned) the beer …" [4] • When the doctors gave my grandmother "platelets, they also gave her bone cancer, and she never came out of the hospital. (She died there.)" [4]	34
Patient	The patient is suffering.	• "(I)t's horrible if a man's in pain—he's dying of cancer, … and he's in terrible pain…" [6] • My sister with leukemia was taken off the ventilator. "She died about two hours later. It was sad, but (she was no longer) suffering. She was better off resting." [7]	57
	The patient accepts death (or at least does not resist it).	• My father "knew he was going (to die), and he was at peace." [4] • A good death "would be one you'd be in control of. You'd know it's coming, and you could … put things in order … and then die." [6]	48
	The patient depends on "artificial" life support to live.	• My friend was two months in a coma. Then "the doctor said, 'He's going to die.' But he was dead already! The machine was pumping, doing the work for him … He couldn't move his arms. (His fingers had) no heat. It was a fraud!" [7]	47

(continued)

Table 11.1 (continued)

Source	Sign	Examples [references]	Percent of sample (%)[a]
Family	The family is emotionally prepared for the death.	• "I had already accepted that (my father) was going to die … (So) I didn't have remorse … You've got to be prepared." [4]	29
	The family is present to witness the death.	• "I saw (my aunt) … in a coma … I went back home (to change clothes) … (Then) the doctors (called and) said, 'You'd better get back here.' She died before I got (back to the hospital) … I should have been there for her (when she died) … That's why I get real(ly) depressed." [4]	29
	The family is suffering due to the patient's illness.	• "I will not be put on machines to live … I would not put my kids through that … (Why) not … make (their) lives easier even if it means ending mine?" [4]	28
	The family has visited the patient before death.	• I urged my brothers and sisters "to go to that hospital and talk to your (dying) daddy for the last time … (to) tell him you love him … They would not go … (but) I did … every day …" [4]	26

[a]Sample percents refer to the study described in Refs. [4, 7]

mentioned each of the following patient-related signs: the patient's suffering, his or her having accepted impending death, and his or her being kept alive on "artificial" life support. And some respondents mentioned *families* as a source for such signs. About 28 % mentioned each of the following family-related signs: the family's being emotionally prepared for the patient's death, their being present to witness it, their suffering due to the patient's dying, and their having already visited the patient (perhaps to say last good-byes). The frequencies of mention for the various signs remained roughly constant across all ethnic and gender subsamples. These frequencies may reflect the importance people attribute to the particular signs for indicating the "right" times to die: That is, circumstantial signs are most important; patient signs, less so; and family signs, least so. Both studies together give a possible mental checklist for helping discern the "right" time for a person to die. Not all considerations, of course, apply to all people.

Case 1

By Mr. D.'s own discernment several considerations suggest the "right" time for him to die has come. Like the first two Britons, Mr. D. has raised his children, completed his career, and seen his spouse die. Mr. D. surely misses his wife. He has also visited his childhood home one last time, attended one last Czech folklife festival, and bequeathed his car to the unemployed daughter. He has lost interest in his hobbies, and his life has generally lost its "zest." He appears to be separating emotionally from his life. His comment to his retirement center companions ("We are all just waiting around to die.") confirms that he believes his "biography is complete."

Other signs also suggest his "right" time to die has come. Though not immediately fatal, his periodic acute illnesses are taking an ever-heavier emotional toll on Mr. D., and his overall physical condition is deteriorating. His heart failure exacerbations come more and more frequently. He has increasing difficulty controlling his fluid weight between exacerbations and perhaps has a decreasing desire to do so. He has also suffered two recent pneumonias, both of which have required airway intubation and mechanical ventilation. Mr. D. hates the ventilator and is less and less willing to tolerate it even for prolonged survival. Furthermore, ever since the second pneumonia he adamantly refuses further hospitalizations. He reasserts this refusal as he regains consciousness from the blackout in the dining room.

The least convincing signs that Mr. D.'s "right" time to die has come concern his family. The caregiving daughter obviously suffers emotionally with Mr. D.'s heart failure exacerbations and pneumonias. But neither daughter appears ready to accept Mr. D.'s impending death. The caregiving daughter cannot bring herself to agree to his refusal of further hospitalizations. And the other daughter does not visit Mr. D. with an urgency suggesting she realizes he may die soon. Such family concerns, however regrettable, must not outweigh Mr. D.'s own concerns in this highly personal matter.

At the critical event in the dining room, all responders recognize the earnestness of Mr. D.'s refusal to go to the hospital. They finally accept Mr. D.'s sense that the "right" time for him to die has come, and honor his request to stay home. His attending nurse practitioner and the paramedics agree to return him to his apartment. They all expect him to die there soon.

Resolving Conflicts Over the "Right" Time to Die

Unlike Mr. D. most critically ill patients do not present such conclusive evidence that the "right" time for them to die has come. The often vague, sometimes conflicting literature on the topic only compounds the uncertainty involved. Still, some life-or-death crises require clinicians and others to make quick, definitive decisions on the matter [8].

How can those people do so responsibly? The patient's own informed view, of course, should usually carry the greatest weight [9]. But, when the patient cannot or does not express a clear, logical view, others (including the attending clinicians) may have to decide on the patient's behalf. Despite the last section's general considerations, no precise calculus exists for making such decisions. The "weights" of those considerations may change from situation to situation, time to time, or person to person, and other considerations not listed in the section (for example, the availability of life-saving transplantable organs) may occasionally emerge to play a dominant role. The decisions, therefore, cannot avoid some subjectivity.

As I have suggested before, exactly that subjectivity can create conflicts at the bedside. Some people may conclude the patient's "right" time to die arrives early on. They may see others as needlessly prolonging the patient's suffering. In contrast, other people may conclude that time arrives only much later. They may see others as trying to "kill" the patient prematurely. The intense emotions underlying these opposing views may escalate disagreements into bitter conflicts. The difficulties with reasoning through such conflicts [10] unfortunately compromise people's ability to find mutually agreeable solutions [11].

Many of the clinical ethics consultations I conducted over the years involved conflicts concerning patients' "right" time to die. Although I detail elsewhere in this book the steps I recommend in resolving end-of-life conflicts in general, I sketch out here the steps particularly relevant to discerning the "right" time to die.

When such conflicts arise, the principal attending doctor should ordinarily take the lead in guiding parties to a resolution. The doctor should begin by reviewing the facts of the case and hearing the conflicting viewpoints firsthand. The doctor might check the patient's medical workup; clarify the diagnostic reasoning behind it (including distinguishing the aspects that can be known with certainty from those that cannot); ask the opinions of patient, family members, doctor consultants, and other health professionals involved in the case; and perhaps notify the hospital ethics committee if the conflict may eventually require an ethics consultation. My consultative experience consistently taught me that such a thorough review avoids unnerving factual surprises during decision-making meetings, helps the doctor approach such meetings tactfully, and allows the conflicting parties the intervening time to resolve disagreements on their own.

If no resolution occurs spontaneously, the doctor may assemble the conflicting parties to try to negotiate one. I urge the lead doctor to have a good idea before such meetings what the range of reasonable resolutions is. (These meetings almost never yield completely unanticipated resolutions, and thinking through the likely range of reasonable resolutions beforehand helps the doctor guide the discussion to a productive outcome.) The doctor might also decide beforehand the resolution he or she tentatively prefers. The meeting itself should allow everyone to speak who wants to speak while others listen respectfully. The lead doctor should summarize comments either after each speaker or at the end of the meeting to assure participants they were heard and to allow corrections for any misunderstandings. I strongly urge the lead doctor not to adjourn the meeting without articulating some follow-up action plan, a rationale behind it, and a timetable for it—all ideally based on consensus. For example, if that plan involves a trial of antibiotics, the doctor might say, "We agree to give antibiotics for seven days, then to stop them, and to reassess the patient's condition." The most discouraging outcome from these meetings is deciding on no particular follow-up action plan. The lead doctor, therefore, should avoid such indecisive outcomes even if he or she must assertively articulate an action plan at meeting's end.

Contrary to popular belief many end-of-life conflicts a doctor now faces concern *patient or family—not doctor*—insistence on aggressive treatments that cannot benefit the patient. The doctor's foremost duty in such situations, of course, is serving reasonable patient interests. If the patient or family consistently demand treatments the doctor believes are causing the patient a needlessly prolonged, uncomfortable death, the doctor may have to object politely but firmly. The doctor should do so only after eliciting the views of other health professionals or the hospital's ethics committee. The doctor should ask those colleagues specifically about the reasonableness of the patient or family views that disagree with his or her own. When colleagues agree with the doctor, their support strengthens his or her position when confronting the patient or family. Then at an appropriate time and place that doctor with at least one other health professional involved in the patient's care should approach those who insist on continued aggressive treatments. The doctor should respectfully explain that continued aggressive treatments produce only net harm for the patient and must be stopped. The doctor should avoid such questions as Do you agree? or Is that acceptable to you? A no answer from patient or family to those questions only reestablishes the impasse. Instead, the doctor should explain the technical rationale behind stopping the treatments, the consensus among the professionals for the stoppage, and everyone's continuing commitment to all beneficial care for the patient.

If the patient is irreversibly dying and these discussions take place with family members, I prefer to say something like "I know you still hope Mr. Jones will recover. I sincerely wish that were possible. But his other doctors and I believe he cannot. We think that he is dying and we cannot arrest the dying process. We, therefore, must stop all treatments that do not benefit him and, in fact, only cause him to suffer. We have no choice. But I promise you we will continue to keep him as comfortable as possible to the end." (The doctor should obviously modify this comment if the other party to the discussion is the patient.) If the stoppage involves life support, the doctor should then suggest a specific date and time to stop the treatments. The date should ordinarily come no more than two or three days later. In my own practice I allow a few close family or friends of the patient to be present for the stoppage if they wish. But I explain beforehand that the patient may not die immediately when life support stops: He or she may live for some minutes, hours, or even days afterward. I say explicitly that no one can predict the exact time of death.

If some people still object to the stoppage before it occurs, I explain that certain legal proceedings can prevent it, but they must be started immediately. I say that, although I do not favor doing so, I will continue the treatments a few more days to give those who object the time to start such proceedings. I also say I will carry out the stoppage on the schedule I outlined unless I receive official notice of legal proceedings to prevent that plan. Notably, in my many years of performing these clinical ethics consultations, no one has ever started those legal proceedings.

Case 1

Mr. D. concludes that his current life is no longer worth living, and the "right" time to die has come. He wants no further aggressive treatments. He just wants to stay home until he dies. Unfortunately, the caretaking daughter cannot bring herself to agree. "The doctors in the hospital say, 'Antibiotics can cure him if he gets another pneumonia,'" she explains. "I can't let Daddy die from suffocation when antibiotics can still cure him." Her objection prevents Mr. D. from ever signing a do-not-resuscitate or do-not-hospitalize order. The other daughter never expresses an opinion on the matter.

Although she never does, the caretaking daughter might have objected when Mr. D. refuses transfer to the hospital after collapsing in the dining room. If she had, the paramedics might have immediately telephoned not only Mr. D.'s attending nurse practitioner, but also her supervising primary care doctor to intervene in the conflict right away. The doctor might then have telephoned the daughter; patiently heard her objections, likely tearful; and explained Mr. D. is voicing a reasonable refusal. The doctor might have explained how he draws this conclusion: Mr. D.'s heart disease is increasingly hard to treat, he has become dissatisfied with his life, it will likely change little in the future, and he appears to be consciously bringing his life to a close. The doctor could have told her that, sad as it is, Mr. D.'s refusal of transfer and hospitalization should be honored. The doctor or the nurse practitioner could have then offered to visit Mr. D. and the daughter later at Mr. D.'s apartment to provide moral support. The doctor could have also suggested immediate hospice admission and a visit from the retirement community's chaplain.

Fates Worse Than Death

Although some people talk jokingly about "fates worse than death," the dying may sometimes actually think of their situations that way. One researcher who has studied the topic explains that the dying constantly weigh the burdens and benefits of living [12]. They, like Mr. D., may occasionally conclude the burdens outweigh the benefits, producing real-life fates worse than death. One empirical research method, the rating-scale method for assessing health states, bolsters the concept. As Chap. 4 explains, this method has respondents rate health states on a line typically running between death and full health. The usual underlying assumption is that any living state is better than death. But respondents may actually want to rate some living states as worse than death. In such cases the rating line must extend below death to include those states.

In perhaps the most insightful study on the topic, that same researcher and his colleagues asked a mix of chronically ill patients, nursing home residents, and well adults to describe fates worse than death. The researchers asked specifically, Have you ever thought of circumstances in which it might be preferable for someone to die rather than to remain alive? and Can you imagine any such circumstances for yourself? The respondents gave unstructured answers, yielding a "universe" of 26 characteristics of fates worse than death.

Half or more of respondents mentioned 16 of the characteristics, divided among three headings, "The Patient's Physical Illness," "The Patient's Mental State," and "The Effects of the Patient's Illness on Others." (I admit that some characteristics might fall logically under more than one heading, and that age does not fit easily under any of the headings.) The Patient's Physical Illness includes eight characteristics: bad prognosis, bad diagnosis, poor bodily function, extreme discomfort, dependence on life-support technologies to live, duration of illness, permanent unconsciousness or inability to communicate, and age. Fifty-five to ninety-six percentage of respondents mentioned each of those characteristics. The Patient's Mental State includes seven characteristics: emotional distress, undesired manner or location for dying, the lost will to live, poor mental function, a sense of worthlessness to society, difficulty adapting to illness, and loss of control over one's circumstances. Fifty to eighty percentage of respondents mentioned each of those characteristics. The Effects of the Patient's Illness on Others includes only one characteristic, burdensomeness to others, and 61 % of respondents mentioned it. Table 11.2 illustrates each characteristic with quotes [13].

The researchers mention two cautions that may limit the generalizability of these data: All respondents lived in Seattle, and 91 % were Euro-Americans. Studies occurring in other locations and involving other American ethnic groups might produce different results.

Table 11.2 Characteristics of fates worse than death

Characteristics	Interviewees mentioning (%)[a]	Example quotes
The patient's physical illness		
Bad prognosis	96	"No chance of recovery." "Death is just outside ..."
Bad diagnosis	82	"Cancer." "Stroke." "AIDS." "Alzheimer's disease."
Poor bodily function	82	"(The) body is incapacitated." "(One's) every ... need (has) to be met by someone else."
Extreme discomfort	73	"I hurt so bad." "(You lie there) and suffer and suffer."
Dependence on life-support technologies to live	66, 70[b]	"Technology keeps (you) alive." "Machinery has to be tied to your body."
Duration or constancy of illness	64	"No let up (in being ill)."
Permanent unconsciousness, inability to communicate	55, 59[c]	"(Living) like a vegetable." "(You're) not even aware ..." "Nothing going in, and nothing coming out." "(I) couldn't convey my feelings to someone else."
Age	55	"My age or older." "(Having too much) quantity of life."

(continued)

Table 11.2 (continued)

Characteristics	Interviewees mentioning (%)[a]	Example quotes
The patient's mental state		
Emotional distress	80	"I (resented) the illness." "Life isn't fun anymore."
Undesired manner or location for dying	79	"(Instead,) he wanted to die at home." "We should have "just put (him) to bed (to die)."
The lost will to live	73	"My life (was) dripping out of my veins." "I would welcome death right now."
Poor mental function	71	"(My) brain has quit working— …" "(I) don't know my family."
Sense of worthlessness to society	61	"I can't do anything good for society anymore." "(People) feel they have no friends (or) nothing."
Difficulty adapting to illness	52	"(It's) hard for me to get used to it." "I've never had to stand any pain (before)."
Loss of control over one's circumstances	50	"(Not having) some control over deciding (one's fate)."
The effect of the patient's illness on others		
Burdensomeness to others	61	"I don't want my children to have to take care of me." "You feel like you're taking away from others' time." "(Being a) drain on society."

Adapted from Pearlman et al. [13], with permission
[a]The table lists only items that at least 50 % of respondents mentioned. Characteristics not listed here include undergoing perpetual treatment (45 %), being blind or deaf (38 %), having no dignity (27 %), living a poor quality of life (23 %), and living in a nursing home (20 %).
[b]The authors originally divided this characteristic into "dependency on machines or technology" (70 %) and "other life-sustaining technologies" (66 %).
[c]The authors originally divided this characteristic into "permanent unconsciousness" (59 %) and "communication" (55 %).

Case 1

I do not know for sure Mr. D.'s thoughts during his last days, but I can imagine those thoughts. His life by then has many characteristics of a fate worse than death. His chronic valvular heart failure is a "bad" diagnosis with a "bad" prognosis. Mr. D. may realize his illness offers no chance for improvement: In fact, his heart condition will likely kill him within a few. Meanwhile it requires a strict diet and conscientious weight control. Both irritate Mr. D. because he cannot eat as many canned soups and bologna sandwiches as he wants. In addition, his heart condition forces him to sleep sitting up so that he can breathe. It also worsens periodically, requiring admissions to the local intensive care unit for stabilization. Mr. D. hates the intensive care unit because its machines make him feel dependent on "artificial" life-support technologies to live and prevent him from sleeping well. Overall, Mr. D. believes the illness, not he, now rules his life. The illness also imposes significant caregiving burdens on the one daughter.

That realization surely depresses Mr. D. because he probably does not want her burdened with his care. Like some of the respondents quoted in Table 11.2, he may "resent his illness," sense his "worthlessness to society," and feel "life isn't fun anymore." He may, therefore, just want to die at home as soon as possible.

Do We Expect the "Ideal Death?"

When Americans think about their deaths, they often despair about ever-achieving "good" deaths. Persistent dissatisfaction with the state of dying in the USA fuels this despair. Society's recent focus on end-of-life issues and the rise of the hospice movement have not been able to dispel that dissatisfaction fully. Americans still sense something wrong [14]. Experts, therefore, continue to urge reforms to ensure "good" deaths [15].

The persistent dissatisfaction with the state of dying seems surprising in light of the long-standing consensus about what comprises a "good" death. One author cites Dame Cicely Saunders, the founder of the modern hospice movement, for her influential guiding ideology. According to this author, Saunders' key point stresses enabling the dying to live as fully as possible until they die [16]. It includes treating each dying person as unique and precious, helping that person live out life within the network of his or her closest relationships, allowing that person as much control and independence as possible, respecting that person's deepest beliefs about life and death, and helping him or her achieve a satisfactory closure for life [16]. Another author cites the Institute of Medicine for its similarly influential definition of a "good" death as being "free from avoidable distress and suffering (for) patients, families, and caregivers; in general accord with patients' and families' wishes; and reasonably consistent with clinical, cultural, and ethical standards" (as quoted in Ref. [17]). And multiple empirical studies have documented the consensus among patients, family members, doctors, and nurses about what constitutes a "good" death: symptom control, honest communication between doctors and patients, clear decision-making, preparation for the end of life, avoidance of unnecessary prolongation of life, and avoidance of excessive burdensomeness on families [17–20].

The concept of a "good" death that emerges from these sources surely appeals to all involved with a death. It aims to provide *patients* physical comfort; competent, attentive, and sensitive care from health professionals; emotional support from family; a sense of accomplishment and completion in life; and peaceful acceptance at its end. In addition, the concept aims to provide *families* protection from physical, economic, and emotional exhaustion; active emotional support; satisfaction with the patient's care; and unregretful remembrances of the death. Furthermore, the concept aims to provide *health professionals* meaningful engagement with patients and families at a key moment in life, and the satisfaction of a difficult job well done. Taken as a whole, the concept has proven a powerful spur to reforming end-of-life care.

So why do Americans continue to be dissatisfied with this care? I believe the reason is that the ideal of a “good” death sets an unattainably high standard [16]. Each person’s death is unique, nuanced, and complex. Care cannot possibly discern and meet all the needs involved. Nonetheless, the end-of-life reform movement has led dying patients and their survivors to expect care that does so, however veiled or complicated the needs. I believe care will always fall short of the publicized ideal. Three factors may explain why.

First, end-of-life care wishes are often varied [6], vague, or conflicting. Two studies show how such wishes can vary. One study about “factors considered important at the end of life by patients” listed symptom management, preparation for death (including funeral arrangements), a sense of completion, consideration as a “whole person,” communication with one’s doctor, mental alertness, helping others, not being a burden to them, and finding peace with God [21]. But a later literature summary, which did not reference the first study, listed six somewhat different goals for end-of-life care: being cured; living longer; improving or maintaining function, quality of life, and independence; being comfortable; achieving life goals; and supporting family members and caregivers [22]. Furthermore, health professionals often do not know *precisely* what people’s expressed care wishes mean. The professionals must, therefore, guess at what care will serve individual dying patients best. Everyone, for example, can imagine unnecessarily prolonged dying. But in actual clinical situations the concept can have different meanings for different observers. Compared to spouses, other family members believe patients should die sooner rather than later when they present physical caregiving problems (such as incontinence or pressure sores), mental problems (such as insomnia or depression), or conflicts with caregivers’ other responsibilities (such as childcare) [2, 18]. In addition, seemingly clear end-of-life care wishes of the same patient may sometimes conflict. A patient may wish to die at home with family around, but also not to burden family physically or emotionally. These two wishes may conflict: Dying at home may put a heavy physical caregiving burden on family, limit their access to professional advice, and magnify their grief when the patient dies [21].

Second, some characteristics of a “good” death require effort from patients or family members to realize. Only patients can designate medical proxies or draw up testamentary wills. And only family members can usually administer medications at home or help with feeding, bathing, or toileting the patient there (even under home hospice care). Yet some patients or family members lack the courage or capacity to perform these tasks. Unfortunately, if patients or family members do not perform them, they often do not get done.

And, third, people’s expectations for end-of-life care may exceed health professionals’ realistic abilities to meet those expectations. As Chap. 8 indicates, many people think hospice, once engaged, provides all physical, psychological, and spiritual care for the dying. Actually, staffing and reimbursement constraints severely limit the time hospice professionals can spend with patients and family members. Patients and families, therefore, must provide much of the care themselves. Patients and families may also have widely varying needs, some difficult for people to express and for health professionals to intuit. The typical hospice admission, lasting less than a week, leaves little time for health professionals to get to know particular patients and families and their needs. Furthermore, as one author says, "Some valued aspects of life do not translate easily into end-of-life preferences." [6] Many people wish, for example, to maintain independence, alertness, and participation in relationships to the very end of life. But dying eventually robs nearly all people of these capacities.

One prominent author assesses this state of end-of-life care and concludes that our current conceptual models of dying are "woefully inadequate at the bedside" (TE Quill as quoted in Ref. [15]). So how might we rethink end-of-life care in a realistic way? A second case introduces suggestions from the professional literature and illustrates them.

Case 2

Mr. E., a man in his mid 60s, presents with an aggressive, widely metastatic prostate cancer only about a year after retiring from a successful career as a non-commissioned military officer and a manufacturer's sales representative. His cancer proves resistant to treatment. The oncologist, the specialist cancer doctor, orders maximal doses of irradiation and powerful systemic chemotherapies to little effect. Meanwhile Mr. E. experiences a steady downhill course. He loses the ability to walk or stand and becomes confined to wheelchair and bed. He eats little and loses weight. Unable to provide singlehandedly all the care Mr. E. needs, Mrs. E. spends some of the E.'s limited savings to hire part-time nurse's aides. Though expensive to pay, they help her perform Mr. E.'s basic living tasks (such as bathing, dressing, and toileting). Mr. E. eventually begins to suffer persistent pains that the oncologist cannot control. The doctor recommends transfer to home hospice care for pain control, and Mr. and Mrs. E. agree.

The hospice transfer occurs late one Friday evening. The hospice admission nurse is tired and harried. She is all-business and short on patience throughout the process. She explains how much of Mr. E.'s personal care Mrs. E. will have to provide. Mrs. E. is dismayed because she has expected hospice to perform all caregiving duties [14]. The nurse also explains that the oncologist will no longer manage Mr. E.'s medical care; the hospice's medical director will do so. Mrs. E. is dismayed again because she and Mr. E. have enjoyed a warm, trusting relationship with the oncologist. She has not expected him "just to disappear" at hospice admission. Nevertheless, as the oncologist intended, the hospice nurse does not leave before calling the hospice doctor to order an intensive morphine regimen for Mr. E.'s pain.

Mr. E. feels better after several days on the new pain regimen. He still has low-level "nagging" pains but considers them "tolerable." Unfortunately, he continues to deteriorate in other ways. He soon stops eating altogether and vomits up black blood several times from his stomach. He slips into and out of a coma. His daughter who is visiting from out of town now holds his hand for hours at the bedside. She finally tells Mr. E., "Daddy, you've fought the good fight. It's OK to die. Just let go." She then leaves briefly for lunch. Moments after she and Mrs. E. return to the bedside, Mr. E. takes one last deep breath and dies. The daughter leaves the room and closes the door behind her, giving Mrs. E. time alone with Mr. E.'s body.

Several weeks after Mr. E.'s death a distraught Mrs. E. calls the hospice nursing supervisor to ask whether Mr. E. suffered excessively in his last days. The nurse briefly reviews the events of those days: Mr. and Mrs. E.'s agreement to hospice; the initiation of the intensive morphine regimen for Mr. E.'s pain; and his subsequent increased comfort enabling valuable family time with him at the end. The nurse also reminds Mrs. E. that Mr. E. probably died from a sudden stomach bleed and that his death likely resembled falling asleep. The nurse reassures Mrs. E. that she, the rest of the family, and the hospice team workers together gave Mr. E. the best end-of-life care possible. The nurse ends by saying, "Mr. E. knew that you were there at the end and that you loved him. You did everything possible to ease his passing. No one could have done better." Mrs. E. feels relieved.

What, Realistically, Is the "Right" Way to Die?

The 1995 landmark Study to Understand Prognoses and Preferences for Outcomes and Risks of Treatments (SUPPORT) rocked the medical world by exposing serious deficiencies in implementing long-established tenets of end-of-life care. The study showed that doctors did not know the care wishes of their critically or terminally ill patients. And, even when given those wishes explicitly, doctors did not modify their care accordingly [23]. These disturbing results galvanized a campaign to reform critical and terminal care in the USA. The campaign unofficially adopted as its goal realizing the ideal of the "good" death described in the last section. But, however well intentioned the campaign, its results have often been disappointing. At least one other author agrees with my belief that its goal—a "good" death for all—is unrealistic in its fullest form. The "inevitable limitations, contingencies, tensions, paradoxes, and accommodations" involved, this author asserts, will always leave the ideal "good" death out of reach. He, therefore, proposes an alternative standard of "good enough" care [16]. While I support realistic end-of-life care practices, I do not support a merely "good enough" standard for them. "Good enough" suggests minimal quality of care and half-hearted efforts at improvement. I believe health professionals must do better. For that reason I prefer the alternative standard of "best possible care under the circumstances." I believe most health professionals already try to deliver end-of-life care to that standard, but improvements are still possible. Those improvements will occasionally be quick and dramatic but more often slow and incremental [24]. They will require long-term effort from health professionals.

Many studies after SUPPORT have produced ideas for improvements in end-of-life care. Grounded in the experiences of patients, family members, and health professionals, these ideas promise to reorient expectations for care from unattainable perfection to realistic, though high, goals. Health professionals providing such care need to know these ideas.

Tables 11.3 and 11.4 summarize the most prominent ideas and call them "steps" that can be taken. Table 11.3 reports steps that health professionals can take; Table 11.4, steps that patients or their caregivers can take. The tables draw on 15 professional journal articles [6, 12–14, 17, 20, 21, 25–32] based on 14 original research projects. These projects studied living elderly or potentially terminal patients (often with cancer, congestive heart failure, cirrhosis, or similarly grave diseases), or family survivors or professional caregivers (including doctors, nurses, social workers, and clergy) of patients who had died. Large majorities of each study's respondents were Euro-Americans. The projects asked about patients' wishes concerning the "right" way to die and about specific steps to fulfill those wishes.

The tables list all ideas, or "steps," that at least three articles report. I categorized the steps two ways: first, under four commonly cited end-of-life care domains—the physical, the psychological, the social, and the spiritual—[15, 28, 33] and, second, under broad patient wishes within those domains. Three review articles [22, 33, 34] served as checks on the comprehensiveness and accuracy of my categorizations. Steps in the physical and psychological domains fall mostly to health professionals to take, steps in the social domain split between health professionals and patients or family caregivers to take, and steps in the spiritual domain fall mostly to patients to take.

I caution readers not to misunderstand these tables as "report cards" or rigid to-do checklists for end-of-life care. I intend the tables merely to suggest ideas for addressing common patient wishes about end-of-life care.

The Physical Domain

The dominant culture of health care tends to focus health professionals' attention on this domain whether in end-of-life care or any other care [20, 34]. The strong clinical values associated with scientifically tested, "evidence-based" physical interventions and the reimbursement regulations based on them promote this singular focus. As a result, health professionals—especially clinicians such as doctors and nurses—attend conscientiously to physical symptoms and disabilities while sometimes overlooking problems in the other domains.

In this domain the one clear wish of dying patients concerns minimizing physical suffering [13, 20, 21, 29]. Accordingly, most of the 15 original research articles and all three review articles urge health professionals to ease patients' pain and other distressing symptoms (such as nausea and shortness of breath) and to help maintain patients' physical functioning to the extent possible.

Table 11.3 The "right" way to die: steps *health professionals* can take to fulfill patients' wishes

Domain	Patients' wish for …	Steps health professionals can take to fulfill patients' wishes	References
Physical[a]	Minimal physical suffering	• Ease pain and other distressing symptoms	[6, 14, 17, 20, 21, 28–31, 45, 46]
		• Maintain patient's physical functioning to the extent possible	[21, 28, 30, 46]
Psychological[b]	Emotional support	• Ease patient anxieties or fears about the future	[21, 26–28, 30, 32, 46]
		• Appreciate "the whole patient"	[14, 17, 21, 28, 29]
		• Support realistic hopes for patient	[27, 29, 32, 45]
		• Care affectively for patient, especially by showing compassion	[14, 17, 25]
		• Help patient prepare for death and accept it	[21, 28, 30]
	Good communications from health professionals to patients	• Talk honestly but supportively about dying	[14, 17, 21, 26, 27, 32]
		• Explain what to expect in the future while dying	[14, 21, 29]
		• Explain clinical matters understandably	[14, 29, 32]
	Some control in dying	• Allow patient some decisions about dying	[17, 20, 26, 28, 29, 32, 46]
		• Help patient provide own care as possible	[6, 26, 30, 32, 45]
	Respectful, personal care	• Maintain continuity of care (especially among doctors)	[14, 21, 25, 27, 29]
		• Make care readily accessible	[14, 25, 27]
		• Be conscientious, trustworthy, and attentive in care	[14, 21, 27]
Social[c]	Putting in order end-of-life matters that affect others	• Encourage as possible easing death's physical, emotional, and financial burdens on others	[17, 20, 21, 29–31]
	Involvement of family and friends in care[d]	• Include family in care discussions	[14, 17, 20, 27]
		• Urge family or friends to visit for patient's emotional support (if not too stressful for patient or others)	[6, 17, 30]

[a]This domain might also be labeled "medical." Health professionals should, of course, also give technically competent care [14, 30], use effective teamwork among health professionals [27], and offer therapeutic touch. [21, 25]

[b]Health professionals might also encourage the patient to "let go" when dying is irreversible and death is close [21, 25]; affirm the patient's accomplishments by communicating that the patient lived a "worthwhile" and, if appropriate, "complete" life [21]; listen attentively to the patient (including asking open-ended questions, inviting the patient's questions, hearing the patient's fears and emotions, and using good eye contact) [21]; and avoid ethnic or cultural stereotypes. [17]

[c]Health professionals should also support caregivers and proxies emotionally [20], and respect the minority patient's cultural rituals. [17]

[d]Health professionals might also encourage prayer [16, 21] and visits or other support from the patient's parish pastor or fellow congregants if the patient wishes [30], but always avoid forcing one's own spiritual or religious beliefs on the patient. [28]

Table 11.4 The "right" way to die: steps *patients* can take to fulfill their own wishes

Domain	Patients' wish for ...	Steps patients can take to fulfill their own wishes	References
Social[a]	Putting in order end-of-life matters that affect others	• Ease death's physical, emotional, and financial burdens on others	[17, 20, 21, 29–31]
		• Make end-of-life plans	[21, 27, 29, 30]
	Involvement of family and friends in care	• Ask family and friends to visit for emotional support	[6, 17, 30]
Spiritual[b]	Support from religious advisors	• Request contacts with clergy	[6, 17, 21, 28]
	Spiritual comfort	• Explore one's own spiritual beliefs and rites	[17, 28, 30]
		• Make peace with God	[21, 28, 31]

[a]Patients might also try to resolve conflicts with others. [28]
[b]Patients might also explore the meaning of life, illness, and death. [21, 29]

Case 2

Mr. E. receives excellent physical care throughout his terminal illness. When the usual analgesic regimens no longer provide adequate pain relief, his oncologist quickly refers Mr. E. to home hospice for specialized analgesic regimens. Mr. E. eventually receives fentanyl skin patches and oral and intravenous morphine. The hospice nurse and nurse's aides reposition him frequently in bed and apply moisturizing cream and cushioned pads to his back and heels to prevent pressure sores. The aides also perform daily bedside range-of-motion exercises to maintain joint flexibility. Once, when Mr. E. is imminently dying, he becomes agitated and cannot find a comfortable position in bed. The hospice nurse notifies the hospice doctor to change the analgesic regimen. She also climbs onto the bed and rubs Mr. E.'s back until he falls asleep.

The Psychological Domain

Patients' wishes in this domain often include emotional support; [17] good communications with health professionals; [14, 32] some control in dying; and high-quality, respectful care [21, 29, 30, 33]. Many of the 15 original research articles and all three review articles mention health professionals' responsibility to try to fulfill these wishes. To aid in emotional support, health professionals should learn something about patients' values and prior lives [21, 29]. I find that doing so need not require extensive questioning or a lot of extra time. Clinicians can ask about the patient's life, for example, while doing other tasks such as changing the bedsheets, taking blood pressures, or checking for pressure sores. Good initial questions include Where did you grow up? Were you in military service? How long have you been married? and Do you have children? Health professionals just need then show interest and to listen attentively. The patient will usually open up and talk freely about his or her life.

Health professionals should also try to ease patients' fears or anxieties about the future. Those fears or anxieties often concern symptoms, functional losses, emotional isolation from others [16, 17, 27], discomfort at the time of death, and the impact of one's death on the family. Health professionals' responses to those fears or anxieties should always be honest but compassionate [17, 19, 28, 32]. If, for example, dying patients fear discomfort in dying, health professionals can honestly reassure them that imminent death is not usually distressing but peaceful like falling asleep.

Good communication with patients, of course, requires health professionals to explain symptoms, care decisions, and other clinically important matters as they arise. The explanations should avoid medical jargon and confusing metaphors [14], and should be repeated as necessary for patients and family members to understand. All the while, health professionals should never dismiss patients' hopes out of hand but treat those hopes respectfully. If they are unrealistic, health professionals should gently help patients reformulate those hopes (such as for comfort when cure is no longer possible). Health professionals should also promise dying patients they will never be abandoned, and should personally deliver on that promise.

Many dying patients also want some control over the end of their lives. The nature and degree of that control vary from patient to patient, and a particular patient's wishes about control may not be immediately obvious to attending health professionals. They, therefore, may need to ask explicitly which decisions patients wish to make. Those decisions might concern how patients will spend their last days, where they will die, or who should be present for their deaths [6, 17, 28]. Self-care is another way patients can exercise control at the end of life [30]. That control may involve only small, everyday tasks, but caregivers can take simple steps to encourage it: They can help patients perform such tasks, say, by placing patients' slippers at the bedside, squeezing the toothpaste onto the toothbrush, or putting a straw into the bedside water glass.

All patients, of course, want high-quality, respectful care from their health professionals [21, 29, 30, 33]. Such care is especially important when patients are dying and vulnerable. Clinicians must ensure that care for these patients is always of high quality. Clinicians must be conscientious, trustworthy, and attentive in fulfilling their terminal care duties. And all health professionals, not just the clinicians, should make a special effort to maintain continuity of care. That continuity provides an anchor for dying patients in a sea of momentous life changes. I urge health professionals—doctors, nurses, clergy, social workers, and others—who have previously attended these patients to visit them periodically even as the patients near the death [25, 33, 35]. Many of these visits may admittedly serve social purposes only. But such visits are important for patient morale, and reimbursement policies should support them [25].

Case 2

Mr. E.'s care demonstrates many of Table 11.3's points about good end-of-life psychological care. Although the hospice's medical director assumes principal responsibility for Mr. E.'s hospice care [25], the oncologist continues to follow him with occasional home visits or calls. That doctor understands much about Mr. E.'s life history and personal values from taking care of him through the prior anticancer treatments. Therefore, most of the oncologist's contacts after the hospice transfer focus on the psychological and social aspects of Mr. E.'s illness. For example, when Mr. E. transfers into hospice, the oncologist talks with the hospice's medical director to brief him on Mr. E. The oncologist describes Mr. E. as a motivated, orderly, and intelligent man who wants to maintain his independence as long as possible, to perform as much of his self-care as possible, and to demonstrate to his children and grandchildren how to die well. The oncologist also explains that Mr. E. needs clear, honest communications from his doctors. The oncologist illustrates by describing a recent conversation with Mr. E. about entering hospice. The oncologist asked Mr. E., "How are you doing?" Mr. E. replied candidly, "I think I'm not doing very well. I'm getting steadily worse." The doctor replied equally candidly, "I agree with you." He then asked, "What are you now hoping for?" Mr. E. said, "I'd like to live another two years." The doctor replied, "I'm afraid that is not likely." But, trying to help Mr. E. reshape his hopes realistically, the doctor added, "As you know, we have given you all the radiation and chemotherapy we can. For that reason I believe the best plan for you now is for us to keep you comfortable and to maintain each functional ability as long as possible. Home hospice can help, and I recommend it to you." Mr. E. agreed after talking with his family.

Once transferred into hospice, Mr. E. receives conscientious and compassionate end-of-life care there, too. The hospice nurse coaches Mrs. E. and the part-time nurse's aides about how to give personalized care directed at Mr. E.'s specific needs. The nurse's aides, for example, learn to monitor his pain and to notify the hospice nurse quickly when Mr. E. needs a change in his analgesic regimen. They also learn to turn him in bed to prevent pressure sores, to change his clothes and bedclothes frequently to keep him clean from incontinence, and to perform range-of-motion exercises to maintain flexibility. The hospice nurse, social worker, and chaplain also conduct informal life reviews to get acquainted with Mr. E. as a "whole" person and to reassure him he is valued [14]. They learn about his World War II military service, his family, and his interest in outdoor sports. At the same time the staff discover some of the core values of Mr. E.'s life: honesty, loyalty, persistence, orderliness, and faith. This knowledge facilitates their communication with him when much of the face-to-face time becomes purely social.

The Social Domain

Patients often have two wishes in this domain: receiving supportive care from the family and putting in order those end-of-life matters that affect others. Health professionals can actively promote both wishes by urging family members to help fulfill the first wish, and patients themselves to fulfill the second.

Health professionals should promote supportive care by family members [3, 15] but only to the extent the patient wants it [6]. (Health professionals might need to ask the patient what care he or she wants and from whom.) [28] If the patient wants, family members might visit regularly for emotional support, participate in care decisions, or give certain kinds of physical care [30]. Health professionals might also encourage the patient and family members to mend any interpersonal rifts; share memories, wisdom, and material keepsakes; and say important good-byes while they can [17, 19, 28, 29]. All the while, health professionals should observe family members carefully for excessive stress due to involvement in the care. Family members usually do not volunteer that stress on their own. But, if it exists, social workers may be able to explore the sources for it and find resources to help.

Health professionals should also encourage patients to put in order any end-of-life matters to ease future physical, emotional, and financial burdens on others [29]. Patients might, for example, plan ahead to stop aggressive treatments in a timely way, or prepare important documents such as testamentary wills, proxy designations, and instructions about burial or cremation [17, 21, 30].

Case 2

When taking the anticancer treatments, Mr. E. mentions to the oncologist that he welcomes care by his wife and children. The doctor tells Mrs. E., the adult daughter, and the adult son about Mr. E.'s wish, and all agree to do what they can. Family involvement becomes especially important as the anticancer treatments fail and Mr. E. weakens. He is relieved as his family assumes ever more decision-making for him. Mrs. E. and both children, for example, decide with Mr. E. to accept home hospice care when the oncologist suggests it for pain relief. Even after hospice transfer Mrs. E. provides as much physical care for Mr. E. as she can despite her own fragile health. She dresses and bathes him every day. The children provide additional physical care for him as needed. The daughter massages and moisturizes Mr. E.'s feet, and the son walks with him short distances in the home and helps toilet him. To provide companionship and encourage intake, Mrs. E. and the children often sit with Mr. E. at meals. Nieces, nephews, and grandchildren also visit from out of town. Mr. E. shows them old family photograph albums and enjoys reminiscing about his early life. Unfortunately, Mr. E.'s care (especially at night) eventually overwhelms Mrs. E. The hospice medical director notices her exhaustion and contacts the social worker for help. She persuades Mrs. E. to engage trustworthy, hospice-experienced nurse's aides, whom the social worker knows well, to attend Mr. E. at night so that Mrs. E. can rest.

In addition, Mr. E. has always fulfilled his family-related duties conscientiously. Therefore, before he becomes too incapacitated from his cancer, he takes steps to minimize any burdens of his future death on the family. He draws up a detailed testamentary will including descriptions, short histories, and assessed values of family heirlooms to be bequeathed to his survivors. He also designates his wife and adult daughter as his primary and secondary proxies, respectively, for medical decisions; makes funeral plans; and arranges for cremation and burial of his ashes in the family's ancestral plot.

The Spiritual Domain

This domain involves the search for meaning in life, illness, and death [3]. I believe virtually all dying patients engage in such a search, [29] and two wishes often play an important part: engaging with spiritual leaders and achieving spiritual peace [21, 30]. When patients profess a particular organized religion, their search naturally focuses on that religion's tenets. Visits and other communications from clergy of that religion may be especially helpful to those patients in their search. Yet patients who profess no particular organized religion may also appreciate visits by clergy to discuss general existential questions in dying [3, 21]. (Clergy, of course, know not to proselytize actively such unchurched patients during these encounters.)

Clinicians—the doctors and nurses who see dying patients first and foremost for their physical problems—have an opportunity to facilitate these spiritual searches. Unfortunately, most clinicians feel unprepared to address patients' spiritual needs and thus do not do so. Even when recognizing such needs, clinicians often refer patients automatically to clergy without providing much spiritual background about the patients. Clinicians then quickly refocus their attention on patients' physical needs [21].

Yet clinicians *can* help address dying patients' spiritual needs. With any clinical consultation the attending clinician's preliminary assessment helps orient the consultant to a patient's problems. A spiritual consultation is no different. In fact, a preliminary spiritual assessment requires little time or special knowledge, mostly just the clinician's willingness to venture out of the familiar physical domain. I recommend the clinician conduct spiritual assessments with all dying patients and structure those assessments according to either of two simple, four-item mnemonics. (I have modified the mnemonics slightly from their originally published versions.) One mnemonic, HOPE, stands for What are your sources of **H**ope and meaning? Do you identify with a particular **O**rganized religion? Do you observe particular spiritual **P**ractices? and What do you wish the **E**ffects of your spiritual or religious beliefs would be on your care? [36] The other mnemonic, FICA, stands for Do you have **F**aith-based beliefs that help you cope with stress? How **I**mportant is faith to you? Are you a member of a particular spiritual or religious Community? and How can we Accommodate your spiritual or religious beliefs in your care? [37, 38] FICA strikes me as the better mnemonic: it is more concrete and to the point than is HOPE. I, therefore, prefer using FICA. With a quick assessment using either mnemonic, though, the clinician can help the hospital chaplain or parish pastor assess a dying patient's spiritual orientation, define his or her reactions to dying, and plan when and how to intervene [28]. (The reader should consult Chap. 14 for further elaboration about spiritual care by clinicians.)

Spiritual consultations by chaplains or parish pastors can give the spiritual support many patients seek, and help them make peace with God at a critical moment in life [28, 33]. Yet, even when clinicians refer patients for consultation with spiritual leaders, I urge clinicians to stay informed about patients' spiritual searches as an integral part of holistic end-of-life care. Of course, when clinicians do not spontaneously request spiritual consultation, patients or family who want one should request it themselves.

Case 2

When Mr. E. first enters the hospital to begin chemotherapy, his oncologist conducts a preliminary spiritual assessment using FICA. The assessment reveals that Mr. E. is a devout Methodist. He attends worship regularly and has served his local church in various voluntary capacities. When the doctor asks whether any particular faith-based beliefs help Mr. E. cope with stress, Mr. E. replies that faith in God's constant companionship gets him through. He then quotes Psalm 23:4, "Even though I walk through the valley of the shadow of death, I fear no evil for Thou art with me." And, when the doctor asks how care might accommodate Mr. E.'s religious beliefs, Mr. E. asks that the minister at his parish be notified about the hospitalization. The minister subsequently visits Mr. E. in the hospital and later at home.

As death nears, Mr. E. struggles with its spiritual implications. He asks the minister such difficult questions as Why has God given me this cancer? Does He still love me? Will He forgive me for what I've done wrong in my life? Will death be difficult? and Will I go to heaven? But Mr. E. asks ever fewer questions as he becomes sicker. He either becomes too weak to consider the questions any more or has already made peace with God. But, shortly before he dies, Mr. E. tells the minister he has resigned himself to death with all its unknowns.

Dealing with Survivor Reservations When Deaths Fall Short of Expectations

We, health professionals, may try continually to improve our end-of-life care by putting into practice these steps and others in the four domains. But, however much we improve our end-of-life care, it will surely never be perfect. It will always fall short, leaving some survivors dissatisfied. All health professionals involved in end-of-life care, therefore, must be prepared to address dissatisfactions—and whatever guilt or regrets accompany them—as part of aftercare for survivors. But the dying patient's principal attending doctor or nurse, chaplain or clergyman, or social worker is usually best positioned to do so.

The health professional who does address such dissatisfactions might start by asking general questions such as How do you feel about the care Mrs. Smith received as she died? and Do you have regrets about it? If survivors might have regrets but are reluctant to disclose them, the health professional might say, "No care is perfect, and we are always trying to improve ours. Telling us your regrets can help us improve future care and may help you feel better, too. For those reasons I genuinely want to hear what you have to say." The health professional must then listen attentively and avoid responding defensively. He or she may then eventually close the conversation by saying something like "I understand the care Mrs. Smith received as she died may not have completely met your expectations. But I thank you for sharing your comments. Our medical team will take them to heart as we try to improve our care in the future." Because this nondefensive, reassuring response may not sink in immediately, the health professional may need to repeat it one or two times.

A case from my internship illustrates the constructive process of addressing survivors' dissatisfactions with end-of-life care. An elderly woman and I had to make the sad decision to stop life support for her brother, who was dying from cirrhosis. For weeks after the man died, the woman returned to the ward for reassurance. Dissatisfaction with that decision plagued her. She always asked the same question, "Did I make the right decision?" And I always gave the same answer: that her brother had been irreversibly dying, that she and I together had made the only decision possible (to stop life support), and that the best gift we could have given her brother under those circumstances was to allow him to die quickly and comfortably. And he had. The woman took months to gain some peace of mind about the decision. Her agony over what she considered unsatisfactory care (but which probably was not) taught me always to give family members considerable reassurance about decisions to withhold life support before, during, and after they are made; and to acknowledge explicitly my own agreement with those decisions.

Physician-Assisted Suicide

I include physician-assisted suicide (PAS) here because it affects both the timing and manner of death. The reader should understand from the beginning that PAS differs from euthanasia, often called "mercy killing." PAS—unlike euthanasia—requires the patient to consent to aid in dying, a doctor to prescribe the lethal oral cocktail to be used, and the patient to ingest it him- or herself [39, 40]. Furthermore, only a few states in the USA (California, Montana, Oregon, Vermont, and Washington) have legalized PAS at this time. Other states may do so soon. But all states still strictly forbid euthanasia [40, 41].

Switzerland and the Netherlands were the first countries effectively to decriminalize PAS. Switzerland decriminalized suicide in 1942, and the Netherlands has openly tolerated PAS (and euthanasia) since the 1980s [39, 40, 42, 43]. Oregon pioneered legalization of PAS in the USA when voters passed the state's Death with Dignity Act in 1994 [26, 42]. That law legalizes PAS but lays out strict conditions for performing it: The requesting patient must be older than 17, residing legally in Oregon, terminally ill with a remaining life expectancy less than six months, and able to make and communicate decisions. The patient must also communicate to the prescribing doctor several requests, oral and written, for assisted suicide. That doctor and a consulting doctor must confirm the patient's diagnosis, prognosis, lack of depression, and ability to make informed, voluntary decisions. The prescribing doctor must also inform the patient about the availability of comfort care and must urge the patient to disclose the PAS request to his or her next of kin [44, 45]. Since its passage the Death with Dignity Act has served as a model for other states considering legalizing PAS.

According to recent studies, 60 % of terminally ill Americans now support PAS in general, 10–11 % "seriously consider" it for themselves, and 4 % actually discuss it with their doctors [18, 28, 42]. But less than 1 % of Oregonians who die request the lethal cocktail, and less than 0.4 % use it to commit suicide [18, 40, 42, 44]. Looking back, some observers declare this experience with the Death with Dignity Act "reassuring." I agree. The catastrophic social consequences some opponents predicted early on have not occurred. Few people from outside Oregon have migrated there solely to use the law, and the vulnerable dying say they do not feel pressure to use it against their wills. Some observers even claim the Act has encouraged improvements in end-of-life care generally and increases in hospice use in particular [41, 44].

Importantly, the Death with Dignity Act requires Oregon's state government to collect records of all lethal prescriptions given. That requirement has prompted follow-up studies of the people who request PAS. Two studies have explored their motivations. These studies involved interviewing patients who were considering PAS or relatives of patients who had used it to end their lives [12, 26]. Nearly all the patients were Euro-Americans, and many first acquired their motivations for PAS by watching the distressing deaths of others. The motivations rated most important included the desire to control certain aspects of one's death (especially dying at home), fears about the future (usually concerning symptoms, poor quality of life, or inability to do self-care), current losses (especially of function, independence, or "the self"), and burdensomeness of one's care to others (Table 11.5). The motivations rated moderately important included current symptoms and readiness to die. Surprisingly, the motivations rated least important included the meaninglessness or poor quality of current life [12, 26, 40, 44, 46].

How, then, should a health professional respond to a patient's request for PAS? Although many terminal patients in the USA support PAS and some may think about it for themselves, the few who act on the idea where it is legal may have misgivings

Table 11.5 Patients' motivations for physician-assisted suicide

Importance	Motivation	Source(s) [references]
High	Desire for control over time or manner of death	[26, 46]
	Fears about future, such as symptoms,[a] poor quality of life, or inability to do self-care	[26, 46]
	Current loss of independence and function (especially the ability to do personal self-care)	[26, 46]
	Burdensomeness of personal care on others	[26]
	Current "loss of self" (e.g., personal identity, essence, or "dignity")	[26, 46]
Moderate	Experiences of prior "bad" deaths of others	[26, 46]
	Current symptoms[a]	[26, 46]
	Worries about future "loss of self"	[26]
	Readiness to die	[26, 46]
Little or none	Meaninglessness of current life: important tasks completed, no pleasures any more	[26]
	Financial drain on others	[26]
	Current poor quality of life	[26]
	Current inability to do self-care	[26]
	Lack of social support	[26]
	Depressed mood	[26]

Data from Pearlman et al. [12]; Ganzini et al. [26]
[a]Symptoms most often mentioned were pain, dyspnea (shortness of breath), weakness, fatigue, incontinence, sadness, and mental confusion.

[43]. Dr. Susan D. Block, the cancer psychiatrist I cited in Chap. 2, believes that thoughts of PAS, even with the misgivings about it, indicate deep dissatisfactions with one's circumstances. Yet few patients volunteer those dissatisfactions to health professionals. Dr. Block believes, therefore, (and I agree) that the principal attending doctor should ask each dying patient about thoughts of suicide. The doctor might say evenhandedly, "Some patients in your situation think about ending their lives early. Have you ever had thoughts like that?" Dr. Block believes that, if the patient responds affirmatively, the doctor should not reply with mere facile reassurances. Rather, the doctor should probe for the patient's dissatisfactions. In general, they will be grounded in psychological, rather than physical, distresses [18, 26, 42, 44, 46]. Table 11.5 can serve as a checklist of possibilities. If correctable dissatisfactions surface, the doctor, of course, should try to remedy them.

In states that have not yet legalized PAS, I urge the doctor who receives a PAS request neither to ignore it nor to accede to it. Instead, the doctor should probe for dissatisfactions that might underlie the request, and try corrective actions [12, 46]. The patient him- or herself may have good ideas. The doctor might ask, for example, "How can I help you, short of ending your life, to get through this difficult time?" [43]. The doctor may need to intensify symptom relief or make referrals to a chaplain, pastoral counselor, or social worker [43]. Above all, doctors in these states must not risk their licenses by performing illegal PAS, even covertly. However well motivated by compassion for the suffering patient, the doctor must consider the duty to care for *all* his or her patients. Performing illegal PAS threatens that care through legal and professional sanctions: specifically, the doctor might lose his license. Doctors who favor PAS in these states, therefore, should channel their advocacy efforts into the political process.

In states that have legalized PAS, doctors should know beforehand the requirements for carrying out PAS legally, and they should honor those requirements punctiliously in practice. Above all, PAS must occur within "an established, meaningful doctor-patient relationship" [43] that offers protections based on doctor–patient familiarity and trust. Other requirements offer additional protections against mistakes and abuses: a clear, established diagnosis and prognosis; the patient's repeated, informed, and voluntary requests; the proven failure of all efforts to relieve the patient's discomforts; an independent second medical opinion; thorough documentation; and a follow-up outside review [39]. Of course, even in these states a doctor who receives a PAS request should carefully explore the causes for it and try to correct them [33, 43]. For example, the doctor should check for depression perhaps by asking simply, "Are you depressed?" [42] The doctor should also remind the patient that important personal growth may occur even during the last weeks or months of life [20, 42]. And the doctor should mention that PAS does not guarantee death: Something might go wrong in the process, leaving the patient suffering more or functioning less than before [40, 43].

PAS can cause people other than the patient to suffer, too. The family will have to live with any sad consequences. Therefore, Dr. Block considers family objections to PAS "a relatively strong contraindication" to it [43]. Doctors can also suffer especially if they feel forced to perform PAS against their consciences. If a doctor cannot comply with a patient's PAS request due to reasons of conscience, the doctor should always feel free to refer that patient to another doctor who can.

Nonetheless, if doctor and patient finally agree on PAS, the doctor should continue as a frequent, reassuring presence for patient, family members, and other health professionals throughout the process and document it meticulously in the patient's medical record.

The Case

As death nears, Mr. E. weakens and becomes bed bound. Brief conversations exhaust him. Even the hospice's intensive analgesic regimen no longer controls his pain. He tires of struggling to live and secretly wants his death hastened. Mr. E. might even think about requesting PAS. But he does not live in a state where it is legal. Therefore, if Mr. E. makes such a request, the doctor must not accede to it. Rather, he or she should explore Mr. E.'s current problems and remedy any that can be remedied. The doctor must accept that as the best he or she can do for Mr. E. under the circumstances [33]. For now, PAS is not a legal option in that state.

Summary Points

1. Americans generally believe a person has at least one "right" time to die, only God knows it for sure, and He ultimately controls it. But everyone has the task of recognizing his or her "right" time to die [3]. Conflicts with others may arise over when it is.
2. The "right" way to die encompasses four domains: the physical, the psychological, the social, and the spiritual. Health professionals who provide end-of-life care should learn and implement steps addressing common patient wishes in each domain.
3. Withholding life support at a patient's request has long, consistent legal support in the USA. However, only five states—California, Montana, Oregon, Vermont, and Washington—have legalized physician-assisted suicide for patients. Doctors should participate in physician-assisted suicide cautiously and adhere strictly to its requirements where it is legal, and should not participate in it at all where it is illegal.
4. No states have legalized euthanasia, that is, killing a patient without his or her consent and ability to perform the lethal act. No health professional should ever perform euthanasia in the USA [39].

To Learn More ...

1. Block SD. Psychological considerations, growth, and transcendence at the end of life: the art of the possible. JAMA. 2001;285(22):2898–905.
2. Kaldjian LC, Curtis AE, Shinkunas LA, Cannon KT. Goals of care toward the end of life: a structured literature review. Am J Hosp Palliat Med. 2009;25 (6):501–11.
3. Perkins HS, Cortez JD, Hazuda HP. Cultural beliefs about a patient's right time to die: an exploratory study. J Gen Intern Med. 2009;24(11):1240–7.
4. Steinhauser KE, Christakis NA, Clipp EC, et al. Factors considered important at the end of life by patients, family, physicians, and other care providers. JAMA. 2000;284(19):2476–82.
5. The SUPPORT investigators. A controlled trial to improve care for seriously ill hospitalized patients: The Study To Understand Prognoses and Preferences for Outcomes and Risks of Treatments (SUPPORT). JAMA. 1995;274(20):1591–8.

References

1. Leget C. Retrieving the *ars moriendi* tradition. Med Health Care and Philos. 2007;10 (nn):313–19.
2. Seale C, Addington-Hall J. Dying at the best time. Soc Sci Med. 1995;40(5):589–95.
3. Hardwig J. Going to meet death: the art of dying in the early part of the twenty-first century. Hast Cen Rep. 2009;39(4):37–45.

4. Perkins HS, Cortez JD, Hazuda HP. Cultural beliefs about a patient's right time to die: an exploratory study. J Gen Intern Med. 2009;24(11):1240–7.
5. Morrison R. Patients' sense of completion. BMJ. 1994;308:1722.
6. Vig EK, Davenport NA, Pearlman RA. Good deaths, bad deaths, and preferences for the end of life: a qualitative study of geriatric outpatients. J Am Geriatr Soc. 2002;50(9):1541–8.
7. Perkins HS, Cortez JD, Hazuda HP. Diversity of patients' beliefs about the soul after death and their importance in end-of-life care. South Med J. 2012;105(5):266–72.
8. Trombatore K. With hopes for a very good death. Tex Med. 1994;90(11):17–21.
9. Karlawish JHT, Quill T, Meier DE for the ACP-ASIM End-of-Life Care Consensus Panel. A consensus-based approach to providing palliative care to patients who lack decision-making capacity. Ann Intern Med. 1999;130(10):835–40.
10. Dubler NN. Conflict and consensus at the end of life. Hast Cen Rep Special Report. 2005;35 (6):S19.
11. Fischer M. Defending compromise. Philos Lit. 2014:38(nn):587–602.
12. Pearlman RA, Hsu C, Starks H, et al. Motivations for physician assisted suicide: patient and family voices. J Gen Intern Med. 2005;20(3):234–9.
13. Pearlman RA, Cain KC, Patrick DL, et al. Insights pertaining to patient assessments of states worse than death. J Clin Ethics. 1993;4(1):33–41.
14. Hanson LC, Danis M, Garrett J. What is wrong with end-of-life care? Opinions of bereaved family members. J Am Geriatr Soc. 1997;45(11):1339–44.
15. Emanuel EJ, Emanuel LL. The promise of a good death. Lancet. 1998;351(Suppl II):21–9.
16. Masson JD. Non-professional perception of "good death": a study of the views of hospice care patients and relatives of deceased hospice care patients. Mortality. 2002;7(2):191–209.
17. Tong E, McGraw SA, Dobihal E, et al. What is a good death? Minority and non-minority perspectives. J Palliat Care. 2003;19(3):168–75.
18. Emanuel EJ, Fairclough DL, Emanuel LL. Attitudes and desires related to euthanasia and physician-assisted suicide among terminally ill patients and their caregivers. JAMA. 2000;284 (19):2460–8.
19. Heyland DK, Dodek P, Rocker G, et al. for the Canadian Researchers, End-of-Life Network (CARENET). What matters most in end-of-life care: perceptions of seriously ill patients and their family members. Can Med Assoc J. 2006;174(5). doi:10.1503/cmag.050626
20. Singer PA, Martin DK, Kelner M. Quality end-of-life care: patients' perspectives. JAMA. 1999;281(2):163–8.
21. Steinhauser KE, Christakis NA, Clipp EC, et al. Factors considered important at the end of life by patients, family, physicians, and other care providers. JAMA. 2000;284(19):2476–82.
22. Kaldjian LC, Curtis AE, Shinkunas LA, Cannon KT. Goals of care toward the end of life: a structured literature review. Am J Hosp Palliat Med. 2009;25(6):501–11.
23. The SUPPORT investigators. A controlled trial to improve care for seriously ill hospitalized patients: The study to understand prognoses and preferences for outcomes and risks of treatments (SUPPORT). JAMA. 1995;274(20):1591–8.
24. Callahan D. Death: the distinguished thing. Hast Cen Rep. 2005;35(6):S5–8.
25. Back AL, Young JP, McCown E, et al. Abandonment at the end of life from patient, caregiver, nurse, and physician perspectives: loss of continuity and lack of closure. Arch Intern Med. 2009;169(5):474–9.
26. Ganzini L, Goy ER, Dobscha SK. Oregonians' reasons for requesting aid in dying. Arch Intern Med. 2009;169(5):489–92.
27. Hanson LC, Henderson M, Menon M. As individual as death itself: a focus group study of terminal care in nursing homes. J Palliat Med. 2002;5(1):117–25.
28. Pierson CM, Curtis JR, Patrick DL. A good death: a qualitative study of patients with advanced AIDS. AIDS Care. 2002;14(5):587–98.
29. Steinhauser KE, Clipp EC, McNeilly M, et al. In search of a good death: observations of patients, families, and providers. Ann Intern Med. 2000;132(10):825–32.
30. Vig EK, Pearlman RA. Quality of life while dying: a qualitative study of terminally ill older men. J Am Geriatr Soc. 2003;51(11):1595–601.

31. Vig EK, Pearlman RA. Good and bad dying from the perspective of terminally ill men. Arch Intern Med. 2004;164(9):977–81.
32. Wenrich MD, Curtis JR, Shannon SE, et al. Communicating with dying patients within the spectrum of medical care from terminal diagnosis to death. Arch Intern Med. 2001;161 (6):868–74.
33. Block SD. Psychological considerations, growth, and transcendence at the end of life: the art of the possible. JAMA. 2001;285(22):2898–905.
34. Hales S, Zimmerman C, Rodin G. The quality of dying and death. Arch Intern Med. 2008;168 (9):912–8.
35. Pantilat SZ. Communicating with seriously ill patients: better words to say. JAMA. 2009;301 (12):1279–81.
36. Anandarajah G, Hight E. Spirituality and medical practice: using the HOPE questions as a practical tool for spiritual assessment. Am Fam Physician. 2001;63(1):81–9.
37. Puchalski CM, Romer AL. Taking a spiritual history allows clinicians to understand patients more fully. J Palliat Med. 2000;3(1):129–37.
38. Puchalski CM, Ferrell B, Viriani R, et al. Improving the quality of spiritual care as a dimension of palliative care: consensus conference report. J Palliat Med. 2009;12(10):885–903.
39. Quill TE, Lo B, Brock DW. Palliative options of last resort: a comparison of voluntarily stopping eating and drinking, terminal sedation, physician-assisted suicide, and voluntary active euthanasia. JAMA. 1997;278(23):2099–104.
40. Emanuel EJ, Onwuteaka-Philipsen BD, Urwin JW, Cohen J. Attitudes and practices of euthanasia and physician-assisted suicide in the United States, Canada, and Europe. JAMA. 2016;316(1):79–90.
41. Petrillo LA, Dzeng E, Smith AK. California's end of life option act: opportunities and challenges ahead. J Gen Intern Med. 2016;31(8):828–9.
42. Bascom P, Tolle SW. Responding to requests for physician-assisted suicide: "These are uncharted waters for both of us." JAMA. 2002;288(1):91–8.
43. Block SD, Billings JA. Patient requests to hasten death: evaluation and management in terminal care. Arch Intern Med. 1994;154(18):2039–47.
44. Okie S. Physician-assisted suicide—Oregon and beyond. N Engl J Med. 2005;352(16):1627–30.
45. Oregon Revised Statutes, Chapter 127.800-897. Accessed 03 June 2015 at https://www.oregonlegislature.gov
46. Ganzini L, Goy ER, Dobscha SK. Why Oregon patients request assisted death: family members' views. J Gen Intern Med. 2008;23(2):154–7.

Chapter 12
Going Home, the Afterlife, and Other Beliefs About Death

> O Lord, support us all the day long of this troublous life until the shad(ow)s lengthen, the evening comes, the busy world is hushed, the fever of life is over, and our work is done. Then, Lord, in your mercy grant us safe lodging, a holy rest, and peace at the last. Amen.
>
> John Henry Newman.
>
> Wilson AN. *John Henry Newman: Prayers, Poems, Meditations*. 1990.

John Cardinal Newman's "Peace at the Last" prayer, which opens both Chap. 1 and this chapter, is a brief but cogent reminder to make sure end-of-life care addresses not only dying but also whatever comes next [1]. Yet death and a possible afterlife are as foreign to scientifically oriented mainstream medicine as any concepts can be.

The recent hospice and palliative care movements, of course, have focused health professionals' (HPs') attention on care in dying as never before. The medical literature on the topic has grown exponentially, funding agencies have underwritten numerous research and demonstration projects about it, palliative care has become an officially acknowledged medical speciality, and end-of-life care delivery has improved greatly. Yet nearly all this attention has concerned dying itself, not death. The reason may lie with the dominance of science in the modern mind-set. Dying yields to science; death does not. Dying, for example, has been subject to scientific inquiries about advance directives, analgesic regimens, decision-making strategies, do-not-resuscitate (DNR) orders, disease trajectories, prognostic predictions, and cost-benefit analyses. But death stubbornly retains its mystery [2, 3]. "Death is always outstanding," one author quotes Ralph Waldo Emerson as saying. "Our (knowledge) never reaches (it), never possesses it. We are always at the beginning, (asking) … What is death?" [4]. In one recent study, a patient named Cissy who was dying from lymphoma described her view of the problem. Whenever I face a challenge, Cissy said, "I read everything I can … about it— … (T)hat's how I deal with things. (But what angered me this time is) I couldn't find anything on the Internet or in books about (death) from lymphoma … What's involved? How does it happen? What's (the) process?" [5]. Cissy was right: Scant professional literature in medicine, psychology, or sociology addresses death. Those disciplines leave the

H.S. Perkins, *A Guide to Psychosocial and Spiritual Care at the End of Life*,
DOI 10.1007/978-1-4939-6804-6_12

topic largely to philosophy and theology [6, 7], and even their literatures say little. The sparseness of all professional literatures on death is surprising given it is a universal, momentous human experience.

What little professional literature about death does exist includes only a few studies on topics that are merely tangentially related to it or are fully grasped only through personal experience. That literature consists of scholars' speculations and ordinary patients' accounts of "death-like" experiences. Clergy, psychologists, and social workers know this literature; scientifically trained clinicians do not. Still, I believe that all end-of-life care professionals should get to know this literature because it may offer therapeutic benefits for dying patients. Like Cissy many of those patients surely wonder—and worry—what will happen to them at and after death [3, 8]. Knowing the literature, however scant, should prepare caregivers as possible to counsel and comfort patients about what death is. And this literature may also help HPs to define their own views on death and to sustain themselves in the arduous task of delivering end-of-life care.

This chapter divides the literature on death into four sections. The first addresses beliefs about what happens to body and soul at physical death. The second describes near-death experiences (NDEs), which may give insights into the transition that death is. The third discusses meanings that people attribute to death. And the fourth describes contacts between the living and the dead. The format of this chapter differs from the format of most other chapters in the book because this chapter does not use just one illustrative case. One case cannot adequately address death, its many nuances, and its significance. Therefore, this chapter uses illustrative quotes from various studies, cases, and informants. Some quotes come from studies by other researchers; and other quotes, previously published or not, come from studies by my colleagues and me. This chapter does conclude with one short case illustrating contacts between the living and the dead.

Beliefs About the Body at Physical Death

Although everyone dies, we know virtually nothing about the transition that is physical death. We do not know exactly when it happens, why it happens, how long it lasts, or what it is like. It remains largely beyond science. Thus, much of what we think about physical death is conjecture. Even clinical practice, which involves witnessing many physical deaths, pays little attention to them. Clinicians consider their duties to the patient ended once a doctor declares the patient dead and notifies the next of kin, the surgeons harvest any organs for transplantation, the nurses prepare the body for transfer to the morgue, and perhaps the pathologists have performed an autopsy. But the clinicians are wrong: The profound impact physical death has on patients and survivors should prompt clinicians and other HPs who provide end-of-life care to understand the variety of beliefs about physical death. HPs owe it to dying patients, survivors, and themselves to appreciate the significance of such beliefs. Sensitive, insightful care depends on doing so [3, 8, 9].

This section addresses beliefs about the body at physical death. Four specific questions have clinical implications: Does some part of the person separate from the body at physical death? How long does physical dying take? Which signs identify physical death? And does the body retain its senses after physical death? A study by my colleagues and me asked those questions of Euro-American (EA), African American (AfA), and Mexican American (MA) elders. The next sections summarize respondents' answers, question by question, and illustrate their answers with quotes.

Does Some Part of the Person Separate from the Body at Physical Death?

Most Americans believe a person's soul or spirit separates from the body and leaves it at physical death. In fact, 90 % of respondents in our study expressed this belief [3]. That percentage remained roughly the same across EAs, AfAs, and MAs and their gender subgroups. Most respondents (79 % overall) and majorities of all three ethnic groups and their gender subgroups except MA women explicitly described the person's soul or spirit as "leaving" or "going" from the body, being "taken" from it, or "passing through" this earthly life. One MA man described physical death as "the spirit comes out of the body … because God wants it" [previously unpublished quote]. Similarly, an EA woman said, "I think (my dead brother) left his body … as soon as (it) died … (His) spirit is in heaven, and (his) body is waiting (here) on earth for the resurrection" [previously unpublished quote]. And an AfA woman said, "I've always thought, when a person died, their body left them (sic), and (their soul) overseed (sic) everything that was going on … how (the bodies) are handled" [previously unpublished quote].

As the first two quotes suggest, the belief about separation of body and soul at physical death sometimes originates in formal church doctrines. Roman Catholic doctrine, for example, permits stopping life support and allowing death to come whenever the body can no longer support God's purpose for the person—that is, maintaining a right relationship with God and with other people [10, 11]. However, other beliefs about this separation originate in cultural traditions or individual experiences independent from formal religious doctrines [3, 11, 12]. Some people believe, for example, in secular ghosts or similar spirits existing outside the dead body. One EA woman in our study said about her dead mother, "We put flowers once a month at the table … where she sat all the time. Where she had breakfast … we keep flowers on the table because we know she's there … she is not in the cemetery" [previously unpublished quote]. And an AfA man said, "After my father's funeral (he) was at my mother's house and … woke her up. He said, 'Look what (the doctors) done to me (during those resuscitations).' He opened his shirt, and he had two big white burns where they was (sic) hitting him with that electric plate. And (then) he buttoned his shirt and said, 'I got to go.' He disappeared, and she's never seen him again … (But) about a month after he died, I would be sitting there in my mother's (house), … I could tell he was there … I'd tell him, 'Hey, old man, come on out where I can see you, … (T)alk to me.' … (H)e never did, but I knew his presence was there in that house" [3].

A few people, however, reject the separability of body and soul at physical death. These people insist the two exist only as one thoroughly intermixed entity. Accordingly, then, the soul dies along with the body.

How Long Does Physical Dying Take?

Views on this question divide into two groups. One group considers physical dying a gradual process and the other, an instantaneous one. One extreme view from the first group claims physical dying occurs lifelong. I have heard this view mostly from theologians. Their idea is the body experiences "dying," or physical losses, throughout life. Thus, as long as the body is living, parts of it are also dying. The losses involved may be physical features (such as skin tone, height, or muscle mass) or capabilities (such as visual or hearing acuity, quick reaction times, or memory and abstract learning). A more moderate view holds that physical dying covers not a whole life span but only some minutes, days, weeks, or years. I mentioned in Chap. 5 one study consistent with this view. That study tracked people's declining abilities to perform everyday activities (such as bathing, dressing, and eating) in the year before death [13]. The abilities of patients who died from severe strokes or heart attacks declined precipitously over seconds to minutes. However, the abilities of other patients who died from cancers, chronic organ system failure (such as gradually progressive heart, lung, or kidney failure), or dementia declined and died more gradually over days to years.

Linda Emanuel, a well-known geriatrician and bioethicist, proposes a variation on the concept of gradual physical dying. She imagines opposing forces acting constantly on the body: degradation and decline versus repair and strengthening. Emanuel believes physical dying begins when an organism's degradation and decline consistently outpace its repair and strengthening. As Chap. 7 notes, Emanuel sees physical dying as approaching, but never quite reaching, the asymptote of total physical death. That asymptote marks the point where the forces of degradation and decline completely overwhelm the forces of repair and strengthening and extinguish any remaining life force. Because the dying organism never quite reaches that asymptote, Emanuel must specify a threshold considered physical death for medical, legal, and cultural purposes. She locates that threshold somewhere on the continuum between the persistent vegetative state and total cessation of cardiorespiratory and neurologic functions [14]. However intriguing her idea, I cannot accept its implication that an organism never reaches actual physical death. That implication strains credibility for me. Others may also object that her idea leaves too indefinite the time of death and might portray the physical dying process as far too long. Still, Emanuel's theory highlights the core implication of all theories of gradual physical dying: Life and death coexist for some part of an organism's life span.

The other group's views about how long physical dying lasts hold that it is not gradual but instantaneous. That is, physical death separates life from death instantly, cleanly, and completely. Such views prompt expressions such as "sudden death" or "the moment of death." Frequent movie portrayals of death as instantaneous (often due to time constraints or dramatic purposes) may fuel this perception. Movies may symbolize such instantaneous deaths by blowing out a candle or turning out a light. Important social and legal procedures depend on this view. Organ harvesting, notification of family, social security benefits, property transfers, and other important matters all depend on considering death as occurring in an instant. I believe most Americans hold that view [15].

The study by my colleagues and me showed only slight differences among ethnic groups about how long physical dying takes. EAs were twice as likely to believe physical dying takes less than a minute (instantaneous) than more than five minutes (gradual), but AfAs and MAs split evenly between the two views [8]. Taking the instantaneous view, one EA woman described her father's death, saying, "We were right outside (the hospital room when my father suffered his cardiac arrest). We knew, when the alarm went off on the heart monitor, it was … the last time we'd see him alive" [8]. Similarly, an MA man explained his mother's soul left her body when "her heart stopped. Once the heart stops, your soul comes out of the body and floats away" [3]. Yet, taking the gradual view, an AfA man described the resuscitation efforts on his father before he died. "(The doctors) started (his heart) three different times," the man said. "So I told them, 'Look, if my dad's heart would quit (again), leave him alone. Don't put his body through that. When a person's mind don't (sic) get oxygen for a certain length of time, he dies … That's what happened to him. He was dead for 25 min before the doctor got his heart going again. They were trying to keep him alive on a machine …" [16]. Similarly, an EA man believed his dad lived for some days after pronounced death. The man, therefore, insisted on riding in the hearse carrying his father to the cemetery. The man explained, "Dad came to me (before he died) and said, 'I get a little scared when I think about dying." (If) I've got to be out in a coffin and buried, will you go the last mile with me?' I said, 'Hell, yes, Dad.' … So on the day of his burial I got in (the hearse) by the driver and lay (sic) my hand on the coffin … I said, 'I'd like to ride here with my dad … Dad hasn't gone. I've still got him with me'" [3].

Which Signs Identify Physical Death?

The cessation of heartbeats, circulation, and respiration has historically identified physical death, prompting doctors' tradition of listening to the patient's heart and lungs before pronouncing death. However, the advent of mechanical ventilators and of cadaveric organ harvesting in recent years has necessitated an additional sign for physical death: the cessation of vital neurologic functions. (Chap. 7 discusses the specific characteristics of this sign, sometimes called "death by brain criteria.") Either sign alone—cessation of cardiorespiratory functions or of vital neurologic functions—is sufficient to diagnose physical death [17]. Either sign also provides solid grounding for meeting society's practical need to differentiate the living from the dead. Medical science understands both signs, and the physical examination detects them easily and reliably. Clinicians have, therefore, incorporated both signs into everyday practice.

Our study, however, suggested the general public look to *various* physical signs to decide for themselves whether physical death has occurred [8]. Some of those signs relate to the two standard clinical signs. For example, referring to cessation of heartbeats and circulation, one MA man told my colleagues and me, "My mother died (as) her heart monitor kept going down little by little" [8]. Similarly, referring to cessation of respirations, an AfA woman said, "Once that breath is going out of the body … (a person's) already dead" [8]. And, referring to cessation of vital neurologic functions, still another MA man exclaimed about a comatose friend, "After 2 months [in the intensive care unit (ICU)] the doctor said, 'He's going to die.' But he was dead already! The (ventilation) machine was pumping, doing the work for him. He couldn't move his arms … It was a fraud!" [3]. Yet the general public may look to other, nonstandard signs, too, to decide for themselves that physical death has occurred. Some people look to cooling of the body, for example. One EA woman told us, "I didn't feel the coldness of (my dead mother in the coffin) … (S)he didn't belong there … I felt like she was asleep. I was telling her to get up" [8]. And other people may decide about physical death based on a person's inability to move, speak, or recognize people, or the person's gurgling, gasping, or turning the eyes back into the head [8].

Conflicts may arise when different people look to different signs to determine physical death. Observers at the deathbed, for example, may disagree about whether a ventilated patient with a heartbeat but few high neurologic reflexes is alive or dead. One of our respondents talked about such a conflict between him and the doctors when his father died. "It was time for (my father) to (die.) He had to go," the MA man said. "He was suffering a very long time. Over 2 years. By the time my dad died, I figured he was already dead. (The doctors) kept him too long on the machines. He was already dead before they announced him (sic) … a long time before" [previously unpublished quote].

Does the Body Retain Its Senses After Physical Death?

Views on this question also vary. Though skeptical myself that the dead body retains its senses, I first encountered the idea years ago when I interviewed a funeral director for a study on attitudes about autopsies [18]. This MA woman expressed dismay that teenaged relatives of a recently dead man had cracked jokes about him at his open casket. She insisted the man could hear the jokes and felt disrespected. A follow-up study by my colleagues and me suggested that 20–33 % of Americans share this woman's belief that dead bodies retain some of their senses [8]. Americans appear to think sight, hearing, and touch are the senses retained most often. For instance, one respondent of ours, an MA woman, said about sight that the dead person's soul "lingers above (the body, watching) to see how the family takes (the death)" [8]. Another respondent, an MA woman, said about hearing, "I think people can hear after they die … for a while. Fifteen minutes at least … They can't answer you, but I'm pretty sure … Like when I talked to my (newly dead) father, I thought he would hear me" [previously unpublished quote]. And a third respondent, an EA man, mentioned touch in objecting vehemently to doctors' practicing medical procedures on a dead body. The man said, "(If the doctors) hurt the guy that's dead (sic), he ain't going to holler … (T)hey wouldn't know they hurt him … (The doctor's) liable to get back too far and hit the bone. He won't know he hit that bone. He (might) break the needle, and … (h)e won't know … 'cause the guy's dead" [3, 8].

The Clinical Implications of These Questions

The answers to these questions can have a dramatic impact on how patients' bodies are handled at physical death. Those who believe that the absence of detectable cardiorespiratory or neurologic functions signals total physical death, that it occurs instantaneously, that a patient's spirit leaves the body immediately, and that the dead body retains no senses will consider the dead body an inanimate, nonsentient object. Those people will treat it no differently than a valuable piece of jewelry or furniture. They tend to approve postmortem invasive procedures such as organ harvesting, autopsies, and practicing of medical procedures. Clinicians tend to fall squarely into this camp. In contrast, those who believe the opposite on all four points will consider the dead body a final refuge for an animate, sentient human spirit. These people tend not to approve postmortem invasive procedures. Many nonclinicians tend to fall into this camp. Thus, when people from both camps witness the same patient's death, conflicts may arise over how to treat the body after physical death.

Beliefs About the Soul at Physical Death

Most Americans believe the soul lives on continuously after bodily death. One study found more than 60 % of university students do so [12]. And the study by my colleagues and me found 90 % of elders do so. This percentage varies little by ethnic group or gender. As one respondent, an MA man, told us, "The spirit is always going to exist" [3]. And a respondent in another group's study said, "The real me (sic) is a spirit that lives in this body. That spirit will live on when my physical body dies" [19]. Likewise, another respondent of ours, an EA man, said,

"I believe that our body is just a place … our actual spirit … use(s) it. The body dies, but the soul never dies. So, when you die, it's actually just your body dying 'cause that's when your soul goes out of 'em. (sic) … All you're keeping is a body … with nothing in it …" [previously unpublished quote]. Like this last respondent many others who hold this belief see the soul as encaged in the body during life and freed at physical death.

A few Americans, however, believe the opposite: They see the soul as inextricably enmeshed within the physical body and dependent on it for existence [14, 20]. Thus, the soul dies when the body dies, and nothing of the person remains afterward [14, 20]. One respondent in our study espoused this belief. "(The soul) stays in the body (after physical death)," this MA man said. "It decomposes with time. It gets destroyed underground … I don't think it leaves (the body ever)" [3].

Even those Americans who believe the soul lives on continuously say it undergoes momentous change at physical death. Many people describe that change as moving away from the living and this earthly life [5]. In our study, for example, 79 % of respondents described the soul as "going," "leaving," or "passing"—a movement away that may occur under the soul's own power [8]. One EA man described the soul to my colleagues and me as "leav[ing] this life, and … go[ing] to the next" [8]. Another respondent, an MA woman, said, "[W]e are born, and we die. Jesus is letting us come to this world … (He) lets us stay here a while. [We're] passing through … to the other side" [8]. Yet other respondents described the soul as "taken," implying a movement away due to some external power, usually God. "We're here on borrowed time," an MA man said. "When God tells you (that) you gotta leave, you do. That's when you die. He's going to take your soul away from you …" [8, 16]. "(W)hen the day comes that there's a medical problem," said another MA man, "I would try for (God) not to take me or … not to take me so easily. (I'd) go to the doctor right away" [previously unpublished quote]. A few respondents, however, mentioned the soul's being taken away by an external force that might not be God. One respondent, an MA woman, said, for example, "I had already accepted that (my father) was going to die, … I knew that one day he had to go. My mother the same thing … Somebody takes them, takes us. We can't do nothing. If it's time, it's time" [previously unpublished quote].

Yet some respondents in this study (14–19 % depending on the particular ethnic or gender group) described the soul not as moving away at physical death but as "resting," "sleeping," or becoming "lost" to the living. One respondent, an MA man, said his sister's "soul went … to heaven (where) she's resting" after her long, exhausting fatal illness [8]. Another respondent, an AfA man, described heaven as "sleep for the believer" [previously unpublished quote]. And still another respondent remembered trying to lift her dead mother's body out of the open coffin at the funeral. "She didn't belong there," this EA woman explained. "I just felt like she was asleep. I was telling her to (wake) up" [8]. And yet another respondent, an MA woman, described her pregnant daughter's death in the childbirth of a grandson who survived. "(My daughter's) death took part of my life away," the woman said. "(But) my grandson … lost his mother" [8].

Near-Death Experiences

Direct study of what happens to the soul at physical death, of course, is impossible. But research has examined the next closest phenomenon, near-death experiences (NDEs). They are the altered states of consciousness that result from life-threatening conditions [7]. The assumption underlying research on NDEs is that they trace what happens to the soul in the earliest, still reversible stages of physical death.

NDEs have been reported for centuries. One author claims Plato's *Republic* reports NDEs as early as 400 B.C.E [21]. The Christian Bible reports Jesus raising Lazarus and others from the dead in New Testament times. And reports of NDEs helped spark the spiritualism fad among Americans during the late 1800s [22]. Yet NDEs attracted little attention from mainstream medicine [21] until American physician Raymond Moody published his book, *Life after Life*, in 1975. In it, Moody reported the experiences of 150 people said to have "died" and then returned to life [4, 6, 23]. The book popularized the now familiar description of death as moving down a tunnel toward a wonderful, bright light. More NDE case reports quickly followed in popular and professional magazines. At the same time new life-saving medical technologies such as cardiopulmonary resuscitation, intensive care, and mechanical ventilation created more opportunities than ever before for patients to be revived from brief death [7]. Formal studies began to document remarkably high prevalences of NDEs. For example, a study of nearly 1600 patients admitted to the University of Virginia's cardiac ICUs found that 2 % of all these patients and 10 % of those suffering cardiac arrests had had NDEs by predefined criteria [24]. Other studies found similar prevalences of NDEs: 4–5 % among the general American population, 6–12 % among cardiac arrest victims, and 15–20 % among critically ill patients [6, 7, 25, 26]. Lingering questions, however, include why other resuscitated cardiac arrest victims or critically ill patients do not report NDEs and what, if not NDEs, these patients *do* experience.

People who do report NDEs consistently describe them as pleasant [6, 7]. These people also describe them using some of 16 other characteristics falling into four domains—the cognitive, relating to thinking and reasoning; the affective, relating to emotions; the paranormal, relating to unexplained phenomena of this material world; and the transcendental, relating to phenomena outside this material world (Table 12.1). No one NDE as described exhibits all 16 characteristics, and no characteristic occurs in all NDEs [25]. Moody's famous all-enveloping, bright light at the end of a tunnel illustrates. Two respondents in a study by colleagues and me spoke about that light in different ways. One respondent, an EA woman, said, "(P)eople who have experiences about death … the light … and that … They say, 'I saw people working on my body, and I went down this bright(ly) lit tunnel, and then I was sent back to my body'" [previously unpublished quote]. But another respondent, an EA man, described his own NDE, saying, "The light at the end of the tunnel. That's what everybody says, but there was no tunnel … and no light (for me) …" [previously unpublished quote].

Table 12.1 "Near-death experiences": domains and prevalences of specific characteristics

Domain	Characteristics[a]	Prevalence rates	
		Schwaninger et al. [26] (%)[b]	Greyson [24] (%)[c]
Cognitive	An altered sense of time: accelerated, decelerated, or complete stoppage	82	67
	Accelerated thinking	NR[d]	44
	A review of specific memories or one's whole life	NR	30
	Sudden, new insightfulness	82–90	30
Affective	Peacefulness	100	85
	Joy, happiness	63	70
	An all-enveloping light	63	67
	Harmonious communion with the whole universe	NR	52
Paranormal	Existence outside of one's body	90	70
	Extraordinarily vivid sensory perceptions	54	15
	Extrasensory perceptions	NR	11
	Visions of future events or states	NR	7
Transcendental	An "other worldly" environment	54	63
	Visions of significant religious figures (e.g., God, Jesus, or Moses) or dead loved ones	72	52
	A boundary or "a point of no return"	NR	41
	A mystical encounter (i.e., a spiritual experience not grasped through the senses or intellect)	63	26

[a]Adapted from Greyson [24], with permission; and adapted from Schwaninger et al. [26], with permission
[b]Schwaninger interviewed 30 of 55 cardiac arrest survivors at one hospital. The prevalences reported here derive from the seven interviewees who experienced near-death experiences (NDEs) during the most recent arrest and from four more who had experienced them before.
[c]Greyson interviewed 1595 patients admitted to a cardiac intensive care unit or a step-down unit at one hospital. Of all these patients, 7 % had an admitting diagnosis of cardiac arrest. Nineteen percent reported a total loss of consciousness, and 27 % more reported only partial consciousness. Two percent of all the patients claimed they had died and then returned to life; the prevalences reported here derive from those patients. The rest of the patients had prevalences of one percent or less for each NDE characteristic except for altered sense of time, which occurred in 2 %.
[d]*NR* = Not reported

Nonetheless, some characteristics occur more frequently than others. Table 12.1 lists all 16 characteristics and their prevalences in two recent studies [24, 26]. Seven characteristics occurred in most NDEs in both studies: an altered sense of time in the cognitive domain; peacefulness, joyfulness, and an all-encompassing light in the affective domain; a sense of being outside one's body in the paranormal domain;

and a sense of an "other worldly" environment and of seeing significant religious figures or dead loved ones in the transcendental domain.

In describing their own NDEs several respondents in the study by my colleagues and me spontaneously mentioned some of the 16 characteristics. (All respondents I quote here happen to be men even though women generally predominate in reporting NDEs [26, 27].) One of these NDEs, which occurred during a cardiac arrest, illustrates an "other worldly" environment, a vision of a dead loved one, and a boundary or "point of no return." "(W)hen I was passed out," said this EA man. "I dreamed … but it didn't seem like a dream. I was going through the chow line … I got my tray full of food … (and) went to a table … (As) I was getting ready to go back (with my tray), I turned around and there stood my (dead) son … He said, 'What are you doing here, Pop?' I said, 'I don't know.' And he said, 'You don't belong here. You go back over there.' I immediately turned around. He was gone … (and) I was back in the (ICU) bed" [3]. Another NDE, which occurred during a car crash with a trailer truck, illustrates an altered sense of time (deceleration), a whole life review, and an all-encompassing light. This AfA man said, "I seen (sic) the truck when he made the turn, and I hit my brakes. My car just started sliding straight towards him. That's when I could see the light underneath the trailer. Real bright. I said (to myself), 'Man, here we are.'… I was seeing my life pass before me underneath that trailer truck. That was the brightest night of my life. I should have been dead, (but) God takes care of me" [previously unpublished quote].

Two more NDEs, one during a "seizure or heart attack" and the other during a coma, both illustrate the characteristics of peacefulness, joy, an "other worldly" environment, a vision of significant religious figures (here, angels, Jesus, and God), and a mystical encounter. "I died for about 45 min," said the first man, an AfA. "It was so sweet down there. I could see the angels. They were all over me … They had wings, … just in black and white … They were the cutest little things … It (all) was beautiful. So peaceful, quiet, and happy" [previously unpublished quote]. And the other man, an MA, explained, "I was … on the other side of … a mountain (from Jesus). I kept asking Him to take me to the other side. I dreamed about myself … I wanted to be over there with Him. I wanted to get down from the mountain … so I could go with Him … I was happy … I was ready … That was when … I asked God, Was He ready to receive me? (Suddenly) I was (back) in intensive care with my kids and the (rest of) the family all crying" [3].

Although the men in the first and fourth examples above said they were dreaming, 82 % of people who have NDEs believe the experiences are "real." I agree about the reality of NDEs for three reasons [23]. First, numerous NDEs have been reported over the years—often by people who surely did not know about them or believe in them beforehand. NDEs, therefore, hardly seem like hoaxes. Second, many NDEs have occurred during similar life-threatening events such as cardiac arrests or accidents. That consistency of circumstance increases the credibility of the events for me. And third, many people reporting NDEs claim the experiences changed dramatically their perspectives on life [4, 25]. Large majorities of such people in one study said NDEs had strengthened their religious beliefs (90 %), increased their appreciation for life (90 %), renewed their sense of purpose (82 %), increased their ability to love others (72 %), and decreased their fear of death (63 %) [6, 9, 26]. Such momentous changes in perspective defy explanation in other ways.

But accepting NDEs as real requires explaining them. Many explanations already proposed tend to fall into three categories: physiological, psychological, and transcendental [23, 24]. The physiological explanations [25] come from mainstream medical science and posit various causative abnormalities that occur as the brain begins to die [4]. The abnormalities may be seizure-like electrical discharges; the presence of exogenous hallucinogenic medications or anesthetics (such as ketamine); or abnormal concentrations of endogenous blood components (such as a low oxygen or a high carbon dioxide, naturally occurring opioid-like compounds called endorphins, or serotonin and other specialized compounds that transmit messages among neurons). The physiological explanations appear to fit best NDEs during cardiac arrests.

The psychological explanations [25] come from psychiatry and posit the mind's expunging frightening real experiences (such as threats to life) and substituting emotionally pleasant experiences. These explanations resemble those for denial (see Chap. 8). They fit best NDEs during accidents or bodily assaults that do not involve catastrophic collapse of vital body functions (as with cardiac arrests).

The transcendental explanations [25] come from philosophy or the spiritual disciplines such as theology. These explanations posit NDEs' direct link to the soul's experience at bodily death. People have suggested three arguments for this link. Some people argue that one of the most common NDE features, out-of-body sensations, proves that a person's mind, spirit, or consciousness can exist independent of the physical brain and continues on after the brain's disintegration at bodily death [6, 19, 23]. Other people argue that NDEs during cardiac arrests bring us as close as possible to experiencing bodily death without actually dying irreversibly [6]. These people reason that because all physical deaths eventually produce cardiac arrests, whatever happens in NDEs during cardiac arrests must resemble what happens in the early dying process [23]. Still others argue that NDEs offer a glimpse into the afterlife [9, 19, 25] because certain prominent characteristics such as an altered sense of time; the "other worldly" ambience of peacefulness, joyfulness, and light; and the communion with significant religious figures and dead loved ones differ so radically from experiences of this earthly world. Other forms of transcendental linkage may also exist.

Despite many proposed explanations for NDEs, none accounts for all their features [25]. Therefore, while I believe NDEs are real, I do not favor any one explanation over all others. I accept that different explanations may fit different circumstances.

Even so, some patients who have had NDEs or heard about them may want to discuss them with health professionals [26]. Dying patients in particular may want to discuss NDEs and their implications. By training, clinicians are typically best prepared to discuss the physiological explanations; psychiatrists, psychologists, and social workers, the psychological explanations; and hospital chaplains and other clergy, the transcendental explanations. Yet I urge all health professionals to learn something about all three kinds of explanations and be prepared to discuss them with patients. The transcendental explanations may be the most difficult to grasp but the most relevant to dying patients. Because the transcendental concepts fall outside both the scientific paradigm and most health professionals' training (including that of hospital chaplains), HPs must make a special effort to learn those concepts on their own. HPs should also try to define their own views and to anticipate other views. Then, if patients broach the topic, HPs are ready to respond. They should then communicate receptivity to the topic, a willingness to listen, and respect for the patient's views [9]. HPs should also avoid harsh, demeaning judgments; try to comfort the patient; and involve other professionals as necessary to help.

Overview: The Meaning of Death

As the end of a person's earthly life story, a death derives much of its meaning from the life that preceded it [4, 5, 28]. Moses guided the Hebrews to the Promised Land, but God forbade him from entering it. The Hebrews needed different leaders for the tasks of conquering and settling their new land. God, therefore, had Moses die in an unknown place in the wilderness to make way for those new leaders. Senators assassinated Julius Caesar after he seized supreme political power in Rome. But Caesar's murder only exacerbated Rome's prior political chaos and accelerated the decline of the Republic. The Senate steadily lost power to other ambitious men who emerged to rule Rome as absolute emperors. Madame Curie discovered radioactivity and studied it lifelong only to die from radiation-induced cancer. Still, her research paved the way for many groundbreaking new diagnostic tests and treatments (ironically including some cancer treatments). And Amelia Earhart died trying to fly her airplane around the world in the 1930s. Earhart is remembered today as not only a fearless aviatrix but also a pioneer of new opportunities for women. As these examples illustrate, life context determines much of the meaning of a person's death.

Of course, certain meanings commonly attributed to a death emerge from the trials of earthly life and from the human hopes about an afterlife. Cardinal Newman's brief "Peace at the Last" prayer, which the first chapter quotes and this chapter repeats as an epigraph, derives much of its consoling power from acknowledging those deeply human trials and hopes [1]. The prayer's first sentence asks God's support throughout "this troublous life until … the fever of life is over, and our work is done." This sentence acknowledges that earthly life with all its trials is short, turbulent, and arduous. The prayer's second sentence asks God's mercy in granting "safe lodging, a holy rest, and peace at the last." This sentence expresses the widespread human hopes for eternal safekeeping, rest, and peace in an afterlife.

Following the structure of Newman's prayer, the next two sections—one per sentence in the prayer—address possible meanings of a death. The first section addresses changes people may expect in disengaging from earthly life and the second, the visions people may have of heaven and an afterlife. Quotes from prior research illustrate these meanings.

Disengaging from Earthly Life

The changes people may expect in disengaging from earthly life include returning the "borrowed" body, resting from life's struggles, being liberated from specific hardships, incurring significant losses from earthly life (especially human relationships), and leaving a legacy that transcends one's earthly life.

Returning the "Borrowed" Body

Some people see their earthly bodies as their own personal property to do with as they wish [10]. But many others see their bodies instead as a loan or trust from God [2]. This body-as-loan belief implies God expects people to repay Him by returning their bodies when they die. In this vein an EA man told my colleagues and me once, "It's a debt we've all got to pay" [previously unpublished quote]. Likewise, an MA man told us, "We're here on borrowed time. When God tells you (that) you gotta leave, you do. That's when you die" [8, 16].

The indebtedness belief may cause unexpected problems for organ donations or autopsies. People who hold that belief may conclude that by removing or damaging body parts, organ donations and autopsies prevent full repayment of one's bodily debt to God at death and, thus, bar the debtor from heaven. As one respondent, an AfA man, asked rhetorically about organ donation, "Would [God] … welcome me into heaven without … (some) parts of my body?" [29]. And another respondent, an MA man, said about autopsies and other invasive procedures done on cadavers, "I don't like it that one day when I go to God, if I sign a paper for (the doctors) to experiment … on my body … How am I going to permit that just for science … (or) practice?" [29]. The indebtedness belief may, therefore, prohibit some beneficial medical uses of cadavers or cadaver parts. It may also logically prohibit cremations.

Resting from Life's Struggles

Newman's "Peace at the Last" prayer also asks God for "a holy rest" when "the fever of (this) life is over, and our work is done." For many of the sick or elderly, the struggle to survive day to day eventually exhausts their will to live. Harboring more memories than dreams, these people question the wisdom of continuing the fight to survive. They begin to look forward to death as rest for both body and soul.

As an MA man told my colleagues and me, "(T)he body gets so tired of struggling. (You) have worked all (your) life … (You've) had enough, and (you're) ready to go … (Y)ou're expiring your soul …" [16]. And another respondent, an AfA man, said of his dead father simply, "His spirit finally went (to heaven) to rest" [previously unpublished quote].

This weariness with life may prompt what Dr. Joanne Lynn calls "taking to bed," the phenomenon whereby people without an obvious life-threatening illness go to bed to await death. Their lives shrink down to the apartment, the bedroom, or the bed, and it becomes their final sanctuary from the demands of the outside world [20].

Being Liberated from Specific Hardships

Death may also liberate the dying and their caregivers from specific hardships. It, for example, may liberate the dying from hardships due to disease or treatment [16]. One MA man in our study, who attributed the suffering of his dying father to disease, said, "He had Parkinson's disease (and) fell down and broke his hip … It was time for him to leave … He was suffering" [16]. And an AfA woman, who attributed the suffering of her dying friend to treatment, said, She "suffered so … with that life-support (ventilator) on … (H)er kids just let her suffer right on through it … (She) died with that thing on" [16].

Death may also liberate the dying from time's own peculiar tyranny, the constant ticking down of the clock during a terminal illness [5, 20]. Short predicted survival times, say, of only weeks to months may cast a chilling, stifling, dispiriting pall over all remaining life. The dying may feel trapped with death as their only escape from the tyrannical clock.

Furthermore, death may liberate the dying from personal problems, social stigmatization, or political oppression. The personal problems may include insurmountable debts, domestic violence, rejection, blame, or shame. One respondent of ours described her daughter's death as simultaneous liberation from terminal disease, domestic violence, and her husband's infidelity. Her husband "was very mean to her," this MA woman recalled. "He hit her when she was lying down in bed. He would just leave for days. She was dying, and … nobody could find him. He was staying with another girl … (My daughter) was so sick of him." The woman then added, "She went to heaven, … (to escape)" [previously unpublished quote]. The social stigmatizations may include the stigmatization of dying itself. Some family and friends may not be emotionally prepared to face a patient's impending death. They may withdraw, leaving the patient isolated and emotionally bereft. The dying person may eventually abandon hope for reconnecting with them meaningfully in this life but maintain hope for reconnecting in the afterlife. And for AfAs, political oppression may include the vestiges of slavery and Jim Crow laws. Death may, therefore, promise AfAs not only release from those earthly burdens but also a spiritual return to their African homelands and cultural roots [30].

Death may even liberate caregivers of the dying from the hardships of physical care duties, the accompanying emotional strains, and the suspension of their own lives. Though saddened at the patient's death, caregivers may find a certain relief at shedding that heavy load. Still, guilt may plague them when they do shed it. As a result, caregivers may need considerable emotional support and explicit reassurance from HPs and others that the feeling of relief is natural and justified.

Incurring Significant Losses from Earthly Life

One significant loss due to death involves human relationships. Both patient and survivors are affected, but the patient more so because he or she loses all human relationships at once. "I just don't want to be away from (those family relationships)," said a respondent of ours, an MA man. "That's the fear I have (about dying)" [previously unpublished quote]. Of course, the patient who believes in afterlife reunions may avoid the distressing thought of losing all human relationships permanently.

But the loss of an earthly relationship with the patient typically distresses survivors, too. That same MA man told my colleagues and me, The family "fear losing you, not having you around. You get so close to ... your (family) ... (They) just don't want to accept that you are going to die" [previously unpublished quote]. And Dr. Greg A. Sachs, an experienced geriatrician and palliative care specialist, described a similar distress in "Sometimes Dying Still Stings," a heartfelt commentary about the death of his beloved father-in-law. "Al was being ripped from our midst," Dr. Sachs remembered. "And it hurt like hell." He then added, "Death means someone is lost to us forever" [31]. Some survivors, of course, do not believe death breaks the relationship evenly temporarily. These survivors claim they maintain active relationships with their dead through regular conversations and postmortem visitations on earth. I discuss both those visitations on earth and reunions in the afterlife in a later section of this chapter.

Even more sobering is the possibility of losing everything to death including any form of existence [2, 20]. Some Americans, perhaps as many as 40 %, take that view [12]. As one of our respondents, an EA man, exclaimed, "Death may be ... no being at all" [previously unpublished quote]. Dr. Ira Byock, a nationally renowned hospice physician, believes so. He says the dying patient faces losing "everything we know and love." Dr. Byock reasons that life and death are opposites: Life is existence; death is nonexistence. "Life involves activity, purpose, and bringing some order out of the universe's (apparent) disorder," he says. Death, in contrast, is total nothingness. Accordingly, Dr. Byock believes many people satisfy themselves with superficial, mere pathophysiologic explanations of individual deaths and avoid any serious conceptual discussions of death itself. Ironically, Dr. Byock considers the nothingness of death the ever-present "backdrop" of life and the essential substrate for its passion, joy, and meaning [20].

Leaving a Legacy

Death for many people means the occasion to leave a legacy when they die. A legacy allows the dead to leave something of themselves and their lives, to "speak from the grave" [30], to influence earthly human events after death. I believe most Americans want to leave some legacy whether or not they intentionally create one.

At least seven kinds of legacies exist [15, 30, 32]. Some are protective: The dead live on in their watching over the living and in keeping them from harm. One MA man in our study hoped to "be an angel (after bodily death) and (to) protect my daughter (and) my son ... I just don't want to be away from (them)" [previously unpublished quote]. Some legacies are biosocial: The dead live on in the genetics and accomplishments of their children [30]. Others are material: The dead live on in letters, photographs, other personal keepsakes [32], and even tombstone inscriptions. Still others are temporal: The dead live on in remembrances on special days such as birthdays, wedding anniversaries, or death days [15]. Some legacies are creative: The dead live on in their works of art, literature, music, or science. They may even live on in their past administrative accomplishments. Still other legacies are philosophical, religious, or spiritual: The dead live on in their wisdom, examples of piety, or presumed powers to intervene with God for survivors. (Jews have an interesting tradition of "ethical will" legacies, that is, documents containing practical, everyday wisdom for survivors.) Perhaps the most basic, most universal legacies are natural: The dead live on in their decayed body matter, which eventually reconstitutes itself in new forms, animate or not [2]. The ancient Greeks called such natural legacies "pneuma," the spirit that permeates all matter [33].

Visions of the Afterlife and Heaven

Sixty-three to seventy-five percent or more of Americans believe in an afterlife. Most anticipate spending it in heaven [2, 3, 12, 34–36]. According to the scant professional literature on the topic [3], people commonly expect to face and pass God's judgment on their earthly lives and then to go to heaven, reunite with dead loved ones, and commune with God forever [11, 28, 36].

Facing and Passing God's Judgment

In the study by my colleagues and me, many respondents expressed the belief that God will judge their earthly lives after physical death. One EA woman explained in our study, for example, The soul of a dead person floats "around somewhere ... until Judgment Day" [3]. And another respondent, an EA man, said, God then tests your soul "to find out what you are" [3]. Many respondents also expressed the belief that God's judgment determines the soul's final, eternal disposition. What happens after you die "depends on how you behaved on earth," said yet another respondent, an MA man. "If you were faithful, you (go) directly to God ..." [previously unpublished quote].

Many of our respondents felt sure they and their loved ones would pass God's judgment and go to heaven [11]. One EA woman insisted to us that God prepares a place in heaven for each person [3], and another EA woman joked that, when her dead mother "got to heaven, she told God to move over. She was going to help Him run heaven" [3]. Some respondents described their expectations for going to heaven in specifically Christian terms [12]. One MA man said his dead mother was "a good Christian. She was a member of the Pentecostal Church. (Church members) would pray for her (as she finally died) ... She was ready, and she went (to heaven. The Lord was waiting for her)" [previously unpublished quote]. Another respondent, an AfA woman, said about a dead friend, "When the resurrection comes, she ... (will) rise and rise in the bosom of Jesus Christ ... (and) will be standing with the King" [previously unpublished quote]. And a third respondent, an EA woman, said about herself, "I may not be the best Christian in the world, but I think I've got as good a chance as anybody else (to make heaven)" [previously unpublished quote].

But what happens to souls rejected for heaven? Respondents suggested some of those souls go to limbo or purgatory. According to Roman Catholic doctrine, limbo is the permanent residence of innocent souls who are unbaptized and thus barred from heaven, or of the righteous souls who lived before Christ. Purgatory, in contrast, is a way station where sinners work off their sins and thereby gain the holiness necessary to enter heaven. Other souls, however, may actually go to hell. Only seven percent of our respondents discussed that possibility. One AfA woman said, "If (people) are good, they go to heaven." But she then added, implying the possibility of going to hell, "You've got to be honest (about your life)" [previously unpublished quote]. Another respondent, an EA woman, said bluntly, The soul "goes to heaven or hell ... (You) are rewarded by the good that you do" [previously unpublished quote]. And the AfA woman who talked about her dead friend's rising in Christ's bosom said that on Judgment Day the Christians will be "standing with the King (Jesus Christ, and) the devils will be standing with the devils" [previously unpublished quote]. Still another respondent, an AfA man, described damnation for evildoers as eternal death without hope [previously unpublished quote].

Thus, many Americans expect God's judgment after physical death, feel sure they and their loved ones will pass that judgment and enter heaven, and worry little about going to hell.

Going Home to Heaven

Many people who talk about going to heaven after physical death describe that transition as "going home." If I am very sick, said one of our respondents, an EA woman, I "just breathe a prayer. If it's God's will for me to die, … He takes me home" [previously unpublished quote]. The idea of going home to heaven expresses a deep, universal human hope. Though seemingly simple, that hope actually incorporates many complex, highly personal meanings. All are emotional, and nearly all are pleasant [37]. Still, the fundamental concept involves returning to comfortable, familiar surroundings to rest forever after the struggles of this "troublous" earthly life. "Going," of course, means movement, either a physical or a mental change of locations. When the soul leaves the dead body, explained one of our respondents, an MA woman, "I don't know if it's in heaven or … in another place … I don't know where it is, but I know that there's a place … for us … (The soul) lingers above, around; and then … goes on. I think it's … a transition … going from one place to another" [previously unpublished quote]. And the heavenly "home" Newman's prayer describes is a place of "safe lodging, a holy rest, and peace." The details of that "heavenly home," however, vary according to each person's earthly life, the environment and experiences of which shape one's expectations for a heavenly home [28, 37].

"Home" in its most concrete earthly sense, of course, refers to a house, a residential structure at a specific address. Yet "home" in its heavenly sense takes on numerous broader cultural, psychological, and philosophical meanings [37]. They include identity, familiarity, rootedness, and continuity [38]; safety, stability, and nurturance [39]; and acceptance, comfort, intimacy, and fulfillment [28]. In such a heavenly "home" the dead can lay down the burdens of earthly life, drop all pretenses, and reveal their true selves to assured positive regard [39].

"Going home" for some may also refer to earthly trips back to the places of one's roots in preparation for physical death. A powerful urge to return to their homelands grips many immigrants who are aging or dying. These immigrants want to end their lives in their original home communities amid familiar surroundings, people, and rituals. Illustrating with Filipino and Cambodian Americans, one author explains that returning to their homelands helps immigrants reassert their core personal and ethnic identities, reconcile various discontinuities in their life stories, prepare for death in culturally meaningful ways, and allow close family and friends to provide end-of-life care [40].

Of course, not only immigrants feel this urge; nonimmigrants do, too. In the movie *Trip to Bountiful*, Carrie Watts, an elderly EA woman, lives with her adult son and daughter-in-law in Houston but has the compelling wish to visit her rural childhood home once more before she dies. Carrie stops at nothing to fulfill that wish. She outwits her overly protective son and angers her petty daughter-in-law by eventually escaping from the Houston house. She travels alone by bus to her childhood hometown, called Bountiful. The audience senses that when Carrie finally sees her childhood home again, dilapidated though it is from years of neglect, she can finally die at peace with herself.

Reuniting with Dead Loved Ones

While physical death seems to sever permanently relationships between the living and the dead, many people believe that severance is only temporary. They expect to reunite eternally with dead family members, friends, and even pets in the afterlife [2]. Some people even expect dead loved ones to greet them at the threshold of death, to accompany them through the transition, and then to guide them on into heaven [4]. Curiously, the hope for reuniting extends only to people the dying person misses, not to others.

Communing with God

This concept expresses Christians' core hope for the afterlife: living anew in God's presence [2, 10]. Some people talk about "entering the church triumphant," taking their places alongside of past saints, or singing God's praises eternally in heavenly choirs. That heaven is total bliss. "Amazing Grace," the famous hymn played at many funerals, expresses those ideas in its last two verses:

> Yes, when this flesh and heart shall fail,
>
> And mortal life shall cease;
>
> I shall profess, within the veil,
>
> A life of joy and peace.
>
> When we've been there ten thousand years
>
> Bright shining as the sun.
>
> We've no less days to sing God's praise
>
> Than when we've first begun [41]

Other faiths may have similar beliefs about heavenly communion with God in the afterlife. Interestingly, among the world's religions, perhaps only Islam envisions such communion in a paradise of sensuous pleasures or a new Garden of Eden [4, 36].

Contacts Between Survivors and the Dead

Survivor contacts with the dead have long had a place in popular cultures. EA beliefs about such contacts trace back at least to medieval times. People in Europe lived then under constant threat of sudden, unexpected deaths from injury or infection. Because life existed in the shadow of death, the two states seemed inextricably linked [15]. The suddenness of many deaths naturally made survivors hope for, even expect, uninterrupted contact with the recently dead. Survivors often claimed to see, hear, or feel the dead much as before their physical deaths. The English language between the twelfth and sixteenth centuries adopted the term, "ghosts," for the dead who actually reappeared in earthly life. The new materialistic science of the seventeenth to nineteenth centuries, however, came to consider physical matter as the only reality, and ghosts and contacts with the dead as mere illusions, deceptions of the mind. According to science, the spirits of the dead could exist only in memory or the imagination [33]. Bertrand Russell, the brilliant twentieth-century English mathematician and philosopher, is quoted as saying in this vein, "When I die, I shall rot. And nothing of my ego will survive" [30]. Nonetheless, widespread belief in visitations by the dead has persisted among EAs into modern times [27]. Marley's ghost, for example, appears in Dickens' still popular nineteenth-century novel, *A Christmas Carol*, to warn Scrooge about the dire consequences in the afterlife if Scrooge continues his miserly ways.

AfAs also have long-standing cultural beliefs in ghosts and contacts with the dead. Dead AfA ancestors, whom one writer calls "the living dead," can present to survivors as "shades or vapors" of human forms. These beneficent ghosts work under God's supervision to serve the living as "moral beacons" in personal crises [15]; guardians of the family; a stabilizing, cohesive influence in the wider community; and witnesses to an afterlife [33].

The percentages of modern-day survivors who report contact with the dead are remarkably high, ranging from 27 % of randomly selected Americans [42] and 28 % of Massachusetts firefighters or policemen who have attended the dying in the field [19] to more than 50 % of English widows from Leicester [32] and 63 % of Nevada adults in a telephone survey [43]. These survivors consistently say the dead initiate the contact [15]; it consists of either a vague awareness or a distinct sensory perception—visual, auditory, tactile, or olfactory—of the dead person's presence [32, 43]. The contact occurs most often during the first year after the death but may recur for many years afterward. Some widows have experienced contact with their dead husbands for as long as 26 years [32]. Yearning for the dead may prompt the contact [43]. One author claims, in fact, that surviving spouses of only happy marriages receive visits from their dead spouses [32]. (I give an exception, though, later.) The vast majority of survivors contacted by the dead react positively to the experience: 86 % find it comforting while only 14 % find it upsetting or frightening [42].

To characterize postmortem contacts further, Klugman and colleagues conducted that Nevada telephone survey [43]. Respondents, mostly EAs and women, answered the question, "Do you have a connection with someone who has died?" Those who answered yes were then asked to describe the connection. The dead person was often a parent or grandparent. The time since that person's death ranged from 1 month to 57 years with a mean of 10 years. Almost all respondents reported two or more contact experiences; the mean number was eight. Majority experiences included dreaming of the dead person, hearing music he or she liked, talking with him or her, sensing his or her presence, and sensing he or she was acting as a guardian angel (Table 12.2) [32, 43]. Higher percentages of unmarried respondents

Table 12.2 Helpful questions for encouraging survivors to describe their contacts with the dead

Questions	Typical prevalences for yes answers (%)[a]
Do you dream about the dead person?	85
Do you talk with the dead person?	70
Do you ever just "sense" or "feel" his or her presence?	55
Do you think the dead person acts like your guardian angel, watching over you to protect you?	54
Do you ever see him or her?	38
Do you ever feel the dead person's touch?	25
Do you ever detect smells (e.g., particular perfumes, foods, flowers, or workplace chemicals) indicating the dead person is near or reminding you of him or her?	15–26
Do you ever hear sounds the dead person is making (e.g., his or her voice or footsteps)? Do you ever hear sounds that remind you of him or her (e.g., favorite songs)?	14–82
Do you ever notice the dead person has moved the objects around you? For example, the dead person might raise or lower the volume dial on the television, switch lights on or off, or open or close a window.	7–18

[a]Adapted from Tables 3 and 4 in Klugman [43], with permission

than married ones experienced each kind of contact, and higher percentages of women than men experienced each kind except for detecting smells associated with the dead person (such as favorite perfumes, foods, or flowers). While more than half of respondents with a bachelor's degree or less education had had such contacts, over a quarter of respondents with a graduate degree had also had them.

In the study of dying and death by my colleagues and me, many respondents described contacts with their dead. AfAs provided most of the examples. One AfA woman was referred to psychiatry for seeing visions of her dead aunt. "(The psychiatrists) ask me all these questions … Do I see things? Yes, I know when (my dead auntie) is present, … she sits on my bed 'cause you can feel the impression of somebody when they (sic) flop down on your bed. I know when she comes back because too many things are falling and dropping around the house. When something's wrong, she will come back." This woman then added, "I have a friend … (whose dead) husband comes back all the time. He lets her know when something ain't right. Her (dead) mother does, too … to warn (her)" [3]. Another AfA woman said about her dead brother, "I think about him a lot especially on Veterans' Day. He was in the service … I talk to him. I say, 'Oh, I wish you were here.' or 'Do something (for me)'" [previously unpublished quote]. And an AfA man described one of his dead father's visits to his mother. "He came and sat down in the chair (next to my mother's bed). He was sitting there watching her; she was laying (sic) there in bed asleep. When she woke up, … she said, 'What are you doing?' He said, '(J)ust seeing if everything's all right.' Then he talked to her (for) a few minutes and left" [3].

MAs and EAs also provided some examples. One MA woman said about her dead mother, "At night I hear her saying, 'Open the door.' I open (it, but) nobody's there … I just get up and pray for her. (T)he Holy Spirit, … (is) calling me to pray for her … because I am probably thinking of her and not letting her rest … (I)f you think on (the dead) too much, you need to pray for them so they may (have) peace …" [previously unpublished quote]. And an EA women said about her dead mother, "We put flowers … at the table where she sat (for breakfast). We know she is (there). She can see (the flowers), and she smells them" [previously unpublished quote]. Still another EA woman said about her dead brother, He "was killed when we were young … I dreamed about him (for years.) … (W)hen I would, I felt like I had had a visit … like we had had some time together" [previously unpublished quote].

While most postmortem contacts are positive, a few are negative. Two respondents gave examples of negative contacts. An AfA woman said, "Only (the ghost of) somebody who gives you problems like an old drunk is going to let you see (him) … I'm not going to live in (some) house(s) 'cause you're liable to hear somebody (like that) walking in there … A haunting. When you die in sin, your ghost will come (back) in sin" [previously unpublished quote]. And the MA woman who described her daughter's being beaten by her husband and dying alone in childbirth said, The ghost of my daughter "is still after him … because he was very mean to her … (N)ow he sees (my dead daughter) on the door every night. He can't sleep because she's there … (H)e knows what he did. He's guilty" [previously unpublished quote].

I believe that with postmortem contacts as with NDEs, the sheer numbers of reported cases make them unlikely to be hoaxes. But medical science remains skeptical. The problem for medical science is its sharp distinction between the worlds of the living and the dead [15]. The living belong strictly to this physical world; the dead belong, if at all, strictly to some other world. By rejecting any meaningful interactions between the two worlds, medical science becomes unable to explain the reports of postmortem contacts. It, therefore, tends to dismiss the experiences as merely disordered thinking [44].

As a result, medical science looks to psychiatry for explanations based on the concepts of hallucinations (experiences due to confused processing of sensory stimuli by the brain) or illusions (imagined experiences falsely believed to be real) [32]. Two explanations predominate. One, favored by Freud, attributes postmortem contacts to severe grief. Freud reasoned that the mind deceives the survivor into thinking the dead person is still alive. The survivor feels compelled to find the dead person and to reestablish the lost relationship. Freud concluded that grief ends only when the survivor accepts the reality of the death, detaches emotionally from the dead person, and relegates him or her to past life [45]. Then the postmortem contacts stop [15, 32, 42]. The other explanation, favored by more recent psychiatrists, attributes postmortem contacts to such psychiatric disorders as neuroticism, anxiety, and postmortem adjustment problems [42]. Some data support these associations, but they may not be causal. Potentially confounding associations exist, for example, between postmortem contacts and liberal religious views [44]. Thus, those religious views, not the psychiatric disorders, may actually cause the perceptions of postmortem contacts.

These medical scientific perspectives obviously conflict with the spiritualist perspective of most survivors who claim to experience contacts with the dead [7]. These survivors are convinced that their contacts are real and need no external validation or scientific explanation [32]. In fact, the contacts may seem so real that survivors sometimes use the present tense to refer to the dead as though they were still alive [32]. The spiritualist perspective offers two explanations for postmortem contacts. First, the continuing bonds theory of grief asserts that the survivor does not delete the dead person completely from memory or relegate him or her strictly to the past. Rather, the survivor uses perceived postmortem contacts, perhaps unconsciously, as a way to extend the relationship with the dead person into the future [43]. Second, death highlights the basic human need for stability [39], continuity, and meaning [4, 40]. One author believes we, humans, are driven to fashion such stability, continuity, and meaning even in the destabilizing, discontinuous, and seemingly meaningless experiences of death. We use perceived postmortem contacts to do so by creating ongoing "dynamic relationships" with our dead. This author, therefore, urges us to incorporate "permanently and deliberately" such relationships into our future lives [5, 15].

Unconsciously facing this task, some dying patients and their survivors may want to discuss with HPs the possibility of postmortem contacts between the living and the dead. HPs should be willing and prepared to do so if only to ease the grief of separation [27]. Yet patients or survivors often hesitate to broach the topic for fear of being considered crazy. HPs might, therefore, invite patients and survivors to talk about the topic if they wish. HPs might introduce it by mentioning that many people have reported such experiences, by giving some examples, and by stating that such experiences should be taken seriously. HPs should also emphasize that such experiences do not indicate that the people who have them are crazy [42, 43]. Then, with a dying patient, HPs might ask, "Do you hope for contact with your family and friends after you die?" and "What form do you think that contact might take?" Or, with a grieving survivor, HPs might ask, "Have you had contact with (the dead person) since he or she died? Would you like to talk about the experience?" The questions used by Klugman and colleagues in their survey can serve as helpful prompts in discussions with survivors. I suggest starting with the topmost questions in Table 12.2 because they have the highest rates of positive responses. Of course, regardless of their own personal views HPs should listen intently and support patients and survivors emotionally as they describe their expectations and experiences [3, 5, 27].

A Case

Mr. T. died from diffuse prostate cancer approximately 18 months after its diagnosis. His disease course followed the typical terminal trajectory of cancer—that is, a slow functional decline over many months followed by a rapid decline over the last few months. He suffered moderate pain (despite hospice care), nausea, and fatigue in the days just before death. He also saw visions of his parents who had died over 25 years before. They talked with him and encouraged him to join them in the afterlife [4]. After the doctor had declared Mr. T. dead, Mrs. T. held and comforted him in bed for about an hour in case he might be afraid or still feel pain.

Late one evening soon after the funeral, Mrs. T. was dozing alone in her easy chair. Mr. T. suddenly called from the next room, "Margaret! Margaret!" His voice was clear and vigorous as it had been before he became ill. Mrs. T. got up to look for him. But, when she reached the next room, Mr. T. had already disappeared. Several weeks later Mr. T. returned. Mrs. T. was again sitting alone in the evening when Mr. T. stood before her. He wore the same sweater and pants as in a photograph on the nearby desk. And he looked healthy and comfortable—contrasting sharply to how he looked when he was dying. He said, "Margaret, I'm OK. Everything is fine. Don't worry. I have no pain." He then disappeared as suddenly as he had appeared. He has not visited her since.

Mrs. T. firmly believes she did not dream Mr. T.'s visits. Yet she has hesitated to tell others about them because she fears being considered "crazy." She did finally mention the visits to her adult children. She described being especially relieved to learn Mr. T. is suffering no more pain. Although her children did not know what to think, they listened supportively and accepted Mr. T.'s postmortem visits as real for her. Mrs. T. later mustered the courage to disclose the visits to her widows' grief group. Several other women surprised her by disclosing similar postmortem visits from their husbands. Those descriptions helped Mrs. T. feel "validated" [27], but several years later she still feels too intimidated to tell her doctor, pastor, or grief counselor about the visits.

How, then, does the T.s' story illustrate beliefs and experiences concerning death and the afterlife? For one thing Mr. T. seemed to reunite with his long-dead parents shortly before he died. He talked with them, and they encouraged him to join them in the afterlife. Furthermore, just when Mr. T. died, Mrs. T. held him and talked with him for a long time after the doctor declared him dead [32]. Mrs. T. obviously believed Mr. T. could still feel and hear perhaps because she still considered him alive (despite the lack of breathing, heartbeats, or responsiveness) or because she thought he retained sentience in death.

As an accomplished, college-educated career woman, Mrs. T. seemed an unlikely recipient of postmortem visits. Of course, she might have been asleep on both occasions and simply dreamt those visits from Mr. T. [45, 46]. But other circumstances of hers matched those of other widows who have such visits [27]. She was grieving Mr. T. after a mostly happy marriage in the later years [32]. His visits occurred early in her bereavement [27, 32]. And each occurred when she probably felt her grief most, that is, when she sat alone reading or watching television in the evenings as she and Mr. T. had done together for years.

Mr. T.'s visits also showed some typical features of postmortem contacts. First, the dead person, Mr. T., initiated each contact [32]. Second, Mrs. T. experienced auditory and visual components, both modestly prevalent in postmortem contacts [27]. Third, Mrs. T. balked at disclosing the contacts for fear of being labeled "crazy" [32]. And fourth, the contacts may have prompted Mrs. T. to reimagine her relationship with Mr. T. in spiritual rather than earthly terms [5, 15, 45, 46].

Finally, Mr. T. left Mrs. T. many legacies [32, 46]: a comfortable inheritance; important keepsakes such as letters and photographs; special days on the calendar such as their wedding anniversary, Memorial Day, and Veterans Day (Mr. T. had served in the military during World War II) for remembering their life together; stories he had written about his life; and, of course, the existence of children and grandchildren. He may have also made postmortem contact to leave Mrs. T. a protective legacy: he relieved her worry that he still suffered pain.

Mrs. T., of course, misses Mr. T. greatly and the close companionship they shared especially after their children had grown and both Mr. and Mrs. T. had retired from their jobs. Mr. T. sometimes still seems alive to Mrs. T. She occasionally even talks about him in the present tense [32, 45]. Once, for example, she wondered aloud whether the late family dog, who had bonded closely with Mr. T., had found him in heaven and was sleeping with him again as the two had done in earthly life. When Mrs. T. is especially lonely or depressed, she often says, "I just want to go be with him." Still, as a devoutly religious woman, Mrs. T. believes she will join Mr. T. in heaven someday, and they will enjoy God's presence together forever [32, 45].

Summary Points

1. Though largely inaccessible to medical science, the experiences of physical death and the afterlife are an important concern for many dying patients and their survivors.
2. The meanings of physical death differ from person to person. Possibilities include returning the "borrowed" body to God, resting from life's struggles, incurring significant losses (especially of human relationships), and leaving a legacy; and facing (and passing) God's judgment, going home to heaven, reuniting with dead loved ones, and communing with God forever [15, 30, 32].
3. All health professionals who provide end-of-life care should be prepared to discuss such beliefs with dying patients and their survivors [11]. To do so effectively, health professionals should know their own beliefs and a broad range of other possibilities.
4. Health professionals should treat respectfully all patient or survivor beliefs about physical death even when those beliefs differ from the health professionals' own. Still, when such beliefs suggest underlying psychopathology, health professionals may need to refer those people to clergy, psychotherapists, or similar professionals [9].

To Learn More ...

1. Greyson B. Incidence and correlates of near-death experiences in a cardiac care unit. Gen Hosp Psych. 2003;25(4):269–76.
2. Perkins HS, Cortez JD, Hazuda HP. Patients' diverse beliefs about what happens at the time of death. J Hosp Med. 2012;7(2):110–6.
3. Perkins HS, Cortez JD, Hazuda HP. Diversity of patients' beliefs about the soul after death and their importance in end-of-life care. South Med J. 2012;105 (5):266–72.
4. Wright K. Relationships with death: the terminally ill talk about dying. J Marriage Family Ther. 2003;29(4):439–54.

References

1. Newman JH. Peace at the last. In: Wilson AN. John Henry Newman: prayers, poems, meditations. New York: The Crossroad Publishing Company; 1990. P. 192.
2. Rain Evans A. Negotiated death: end of life issues and Christian faith. Insights: Fac J Austin Semi. 2005;121(1):36–40.
3. Perkins HS, Cortez JD, Hazuda HP. Diversity of patients' beliefs about the soul after death and their importance in end-of-life care. South Med J. 2012;105(5):266–72.
4. Haddow AH. Dying, death and after death. J Rel Psych Res. 2000;23(3):133–40.
5. Wright K. Relationships with death: the terminally ill talk about dying. J Marriage Family Ther. 2003;29(4):439–54.
6. Parnia S, Fenwick P. Near death experiences in cardiac arrest: visions of a dying brain or visions of a new science of consciousness. Resuscitation. 2002;52(1):5–11.
7. Facci E, Agrilio C. Near-death experiences between science and prejudice. Front Human Neurosci. Accessed online at http://journalfrontiersin.org/Journal/10.3389/fnhum.2012.00209/full.
8. Perkins HS, Cortez JD, Hazuda HP. Patients' diverse beliefs about what happens at the time of death. J Hosp Med. 2012;7(2):110–6.
9. Morris LL, Knafl K. The nature and meaning of the near-death experience for patients and critical care nurses. J Near-Death Stud. 2003;21(3):139–67.
10. Paris JJ. Terri Schiavo and the use of artificial nutrition and fluids: insights from the Catholic tradition on end-of-life care. Palliat Support Care. 2006;4(2):117–20.
11. Leget C. Retrieving the *ars moriendi* tradition. Med Health Care Philos. 2007;10(3):313–9.
12. Holcomb LE, Neimeyer RA, Moore MK. Personal meanings of death: a content analysis of free-response narratives. Death Stud. 1993;17(4):299–317.
13. Lunney JR, Lynn J, Foley DJ, Lipson S, Guralnik JM. Patterns of function decline at the end of life. JAMA. 2003;289(18):2387–92.
14. Emanuel LL. Reexamining death: the asymptotic model and a bounded zone definition. Hast Cen Rep. 1995;25(4):27–35.
15. Howarth G. Dismantling the boundaries between life and death. Mortality. 2000;5(2):127–37.
16. Perkins HS, Cortez JD, Hazuda HP. Cultural beliefs about a patient's right time to die: an exploratory study. J Gen Intern Med. 2009;24(11):1240–7.
17. Jonsen AR, Siegler M, Winslade WJ. Clinical ethics: a practical approach to ethical decisions in clinical medicine. 7th ed. New York: McGraw-Hill; 2010. p. 40–3.

18. Perkins HS, Cortez JD, Hazuda HP. Autopsy decisions: the possibility of conflicting cultural attitudes. J Clin Ethic. 1993;4(2):145–54.
19. Kelly EW, Greyson B, Stevenson I. Can experiences of near death furnish evidence of life after death? Omega. 1999–2000;40(4):513–9.
20. Byock I. The meaning and value of death. J Palliat Med. 2002;5(2):279–88.
21. Owens JE, Cook EW, Stevenson I. Features of "near-death experience" in relation to whether or not patients were near death. Lancet. 1990;336(8724, 10 November):1175–7.
22. Greyson B. Dissociation in people who have near-death experiences: out of their bodies or out of their minds? Lancet. 2000;355(9202):460–3.
23. Parnia S, Waller DG, Yeates R, Fenwick P. A qualitative and quantitative study of the incidence, features and aetiology of near death experiences in cardiac arrest survivors. Resuscitation. 2001;48(2):149–56.
24. Greyson B. Incidence and correlates of near-death experiences in a cardiac care unit. Gen Hosp Psych. 2003;25(4):269–76.
25. Greyson B. Biological aspects of near-death experiences. Perspect Bio Med. 1998;42(1):14–32.
26. Schwaninger J, Eisenberg PR, Shechtman KB, Weiss AN. A prospective analysis of near-death experiences in cardiac arrest patients. J Near-Death Stud. 2002;20(4):215–32.
27. Barbato M, Blunden C, Reid K, Irwin H, Rodriguez P. Parapsychological phenomena near the time of death. J Palliat Care. 1999;15(2):30–7.
28. Dekkers W. On the notion of home and the goals of palliative care. Theor Med Bioeth. 2009;30(5):335–49.
29. Perkins HS, Shepherd KJ, Cortez JD, Hazuda HP. Exploring chronically ill seniors' attitudes about discussing death and postmortem medical procedures. J Am Geriatr Soc. 2005;53 (5):895–900.
30. Hood RW, Morris RJ. Toward a theory of death transcendence. J Sci Stud Rel. 1983;22 (4):353–65.
31. Sachs G. Sometimes dying still stings. JAMA. 2000;284(19):2423.
32. Bennett G, Bennett KM. The presence of the dead: an empirical study. Mortality. 2000;5 (2):139–57.
33. Hood RE. Ghosts and spirits in Afro culture: Morrison and Wilson. Anglican Theol Rev. 1991;73(3):297–314.
34. Harley B, Firebaugh G. Americans' belief in an afterlife: trends over the past two decades. J Sci Stu Rel. 1993;32(3):269–78.
35. McClain-Jacobson C, Rosenfeld B, Kosinski A, et al. Belief in an afterlife, spiritual well-being and end-of-life despair in patients with advanced cancer. Gen Hosp Psych. 2004;26(6):484–6.
36. Flannelly KJ, Ellison CG, Galek K, Koenig JG. Beliefs about life-after-death, psychiatric symptomology and cognitive theories of psychopathology. J Psych Theol. 2008;36(2):94–103.
37. Moore J. Placing *home* in context. J Environ Psych. 2000;20(3):207–17.
38. Kontos PC. Resisting institutionalization: constructing old age and negotiating home. J Aging Stud. 1998;12(2):167–84.
39. Wright G. Prescribing the model home. Social Res. 1991;58(1):213–25.
40. Becker G. Dying away from home: quandaries of migration for elders in two ethnic groups. J Geront. 2002;57(2):S79–95.
41. Newton J, Excell EO. Amazing grace. In: The Presbyterian hymnal. Louisville, Kentucky: Westminster John Knox Press, 1990. p. 280.
42. Datson SL. Personality constructs and perceived presence of deceased loved ones. Death Stud. 1997;21(2):131–47.
43. Klugman CM. Dead men talking: evidence of post death contact and continuing bonds. Omega. 2006;53(3):249–62.
44. Simon Buller S, Christopherson VA, Jones RA. Correlates of sensing the presence of a deceased spouse. Omega. 1988–9;19(1):21–30.

45. Hines Smith S. "Fret no more, my child. for I'm all over heaven all day": religious belief in the bereavement of African American, middle-aged daughters coping with the death of an elderly mother. Death Stud. 2002;26(4):309–23.
46. Vickio CJ. Together in spirit: keeping our relationships alive when loved ones die. Death Stud. 1999;23(2):161–75.

Chapter 13
Bereavement and Grief

About 3.5 million Americans, or 1 % of the population, die every year. Nearly every death disrupts multiple relationships—relationships the decedent had with a spouse [1], parents, children, other relatives, friends, or colleagues. As a result the death leaves multiple grieving survivors, who must reconfigure their futures without the decedent. Some survivors can work through that process on their own, but others need help.

Seeking to address that need, the hospice movement introduced grief care for survivors as a key part of comprehensive end-of-life care. The impetus for this effort was recognition of the major life traumas a death and the resulting grief can cause survivors. They may have to arrange immediately for body disposition, plan for memorial services, and attend to important legal or financial duties such as wills. They may also suddenly have to assume roles the dead person once filled in the family or workplace. Survivors must simultaneously tend to their own emotional and spiritual wounds caused by the death. The grief care that hospice imagines aims to help survivors navigate the difficult life adjustments while healing from the grief.

This chapter describes such grief care. The chapter reviews recent theories about grief, offers a simple taxonomy for it, differentiates depression from grief, and suggests ways health professionals should monitor (and occasionally treat) survivors' grief. The case of Mrs. J., a recently widowed elderly woman, illustrates many of the points.

Throughout the chapter, I use the terms *bereavement*, *grief*, and *coping* in distinct ways. Bereavement means a loss expected to produce distress [2]. The loss need not be a death but may be a divorce, a lay-off, or a competitive defeat [3]. Bereavement is an objective, historical fact. Grief, in contrast, means a person's emotional distress precipitated by such a loss [4]. Grief is a subjective reaction. And coping means one's efforts to manage the stresses provoked by fears, anxieties, threats, or even challenges—some of which may result from significant losses—to achieve "a livable balance" for the future [5]. Coping includes both objective

H.S. Perkins, *A Guide to Psychosocial and Spiritual Care at the End of Life*,
DOI 10.1007/978-1-4939-6804-6_13

behaviors and subjective feelings. This chapter focuses mainly on grief but occasionally also mentions bereavement and coping.

The Case

Mr. J., a 77-year-old man, is diagnosed with leukemia. The hematologist tells Mr. and Mrs. J., who have been married 53 years, his expectation for Mr. J. to live less than a year. Still, the doctor offers chemotherapy, and the J.s accept. Mrs. J. says she wants five more years with her husband. Mr. J. improves briefly with the chemotherapy but then declines steadily.

Over the months Mr. J. declines, Mrs. J. is too feeble to care for him herself. The J.s' adult children and several paid attendants provide home care for him instead. Mr. J.'s function deteriorates by fits and starts. One day he can walk with his walker down the hall; the next day he cannot. One day he can toilet himself; the next day he cannot. Once lost, a function never returns. Desperately wanting Mr. J. to return to his old self, Mrs. J. grieves every lost function. Her discouragement sometimes boils over in accusations that Mr. J. feigns disability to gain attention. The hematologist urges referral to hospice when Mr. J. begins to suffer severe back pain. Mrs. J. resists the idea at first. She relents only when Mr. J.'s back pain exceeds the ability of the current treatments to relieve it. Mr. J. dies six weeks later in hospice care at home.

In the first moments after he dies, a distraught, weepy Mrs. J. asks everyone else to leave his room. She closes the door and stays several hours. She emerges "dazed and numb." Her pastor visits the next day. At one point he gathers Mrs. J. and her adult children to talk about Mr. J. and his life. This review comforts Mrs. J. because she remembers many of Mr. J.'s accomplishments in life. The pastor mentions some of those accomplishments during the memorial and burial services he conducts for Mr. J. the following week.

When the frenzy of events immediately surrounding the death subsides, Mrs. J. finds herself alone at home, badly missing Mr. J. She cannot bring herself to leave the apartment for several weeks. She eats little and loses weight. She lacks interest in activities she ordinarily enjoys such as socializing with friends, attending church, and knitting. She appears "emotionally distant." Once when others express sympathy, she snaps back, "No one really knows how I feel." She often just sits, ruminating sadly about Mr. J. Certain items in the apartment remind her of him: the pictures on his desk, the empty bed, the clothes in his closet. She tells her children that she regrets not having taken better care of him. When the pastor visits again, she asks him why Mr. J. died and what has happened to him since. She often senses Mr. J.'s "presence," and says he appeared to her twice, assuring her he is "fine." She often worries that her children will abandon her to a lonely widowhood, and sometimes just wants to go join Mr. J. in heaven.

Mrs. J. tries several kinds of grief counseling. She meets with the hospice's bereavement counselor but cannot express her feelings candidly to "a stranger." She fares only somewhat better with her pastor. Mrs. J.'s daughter finally convinces her to attend a grief support group at the nearby senior center. That group proves key to Mrs. J.'s recovery. The group's expert curriculum, devoted attendees, and ample open discussion help Mrs. J. learn to cope with her grief, to find new meaning in life, and to redirect her future.

Mrs. J.'s grief lifts over the next six months. She begins to meet friends again for dinner. She resumes knitting and attending events such as concerts, movies, and church services. She even sorts through Mr. J.'s belongings, keeps a few significant items (such as his diplomas, his military uniform, and a stuffed hippopotamus he won for her at a carnival), and donates much of the rest to charity. Of course, in later years she still misses Mr. J. on special days such as their wedding anniversary, his birthday, and Memorial Day. She also sometimes still speaks of him in the present tense as though he were still living.

Recent Theories About Grief

Successful living requires adaptation to many personal losses including those of youth, beauty, strength, memory, dexterity, dreams, opportunities, and, of course, people [1]. Vigorous discussion over the last 100 years has tried to specify the role of grief in that adaptation process. In the early twentieth century, Freud's theories of personal attachment produced the idea of "grief work" after deaths. Psychiatrists understood this idea to mean that a return to mental health required the bereaved to express their grief, to work through it somehow, and eventually to detach emotionally from decedents. The idea of working toward emotional detachment dominated the clinical approach to grief for years. Yet no research ever validated the idea or the clinical approach using it.

In the 1960s Parkes modified Freud's original idea [6]. While retaining the goal of eventual emotional detachment from the dead, Parkes proposed an orderly progression of grief through four stages: shock–numbness, yearning–searching, disorganization–despair, and reorganization. Kuebler-Ross later expanded those stages into her five famous stages of denial, anger, bargaining, depression, and acceptance. A recent empirical study now largely confirms the Kuebler-Ross idea not as distinct stages in a predictable progression but as *nearly coinciding elements* of early grief. Bereaved subjects in the study rated themselves several times on the five Kuebler-Ross elements during the two years after a family member's death [6] Ratings for the elements overlapped considerably during the first six months and peaked in roughly the order, first to last, that Parkes and Kuebler-Ross predicted: disbelief, yearning, anger, depression, and acceptance (Fig. 13.1). Specifically, mean scores for the first four elements all peaked during the first six months and then fell. In contrast, mean scores for the last element, acceptance, rose consistently over the two years and then remained high. These data suggest the distress of grief usually concentrates in the first six months of bereavement and then decreases. Acceptance meanwhile increases steadily. These data appear to refute the idea of

Fig. 13.1 Stages of grief according to Parkes and Kuebler-Ross: disbelief, yearning, anger, depression, and acceptance (From Wortman and Silver Cohen [7], with permission)

delayed grief whereby initially suppressed grief simmers over months or years and eventually boils over in full-blown distress.

Many authors commend the Parkes and Kuebler-Ross ideas as important advances in describing grief. But some authors point out apparently false conclusions that have resulted [7, 8]. One such conclusion asserts that grievers always suffer significant distress after a death and, therefore, must confront and "process" their true feelings of loss. This processing, or "working through" the grief, requires conscious, concerted effort to understand why the death occurred, whether it could have been avoided, and what it means for survivors. According to this idea, any denial of the loss or avoidance of the "working through" process is considered pathologic and indicates the need for psychotherapy. Without such therapy the unresolved distress of grief remains a threat to future physical or mental health. Critics of this conclusion note that distress may not always accompany grief, that little empirical evidence demonstrates such conscious "working through" actually occurs in most cases, and that psychotherapy is rarely necessary and may actually harm grieving survivors when used indiscriminately.

Another apparently false conclusion is that the hypothesized stages of grief serve as useful indicators of progress along the road to mental recovery. As the empirical study I describe above shows, grief's distressing elements often coexist in the first 6 months after a death [6]. A person's grief at any particular time cannot be pigeonholed into one neat stage or another. Furthermore, grief rarely progresses steadily toward resolution. Grief is idiosyncratic: its distresses may come and go in a pattern that defies steady progress. Exhortations to "get over it" or "to move on with your life," therefore, may not address grief accurately and may simply ignite a griever's resentment. Still another apparently false conclusion is that grief always resolves. It does in most cases, but not all. For example, complicated grief—an unusual, prolonged, and especially distressing kind—may require intense psychotherapy and occasionally may never fully resolve. I discuss this kind of grief later in the chapter.

Taxonomies of Grief

Experts proposed many clinically oriented taxonomies for grief late in the twentieth century. These taxonomies have created confusion for two reasons. First, different taxonomies use different terms for similar concepts, or similar terms for different concepts. The resulting confusions complicate comparisons across taxonomies. Second, the earliest taxonomies were based on theories, not research. When empirically tested later, some key aspects of these early taxonomies could not be verified. Nonetheless, mental health professionals and the general public still use some confusing or dubious concepts from these taxonomies.

I prefer the taxonomy proposed by Bonanno and Kaltman [2]. It has the advantages of simplicity, empirical verification, and sensible therapeutic implications. It identifies three kinds of grief, two of which often do not require active treatment but one of which does. This differentiation, therefore, can be useful to professionals in their grief care.

Two Mild Kinds of Grief: Minimal and Typical

Although considerable professional and popular literature emphasizes the problems grief can cause, the great majority of grieving survivors experience a relatively mild, self-limited course (Table 13.1) [2]. Fifteen to fifty percent experience only *minimal grief*. Though sad, they have little disabling distress. Some psychotherapists consider such minimally symptomatic grief as pathologic perhaps due to the old grief-work-and-detachment philosophy. They recommend active therapy with the goal of bringing distresses to the surface and thereby preventing a long-standing, subconscious grief that finally bursts out "as fresh(ly) and intense(ly) as if the loss had just occurred." [2-Humphrey and Zimpfer as quoted by Bonanno and Kaltman]. Yet multiple studies have failed to identify such delayed grief. These studies convince me (and others) that it does not exist, and that lack of grief distress actually indicates a healthy adjustment to loss.

Another 35–70 % of grieving survivors experience *typical grief*. These grievers suffer moderate functional disruption during the first 6 months after the death but return to near-normal function during the next 6–18 months. The early functional disruptions may affect the emotional, physiologic, cognitive, social, or occupational aspects of life. As the ideas of Parkes and Kuebler-Ross suggest, the emotional

Table 13.1 Differences among minimal, typical, and complicated griefs

Kinds of grief	Approximate prevalences (%)	Distress at various times during bereavement		
		6 Months	12 Months	24 Months
Minimal	15–50	Low	Low	Low
Typical	35–70	High	Moderate	Low
Complicated	15	High	High	High

Adapted from Bonanno and Kaltman [2]

aspects dominate the first months after a death. Pining or yearning for the dead person is a key feature, but sadness, irritability, anger, fear, and guilt may also occur. A griever's loneliness is more emotional (loneliness even in the company of others) than social (marginalization or exclusion from social networks). The physical aspects may include temporary impairment of the immune system and increased mortality. Infections occur more frequently than usual among grievers of any age while deaths by accidents or suicide occur especially frequently among young grievers.

The cognitive aspects involve difficulty with making sense of the death and refashioning one's identity in the absence of the dead person. The death may be hard to believe or accept, giving grievers a lingering sense of unreality about it. They may struggle to know why and how the death occurred. They may even discover their confidence eroding in the predictability, orderliness, and meaning of the future. That future without the dead person may seem foreshortened, uncertain, and hopeless. By identifying closely with the dead person, some grievers may feel "a piece of (themselves) missing." Yet others may actually experience *added* connection to the dead person. They may feel physical symptoms like the dead person's, do things the dead person would do, or imagine their own deaths like that person's death. Persistent, "unbidden" thoughts or occasional appearances of the dead person may occur, [2-Horowitz as quoted by Bonanno and Kaltman] worrying grievers that they are "going crazy." Social and occupational disruptions may also affect grievers. "Emotional distance" may cause grievers to avoid friends, social events, and new romantic relationships. This "distance" may alienate others. Work performance may also suffer, eroding grievers' confidence in their abilities and creating dissatisfaction with their work for them and their coworkers.

Both mild kinds of grief run a short, self-limited course. As a result they need no treatment besides patient listening and reassurance. In fact, they may bring unexpected benefits. Grieving survivors may experience increased affection for the dead person or increased pride in his or her accomplishments. Grievers may also find solace in the belief that the dead person has escaped earthly sufferings and finally resides comfortably in heaven. Grievers themselves may also feel liberated from old living restrictions, say, caused by caregiving duties, and see opportunities to pursue new interests and experiences.

Complicated Grief

The remaining 15 % of grieving survivors suffer a chronic, more complicated grief than either the minimal or typical griefs. Prigerson and her colleagues characterize this complicated grief in terms of major and minor symptoms [3, 4, 9, 10] (The distinction between "major" and "minor" here refers not only to how much distress the symptoms cause the griever but also how important they are in establishing the diagnosis.) The key major symptom, these authors believe, is yearning for the dead person. But three other major symptoms also exist: having intrusive thoughts about the dead person, being compelled to search for that person, and experiencing excessive loneliness without him or her (Table 13.2). The authors also list nine minor symptoms: being unable to accept or even to acknowledge the death; having excessive bitterness, anger, irritability, or "edginess"; feeling that a part of oneself died with the decedent; taking on some symptoms or behaviors of him or her; viewing current life as empty or meaningless; viewing the future as unfulfilling, purposeless, or hopeless; feeling emotionally "numb" or "detached"; seeing one's worldview shattered through a sense of lost security, control, or trust in others; and being unable to "move on" with life. Diagnosing complicated grief requires three of the four major symptoms daily, four of the nine minor symptoms daily, and resulting dysfunction in social, occupational, or other important domains of life for at least six months.

Recent research has begun to identify predisposing, exacerbating, and ameliorating factors for complicated grief. Predisposing factors include close kinship relationships to the dead person (such as being a spouse, parent, or child); a psychologically intimate marriage (for a widow or widower); a lack of mental preparedness for the death; and a history of abuse, neglect, or separation anxiety during the griever's childhood [3]. Exacerbating factors include the griever's being young, a spouse to the decedent, and mentally unprepared for the death [11]. The one identified ameliorating factor is dependable social supports [12–15].

Having gathered evidence for complicated grief as a distinct disorder and offered diagnostic criteria for it, Prigerson and colleagues then suggest a possible psychopathologic mechanism behind it. (The reader should note that Prigerson and her colleagues describe this mechanism by referring specifically to widowhood. They may have done so, assuming (as others do [16]) that the death of a longtime spouse is life's worst bereavement.) The mechanism these authors propose involves a grieving survivor's inability to believe the death is permanent. The griever becomes "frozen or stuck" in yearning for the dead person. The griever sees current life without that person as meaningless, and future life as unfulfilling and hopeless. In the case of a dead spouse, the surviving spouse feels cheated out of a fulfilling or complete married life and desperately wants it back. The griever mounts a "psychological protest" against the absence of the decedent. The griever may experience bitterness, anger/irritability, emotional "numbness," and psychological detachment from others. The result is that the grieving survivor cannot develop a new identity without the dead person [3].

Table 13.2 Proposed criteria for diagnosing complicated grief after a death

Types or effects of symptoms	Description	Duration/intensity	Required?
Major	· Yearns excessively for the dead person · Has intrusive thoughts about the dead person · Searches for the dead person · Feels excessively lonely since the death	Daily or "to a marked degree"	Yes for at least three
Minor	· Cannot accept or even acknowledge the death · Feels excessively bitter, angry, irritated, or "on edge" · Feels that part of oneself died with the dead person · Assumes symptoms or behaviors of the dead person · Views current life as empty or meaningless · Views future life as unfulfilling, purposeless or hopeless · Feels emotionally numb or detached from others · Sees own worldview shattered through a sense of lost security, control, or trust in others · Cannot "move on" with a new life	Daily or "to a marked degree"	Yes for at least four
Effects on function	Has marked dysfunction in social, occupational, or other important domains of life as caused by the symptoms	At least 6 months	Yes

Adapted from Prigerson [10], with permission

Prigerson and colleagues now take the next logical step in pressing to establish complicated grief as a valid psychiatric diagnosis. They are urging the *Diagnostic and Statistical Manual V*, the most authoritative diagnostic reference in psychiatry, to list complicated grief as a diagnosis and to give the criteria in Table 13.2 as its defining criteria [3] Other experts support the idea [17] Its importance derives from the need to treat most complicated griefs with active psychotherapy.

The Case: Monitoring Mrs. J.'s Grief

Monitoring Mrs. J.'s grief takes several forms. The hospice's grief counselor meets with her several times as part of the Medicare hospice benefit. Mrs. J.'s pastor and the church's deacons visit her at home. And her primary care doctor schedules several office visits beginning about a month after Mr. J. dies. The doctor checks for symptoms that might indicate a complicated grief and, therefore, require psychiatric referral and formal psychotherapy. Mrs. J. also attends the senior center's grief group where the other widows informally observe her grief.

Mrs. J. has varying affinities for the different monitoring methods. While the hospice's grief counselor treats her with great sensitivity and respect, Mrs. J. has always been a "private" person. She feels uncomfortable revealing her innermost vulnerabilities and feelings to strangers in unfamiliar places. As a result she likes least talking with the hospice's grief counselor who meets with Mrs. J. in the unfamiliar setting of the hospice's administrative offices. She likes more the visits with her pastor, the church's deacons, and her doctor, all of whom she already knows and likes. But, surprisingly to her, she likes most the grief group at the senior center. She especially likes the camaraderie. The group meets once a week, follows a videotaped curriculum, and encourages open discussion at the end of each meeting. Mrs. J. just listens at first but eventually begins contributing comments. She finds the other participants, mostly widows like her, interested in her reactions to Mr. J.'s death. The others encourage her to speak up and give her positive support. That experience helps her overcome her anxiety about talking about her true, raw feelings. In fact, she finds that revealing them gives an unexpected healing benefit for her distress at Mr. J.'s death [12–15].

The Case: Categorizing Mrs. J.'s Grief

Process of elimination reveals Mrs. J.'s grief as typical. The key feature distinguishing her grief as typical, not minimal or complicated, is that she suffers *multiple, self-limited symptoms that only rarely disrupt her daily functioning after the first few months*. Her grief is not minimal because she suffers more than just sadness. She also endures some disruption in her life activities and interests at first. Yet her grief is not severe enough to qualify as complicated because she presents just one of four major symptoms: She feels lonely for many days but only occasionally yearns for Mr. J.; experiences few disruptive, intrusive thoughts about him; and goes searching for him only once after his death when he calls to her from around the corner. She also acknowledges Mr. J.'s death immediately and presents no other minor symptoms. None of her symptoms except for the loneliness occur every day, and all resolve within six months.

Table 13.3 lists the symptoms of Mrs. J.'s typical grief, classifying them as either emotional, medical, cognitive, or social/occupational. For the emotional, Mrs. J. feels sad for parts of many, but not all, days during the first year after Mr. J. dies (especially on significant days such as their wedding anniversary and Memorial Day). She misses him most often during the evenings as she sits alone in their apartment and occasionally yearns for him there. But she also sometimes feels angry at him for dying so soon: She once angrily turns down all photos of him in the apartment because she suddenly does not want "that dead man staring at me." But mostly Mrs. J. just wants Mr. J. to come back again. (According to her, he does so twice for a few minutes each time. I discuss those visits later.) Mrs. J. also feels guilty about Mr. J.'s terminal pain [10] She blames herself for not doing "more" to relieve it in his last days. Friends' insensitive comments such as "I know just how you must feel" and "It must be God's will" irritate her: Mrs. J. does not believe others can know how she feels or what God's will is. Their attempts at empathy and support, therefore, fall flat [12] Mrs. J. also worries that her children will abandon her now that the family crisis has passed. And she wonders what reason she has to live (although she never seriously considers suicide).

For the medical, Mrs. J. sometimes has a malaise—a vague feeling of illness or debility—which keeps her in bed much of an occasional day. She can never specify her symptoms and tells her doctor only, "I don't feel good." Meanwhile she shows no evidence for impaired immunity, increased infections, or other medical risks for premature death.

For the cognitive, Mrs. J. has difficulty grasping Mr. J.'s death. She is not mentally prepared when he dies even though his death occurs after a months-long fatal illness. She holds to the very end the unrealistic hope that his illness will miraculously reverse itself and give them "another five years together." Then, after he dies, she repeatedly asks herself why Mr. J. died before *she* was ready and what has happened to him since. She also struggles to imagine her future without him in this life. Fifty-three years of marriage has melded their lives and identities more than she realized. Building a new future without her lifelong companion now seems an overwhelming task, and she is afraid. Despairing at times, Mrs. J. says, "I just want to be with Bill."

And for the social and occupational, Mrs. J. avoids friends and social events for a long time after Mr. J. dies. She eats meals alone in her apartment and initiates no contact with friends. She refuses many of their requests to visit. New romance does not interest her. Her difficulty at accomplishing everyday tasks frustrates her. Cleaning the apartment, knitting, and shopping for clothes—ordinarily sources of satisfaction or fun—and keeping her appointment calendar require more effort than she can muster for months.

As expected, Mrs. J.'s typical grief resolves gradually over six months. She requires no special treatment for it (although she does restart medication for an old, mild depression). Now several years later she still has an occasional sad day when she thinks a lot about Mr. J. But she has generally reconfigured her life successfully for widowhood. She has resumed social activities with friends, attends church services, and keeps up active contact with her grandchildren. Overall, she enjoys life again.

Distinguishing Complicated Grief, Depression, and Anxiety in Grievers

Complicated grief, depression, and generalized anxiety seem at first to share many of the same presenting symptoms among grieving survivors. Some survivors, in fact, seem to have two or all three of these diagnoses at once. HPs can justifiably wonder, Do these diagnoses represent the same disorder or different ones? Researchers tackled this question by interviewing 135 elders whose spouses had died from a critical illness within the prior six months. The researchers asked the symptoms of these widows and widowers and used extensive, validated research

Table 13.3 Symptoms of Mrs. J.'s typical grief

Domain	Symptom	Present?	Illustration
Emotional	Sadness	Yes	Low mood for many days especially on significant dates
	Yearning or pining for the dead person	Yes	Wishes to "be with" Mr. J
	Anger/irritability	Yes	Anger at Mr. J. for dying. Irritability at others' facile sympathetic comments
	Emotional unavailability, emotional distance, or loneliness	Yes	Comment: "You can't possibly know how I feel."
	Intrusive thoughts about the dead person or appearances by him or her	Yes	Two postmortem visits by Mr. J. to Mrs. J
	Guilt over past experiences with the dead person	Yes	Self-accusations about not having done "more" to relieve Mr. J.'s pain in his last days
	Fear for the future: lack of confidence in life's predictability, orderliness, or meaningfulness	Yes	Worry that her adult children will abandon her
Medical	Malaise (i.e., a vague feeling of being ill or debilitated)	Yes	Episodes of "just not feeling good" without ability to specify further
	Symptoms resembling the dead person's	No	Not applicable
	Disrupted immunologic function	No	Not applicable
	Increased mortality	No	Not applicable
Cognitive	Difficulty accepting the death	Yes	Comment: "I thought Bill and I would have 5 more years together."
	Difficulty making sense of the person's death	Yes	Discussions with pastor about why Mr. J. died before she was ready and what has happened to him since death
	Tendency to fear the future without the dead person	Yes	Worry that she must face life alone in widowhood
	Tendency to merge one's own identity with the dead person's	Yes	Comment: "I don't want to live. I just want to die to be with Bill."

(continued)

Table 13.3 (continued)

Domain	Symptom	Present?	Illustration
Social and occupational	Avoidance of social events	Yes	Eating alone in apartment for first few weeks. No interest in social events. Inability to initiate contact with friends
	Avoidance of new romantic relationships	Yes	Lack of interest in other men for at least 18 months after Mr. J.'s death
	Dissatisfaction with work performance	Yes	Frustration at lack of motivation to clean, knit, shop, or keep her calendar

tools to screen them for complicated grief, depression, and anxiety [4, 9] Distinctive clusters of symptoms characterized each disorder. Symptoms characterizing complicated grief included sensing the dead spouse's presence (such as through sightings, sounds, or dreams), feeling pain in the same area of the body as did the spouse, needing to call the spouse's name, thinking persistently about the spouse, denying the spouse's death, and yearning for the spouse as he or she was before the fatal illness. Symptoms characterizing depression included feeling "blue," not being happy or not enjoying life, and experiencing everything as an effort. And symptoms characterizing anxiety included nervousness, worry, and restlessness [10]. The distinctiveness of these symptom clusters convinced the researchers that complicated grief, depression, and generalized anxiety are distinct disorders.

Such evidence raises the question, How can health professionals differentiate the three disorders in grieving survivors? Perhaps the most important distinction for treatment purposes lies between grief and depression: Grief does not usually require medications or intensive psychotherapy; major depression does. (Anxiety often accompanies depression and responds to treatment for it.) Identifying major depression may simply require asking, Are you depressed? However, Susan D. Block, M.D., the prominent cancer psychiatrist I first cited in Chap. 2, has detailed the many particular symptoms of grief and depression. Those two disorders share some symptoms but differ on others (Table 13.4). (Although Dr. Block describes these disorders for dying patients, I believe her description applies equally well to grieving survivors). Decreased concentration, social isolation or loneliness, anger or irritability, appetite disturbances, and sleep disturbances can occur with both grief and depression. However, the capacity to experience pleasure and to see a positive future occurs with grief alone. In contrast, feelings of hopelessness, helplessness, worthlessness, meaninglessness, and guilt occur with depression alone. Suicidal thoughts also occur with depression alone. Feelings of grief come intermittently; feelings of depression remain constant. In addition, because grief is so common after a death, prior occurrence does not help predict later recurrence. Yet prior occurrence for depression (and anxiety, too) *does* help predict later recurrence [13, 18–20].

Table 13.4 Distinguishing grief and depression

Characteristic	Specific item	Grief	Depression
Definition		Feelings and behaviors caused by a major loss	Cluster of specific symptoms as outlined below
Psychosocial symptoms	Decreased concentration	Yes	Yes
	Social isolation or loneliness	Yes	Yes
	Anger or irritability	Yes	Yes
	Feelings of hopelessness, helplessness, worthlessness, or meaninglessness	No	Yes
	Guilty feelings	No	Yes
	Suicidal thoughts	No	Yes
	Capacity for pleasure	Yes	No
	Able to foresee a positive future	Yes	No
Physical symptoms	Appetite disturbances	Yes	Yes
	Sleep disturbances	Yes	Yes
Persistence of symptoms		Intermittent	Constant
Treatment		Expressing feelings informally or attending formal psychotherapy and/or support groups	Taking medications with or without psychotherapy

Adapted from Block [21], with permission

The Case

Mrs. J. suffered a depression earlier in her life and was successfully treated with medication. She had not suffered a recurrence or needed antidepressant medication since. Still, she has some risk for recurrence when Mr. J. dies. She certainly presents all the symptoms shared between grief and depression. She tries to read mystery novels but cannot concentrate on them. She also tries to watch televised sports events but cannot concentrate on them, either. Despite feeling lonely, she avoids social contact. She easily gets angry at friends' comments meant as consolation. And she finds herself occasionally angry at Mr. J. for dying. Disturbances of appetite and sleep plague her early on: she eats little and sleeps a lot.

I do not know whether any of the professionals attending her ask Mrs. J. the simple question, Are you depressed? But she does present a few of the symptoms distinctive for depression. She feels "flat" or "down" most days and cannot muster interest in activities she ordinarily enjoys (such as knitting). She expresses some meaninglessness about her new life, saying she does not know how she can live without Mr. J. And she feels guilt over "not having done more" to relieve his pain during his last days. She occasionally takes to bed for a day or two and says she wants to die. Several times her primary care doctor rules out urinary tract infections, medication side effects, blood chemistry abnormalities, and other medical causes for mental status changes. Yet Mrs. J. never expresses thoughts of hopelessness, helplessness, or worthlessness and often foresees a positive future reunited with Mr. J. someday in heaven.

She, therefore, presents a mixed picture of both grief and depression. Remembering her prior depression, her primary care doctor makes a psychiatric referral. The psychiatrist begins an antidepressant and conducts brief psychotherapy. Mrs. J. "perks up" over the next few months.

Belief in Contacts Between the Living and the Dead

Many people believe in life after earthly death. Continuing contact between the living and the dead may be part of their belief. Scientists, of course, tend to discredit such beliefs as purely speculative and objectively unverifiable. Psychotherapists and other scientifically oriented clinicians may interpret them as psychopathology, maladaption, or "semirealistic perspectives that compromise objective reality." [22]. One researcher even calls reported contacts with the dead "hallucinations." [23].

Nonetheless, the prevalences of these beliefs are likely high among Westerners. Ninety percent of hospitalized patients in a San Antonio sample [24], and 68 % of widowed spouses in a Detroit sample believed in an afterlife [16]. In addition, 33 % of patients in the San Antonio sample [24], 44 % of urban residents in a Los Angeles sample [25], and 47 % of widowed spouses in a Wales sample [23] claimed direct, personal contact with the dead. By their own descriptions, 39 % of the Welsh widows and widowers "felt the presence" of the dead, 14 % saw them; 13 % heard them; and 3 % were physically touched by them [23].

People often interpret contacts with the dead as either dreams or supernatural visits. As I noted earlier in this book, one Euroamerican man in the San Antonio study described such a "dream" during his cardiac arrest. "I remember going through a chow line," he explained. "I turned around, and there was my (dead) son. He said, 'What are you doing here, Pop? You don't belong here.' Suddenly he was gone, and I was (back) in the (intensive care) bed." And an African-American woman in that same study described supernatural "visits" from her dead aunt. "I know when she is there (in the room) because too many things (are) falling or dropping," the woman said. "When something's wrong, she comes back." Grieving survivors may see these dreams or visits as giving warnings, reassurance, or guidance [24]. Of course, contemporary religion, literature, and popular culture may reinforce such beliefs. In any case, because those beliefs are so prevalent, professionals providing end-of-life care should anticipate them among survivors.

The Case

Mrs. J. illustrates those beliefs. Her yearnings "to die and go be with Bill" certainly demonstrate a belief in an afterlife. Her Protestant faith provides a basis for that belief: the resurrection of the dead to eternal life with God. Troubled or perhaps just curious about what happened to Mr. J. after he died, Mrs. J. asks her pastor many questions about the afterlife during his follow-up visits. (Chap. 12 in this book discusses beliefs about the afterlife in detail.)

Mrs. J. believes firmly, too, that Mr. J. came to her twice after he died. She insists he came not in dreams, but actual visits. Both occurred late at night as she sat reading in her apartment. The first visit occurred about two weeks after his death. He stayed out of sight in another room and called, "Mary! Mary!" Mrs. J. explains later to others, "Bill called me in the same tone he always used. I heard him even without my hearing aids!" She got up and looked around the corner, but he had already disappeared. The second visit occurred about a month later. Mr. J. appeared that time for Mrs. J. to see. He wore his favorite plaid golf shirt and khaki slacks. He assured her that he had no pain and was having fun. He was back to playing golf. He then disappeared and has not visited since.

Mrs. J. waits several weeks after these visits to tell anyone about them. She worries what others will think of her. She finally tells her adult children and is pleased that they take the experiences seriously. The children make no dismissive comments and give no facile explanations about "psychological stress." Their interest encourages Mrs. J. to mention the visits to her closest friends. She is surprised when those friends also take an interest. Some even describe postmortem contacts of their own.

The Psychological Significance of Beliefs in Postmortem Contacts with the Dead

Nearly all religions have beliefs about such contacts between the living and the dead [16, 22], but empirical research about the prevalence, content, and purpose of such beliefs is sketchy and inconsistent. In particular, data relating the beliefs to age group or education are sketchy, and data relating them to ethnic group or gender are inconsistent. Many studies that do exist are merely descriptive anecdotal case series or cross-sectional opinion surveys. Neither address the important cause-or-effect question, Do certain psychological needs promote religious and other beliefs in these contacts, or do certain preexisting religious beliefs promote psychologies receptive to beliefs in these contacts?

In light of all the theoretical speculation and the limited empirical research on the topic, what tentative understanding might exist for the psychological significance of beliefs in postmortem contacts? Although some authors think such beliefs are neither uniformly comforting nor distressing [16], one organizing concept that suggests an overall comforting effect appeals to me [22]. Two kinds of meaning—global and situational—form the heart of the concept. Global meaning involves understanding one's basic perspectives on life (including ideas about justice, order, hope, and God's love); situational meaning involves understanding a particular event. A death, of course, is a traumatic event for grieving survivors. They must try to reconcile the psychic trauma of the death (the situational meaning) with their faith in life's orderliness and goodness (the global meaning).

Beliefs in postmortem contacts help to accomplish that reconciliation by producing two comforting thoughts for grieving survivors. First, the dead person lives on [26] or will live again in the future. And second, the grievers themselves will live on after their own deaths. Together the two thoughts support survivors' hope for a future reunion with dead loved ones. That hope helps heal any anger at the dead person for dying or at God for allowing any perceived injustice in the death. Belief in postmortem contacts also reassures grieving survivors that their relationship with the dead person has not necessarily ended, only changed. Grievers can continue the relationship (if they wish) [22] even as they prepare for life without an earthly version of the dead person. One African-American woman in the San Antonio study said, for example, she maintains a relationship with her dead brother by thinking about him, speaking with him, and sometimes asking favors from him in this life [24].

Even if such contacts with the dead subside over time, grieving survivors can use other, less dramatic ways to maintain the relationship. They can perform memorial rituals including visiting the dead person's grave, sharing stories or recollections about him or her, adopting or teaching others some of the person's values, or bequeathing important objects from his or her life. Mrs. J. still attends Memorial and Veterans Day services at the senior center to remember Mr. J.'s World War II military service. She visits his grave. She also shares with her family and others stories about his idiosyncrasies (often funny ones) and his most important life values including "honesty to a fault," dedication to family, frugality, and service to community. She realizes those stories and values create a philosophical legacy from Mr. J. to others. She also contributes to the material legacy from him to them. For example, she has given to Mr. J.'s namesake grandson, a middle school band teacher and accomplished amateur musician, Mr. J.'s flute from his high school band days.

Though not systematic, observations of such rituals convince me that ongoing relationships with the dead provide grieving survivors important solace. Many probably need to talk with others about their beliefs and experiences. Grievers should certainly not have to fear clinicians' ridicule for doing so particularly because the disclosures may bring important psychological benefits (as they do for Mrs. J. with the disclosures to her adult children and friends). Clinicians should suspend their usual, scientifically based skepticism long enough to encourage the disclosures and to listen supportively to them. When open, sensitive, and non-judgmental, discussions of postmortem contacts have a chance to provide grievers an important emotional healing.

Professionals' Care for Grieving Survivors

Professionals who attend the dying patient can take several steps to help survivors cope with their grief. Before the death those professionals can encourage survivors to prepare mentally. For instance, health professionals can urge family members and friends to visit the dying patient, say their good-byes, and help the patient as appropriate with any unfinished legal or other personal business. Professionals, especially the health professionals, can facilitate early hospice referral when it is indicated. The hospice philosophy includes attention to survivors' grief.

After the death, professionals—doctors, nurses, social workers, or clergy—should continue to monitor grieving survivors' reactions for at least six months. This monitoring should distinguish for treatment purposes self-limited minimal or typical grief from complicated grief. Survivors experiencing minimal or typical grief generally need no formal treatment [1]. In fact, psychotherapy for them may waste time and money and may do harm by incorrectly labeling them "psychiatric." In contrast, survivors with complicated grief, defined according to Table 13.2, should receive formal treatment. The professional should disclose to the affected survivor his or her complicated grief and explain it. Already worried about the distress and inability to function, the affected survivor is often relieved to learn that the professional recognizes the problem, that it occurs sometimes, and that treatment exists. Giving a specific diagnosis may also benefit the affected person by alerting others to a recognized problem and by discouraging such hurtful comments as the survivor is merely exaggerating his or her grief, feeling sorry for him- or herself, or using grief to gain attention.

The professional should ensure the survivor with complicated grief gains access to psychotherapy [3]. I believe a trained professional counselor experienced in grief counseling should conduct this therapy. Overall, it should aim to relieve symptoms, increase function, and restore hopefulness. Of course, the strategies the counselor uses should suit the particular griever's needs [1]. (One expert believes men generally need help in addressing emotions while women generally need help in solving problems [27]) One strategy the counselor might use educates the griever about the symptoms and course of complicated grief. Another strategy tries to help the griever accept or "mentally process" the death. A specific technique to that end encourages the griever to "talk" with the dead person about the griever's loss and adjustment difficulties [3] Still another strategy suggests specific healthy coping skills aimed at relieving everyday life stressors. And yet another encourages the griever to imagine and articulate a new future without the dead person. That future usually entails resuming some old functions and assuming some new ones [26]. A commonsensical method to doing so starts with taking on the easiest functions first. Early success with them can then bolster the griever's confidence for taking on more difficult functions later.

The professional must take a different approach if he or she recognizes a major depression or a generalized anxiety rather than a complicated grief in the survivor [19, 20]. A violent death for the dead person or a prior history of these disorders for the survivor predisposes the survivor to depression or anxiety after a death. Depression in bereavement is especially common and well studied: Its incidence among grievers reaches about 32 % by one year after a significant death. Unlike complicated grief, bereavement-related depressions and anxieties should be treated with psychotherapy *and* medications. The attending psychotherapist must be alert for suicidal ideation among these survivors. Despite good responses among many depressed grievers, 10 % remain depressed at four years after a significant death. These people may suffer chronic depressions even under ordinary circumstances. Mrs. J., for example, had a major depression in the past. Mr. J.'s death likely reignites it. But the symptoms from her combined typical grief and depression respond well to individual counseling, group support, and aggressive medication management. She now has no psychological symptoms.

Summary Points

1. Three kinds of grief exist. *Minimal grief* presents primarily as sadness and rarely disrupts everyday activities. *Typical grief* presents with multiple emotional, medical, cognitive, or social/occupational symptoms; intermittently disrupts everyday activities; and lasts no more than six months. *Complicated grief* presents with daily excessive yearning or searching for the dead person, intrusive thoughts about him or her, or loneliness; involves other symptoms, too; and disrupts important everyday activities for more than six months.
2. Correct treatment for grief depends on monitoring a survivor for at least six months and categorizing the grief correctly. Minimal or typical grief requires no specific treatment except perhaps informal counseling and reassurance. Complicated grief, in contrast, often requires psychotherapy aimed at developing healthy coping skills.
3. Depression and anxiety differ from grief but may coexist with it during bereavement. If severe, these two illnesses deserve treatment with medications and/or psychotherapy.

To Learn More …

1. Bonanno GA, Kaltman S. The varieties of grief experience. Clin Psychol Rev. 2001;21(5):705–34.
2. Prigerson HG, Bierhals AJ, Kasl SV, et al. Complicated grief as a disorder distinct from bereavement-related depression and anxiety: a replication study. Am J Psychiatry. 1996;153(11):1484–6.
3. Prigerson HG, Jacobs SC. Caring for bereaved patients: "All the doctors just suddenly go." JAMA. 2001;286(11):1369–76.

4. Wortman CB, Silver Cohen R. The myths of coping with loss. J Consult Clin Psychol. 1989;57(3):349–57.
5. Zhang B, El-Jawahri A, Prigerson HG. Update on bereavement research: evidence-based guidelines for the diagnosis and treatment of complicated bereavement. J Pall Med. 2006;9(5):1188–1203.

References

1. Boerner K, Wortman CB, Bonanno GA. Resilient or at risk? A four-year study of older adults who initially showed high or low distress following conjugal loss. J Gerontol. 2005;60(2):67–73.
2. Bonanno GA, Kaltman S. The varieties of grief experience. Clin Psychol Rev. 2001;21(5):705–34.
3. Zhang B, El-Jawahri A, Prigerson HG. Update on bereavement research: evidence-based guidelines for the diagnosis and treatment of complicated bereavement. J Pall Med. 2006;9(5):1188–203.
4. Prigerson HG, Bierhals AJ, Kasl SV, Reynolds CF III, Shear MK, Newsom JT, Jacobs S. Complicated grief as a disorder distinct from bereavement-related depression and anxiety: a replication study. Am J Psychiatry. 1996;153(11):1484–6.
5. Corr CA. A task-based approach to coping with dying. Omega. 1991–2;24(2):81–94.
6. Maciejewski PK, Zhang B, Block SD, Prigerson HG. An empirical examination of the stage theory of grief. JAMA. 2007;297(7):716–23.
7. Wortman CB, Silver Cohen R. The myths of coping with loss. J Consult Clin Psychol. 1989;57(3):349–57.
8. Corr CA, Doka KJ, Kastenbaum R. Dying and its interpreters: a review of selected literature and some comments on the state of the field. Omega. 1999;39(4):239–59.
9. Prigerson HG, Frank E, Kasl SV, et al. Complicated grief and bereavement-related depression as distinct disorders: preliminary empirical validation in elderly bereaved spouses. Am J Psychiatry. 1995;152(1):22–30.
10. Prigerson HG, Jacobs SC. Caring for bereaved patients: "All the doctors just suddenly go". JAMA. 2001;286(11):1369–75.
11. Barry LC, Kasl SV, Prigerson HG. Psychiatric disorders among bereaved persons: the role of perceived circumstances of death and preparedness for death. Am J Geriatr Psychiatry. 2002;10(4):447–57.
12. Lehman DR, Ellard JH, Wortman CB. Social support for the bereaved: recipients' and providers' perspectives on what is helpful. J Consult Clin Psychol. 1986;54(4):438–46.
13. Norris FH, Murrell SA. Social support, life events, and stress as modifiers of adjustment to bereavement by older adults. Psychol Aging. 1990;5(3):429–36.
14. Sherkat DE, Reed MD. The effects of religion and social support on self-esteem and depression among the suddenly bereaved. Social Ind Res. 1992;26(3):259–72.
15. Bisconti TL, Bergeman CS, Boker SM. Social support as a predictor of variability: an examination of adjustment trajectories of recent widows. Psychol Aging. 2006;21(3):590–9.
16. Carr D, Sharp S. Do afterlife beliefs affect psychological adjustment to late-life spousal loss? J Gerontology. 2013;69B(1):103–112. Published online, 29 June 2013; doi:10.1093/geronb/gbt061.
17. Horowitz MJ, Siegel B, Holen A, et al. Diagnostic criteria for complicated grief disorder. Am J Psychiatry. 1997;154(7):904–10.
18. McHorney CA, Mor V. Predictors of bereavement depression and its health services consequences. Med Care. 1988;26(9):882–93.

19. Zisook S, Shuchter SR. Depression through the first year after the death of a spouse. Am J Psychiatry. 1991;148(10):1346–52.
20. Zisook S, Paulus M, Shuchter SR, Judd LL. The many faces of depression following spousal bereavement. J Affect Dis. 1997;45(1–2):85–95.
21. Block SD. Psychological considerations, growth, and transcendence at the end of life: the art of the possible. JAMA. 2001;285(22):2898–2905.
22. Benore ER, Park CL. Death-specific religious beliefs and bereavement: belief in an afterlife and continued attachment. Int J Psychol Relig. 2004;14(1):1–22.
23. Rees WD. The hallucinations of widowhood. BMJ. 1971;4(5778):37–41.
24. Perkins HS, Cortez JD, Hazuda HP. Diversity of patients' beliefs about the soul after death and their importance in end-of-life care. South Med J. 2012;105(5):266–72.
25. Kalsh RA, Reynolds DK. Phenomenological reality and post-death contact. J Sci Study Relig. 1973;12(2):209–21.
26. Perkins HS, Cortez JD, Hazuda HP. Patients' diverse beliefs about what happens at the time of death. J Hosp Med. 2012;7(2):110–6.
27. Ott CH, Lueger RJ. Patterns of change in mental health status during the first two years of spousal bereavement. Death Studies. 2002;26(5):387–411.

Chapter 14
Giving End-of-Life Spiritual Care

For many years I taught a month-long medical humanities seminar for internal medicine residents. One seminar session addressed doctors' personal and professional growth during the practice years. To stimulate discussion during that session, I played video clips from a television documentary that interviewed doctors at several points early in their careers—entry into medical school, graduation, the first day of residency, entry into practice, and 10–15 years later [1]. One clip showed an established primary care internist, speaking at the funeral of a long-term patient of hers. This doctor stood before the patient's open casket and described how much the personal bond between her and this patient had meant to them both during the patient's dying months. Seminar participants sometimes responded to that clip by debating whether, as doctors, they should attend the funerals of their patients who die. Consistently troubled by the idea, participants always concluded they should not. Their reasons included the belief that a doctor's professional duties to a patient end when the patient dies, the regrets they would feel when curative treatment fails, the awkwardness of encountering survivors soon after a death, the unfamiliarity with the patient and his or her life outside the medical context, and the pressures to stay focused on medical tasks for still-living patients. I believe the seminar participants' aversion to attending patient funerals reflected a discomfort of most present-day clinicians—doctors in particular—toward spiritual matters in clinical settings. Yet those matters surely influence in important ways patient responses to end-of-life situations.

Medicine has not always been so uncomfortable with the spiritual dimension of illness. Both medicine and spirituality in ancient times played interrelated roles in life's major transitions such as birth, puberty, procreation, serious illness, and death. The ancients believed that medicine and spirituality together were needed for living healthfully and healing from illness [2]. The two domains shared twin guardian status over a person's welfare [2, 3]. But, as the scientific revolution of the nineteenth century swept over it, medicine focused more and more on the seemingly objective problems of the physical body [4] and less and less on the nonobjective problems of the spirit [5–10]. As ever more physical problems yielded to treatments based in

H.S. Perkins, *A Guide to Psychosocial and Spiritual Care at the End of Life*,
DOI 10.1007/978-1-4939-6804-6_14

science [3], medicine felt justified in concentrating on the body. Unfortunately, doctors often came to think of patients as only the physical problems they presented, and patients began to feel like mere objects of doctors' scientific curiosity [2, 11]. In addition, medicine came to view spiritual problems as either not real [8, 12], not important [12, 13] or merely emotional [14] and, thus, outside its responsibility to patients [3, 12, 15]. The longer the division between body and spirit persisted [7], the wider and firmer the gap between medicine and spirituality became.

Two prominent contemporary authors, however, argue convincingly that a person's physical and spiritual natures are inseparable [2], and thus, any physical problem has spiritual effects. According to one of these authors, all illness results from disruptions in relationships [2]. The disruptions may originate in the body's internal physical structures or biological processes, or in the patient's external relationships with other people or even God. But in each case the disruptions affect the spirit. The other author insists that physical "illness strikes a person in his (or her) totality," [16] both body and spirit, and especially powerfully if the illness is life threatening. Care must, therefore, address both [2, 17]. For these reasons the first author concludes that, while uncommon in modern, scientifically oriented medicine, attending to patients' spiritual needs is still "integral" to good health and good health care [18].

I agree. In particular, I believe that all health professionals (HPs)—especially those providing end-of-life care—have a duty to ensure patients receive the spiritual care they want and need [14, 19]. Spiritual care is a key part of complete health care [20]. And the more serious the illness, the more important is the spiritual care. Most circumstances allow spiritual care specialists such as chaplains, social workers, or pastoral counselors to provide this care, but circumstances occasionally require clinicians to do so. They should be prepared for it. But even when the specialists give spiritual care, the medically trained clinicians attending the patient should follow it as part of their responsibilities for coordinating all care.

Several medical organizations have published guidelines to that effect. The American Association of Medical Colleges expects graduating medical students to be able to elicit a patient's spiritual history, to understand the patient's spiritual beliefs in a clinical context, and to understand their own spiritual beliefs and the possible impact of those beliefs on their clinical care [21]. In addition, the World Health Organization recognizes the spiritual as an important dimension of quality of life [2]; the Institute of Medicine, the National Hospice and Palliative Care Organization, the National Consensus Project for Quality Palliative Care, and the Joint Commission on Accreditation of Hospital Organizations all emphasize attention to patients' spiritual needs as a dimension of good-quality health care [22, 23]. Therefore, all HPs—clinicians and nonclinicians alike—should have the training to recognize spiritual issues and to address them.

This chapter lays a foundation for that training as related to end-of-life care. The chapter opens by reviewing breaking bad news, the clinical event that often raises spiritual issues most starkly. The chapter then defines the spiritual and its two key components, the religious and the philosophical. The chapter next discusses why spiritual matters are important to dying patients and what spiritual needs many of them have. The chapter also summarizes and critiques some empirical studies that have tested for the health benefits of spirituality and spiritual care. The chapter ends by suggesting practical steps for addressing spiritual needs in everyday clinical care.

Admittedly, I rarely addressed patients' spiritual needs and never attended patients' funerals early in my own medical career (though some patients presented opportunities for me to do so). I made the same excuses as did the participants in my seminar. Yet one memorable patient taught me the importance of spirituality in physical illness and thereby caused me to change my practice. I use her case to illustrate many points in this chapter.

The Case

Mrs. R., a twice divorced, middle-aged African-American nurse administrator, managed the public hospital's clinic where I practiced most of the time. But one month when I just happened to be supervising inpatient care on the hospital's teaching wards, Mrs. R. was admitted for trashing her bedroom at home and not remembering the incident afterward. Her admission history revealed she used to smoke heavily but had quit several years before. A chest X-ray showed an abnormal mass. The housestaff and I suspected cancer. Computerized axial tomography (a CAT scan) the next day revealed another mass in Mrs. R.'s brain. That mass was likely a seeding from a primary lung cancer. Mrs. R. underwent a bronchoscopy (a direct visual examination of her large airways with a flexible, lighted scope) and a biopsy of the chest mass. The biopsy reading confirmed our suspicions of cancer several days later. As I attended Mrs. R. while she was dying over the next few months, her case would teach me important lessons about spiritual care at the end of life.

Breaking Bad News

Most people immediately think of medical bad news as a new diagnosis of cancer or other fatal disease. But other kinds of medical bad news exist, too. As Chap. 4 says, bad news is "any information which adversely and seriously affects (one's) view of the future." [24]. Along with fatal diagnoses, examples of medical bad news include the progression or reappearance of a chronic disease; the failure of treatment aimed at cure; the new onset of irreversible side effects from treatment; a positive genetic test for a serious disease; and the indications to begin hospice care [24]. In this chapter, however, I limit my discussion of medical bad news to fatal diagnoses. Such a diagnosis inevitably raises difficult medical, emotional, and spiritual issues for the patient. But the doctor must first break the bad news. Any further considerations, consultations, or treatment plans await that key step.

Doctors never disclosed fatal diagnoses to patients until recently. Instead, doctors concealed such diagnoses in trying to save patients from emotional distress and hopelessness. However, the Western public's improving education and increasing demands for a say in their medical care eventually convinced doctors (especially in the USA) to abandon this tradition. Surveys of American doctors documented the seismic change in disclosure practices between 1961 [25] and 1979 [26]. For fatal diagnoses, then, disclosure has replaced concealment as the standard of care.

But that change presented a difficult new question for doctors: How should they break such bad news to patients? Little formal research or training existed at first to guide end-of-life communications in general and breaking bad news in particular. Doctors had to "figure it out on their own." Unfortunately, many developed bad disclosure habits. Public dissatisfaction with those habits soon prompted directed research and education on the topic.

Of all end-of-life communications, breaking bad news has consistently commanded the most attention in the professional literatures. Most doctors dread breaking bad news. Several reasons may explain this dread: doctors' sense of being unable to anticipate or manage patient emotional reactions to bad news [24, 27]; fears about dying themselves [28]; self-doubts from possibly overlooking a key diagnosis or treatment; discouragement at the inability to cure an irreversible, fatal disease [24, 28]; embarrassment at having given earlier overly optimistic outcome predictions [24]; fears of dashing patients' hopes for the future [28]; and fears of being blamed for patient deaths [29]. Yet patients, family members, and other HPs now consider good doctor communications the most important part of giving care during a critical or terminal illness [28]. Patients and family members expect doctors to communicate in clear, honest, compassionate, and timely ways at those times [22, 28]. Doctors, however, may lack the skills or confidence to do so [27]. Even so, doctors must not avoid or delegate to others the duty to break bad news [30, 31] (unless, of course, the patient wishes otherwise). Providing that news to patients who want it (as almost all Americans do) is essential for respecting patients as persons and maximizing their control over their lives. Though sad, the news is essential to communicate [30].

The research and education efforts on breaking bad news have yielded two points of agreement about how to conduct the delivery. One involves goals. The goals for the delivery are not mutually exclusive and may include eliciting the patient's initial understanding of the situation, providing critical medical information tailored to the patient's wishes and needs, achieving a shared perspective between doctor and patient, supporting the patient emotionally, and formulating a preliminary care plan. The doctor should identify beforehand the main goals he or she wants the breaking-bad-news procedure to achieve. The other point of agreement involves the process of breaking the bad news. While several versions of that process exist [24, 31], I prefer one I call SPINES. I introduced it in Chap. 4. SPINES is an acronym indicating the six steps in disclosure: **S**et an overall approach, **P**lan for the disclosure, **I**nquire about what the patient already knows, deliver the bad News, Empathize, and Summarize the discussion and make follow-up care plans. I review the SPINES approach briefly here because so much of the spiritual reaction patient, family, and HPs have to bad medical news depends on the style and content of the disclosure. The reader should also review the more detailed discussion of SPINES in Chap. 4.

Set the Approach

Breaking bad news deserves careful prior planning to avoid significant problems later on. Experts emphasize a few particularly important arrangements. First, one doctor, typically the main, senior-most one attending the patient at the time, should take responsibility for disclosing the bad news. If that doctor does not have a long relationship with the patient, the patient's primary care doctor might also participate in the disclosure. The primary care doctor can help explain the clinical situation to the patient and ask questions on the patient's behalf. Second, as early as possible in the illness the lead doctor should ask how much information the patient wants and who else among the patient's family or friends should hear it. Third, before the breaking-bad-news procedure the lead doctor should review the patient's medical record and hear the views of other doctors on the case. Identifying controversies and the reasons behind them can minimize awkward surprises during the disclosure. Fourth, the lead doctor should consider where, when, and how the bad news should be delivered. Unfortunately, practice patterns in many hospitals often make the patient's hospital room, which is neither quiet nor private, the only practical place to hold the discussion. The disclosure should also be prompt (but not unwisely hasty): The patient usually already suspects bad news and should not be left waiting unnecessarily. The doctor should also think ahead about how the patient might react emotionally and how the doctor might respond supportively.

Prepare for the Disclosure

I recommend that the lead doctor invite to attend the disclosure another health professional such as a nurse or chaplain currently involved in the patient's care. That health professional can help support the patient, the family, and the doctor through the process. The lead doctor should, of course, dress professionally and compose himself or herself before entering the patient's room [32]. The doctor should then introduce him- or herself and the other HPs present, and briefly explain their roles in the patient's care. The doctor might say simply, for example, "I am Dr. Smith. I am the senior doctor supervising Mrs. Jones' care." That doctor should also create an atmosphere of compassion and trustworthiness. Body language can help. I favor sitting down near the patient, making eye contact, and at times either holding the patient's hand or resting a hand gently on the patient's shoulder [17].

Then the doctor might briefly review the patient's medical course in lay terms [24]. The patient may sometimes sense bad news coming, become impatient, and insist the doctor get on with the disclosure. If so, the doctor might explain the importance of taking a moment to get "on the same page," that is, to gain the same perspective as the patient [31].

Inquire About the Patient's View

A technique called asking before telling works well here. The asking part of the process need not be lengthy. The doctor might ask what the patient understands about his or her medical condition and what worries the patient most [31]. The doctor might encourage responses [33] by liberally interjecting such comments as "Tell me more" [30] or "Sounds like you're telling me …" [34].

Deliver the Bad News

Having heard the patient's views on the situation, the doctor should then warn him or her that bad news is coming. The doctor might say, "I'm afraid I have sad news for you," and then deliver it with sensitivity and compassion. One author urges the doctor to avoid "truth-dumping"—a callous unloading of alarming medical facts that overwhelms the patient. This author notes Plato's teaching about the virtue of truthfulness as telling the truth in ways people are able to hear it [33]. The doctor, therefore, should follow the patient's lead in deciding the manner, rate, and direction of delivery [31]. It should always focus on the patient's wishes and needs [24, 30], check for his or her understanding and emotions, and invite questions frequently [28]. The doctor's comments should try as possible to use the patient's own prior words and ideas in giving follow-up explanations. Because the typical patient retains only about half of the information a doctor gives at any one time, the doctor should repeat the information at different times and in different ways as necessary until the patient can grasp it [27, 29]. Adequate patient comprehension, therefore, may require several subsequent discussions.

Empathize

Once the doctor has delivered the bad news, he or she should address its emotional impact. The doctor has already observed participants for their emotional reactions during the disclosure. He or she might then expressly ask the patient and others to describe their reactions. The doctor might say, "I know this news is not what you wanted to hear. I wish it were better," [24] and then ask, "What feelings are you having now?" or "How are you reacting emotionally to this news?"

People may react to the news with disbelief, anger, fear, hopelessness, or a sense of isolation [24, 31]. If people react with denial, the doctor should not argue against it: It may be an important coping strategy for the moment. Some people may even react with angry outbursts [29], tears, or powerful silences. If so, the doctor should patiently wait out such reactions. All eventually pass. In the meantime the doctor might give a gentle, reassuring touch.

All the while, the doctor should strive to make a meaningful emotional connection with the patient. Though perhaps never having received similar bad news, the doctor should try to imagine the patient's feelings [34]. The doctor might try to validate those feelings by saying something like "I can understand how you are feeling. You are scared and confused. Anyone in this situation would feel the same way." or "If I had just heard this news about myself, I would feel angry (or weepy), too." The doctor may even feel genuine sadness for the patient [35]. If so, the doctor might mention it.

Such an emotionally intense discussion is exhausting. I recommend the lead doctor decompress afterward by discussing the emotional trauma with a friend, a colleague, or the other health professional(s) who attended the disclosure. Simply acknowledging the intense feelings and the resulting exhaustion can ease their effect and prevent mental paralysis. The doctor must be able to refocus on other patients' needs and resume professional duties [32, 36].

Summarize the Discussion and Make Follow-Up Plans

The doctor must never end the bad news discussion by saying, "There is nothing more we can do." There is *always* something more to do even if comfort oriented, noncurative care alone. Therefore, the doctor might ask the patient whether he or she wants to talk about future care at that time [24]. The patient may want to do so as an antidote to the immediate discouragement. The doctor might then suggest initial steps in care and invite the patient's reaction to them [24, 27]. If the patient needs additional views on diagnosis or treatment, the doctor might suggest the patient seek another opinion, contact other patients with the same diagnosis, or ask the opinions of important others in the patient's life [31]. And the doctor should invite additional questions by giving his or her contact information and saying, "Please ask questions." [28]. The doctor might also promise to return within a day or so to review the situation. The doctor should, of course, leave the patient with hope, but always a realistic one [28]. That hope can at least be for continuing connection with the doctor, as much comfort as possible [27], and some additional meaningful life experiences.

I believe training, practice, and reflection can improve a doctor's approach for delivering bad news. I also believe that despite the initial discouragement due to bad news most patients eventually adapt to it [27] and may even grow from it. They may gain courage, peace of mind, or a new sense of purpose they never had before [27]. Humans are resilient in even the most difficult circumstances.

The Case

How did the housestaff and I break to Mrs. R. the bad news about her lung cancer? We had suspected it since her admission but delayed any disclosure until the official biopsy reading came back proving the diagnosis. But then we faced the hard task of telling her. We unfortunately blundered our way through, illustrating in many ways how *not* to break bad news.

Despite our strong suspicions of cancer, the housestaff and I did not discuss a disclosure plan early in Mrs. R.'s hospitalization. I did not tell the housestaff that I had worked with Mrs. R. for many years in the outpatient clinic and surely knew her best among all the ward team's doctors. For that reason and for being the senior-most doctor on the team I was surely the logical person to take responsibility for disclosing the bad news to her if the biopsy showed cancer.

The biopsy reading came back late one afternoon when only the intern saw it. The next morning he gave the diagnosis to Mrs. R. during his early-morning rounds. With little time for discussion at that moment, he left her despondent and weepy as he hurried on to his next patient.

I learned the diagnosis only when I arrived several hours later to make my own rounds. I was surprised to learn the housestaff had already disclosed it to Mrs. R. I went immediately to her wardroom and found her alone and still weepy. Unprepared and inadequately focused, I stood nervously at her bedside, probably giving the impression I was rushed. (I was.) Mrs. R. was obviously struggling emotionally. Taking her hand, I acknowledged the bad news [17], said I was sorry about it, and asked clumsily whether she had any questions. She remained silent for some time as she looked away. I finally broke the silence by saying we would consult all the relevant subspecialty doctors about how best to treat her, and we could always offer pain relief. Furthermore, we would try to maximize whatever quality time she had left. I then promised to return to talk with her more once I had finished my rounds. I left dazed by the suddenness and gravity of the conversation.

Upon returning later, I again mentioned what "lousy" news Mrs. R. had received and how genuinely sorry I was about it. I asked what she had thought about since I had visited earlier. She replied that she was mentally numb and not really thinking at all. I asked whether she wanted me to call the chaplain. She said no, explaining that some of her family and her "church people" would stop by later. Because I did not know about spiritual histories at the time, I did not take one. But Mrs. R.'s African-American background should have suggested to me she likely had a significant religious life [37, 38]. I said something about how important it is to trust God in such difficult times. I also offered to inform her family however and whenever she wished. She asked that I come back late that afternoon when her family would be visiting and explain the situation to them then. I agreed. I also offered to continue serving as her primary care doctor throughout the hospitalization and after discharge if she wished. She tearfully nodded agreement.

I look back now on the conversations with Mrs. R. that day and see a few positives. As a senior general internist and the ward team doctor who knew Mrs. R. best, I surely was right to assume primary responsibility for disclosing the cancer to her. Furthermore, when she and I first discussed the news, I purposefully held her hand in nonverbal support. I also offered the realistic hope for pain relief and future quality time; I could not offer hope for cure. And I agreed to return later that same day to inform her family and to answer any questions then in an unhurried way.

Yet those conversations admittedly also had important negatives. One was the lack of advanced planning for breaking bad news even though the housestaff and I suspected a cancer diagnosis from the start. We should have discussed early on who would disclose such news, when, how, and to whom. That lack of advanced planning undoubtedly caused everyone unnecessary distress at the time of eventual disclosure. Another negative was my trying to squeeze those first conversations with Mrs. R. about the cancer into my busy morning rounds. She surely noticed my rushed, distracted demeanor at the time when she needed most my full, unhurried attention. Still another negative was my initial "I'm sorry" comments about her diagnosis. Those comments were reflexive and trite. Worse yet, they may have struck Mrs. R. as expressing culpability for the diagnosis or pity for her. Those comments may also have cut off prematurely further meaningful discussion about her emotions. (Chap. 4 addresses in detail other common reactions to "I'm sorry" statements.) Now I wish I had said something different such as "I wish I had better news" or "I wish things were different." [35]. Another negative was my failure to take a spiritual history. And still another was my potentially misleading statement about seeking consultations for treatment. Mrs. R. may have gathered that I was still hoping for a cure when I was not.

However regrettable these negatives were in Mrs. R.'s case, they yielded benefits for subsequent patients. The negatives, once I recognized them, prompted me to study up about breaking bad news and, I think, to improve my approach. Other HPs might similarly improve their own approaches through study and practice.

The Importance of Spirituality in Serious Illness

Once Mrs. R. had heard the bad news of her lung cancer diagnosis, I tried to address its spiritual implications with her. By referring to God and trust in Him, I cast my attempt in specifically religious terms. That attempt proved awkward (and probably ineffective) because I had little knowledge about how to address those matters with patients. Almost no resources existed at the time to guide me. Since then, however, research and resources about spiritual care have proliferated, highlighting the importance of such care and guiding HPs in providing it.

The importance of spiritual care may surprise some scientifically trained HPs but not patients who experience a new fatal diagnosis, a recurrence of serious disease, or a failure of once-promising treatment. The despair that accompanies such devastating events invariably prompts spiritual questions. Patients may question the purpose of their lives [39], the validity of their hopes [39, 40], or the future of their relationships with other people or even God [39].

But people are fundamentally spiritual creatures [41], and their spirituality often emerges prominently in illness crises. While spirituality encompasses more than just religion, Americans typically equate the two. And most Americans consider themselves religious. About 90 % believe in God [3, 15, 38] and pray at least weekly [15]. Most attend worship at least monthly [15], and most say religion is a major influence in their lives [7, 15].

Furthermore, many American patients admit that spiritual, usually religious, concerns arise during serious illness and have an important impact on their outlook. Nearly all the hospitalized patients in one study rated spiritual health as equally important to physical health [42], and high percentages of advanced cancer patients in another study recalled important spiritual issues that had arisen during their illnesses [43]. About half of patients in each of several other studies said spirituality would be a key coping mechanism for them during serious illness [44, 45] and would influence their treatment choices [46]. Nearly all patients in yet another study said they prayed often [47]. Though few made medical choices based *only* on prayer [48], many patients made those choices relying on both prayer and advice from their doctors [14, 46, 49]. The patients most attuned to spiritual issues in these studies were seniors [44, 49, 50], minorities [37, 38, 44, 45, 49], and women [44, 49].

Despite a current doctors' debate on the topic [51], most American patients definitely want doctors to take patient spirituality into account when they treat serious illness. Over three quarters of hospitalized patients in one study, for instance, wanted doctors to consider patients' spiritual beliefs in making important medical decisions [23, 42]. Patients believe such consideration helps doctors understand patients, give personalized medical advice, provide compassionate care, and encourage realistic hopes [52]. Of course, not all patients wish to discuss those beliefs: Two-thirds of patients do, one-sixth do not, and another sixth express no opinion [46]. But illness may unexpectedly prompt such wishes. The prevalence of patients with such wishes increases as the seriousness of illness increases until over half of the critically ill or dying patients want their spiritual beliefs to be addressed [52, 53]. Still, doctors hesitate to do so.

As a result even patients who recognize spiritual issues during serious illness and want help with them often do not get it. Between one-quarter and one-half of cancer patients in one study said they had had unmet spiritual needs during their illnesses [54]. Those needs often involved overcoming the fear of death [9, 55]; or finding spiritual resources, hope, or meaning in life [9]. Many patients would have welcomed doctors' help with such needs [13, 56, 57], but only about one-tenth of patients had actually received that help [49, 52, 58–60].

Many doctors and other HPs may lack the confidence to address spirituality with patients who wish them to do so. HPs may simply need a strong knowledge base in the area. What might that knowledge base include? The remaining sections of this chapter discuss some key aspects of it.

Spirituality, Religion, and Personal Philosophies of Living

To help dying patients with their spiritual needs, HPs must have a clear idea of how illness, suffering, and death relate to those needs. Many understandings of that relationship undoubtedly exist, but I present one here that makes sense to me. The core concept of this understanding is the meaning of illness.

A person considers himself or herself *ill* when physical or mental problems disrupt the normal pattern of life. A person is *seriously ill* when the disruption affects all aspects of a person's life. Physical appearance, functional abilities, attitudes, relationships, and vocational and social roles may all be affected, sometimes dramatically [2, 8].

Such disruptions naturally distress the ill person and his or her loved ones. They all cope with the distress by seeking meaning for it. They need assurance that the disruption is not arbitrary: That is, life makes sense despite its changes; it coheres [14]. Two authors offer different explanations for people's reflexive search for meaning in circumstances of illness. One author claims a person—as the locus of feeling, thinking, and acting in this world—is inwardly compelled to find meaning in experiences, especially the distressing ones such as illness [4]. The other author claims the human psyche needs to find meaning in those distressing experiences for the emotional stability that meaning brings [61, 62].

From either viewpoint a person's search for meaning becomes especially compelling during an illness that might prove fatal. A conscientious search for meaning at those times requires courage to face unfamiliar and potentially frightening outcomes [63]. They are surely life's greatest challenge [23]. "The proximity of death," says one author, "clears away 'the dust of our daily lives' and silences 'the noise of our (everyday existence).'" [63]. By stripping away life's usual, superficial façade and laying bare life's inner core, dying may reveal truths about life that have lain hidden before [64, 65].

Spirituality, as I use the word here, refers to a meaning, purpose, or truth that transcends the physical world and everyday experience [6, 15, 18, 41, 42, 58, 65–67]. Spirituality's insights come often as spontaneous revelation but occasionally as logical deduction. They have either of two sources. One source, religion, is what most people immediately associate with spirituality. Religion is an organized set of beliefs, practices, symbols, and texts shared by a community searching together for transcendent meaning, purpose, or truth. The community's search often focuses on belief in a deity [2, 9, 39, 41, 58, 68], whom neither the natural world nor human science fully encompasses [68]. Religion provides access to that supernatural deity. Religion also provides a foundational perspective on life in general and ethical precepts for living in community in particular [44, 50, 69].

Because most Americans have some personal connection to a religion, they tend to use it as their principal coping mechanism during life-threatening illness [44, 50]. Yet religious coping is double-edged, either positive or negative. Positive religious coping may help people find strength and emotional stability inside themselves and support from a trusted outside community of like-minded believers. Positive religious coping may also provide admirable examples of coping in life-threatening situations (such as brave crusader-warriors or martyred saints). Negative religious coping, in contrast, may create doubts about the deity's goodness or attribute the believer's suffering to internal moral defects, alienation from others, or divine punishment by the deity [18, 50]. Some call negative religious coping "the baggage of religion." Thankfully, positive religious coping occurs more often than negative religious coping.

The other source of spiritual insights is nonreligious personal philosophies of living. Everyone has such a philosophy. While religion may address some transcendent questions about living (for example, What is the human being's relationship to God? Does suffering strengthen the believer? and Is there life after death?), religion does not address them all. Personal philosophies of living, though often not systematically thought out or articulated, address transcendent questions religion may not. Examples include What significance has my life had? How should I spend my remaining time on earth? and What should I plan for my legacy after I die?

To help the dying as they search for meaning, spiritual care must be able to engage religion and personal philosophies of living alike [64]. Overtly religious people may find their spiritual insights in either religion or personal philosophies of living while nonreligious people must find those insights in personal philosophies of living alone. HPs must, therefore, be able to address whichever source of spiritual insights dying patients and their survivors use. The next section helps HPs do so by taking up questions dying patients commonly ask themselves.

The Spiritual Needs of Dying Patients

Fatal illnesses confront patients with difficult but important questions about life's value and meaning [15]. The questions arise sporadically—not all at once or in a steady stream. Their timing often depends on the underlying illness. Cancer, for example, tends to raise the questions when it is first diagnosed, treatments fail, or the terminal decline in function sets in. Congestive heart failure and the other kinds of organ-system failure, in contrast, tend to raise the questions as exacerbations increase in seriousness or frequency and gradually erode the patient's will to live. But, whenever they arise, the questions may include some of the following:

Why is this happening to me? [60, 70]. Fatally ill patients may ask this question about the disease, the experience of the illness, or the meaning of the illness. Clinicians are most comfortable answering the first two versions of this question, which have typical medical answers. The version about the disease calls for a biologic explanation of pathology; the version about the experience of the illness experience, a clinical explanation of symptoms or disabilities. A common form of the second version just after diagnosis is What can I expect for survival? or How long do I have to live? For many fatal diseases clinicians can answer that question with well-established survival data. I address such prognoses in Chap. 5.

Clinicians often do not feel comfortable with the third version of this question because it may require answers based on personal philosophy. This version can be cast as the question, "What does this illness mean for …?" Patients fill in the blank in various ways such as with "my life goals," "my self-image," "my family," or "my work." The answers vary accordingly. But one answer applies universally: Any fatal illness creates for itself a permanent presence in the patient's consciousness. Awareness of death then lurks everywhere, affecting the patient's entire being.

Several other forms of the question may have great importance for specific patients. One of these forms is "What does this illness mean about who or what is to blame for my disease?" Some patients accept blame themselves for prior unhealthy behaviors such as smoking or overeating that may have caused the fatal illness [71]. Other patients may blame God for inadvertently allowing the illness or purposefully causing it, say, as punishment for sins [18]. The answers blaming God invariably erode patients' trust in Him [15], and their dedication to religious beliefs. Still other patients may blame simple bad luck such as with harmful gene mutations or freakish vehicle accidents.

Yet another form of the question asks, What does this illness mean for my identity? The illness, of course, may change a patient's fundamental identity by changing his or her appearance, functional abilities [71], life values, or future prospects [66]. But the patient may make added changes *in response to the illness*. One author calls this process "reorienting one's life." The patient, for example, may decide to live more in the present [66], to take better care of his or her physical body [72], or to live more harmoniously with others or more spiritually than before [66].

Why do I react as I do? Fear, anger, or other deeply disturbing emotions surface when a patient faces a life-threatening illness. The patient feels vulnerable and insecure. The patient may dread the future he or she envisions: increased physical suffering, functional loss, dependence, and burdensomeness to others [73]. Life seems suddenly out of control. Clinical medicine's arcane diagnostics and therapeutics only exacerbate the patient's feelings of powerlessness.

Some patients react to such disturbing feelings by retreating into a passive role [74]. They relinquish their decision-making authority in medical matters to relatives, close friends, or doctors. Other patients react by actively taking charge for themselves. They demand decision-making authority and exercise it. Either way, primal fears fuel an impulse for quick action in dangerous circumstances [71, 74]. Scared and vulnerable, patients "hold their noses and jump." [74].

How will I cope? Patients must also try to cope emotionally with the fatal illness. Successful coping demands finding sustaining meaning in the crisis [75], gaining some control over it, and thereby bolstering one's self-esteem [70]. Key is living as fully as possible despite the crisis. Patients may do so by affirming still-sound previous priorities, discarding worn-out ones, or embracing completely new ones. As a result patients may spend more time with family and friends, remember past accomplishments or satisfactions, share life experiences, laugh about foibles, forgive past wrongs, deepen their spiritual lives, or find peace with God [67, 76]. Patients may also exert control over the crisis by choosing their attitudes toward it (even when treatments offer no cure). Choosing healthful attitudes bolsters patients' self-esteem by preventing feelings of victimization and helplessness [70]. Some authors describe this process of developing good coping strategies as gaining emotional health amid serious physical illness [75].

How do I understand this illness in the context of my whole life? [71] Finding worth in one's whole life is critical for meeting two challenges—enduring a specific ordeal and gaining satisfaction from life. Both challenges occur together during a fatal illness. They force patients to find meaning in the experience. (*Meaning* here refers to the awareness that an experience has specific, identifiable implications for a person and his or her future. Those implications may affect one's image (internal or external), life goals, and relationships with others [62, 75].) Finding meaning uses at least four guideposts [75]: comprehensiveness, coherence, continuity, and completion. Meaning must account for all relevant aspects of an experience (comprehensiveness) [77, 78], illuminate connections among those aspects and integrate them into a logical whole (coherence) [9, 75, 77], make connections between past experiences and the present one (continuity) [75], and (usually) define an end to the experience (completion) [77]. Notably, completion does not require fulfilling all expectations for the experience: some may ever remain unfulfilled [71]. Furthermore, the meanings of a particular experience can vary. Any one person may see different meanings in the same experience at different times, and different people may see different meanings in the same experience at the same time. The various meanings need not be logically compatible. Disagreements may, therefore, arise.

I do not know for sure the guideposts for meaning Mrs. R. used for her lung cancer, but several possibilities occur to me. She may have seen the cancer's meaning as simply the result of her long-term smoking habit. She had smoked intensely for years, and cigarette smoke contains many cancer-causing substances (comprehensiveness). Smoking had exposed her lungs to enough cancer-causing substances eventually to cause a cancer. That cancer attacked her body's defenses, finally overwhelmed them, and spread (coherence). This meaning would draw from what Mrs. R. knew about lung cancer from her nursing education and had seen with many lung cancer patients during her career (continuity). In addition, firmly rooted in pathophysiology, this meaning defined physical death as the end of her illness (completion).

Alternatively, Mrs. R. may have attributed her cancer to whatever "sins" she thought she had committed in the past. They might have included her smoking habit and two divorces (comprehensiveness). She may have decided God was punishing her for those "sins" (coherence). The Bible describes God's doing just that to the Hebrews' ancient Egyptian captors, the Babylonian King-captor Belshazzar, and others. Some in Mrs. R.'s church might have held similar views (continuity). Such a meaning suggested God's punishment might end when Mrs. R. repented and received God's forgiveness (completion).

Yet again, Mrs. R. might have understood the meaning of her cancer as a test of faith. God might have been testing her to determine whether her faith was strong enough to warrant admission to heaven (comprehensiveness). Mrs. R. might have believed God would reward her if she steadfastly kept her faith through the suffering (coherence). God did so with Noah, Abraham, Moses, Esther, David, and many other Biblical characters (continuity). This meaning implied Mrs. R.'s test of faith would end when God accepted her into heaven (completion).

Each of these meanings could have shaped quite differently how Mrs. R. thought about her cancer and lived out her life with it.

How will dying affect my relationships? [60] All people live their lives in relationships—with themselves, other people, and usually God or other deities. Physical death affects all these relationships, often causing dying patients to worry about forgiveness and disconnection in each one.

Worries about forgiveness arise from regrets due to past disappointments or injuries patients believe they have caused. The disappointments or injuries may have occurred in any of the relationships [58]. A common source of disappointments, injuries, and regrets is unfaithfulness, in a broad sense, to oneself, others, or God. Memories of such disappointments or injuries often prompt wishes for forgiveness.

Disconnection in dying may also affect any of the relationships. Disconnection from oneself may come for the dying person by alienation from one's old healthy self or by denial of one's present reality. Disconnection from others may come in various ways such as by emotional distancing between parties, breakdown of the relationship during life, or the extinction of the relationship at the dying person's death. Distancing can occur when people cannot communicate comfortably with the dying person [71]. People avoid substantive communication by denying the dying person's reality ("Everything will be 'all right.'") or avoiding that person altogether [74]. Another disconnection from others may come through a breakdown of the relationship perhaps because either party to it cannot express or receive genuine love [66] when communication of love is critical to sustaining the relationship. And yet another disconnection from others may come, of course, if at death the dead person leaves this earthly home [79] and loses all means to communicate with people back here. (Chap. 12 addresses the opposite belief of continuing communication with the dead after death.) Disconnection from God probably can also occur in several ways. Some dying patients may feel abandoned by Him; others may lose faith in Him [58]. As a result these patients may decide God is not as faithful, beneficent, or powerful as they thought [43].

Worries about not being forgiven or suffering disconnection along with fears of the unknown may motivate people in powerful ways to avoid death for as long as possible. Such motivations may have influenced some medical ethicists when they faced their own cancers or other life-threatening illnesses. These ethicists had previously taught reason for making treatment decisions in such situations but found themselves using emotion in their own. Despite their prior teachings based on reason, the ethicists accepted many highly burdensome treatments for even low probabilities of life extension: These ethicists, like so many other patients, consistently chose quantity over quality of life [74].

What can I still hope for in life? When fatal illness strikes, patients naturally hope immediately for cure. The patients' hope hinges on faith in medical treatments or God's eventual rescue. Yet, as treatments fail and God does not rescue, that hope dims "haltingly," as one author notes [72]. Disappointment sets in [71]. Patients, family members, and HPs begin to resign themselves to the approaching death but always retain some hope, however diminishing, for miraculous cure. Some people, however, may eventually assume more modest but more realistic hopes for patients that include physical comfort, familiar surroundings to die in, and meaningful legacies to leave survivors [60].

What will happen to me after I die? [60] Patients' answers to this question can profoundly affect their attitudes about dying. Probably everyone has some fear of death because it ends the physical life we have always known [77]. Yet many people find comfort in the belief that spiritual existence continues after physical death. That belief often promises eternal life of one's spirit in God's paradise. But disquieting possibilities exist, too, including judgment and punishment by God, reincarnation into a demeaning future life on earth, and total annihilation of the self [55, 73]. I do not discuss these perplexing ideas further here; I have discussed some already in Chap. 12.

Obstacles to Giving Spiritual Care

Most patients and family members think clinicians should pay increased attention to spiritual issues in illness. Many palliative care or hospice experts agree. Yet clinicians in general (and doctors in particular) are reluctant to do so [7, 8, 39]. Why this reluctance? I believe two sources exist for it.

One source is the tenets of core modern medical culture. One tenet relevant here is medicine's firm distinction between body and spirit. That distinction traces back at least as far as Plato and other ancient Greeks who espoused the separability of body and spirit. Those ancients believed the body traps the spirit inside during waking hours but allows it to escape during sleep and at death [80]. Descartes and other philosophers of the early scientific age espoused a related idea, body–spirit distinctiveness, at the beginning of the seventeenth century. They believed body and spirit are distinct but interdependent kinds of "matter." [81]. Medicine, which was just realizing science's potential power in the physical world at the time, embraced the body–spirit distinction then and still embraces it today. As a result modern scientifically based medical care focuses its attention on the body and away from the spirit. That care often appears to distill, or "reduce," the patient to a body alone [11]. Physical distresses are considered "real" [9]; spiritual ones are not [14]. Doctors, therefore, typically perceive their care duties as limited to the physical body. As one doctor observes, "Contemporary medicine … has reduced human life (to bodily) function; doctors simply replace the broken or dead parts of the human machine … (Human life is) mere physiology." [11] And another laments similarly, We have allowed the physical body "artificially (to) circumscribe our tasks of caring." [8] Thus, modern medicine has generally accepted responsibility for the body but relinquished responsibility for the spirit.

Another relevant tenet is modern medicine's insistence on sound scientific data to justify diagnostic and therapeutic interventions. Unfortunately, the scientific studies about spiritual interventions in clinical situations have consistently suffered serious weaknesses: imperfect quantitative markers for spirituality (such as church attendance or self-assessed spirituality) [82, 83]; uncontrolled comparisons [57, 83]; weak or inconsistent associations between spiritual measures and health outcomes [10]; and the inability to produce clear cause-and-effect conclusions [13, 57, 83, 84]. Furthermore, publication bias may exist. Studies that are published are more likely to show associations between spirituality and health outcomes [13, 85] than are studies that are not published [83]. Negative studies just do not make for exciting journal articles. So no one knows how many unpublished, negative studies about spirituality and health outcomes exist.

Even the published studies show few robust associations between spirituality and either the process of care or physical health outcomes. Two studies about the process of cancer care, for example, report conflicting results. One study reports religiousness associated with increased use of life support [45] while the other study reports religiousness associated with increased use of hospice [58].

The published studies about spirituality and physical health outcomes divide into studies about general healthfulness outcomes and studies about particular disease outcomes. Problems exist with both. Some, but not all, studies about general healthfulness outcomes show a positive association with spirituality [49, 83]. For example, some studies show religiousness associated with increased quality of life [86] or decreased mortality [52, 87]. However, these associations may not be cause-and-effect. And, even if they are, the cause-and-effect direction may go either way. That is, religiousness may cause better or longer life or vice versa.

The published studies of particular disease outcomes also give confusing results. A 2002 review covering five randomized and four nonrandomized studies suggests that intercessory prayer may improve physical health outcomes overall for serious heart diseases, high blood pressure, immune deficiency states, and depression but not for alcoholism [6]. Two of the nine studies reviewed suggest a survival benefit. An epidemiologist who has reviewed this literature extensively sees a general positive association between spirituality and physical health outcomes [10]. I, however, remain unconvinced due to the small number of studies overall, their many methodological weaknesses, their weak or inconsistent findings, and the lack of plausible biological explanations for a cause-and-effect relationship between spirituality and physical health outcomes.

The current state of research, therefore, leaves considerable uncertainty about associations between spirituality and physical health outcomes.

In contrast, the current state of research about spirituality and emotional health outcomes is more consistent and more convincing [47]. Even nonpsychiatric doctors now believe that existing studies show clear emotional benefits from spirituality [64]. And studies with that implication continue to accumulate [64]. For example, two recent studies strongly suggest positive effects of spirituality with advanced cancer patients: One study suggests an effect increasing quality of life [86, 88] while the other study suggests an effect decreasing end-of-life despair [64]. The second study even suggests spirituality affects end-of-life despair more powerfully than does depression, implying that spirituality works independently from depression [44]. Curiously, though, the "meaning component" of spirituality (comprised of inner harmony and peace based on religious *or* philosophical ideas) exerts a larger, more consistent effect decreasing end-of-life despair than does the "faith component" (comprised of comfort or strength based strictly on religious beliefs) [64].

As I mentioned before, the associations of spirituality with health outcomes vary by a few demographics but not by many other common epidemiologic factors. Specifically, strong positive associations occur for senior age [37], African-American ethnicity [44] and female gender [87]; but none occur by patients' symptoms [87], functionality [64], or social support [64, 87]. Several authors try to explain the positive demographic associations by saying that, compared to high social-status Americans, relatively low social-status Americans (i.e., the elderly, African Americans, and women) tend to rely heavily on personal spirituality for internal coping and on churches for external support during serious illnesses [39, 87].

Unlike spirituality's effects on physical health outcomes, its effects on emotional health outcomes have obvious, plausible mechanisms. Spirituality may encourage meditation, conscious relaxation, or other personal habits that promote healthful emotions throughout life [39]. Spirituality may also provide added emotional comfort during terminal illness by helping to make sense of it or of treatment decisions about it [89]. And spirituality may nurture hopefulness [41], a positive will to live, good coping strategies [90], and a sense of personal dignity even in dying [9].

However strong the tenets of the body-spirit distinction and of scientific proof are to medicine, I do not think they explain most of clinicians' reluctance to provide spiritual care. Many clinicians already accept the importance of such care despite those tenets [91]. Eighty-one percent of American doctors in a nationwide study believed spirituality positively affects patient outcomes [9], and 75–90 % of nurses in another study agreed [92]. Furthermore, nearly all doctors in that nationwide study said doctors should know something of their patients' spiritual beliefs [9, 93, 94], and most said doctors should "encourage" patients' spiritual practices that promote health [21]. Yet, while gravely ill patients in other studies said they have spiritual needs [54, 67] and want to discuss such needs with clinicians [9, 46], those patients also said few doctors initiate spiritual discussions [7, 15, 18, 59]. Nurses do little better [41, 92]. I, therefore, conclude that the gap between clinicians' support for spiritual care in theory and their reluctance to provide it in practice can be attributed only to clinicians' personal reservations about doing so.

Six specific reservations may work to discourage such care. First, clinicians may fear overstepping their professional boundaries. Clinicians want to avoid extending their care beyond what patients want, or trespassing on the "turf" of spiritual care specialists such as chaplains or pastoral care professionals [15, 21, 23, 52, 94]. Second, clinicians may feel inadequately prepared to address patients' spiritual needs [15, 18, 21, 41, 53, 71, 82, 94–97]. Clinicians' professional training makes them comfortable with medical science and its relatively neutral, data-based perspective [39, 89], but not with spirituality and its belief-based perspectives [21, 60, 97]. Discomfort with spirituality may actually explain why doctors as a group are less religious than the general American population [18, 98, 99]. Third, clinicians may have difficulty identifying the particular patients who want to discuss spiritual matters [15, 52, 53, 82, 94]. After all, broad demographic associations (such as ethnic group and gender) give only general tendencies, which may not hold for individual patients.

Fourth, clinicians may fear that in discussing spiritual matters they will inadvertently force their beliefs on patients [2, 9, 15, 82]. Clinicians surely do not want to take advantage of vulnerable or unsuspecting patients [18, 52, 53, 94] or to violate patients' privacy [39, 52, 89]. Fifth, clinicians—especially doctors—may feel the physical part of end-of-life care already overburdens them [39]. They may worry that spiritual discussions will take too much time and effort away from the already-overwhelming tasks of physical caregiving [15, 18, 41, 52, 53, 60, 71, 82, 89, 94, 97]. And sixth, clinicians may worry that discussions about spirituality may actually harm patients [52] by creating ethical conflicts [52, 93], religiously based fears, or unwise responses to illness. Some patients, for instance, may worry that religion requires them to use all life support regardless of their wishes or its burdensomeness. Other patients may worry that God has caused their illnesses due to their moral lapses or inadequate devotion [84] or that God no longer loves them or has abandoned them [100]. And still other patients may discontinue sound medical regimens out of the conviction that God will always provide a cure [89].

I think that, however common these reservations among clinicians, convincing counterarguments emerge from formal research and informal clinical experience (Table 14.1). Regarding worry about overstepping professional boundaries, many patients, family members, and HPs rightly believe that spiritual care is a key part of total end-of-life care [101, 102], and that dying patients deserve spiritual care if they want it. Furthermore, many specialists in spiritual care appreciate clinicians' interest in it and willingness to offer it to patients. Accordingly, clinicians should be able to screen patients for spiritual needs and should follow up with prompt spiritual care referrals as necessary. Unfortunately, most clinicians think to request spiritual care referrals only quite late when patients are imminently dying. (I am certainly guilty here.) Spiritual care specialists then have the difficult task of comforting distressed, dying patients and their survivors when the specialists do not already know these people. Early spiritual care referrals for seriously ill patients would be helpful.

Although many clinicians feel ill-prepared to address spiritual matters, they can consciously acquire relevant skills through training and experience. I admit that much of what I now think I know about spiritual care came by "learning through

Table 14.1 Clinicians' reservations about addressing spiritual matters with dying patients and some replies to those reservations

Clinicians' reservations	Replies to reservations
Worry about overstepping professional boundaries	• Many people believe that spiritual care is a key part of total end-of-life clinical care and that dying patients deserve spiritual care if they want it. • Spiritual care consultants appreciate involvement by clinicians (including taking initial spiritual histories and giving early notice about patients who may want spiritual care).
Feeling ill-prepared to give spiritual care	• Formal training exists. • Learning can occur through experience and reflection on it (that is, "learning through doing").
Having difficulty distinguishing patients who do and patients who do not want spiritual matters addressed	• Most seriously ill patients want discussions about spiritual matters. • Even many patients who do not want such discussions encourage doctors to ask *all* patients about wishes to discuss spiritual matters.
Worry about forcing one's own beliefs on vulnerable patients	• Patients want their doctors to be comfortable with discussing end-of-life spiritual matters. • After inquiring about spiritual matters, doctors should follow patients' lead about if, when, and how to go further.
Feeling overburdened already with the physical care of terminal illness	• Addressing spiritual matters may take little time or extra effort. • Doing so parallels the overall satisfaction of patients and family members with end-of-life care. • Doing so may provide clinicians important professional satisfactions and life lessons.
Worry about causing harm to patients	• The harm from addressing spiritual matters at the end of life appears to be rare.

doing." Reflection on Mrs. R.'s case and similar cases certainly gave me ideas for improvement. I then practiced those ideas until they became habits. And regarding difficulties with distinguishing patients who do want to discuss spiritual matters and patients who do not, HPs in the USA have great odds in their favor for guessing that a seriously ill patient wants to talk about those matters. Up to 90 % of Americans consider spirituality important in their lives [15, 100], and up to 79 % find it helpful in coping with illness [44, 45]. Similarly high percentages say they want their doctors to address spiritual matters with them specifically during serious illness [46, 52]. Thus, brief, respectful inquiries by HPs about spiritual matters should offend few patients. In fact, many patients who themselves do not want spiritual discussions encourage blanket inquiries by HPs to identify the patients who *do* want such discussions. The patients who do not want them can simply say no.

Of course, if a patient wants to discuss spiritual matters, a clinician must proceed with sensitivity. I urge the clinician to follow the patient's lead about if, when, and how to go further [60]. Many clinicians worry about forcing their spiritual beliefs on vulnerable patients. Yet many dying patients want their clinicians to be comfortable in spiritual discussions if they occur. To be so, clinicians must understand their own beliefs beforehand and be able to respect contrasting beliefs from patients. Many clinicians also feel already overburdened with the physical care of terminal illness. That care is certainly intense, time consuming, and exhausting. Nonetheless, I find spiritual care takes surprisingly little time or extra effort. Giving that care can sometimes even occur during routine physical care tasks (such as taking vital signs or checking intravenous infusions). Spiritual care also has clear benefits. People generally appreciate clinicians' efforts at it, and that care parallels the satisfaction of patients and family with overall end-of-life care. Clinicians, too, may gain unexpected professional satisfactions from spiritual care and learn important life lessons from it. Finally, regarding worry about harms to patients from discussing spiritual matters, those harms are probably rare in reality. I have never detected ethical conflicts or religiously based fears arising due to spiritual care when those conflicts or fears did not already exist for patients. (I grant, of course, that patients may not always have divulged those conflicts or fears to me.) I *have,* however, witnessed some unwise treatment decisions by patients or proxies based on exaggerated expectations of divine rescue. When discussing spiritual matters with patients, we clinicians may need to ask explicitly about such conflicts, fears, and exaggerated expectations.

Giving Spiritual Care at the End of Life

While these arguments may not convince everyone about the importance of end-of-life spiritual care, three arguments certainly convince me. First, I believe every human being consists of commingled physical and spiritual parts. As the physical part deteriorates near the end of life, the spiritual part gains in relative importance. Care should change accordingly. Second, dying patients and their survivors do perceive spiritual needs and want HPs to help address them. Providing care to address such needs surely increases the satisfaction of all with end-of-life care. And, third, the available research suggests spiritual care *does* provide benefits, especially emotional ones, for patients, survivors, and even HPs. While those benefits may sometimes be difficult to measure, I believe just the likelihood they exist justifies efforts to pursue them. I, therefore, strongly favor consciously providing spiritual care for those dying patients and their survivors who want it.

Unfortunately, the medical literature usually helps only a little, and sometimes actually complicates, efforts to give such care. The literature, for example, describes many spiritual care interventions with a baffling vagueness for those trying to implement them. The interventions include "being present"; listening to the patient; "sharing oneself"; or laughing, empathizing, and praying with the patient [9, 18, 56, 60, 85]. Many published measures of successful implementation are also vague or difficult to compare across studies. Those measures include increases in "personalized" care [86], quality of life at the end of life [7], or satisfaction with care [97, 103]; increases in self-perceived "personal growth," self-esteem, self-acceptance, hopefulness, or meaningfulness; increases in buffering against stressors or mastery over the circumstances of dying; and increases in making peace with God or others [16, 48, 56, 69, 100]. The literature, therefore, may leave HPs wondering exactly what is spiritual care, how to give it, and how to evaluate its efficacy.

Nonetheless, each HP will surely encounter dying patients with spiritual needs [22]. Whether doctor, nurse, social worker, or other, a HP must be willing to attempt spiritual care when necessary. Perhaps the most important elements of such an attempt are a belief in the importance of transcendent matters, a focus on the individual patient's needs, and a willingness to ask sensitively about spiritual needs and then to listen attentively to the patient's responses. Within those broad guidelines the HP can plan an approach that fits his or her personality.

How might HPs, then, plan to give end-of-life spiritual care? Obviously no detailed formula exists, but I can offer some general suggestions here. First, because attitude is important, all HPs who attend dying patients should take the attitude that spiritual care is central to good end-of-life care. I, in fact, believe HPs should consider as a basic professional duty seeing that all seriously ill patients, both the imminently dying and others, get spiritual care if they want it [18, 76]. The principal clinicians—both doctors and nurses—attending such patients should take responsibility for ensuring that care [18]. Second, those clinicians should be prepared to give some of that care as needed. Chaplains, other clergy, social workers, and pastoral care counselors will usually, of course, give it, but some dying patients may prefer to receive spiritual care from their doctors or nurses or may need it when no spiritual care specialists are available. Clinicians must then be willing to give it. Third, all HPs should have specific training in giving spiritual care. Spiritual care skills do exist but do not come naturally: they must be learned and practiced.

Fourth, all HPs should be able to take at least a brief spiritual history and to deduce an initial spiritual assessment from it. That assessment should begin to identify patients' spiritual needs and resources. As I mentioned in Chap. 11, I favor the four-question FICA mnemonic for taking a spiritual history. FICA stands for Do **F**aith-based beliefs help you cope with stress? Is faith **I**mportant to you? Are you a member of a particular spiritual **C**ommunity? and How can we **A**ccommodate your spiritual beliefs in your care? [16, 104] I, however, ask the FICA questions in a different order, I-C-F-A, which is more comfortable for me. One expert simplifies the spiritual history-taking even further by boiling it down to one question, What role does spirituality play in your life now? [13]. Some HPs may prefer to use just that one question.

Fifth, HPs should stay alert for indications that patients or survivors have spiritual needs or want to discuss spiritual matters. Because overlooking such matters is easy to do, HPs must keep “a high index of suspicion” for them. Therefore, when taking a patient’s initial medical history, when suspecting a patient’s spiritual needs at other times, or when realizing a patient has become imminently terminal, a HP (preferably the patient’s principal doctor) might ask the patient explicitly, Do you want to talk about spiritual matters? [93, 94]. Many chronically ill patients in one study said that, if a doctor asked this question, their trust in him or her would increase [46]).

Sixth, HPs should respect sensible limits from their time and training for giving spiritual care. Clinicians, in particular, should not routinely serve as patients’ primary spiritual counselors [13]. Rather, clinicians should know the available spiritual resources [59] such as hospital chaplains or pastoral counselors and should call readily on them to help patients with spiritual needs [16, 58, 60].

And seventh, even if not delivering patients’ spiritual care directly, the principal doctors attending patients should monitor that care as part of total patient care coordination [18]. Those doctors should document patients’ spiritual histories in the medical records, request needed consultations directly from chaplains or other spiritual care specialists, mesh spiritual care consciously with physical care, and track the progress of both. All HPs, of course, must respect patient confidentiality as much in spiritual matters as in physical matters. HPs should disclose the details of spiritual care to others only when those others must know the details of it for care purposes or when patients agree to the disclosure.

The Case

If I had known them at the time, some of these suggestions would have helped me give Mrs. R. the spiritual care she deserved. But even as her case was, I should have realized early on how much she relied on religion to cope with her cancer. There were two clues, both of which I did not immediately recognize. First, Mrs. R. had two of the three demographic characteristics most often associated with religious coping: African-American ethnicity and female gender [37, 38]. And, second, she told me the day she received her cancer diagnosis that her “church people” would visit her in the hospital. I did mention briefly at that time trusting in God but did not broach religion or religious coping again over many weeks until just before she died. During that time I completely overlooked religion’s significance for her. Perhaps I subconsciously prevented myself from recognizing its significance due to my own discomfort with addressing patients’ spirituality. I certainly felt inadequately trained to do so at the time and may have foreseen awkward emotional discussions that I secretly wanted to avoid. I may have also worried about my own feelings or professional comportment during such discussions.

Looking back now, I can identify several times when I could have addressed Mrs. R.’s spirituality in general or her religious coping in particular. One of those times occurred at hospital admission. After she had blacked out, trashed the bedroom of her new home, and remembered nothing of the incident, Mrs. R. had to be frightened. Her nursing education surely fueled the fright by suggesting serious medical causes for this history. Her life must have seemed out of control. She probably needed explicit reassurances that God was still in control, that He intended good for her, and that He would help our ward team investigate the cause for her strange behavior and determine how best to treat her. Another time occurred several days later at the time when, having just learned her cancer diagnosis, Mrs. R. said through tears that her "church people" would be visiting. That comment gave me an opening not just to mention facilely trust in God but to broach seriously the topic of religious coping. Although the FICA method had not yet been published, I could have asked a few intuitive questions such as Is religion important to you? Do you attend a particular church? Which one and for how long? and Does your faith help you in difficult times? How? Mrs. R.'s answers could have told me something about the substance of her faith. For example, they may have revealed whether she participated in church mostly for its social benefits (extrinsic religiousness) or for its guidance in actually living out her faith (intrinsic religiousness) [2]. Furthermore, when the "church people" visited, I could have purposefully stopped at the bedside to meet them and to gauge their supportiveness for her.

Still other instances occurred during Mrs. R.'s office visits to me after hospital discharge. She was taking radiation and chemotherapy treatments by then. She had lost considerable weight and looked ever more frail. She also had begun wearing a headscarf to cover the bare patches of scalp where her hair had fallen out. Because Mrs. R. had always dressed stylishly, I might have sensed that body image was important to her, and it was under severe attack. I ordinarily consider my own religious beliefs a private matter and, thus, rarely bring them up in patient care [51]. But on this occasion I might have taken motivation from the scripture, "(B)e not anxious about your body, … Consider the lilies of the field, … even Solomon in all his glory was not arrayed like one of these … Will not (your Father in heaven) much more clothe you than these?" (Matthew 6:25, 29–30). I could have, therefore, discussed with Mrs. R. how she felt about her hair loss and the other bodily changes happening to her. Still another time occurred when she said overwhelming fatigue kept her from attending church anymore. I could have asked how much support she was getting from her faith and the "church people." If she needed more, I could have called her pastor (with her permission, of course). He might have arranged increased visits to Mrs. R. and even helped her accept hospice. As her case played out, I asked her about hospice at nearly every office visit while she pursued ever less promising chemotherapies. She always refused.

When Mrs. R. missed two appointments in a row without explanation, I became concerned and telephoned her house. Her son answered, and I introduced myself as Mrs. R.'s principal doctor. He explained the cancer had spread despite all the chemotherapies. The oncologist, the cancer specialist, had finally persuaded Mrs. R. to accept hospice care. A local hospice was caring for her at home. I asked him to tell her I had called. Several days later I called back to check how Mrs. R. was doing. Her son answered again and explained that she was declining steadily. I said how sorry I was to hear that and asked whether I could stop by to see her at home. He agreed.

I came the next day. Mrs. R.'s house was located in a new housing development. Mrs. R. had bought the house and moved in just before her cancer diagnosis. She had intended to live there into retirement. I knocked on the front door and introduced myself when Mrs. R.'s son and sister answered. They ushered me into the living room where Mrs. R. lay sleeping in a hospital bed. The atmosphere was quiet and somber. I sat down next to the bed and held her hand. She opened her eyes, gave a flicker of recognition, and then closed them again. I finally said to her in front of the others, "Your family and the hospice people are doing everything possible to take care of you. I am sure they are doing a good job. But mostly you are in God's hands. We all need to trust Him now." Mrs. R. nodded weakly. As I left the house, I asked the family's permission to visit in another two days or so. They agreed. Unfortunately, Mrs. R. died before I could return.

Mrs. R.'s death ended her physical existence on this earth, but questions about her spiritual care lingered for me. I remembered Mrs. R.'s suffering mightily with her cancer and gradually being physically and psychologically worn down by it. I also remembered how frail and pitiful she looked at the end. I wondered whether more conscientious, more skilled spiritual care than I had given might have somehow helped her.

The local newspaper announced Mrs. R.'s funeral scheduled for several days later. I decided to attend to gain some closure for myself. That funeral focused my attention on the spiritual dimension of dying as never before. The funeral took place at a modest evangelical Baptist church in the African-American area of town. The pews were packed. I was the only Euro-American and the only person from the hospital or clinic there. The service celebrated Mrs. R.'s life and proclaimed her triumphal entry into heaven. The choir sang heartfelt gospel music, and the preacher fervently eulogized their "sister." At one point he announced that Mrs. R. had already entered God's eternal presence. The congregation punctuated the preacher's statements with enthusiastic amens. The tone of the funeral surprised me: It was joyful despite the difficult death Mrs. R. had experienced. The congregation was celebrating her return home to God. At that moment I realized how important Mrs. R.'s religious spirituality and these "church people" had been to her throughout her illness.

Since then, Mrs. R.'s case has prompted me to try to improve my spiritual care for dying patients. I now ask specifically about their spirituality. FICA gives me a sensible guide for doing so. I sit, not stand, at the bedsides of dying patients and allow plenty of time to talk. I try to avoid giving the impression of being rushed. And I try to use therapeutic touch liberally (such as by holding their hands). Furthermore, I involve chaplains and other pastoral care professionals in the spiritual care process whenever possible and try to learn from their approaches. I also make telephone calls or write sympathy notes to commiserate with survivors after patients have died.

My commitment to improving my spiritual care stands. And the effort continues. I feel that giving such care to the dying is a professional obligation, and I must not ignore it.

Summary Points

1. Spirituality involves a search for meaning that transcends physical life. That search may have both religious and philosophical components.
2. Total end-of-life care includes spiritual care. Every health professional who attends dying patients should be willing and prepared to give it.
3. Breaking bad news, typically the first step in giving spiritual care, requires the disposition to tell the truth so people can hear it. The process of breaking bad news should include prior planning, listening carefully to patients, and delivering the bad news with sensitivity.

To Learn More …

1. Cassel EJ. The nature of suffering and the goals of medicine. N Engl J Med. 1982;306(11):639–45.
2. Puchalski C, Ferrell B, Virani R, et al. Improving the quality of spiritual care as a dimension of palliative care: the report of the consensus conference. J Palliat Med. 2009;12(10):885–904.
3. Quill TE, Townsend P. Bad news: delivery, dialogue, and dilemmas. Arch Intern Med. 1991;151:463–8.
4. Sulmasy DP. A biopsychosocial-spiritual model for the care of patients at the end of life. Gerontologist. 2002;42(Special Issue III):24–33.
5. Sulmasy DP. Spiritual issues in the care of dying patients: "… It's okay between me and God." JAMA. 2006;296(13):1385–92.

References

1. Doctors' Diaries. Public Broadcasting Service NOVA. http://video.pbs.org/video/1114402491/. Accessed 12 Oct 2015.
2. Sulmasy DP. A biopsychosocial-spiritual model for the care of patients at the end of life. Gerontologist. 2002;42(special issue III):24–33.
3. Marwick C. Should physicians prescribe prayer for health? Spiritual aspects of well-being considered. JAMA. 1995;273(20):1561–2.
4. Kontos PC. Resisting institutionalization: constructing old age and negotiating home. J Aging Stud. 1998;12(2):167–84.
5. Jones AW. A survey of general practitioners' attitudes to the involvement of clergy in patient care. Br J Gen Pract. 1990;40(nn):280–3.
6. Townsend M, Kladder V, Ayele H, Mulligan T. Systematic review of clinical trials examining the effects of religion on health. South Med J. 2002;95(12):1429–34.
7. Balboni TA, Vanderwerker LC, Block SD, et al. Religiousness and spiritual support among advanced cancer patients and associations with end-of-life preferences and quality of life. J Clin Oncol. 2007;25(5):555–60.
8. Cassel EJ. The nature of suffering and the goals of medicine. N Engl J Med. 1982;306(11):639–45.

9. Chochinov HM, Cann BJ. Interventions to enhance the spiritual aspects of dying. J Palliat Med. 2005;8(suppl 1):S103–15.
10. Levin JS. Religion and health: Is there an association, is it valid, and is it causal? Soc Sci Med. 1994;38(11):1475–82.
11. Dugdale LS. Medicine's metaphysics. Hast Cen Rep. 2013;43(2):7–8.
12. Levin JS, Larson DB, Puchalski CM. Religion and spirituality in medicine: research and education. JAMA. 1997;278(9):792–3.
13. Sulmasy DP. Addressing the religious and spiritual needs of dying patients. West J Med. 2001;175(4):251–4.
14. Daaleman TP. Religion, spirituality, and the practice of medicine. J Am Board Fam Pract. 2004;17(5):370–6.
15. Hebert RS, Jenckes MW, Ford DE, O'Connor DR, Cooper LA. Patient perspectives on spirituality and the patient-physician relationship. J Gen Intern Med. 2001;16(10):685–92.
16. Puchalski C, Ferrell B, Virani R, et al. Improving the quality of spiritual care as a dimension of palliative care: the report of the consensus conference. J Palliat Med. 2009;12(10):885–904.
17. Wenrich MD, Curtis JR, Ambrozy DA, et al. Dying patients' need for emotional support and personalized care from physicians: perspectives of patients with terminal illness, families, and health care providers. J Pain Symptom Manage. 2003;25(3):236–46.
18. Sulmasy DP. Spiritual issues in the care of dying patients: "… It's okay between me and God." JAMA. 2006;296(13):1385–92.
19. Mohrmann ME. The practice of the ministry of medicine. Update. 1998;14(3):1–7.
20. Meador KG. Spiritual care at the end of life: what is it and who does it? North Carolina Med J. 2004;65(4):226–8.
21. Curlin FA, Chin MH, Sellergren SA, Roach CJ, Lantos JD. The association of physicians' religious characteristics with their attitudes and self-reported behaviors regarding religion and spirituality in the clinical encounter. Med Care. 2006;44(5):446–53.
22. Hanson LC, Dobbs D, Usher BM, et al. Providers and types of spiritual care during serious illness. J Palliat Med. 2008;11(6):907–14.
23. Puchalski CM, Lunsford B, Harris MH, Miller T. Interdisciplinary spiritual care for seriously ill and dying patients: a collaborative model. Cancer J. 2006;12(5):398–416.
24. Baile WF, Buckman R, Lenzi R, et al. SPIKES—a six-step protocol for delivering bad news: application to the patient with cancer. Oncologist. 2000;5(4):302–11.
25. Oken D. What to tell cancer patients: a study of medical attitudes. JAMA. 1961;175(13):86–94.
26. Novack DH, Plumer R, Smith RL, et al. Changes in physicians' attitudes toward telling the cancer patient. JAMA. 1979;241(9):897–900.
27. Girgis A, Sanson-Fisher RW. Breaking bad news: consensus guidelines for medical practitioners. J Clin Oncol. 1995;13(9):2449–56.
28. Wenrich MD, Curtis JR, Shannon SE, et al. Communicating with dying patients within the spectrum of medical care from terminal diagnosis to death. Arch Intern Med. 2001;161(6):868–74.
29. Fallowfield L. Giving sad and bad news. Lancet. 1993;341(20 February):476–8.
30. Clayton JM, Butow PN, Tattersall MHN. When and how to initiate discussion about prognosis and end-of-life issues with terminally ill patients. J Pain Symptom Manage. 2005;30(2):132–44.
31. Quill TE, Townsend P. Bad news: delivery, dialogue, and dilemmas. Arch Intern Med. 1991;151(3):463–8.
32. Fox E. The Yale curriculum in ethical and humanistic medicine, 1995.
33. Drane JF. Becoming a good doctor: the place of virtue and character in medical ethics. Kansas City: Sheed & Ward; 1995. p. 53.
34. Coulehan JL, Platt FW, Egener B, et al. "Let me see if I have this right": words that help build empathy. Ann Intern Med. 2001;135(3):221–7.

35. Quill TE, Arnold RM, Platt F. "I wish things were different": Expressing wishes in response to loss, futility, and unrealistic hopes. Ann Intern Med. 2001;135(7):551–5.
36. Weaver R. Terminal (mis)diagnosis and the physician-patient relationship in LM Montgomery's The Blue Castle. J Gen Intern Med. 2010;25(10):1129–31.
37. Johnson KS, Elbert-Avila KI, Tulsky JA. The influence of spiritual beliefs and practices on the treatment preferences of African Americans: a review of the literature. J Am Geriatr Soc. 2005;53(4):711–9.
38. Karches KE, Chung GS, Arora V, Meltzer DO, Curlin FA. Religiosity, spirituality, and end-of-life planning: a single-site survey of medical inpatients. J Pain Symptom Manage. 2012;44(6):843–51.
39. Astrow AB, Puchalski CM, Sulmasy DP. Religion, spirituality, and health care: social, ethical, and practical considerations. Am J Med. 2001;110(4):283–7.
40. Silvestri GA, Knittig S, Zoller JS, Nietert PJ. Importance of faith on medical decisions regarding cancer care. J Clin Oncol. 2003;21(7):1379–82.
41. Hermann CP. Spiritual needs of dying patients: a qualitative study. Oncol Nurs Forum. 2001;28(1):67–72.
42. Anandarajah G, Hight E. Spirituality and medical practice: using the HOPE questions as a practical tool for spiritual assessment. Am Fam Physician. 2001;63(1):81–9.
43. Alcorn STR, Balboni MJ, Prigerson HG, et al. "If God wanted me yesterday, I wouldn't be here today": religious and spiritual themes in patients' experiences of advanced cancer. J Palliat Care. 2010;13(5):581–8.
44. Koenig HG. Religious attitudes and practices of hospitalized medically ill older adults. Int J Geriatr Psychiatry. 1998;13(4):213–24.
45. Phelps AC, Maciejewski PK, Nilsson M, et al. Religious coping and use of intensive life-prolonging care near death in patients with advanced cancer. JAMA. 2009;301 (11):1140–7.
46. Ehman JW, Ott BB, Short TH, Ciampa RC, Hansen-Flaschen J. Do patients want physicians to inquire about their spiritual or religious beliefs if they become gravely ill? Arch Intern Med. 1999;159(15):1803–6.
47. Koenig HG, George LK, Titus P. Religion, spirituality, and health in medically ill hospitalized older patients. J Am Geriatr Soc. 2004;52(4):554–62.
48. Ellison CG, Taylor RJ. Turning to prayer: social and situational antecedents of religious coping among African Americans. Rev Religious Res. 1996;38(2):111–31.
49. McCaffrey AM, Eisenberg DM, Legedza ATR, Davis RB, Phillips RS. Prayer for health concerns: results of a national survey on prevalence and patterns of use. Arch Intern Med. 2004;165(8):858–62.
50. Pargament KI, Koenig HG, Perez LM. The many methods of religious coping: development and initial validation of the RCOPE. J Clin Psychol. 2000;56(4):519–43.
51. Perkins HS. Treating both dimensions of the body: commentary on "Providing guidance to patients: physicians' views about the relative responsibilities of doctors and religious communities." South Med J. 2013;106(7):407–8.
52. McCord G, Gilchrist VJ, Grossman SD, et al. Discussing spirituality with patients: a rational and ethical approach. Ann Fam Med. 2004;2(4):356–61.
53. MacLean CD, Susi B, Phifer N, et al. Patient preference for physician discussion and practice of spirituality: results from a multicenter patient survey. J Gen Intern Med. 2003;18 (1):38–43.
54. Moadel A, Morgan C, Fatone A, et al. Seeking meaning and hope: self-reported spiritual and existential needs among an ethnically-diverse cancer patient population. Psycho Oncol. 1999;8(nn):378–85.
55. Cicirelli VG. Personality and demographic factors in older adults' fear of death. Gerontologist. 1999;39(5):569–79.
56. Daaleman TP, Usher BM, Williams SW, Rawlings J, Hanson LC. An exploratory study of spiritual care at the end of life. Ann Fam Med. 2008;6(5):406–11.

57. Sloan RP, Bagielia E, Powell T. Religion, spirituality, and medicine. Lancet. 1999;353(20 February):664–7.
58. El Nawawi NM, Balboni MJ, Balboni TA. Palliative care and spiritual care: the crucial role of spiritual care in the care of patients with advanced illness. Curr Opin Support Palliat Care. 2012;6(2):269–74.
59. Hart A, Kohlwes RF, Deyo R, Rhodes LA, Bowen DJ. Hospice patients' attitudes regarding spiritual discussions with their doctors. Am J Hospice Palliat Care. 2003;20(2):135–9.
60. Puchalski CM. Spirituality and end-of-life care: a time for listening and caring. J Palliat Med. 2002;5(2):289–94.
61. Wright G. Prescribing the model home. Soc Res. 1991;58(1):213–25.
62. Fife BL. The measurement of meaning in illness. Soc Sci Med. 1995;40(8):1021–8.
63. O'Connell LJ. Religious dimensions of dying and death. West J Med. 1995;163(3):231–5.
64. McClain CS, Rosenfeld B, Breitbart W. Effect of spiritual well-being on end-of-life despair in terminally-ill cancer patients. Lancet. 2003;361(10 May):1603–7.
65. Reed PG. Spirituality and well-being in terminally ill hospitalized adults. Res Nurs Health. 1987;10(nn):335–43.
66. Fryback PB. Health for people with a terminal diagnosis. Nurs Sci Q. 1993;6(3):147–59.
67. Hampton DM, Hollis DE, Lloyd DA, Taylor J, McMillan SC. Spiritual needs of persons with advanced cancer. Am J Hosp Palliat Med. 2007;24(1):42–8.
68. Vanderpool HY, Levin JS. Religion and medicine: how are they related? J Relig Health. 1990;28(1):9–20.
69. Daaleman TP, VandeCreek L. Placing religion and spirituality in end-of-life care. JAMA. 2000;284(19):2514–7.
70. Taylor SE. Adjustment to threatening events: a theory of cognitive adaptation. Am Psychol. 1983;38(11):1161–73.
71. Murray SA, Kendall M, Boyd K, Worth A, Benton TF. Exploring the spiritual needs of people dying of lung cancer or heart failure: a prospective qualitative interview study of patients and their carers. Palliat Med. 2004;18(1):39–45.
72. Finucane TE. How gravely ill becomes dying: a key to end-of-life care. JAMA. 1999;282(17):1670–2.
73. Fry PS. A factor analytic investigation of home-bound elderly individuals' concerns about death and dying, and their coping responses. J Clin Psychol. 1990;46(6):737–48.
74. Dresser R. Bioethics and cancer: when the professional becomes personal. Hast Cen Rep. 2011;41(6):14–8.
75. Fife BL. The conceptualization of meaning in illness. Soc Sci Med. 1994;38(2):309–16.
76. Steinhauser KE, Christakis NA, Clipp EC, et al. Preparing for the end of life: preferences of patients, families, physicians, and other care providers. J Pain Symptom Manage. 2001;22(3):727–37.
77. Becker G. Dying away from home: quandaries of migration for elders in two ethnic groups. J Gerontol. 2002;57B(2):S79–95.
78. Motzer SU, Stewart BJ. Sense of coherence as a predictor of quality of life in persons with coronary artery disease surviving cardiac arrest. Res Nurs Health. 1996;19(nn):287–98.
79. Mako C, Galek K, Poppito SR. Spiritual pain among patients with advanced cancer in palliative care. J Palliat Med. 2006;9(5):1106–13.
80. van der Eijk P. Body and spirit in Greek medicine and philosophy. An unpublished lecture accessed online at www.rcpsych.ac.uk/pdf, 28 Oct 2015.
81. The editors. Mind-body dualism. Encyclopedia Britiannica. Accessed online at www.britannica.com/topic/mind-body-dualism, 28 Oct 2015.
82. Guilfoyle J, St. Pierre-Hansen N. Religion in primary care: let's talk about it. Can Fam Physician. 2012;58(3):249–51.
83. Levin JS, Vanderpool HY. Is frequent religious attendance really conducive to better health? Toward an epidemiology of religion. Soc Sci Med. 1987;24(7):589–600.
84. Sloan RP, Bagiella E. Spirituality and medical practice a look at the evidence. Am Fam Physician. 2001;63(1):33–5.

85. Koenig HG. Spiritual assessment in medical practice. Am Fam Physician. 2001;63(1):30–3.
86. Astrow AB, Wexler A, Texeira K, Kai He M, Sulmasy DP. Is failure to meet spiritual needs associated with cancer patients' perceptions of quality of care and their satisfaction with care? J Clin Oncol. 2007;25(36):5753–7.
87. Strawbridge WJ, Cohen RD, Shema SJ, Kaplan GA. Frequent attendance at religious services and mortality over 28 years. Am J Public Health. 1997;87(6):957–61.
88. Fisch MJ, Titzer ML, Kristeller JL, et al. Assessment of quality of life in outpatients with advanced cancer: the accuracy of clinician estimations and the relevance of spiritual well-being—a Hoosier Oncology Group Study. J Clin Oncol. 2003;21(14):2754–9.
89. Curlin FA, Hall DE. Strangers or friends? A proposal for a new spirituality-in-medicine ethic. J Gen Intern Med. 2005;20(4):370–4.
90. Curlin FA, Roach CJ, Gorawar-Bhat R, Lantos JD, Chin MH. When patients choose faith over medicine: physician perspective on religiously related conflict in the medical encounter. Arch Intern Med. 2005;165(1):88–91.
91. Daaleman TP, Frey B. Prevalence and patterns of physician referral to clergy and pastoral care providers. Arch Fam Med. 1998;7(6):548–53.
92. Grant D. Spiritual interventions: how, when, and why nurses use them. Holist Nurs Pract. 2004;18(1):36–41.
93. Luckhaupt SE, Yi SY, Mueller CV, et al. Beliefs of primary care residents regarding spirituality and religion in clinical encounters with patients: a study at a Midwestern U.S. teaching institution. Acad Med. 2005;80(6):560–70.
94. Monroe MH, Bynum D, Susi B, et al. Primary care physician preferences regarding spiritual behavior in medical practice. Arch Intern Med. 2003;163(22):2751–6.
95. Larimore WL. Providing basic spiritual care for patients: Should it be the exclusive domain of pastoral professionals? Am Fam Physician. 2001;63(1):36–41.
96. Lo B, Ruston D, Kates LW, et al. for the working group on religious and spiritual issues at the end of life, American Medical Association. Discussing religious and spiritual issues at the end of life: a practical guide for physicians. JAMA. 2002;287(6):749–54.
97. Williams JA, Meltzer D, Arora V, Chung G, Curlin FA. Attention to inpatients' religious and spiritual concerns: predictors and association with patient satisfaction. J Gen Intern Med. 2011;26(11):1265–71.
98. Behan J, Carmichael S, Edeen R, et al. Understanding the role of spirituality in medicine: a resource for medical students. Accessed online at https://www.aamc.org/download/300720/data/osrspiritualityresources, 12 Nov 2015.
99. Curlin FA, Lantos JD, Roach CJ, Sellergren SA, Chin MH. Religious characteristics of U.S. physicians: a national survey. J Gen Intern Med. 2005;20(7):629–34.
100. Koenig HG. A commentary: the role of religion and spirituality at the end of life. Gerontologist. 2002;42(special issue III):20–3.
101. Steinhauser KE, Clipp EC, McNeilly M, et al. In search of a good death: observations of patients, families, and providers. Ann Intern Med. 2000;132(10):825–32.
102. Steinhauser KE, Christakis NA, Clipp EC, et al. Factors considered important at the end of life by patients, family, physicians, and other care providers. JAMA. 284(19):2476–82.
103. Wall RJ, Engelberg RA, Gries CJ, Glavan B, Curtis JR. Spiritual care of families in the intensive care unit. Crit Care Med. 2007;85(4):1084–90.
104. Puchalski CM, Romer AL. Taking a spiritual history allows clinicians to understand patients more fully. J Palliat Med. 2000;3(1):129–37.

Chapter 15
A Look Back, A Look Now, and A Look Forward

This book has presented many concepts and techniques of end-of-life care. While addressing some of the physical domain of that care, the book has emphasized the psychological, social, spiritual, and familial domains. I believe those last four domains deserve more attention than they ordinarily get. This concluding chapter steps back from the details of the five domains to gain an overview of the whole end-of-life care field. This overview aims to review where the field has come from in recent years and to suggest where it might go in the near future. The health professionals (HPs) delivering that care have the responsibility to attend both to the details of current care and to the future of the whole field.

The chapter begins by reviewing how modern end-of-life care has developed and how this book fits into that process. The chapter then summarizes the answers of HPs and patients to two questions key to that care, What is dying well? and What is quality end-of-life care? Those answers suggest, I believe, that modern end-of-life care does a better job of addressing the physical and psychological domains of dying than of addressing the social, spiritual, and familial domains. Accordingly, the answers suggest avenues for future research. The chapter closes by presenting a new vision for end-of-life care. That vision not only differs from mainstream curative medicine but also encourages changes in current end-of-life care. The vision emphasizes *all five* domains of care and universal responsibility for them among HPs. Quality end-of-life care under this new vision considers all five domains interconnected and equally vital. I offer a few suggestions at the end of the chapter for how to realize this new vision.

A Look Back: A Recent History of Developments in End-of-Life Care

I imagine the modern development of end-of-care as divided into three phases. I have sketched some of that development in several places in this book already but review the development again for purposes of this overview.

H.S. Perkins, *A Guide to Psychosocial and Spiritual Care at the End of Life*,
DOI 10.1007/978-1-4939-6804-6_15

The First Phase

Infections and injuries killed most people for many centuries [1]. The medicine of this phase initially offered little palliation and almost no cures, and dying typically took a rapid downhill course. However, the explosion of medical scientific discoveries in the mid- to late twentieth century dramatically changed dying in the developed world. New treatments (including antibiotics, vaccines, and modern surgical techniques) allowed people to survive infections and injuries that would have killed them before. People lived longer only to face the killers of old age including atherosclerotic heart disease, cancer, chronic lung diseases (such as emphysema), and dementia. Medical science responded by developing powerful life-prolonging treatments such as intensive care units, vascular bypasses and stents, ventilators, cancer chemotherapy, and major organ replacements. For the first time, people could survive a long time treated with these diseases. But people, of course, still eventually died [2]. Furthermore, the initial awe at the new life-prolonging treatments waned as the public began to see the "miraculous" treatments not as saving lives but as prolonging deaths. The prospects of dying on life-prolonging treatments, especially the machines, alarmed many.

The Second Phase

Alternative ideas to using fully aggressive life support to the end initiated the next phase in the development of modern end-of-life care. The right-to-die and death-with-dignity movements arose in the 1960s and 1970s to counterbalance modern medicine's "technologic imperative," the powerful inducement to use life-prolonging technologies even for patients intolerably suffering or irreversibly dying. These movements, paralleling the rise of modern medical ethics, championed a strong say for patients in their end-of-life care. Courts soon followed by recognizing patients' universal right to refuse life-saving treatment even if they died by exercising that right.

In the mid-1970s the modern hospice concept spread from England to North America. Grassroots support from outside mainstream medicine championed the early hospices here. They offered a style of end-of-life care radically different from technology-driven hospital care. They promoted aggressive pain control; attention to patients' illness experiences beyond just the physical; and death at home if possible [2]. Once hospice care took firm root in the USA, Congress agreed to pay for it through Medicare. (This important milestone, however, occurred for the wrong reason: Congress passed the Medicare hospice benefit not primarily to humanize care but to save on federal medical expenses.) Public demand and established reimbursement soon persuaded mainstream American medicine to adopt many hospice practices.

Twenty years later the landmark Study to Understand Prognosis and Preferences for Outcomes and Risks of Treatments (SUPPORT) [3, 4] formally reassessed critical and end-of-life care in the USA. The investigators monitored the care of seriously ill and dying patients in American hospitals (where most Americans still die). Beginning with its first publications in 1995, SUPPORT provided disappointing evidence that many patients were still receiving "unsatisfactory" end-of-life care. Of all patients who died within six months, 38 % spent more than 10 days in intensive care units, and more than 50 % of the patients who died in the hospital suffered at least moderate pain for much of their last three days of life. Furthermore, for the 31 % of study patients who wanted no cardiopulmonary resuscitation (CPR), doctors knew about those wishes less than half the time.

These findings prompted the SUPPORT investigators to design a follow-up intervention that informed doctors about the prognoses and treatment wishes of their seriously ill patients. Though logical, that intervention produced no measurable improvement in any outcome including doctor–patient communication about resuscitation status, doctor knowledge of patient wishes for do-not-resuscitate (DNR) orders, timing of those orders, level of reported pain, or medical resource use.

Another study published in 1995 provided additional disappointing results, this time about actual patient dissatisfactions with critical and end-of-life care. Guyatt et al. [5] solicited opinions about such care from elderly patients who had survived life-threatening illnesses or from relatives of patients who had died from such illnesses. The patients had usually had "moderate to severe" pneumonias or exacerbations of chronic heart or lung disease. The researchers categorized the deceased patients by whether they had or did not have decision-making capacity before they died. While respondents expressed many satisfactions with their care, they also expressed some important dissatisfactions. Table 15.1 shows the dissatisfactions cited by more than 30 % of any group. Two dissatisfactions concerned the arduousness of critical care or the dying experience. ("Patient went through a lot." and "Patient went through more than expected.") Better care may not have been able to alleviate those dissatisfactions. However, better care might have alleviated other dissatisfactions. Two concerned the patients' having a say in care ("Health professionals made decisions without involving the patient" and "The healthcare system controlled the level of care.") and occurred for all the kinds of patients. Still other dissatisfactions occurred for only deceased patients with decision-making capacity or for their relatives. According to the relatives, many of these patients felt that they did not know what was going on with their care, that they were insufficiently listened to by HPs, and that the care situations were spinning out of control. Many of the relatives felt that they also did not know what was going on with the patients' care and that the patients did not receive enough comfort measures or best-quality care soon enough.

The Third Phase

Results from the SUPPORT study, the Guyatt study, and similar studies published shortly afterward severely undercut the optimism of the time that dying patients were finally getting some say in their care and that care was generally improving.

Table 15.1 Patient and family dissatisfactions with critical or end-of-life care

Sources of dissatisfaction	Elderly[a] patients who survived life-threatening illnesses (n = 102) (%)[b]	Family of deceased, capacitated patients (n = 75) (%)[b]	Family of deceased, incapacitated patients (n = 78) (%)[b]
Patient went through a lot.	50	77	54
Patient went through more than expected.	46	57	–[c]
Health professionals made decisions without involving patient/Healthcare system dictated level of care.[d]	46	44	40
Patient did not know what was going on with care.	–	49	–
Patient felt the care situations were out of control.	–	43	–
Family member did not know what was going on.	–	41	–
Patient felt the health professionals were not listening to him or her.	–	37	–
Patient did not receive enough comfort measures.	–	32	–
Patient did not receive the best care soon enough.	–	31	–

Calculated from frequencies in Tables 1, 2A, 2B, and 2C in Guyatt et al. [5]
[a]"Elderly" here means older than 64 years.
[b]The table lists only those items mentioned by more than 30 % of respondents of a group.
[c]A dash means the datum did not meet this threshold.
[d]The tables from the original article pegged "Health professionals made decisions without involving patient" only to elderly patients who survived life-threatening illnesses and "Healthcare system dictated level of care" only to family members of deceased patients, both capacitated and not. I combined these two items as virtually identical.

The results of these studies dismayed patients, HPs, and the general public alike. Professional healthcare organizations responded by redoubling their efforts to examine, evaluate, and improve end-of-life care. The Institute of Medicine reviewed the state of the field and published *Approaching Death: Improving Care at the End of Life*, something of a manifesto calling for further, widespread reforms. Prominent funding organizations including the National Institutes of Health, the Robert Wood Johnson Foundation, and The Open Society underwrote additional empirical research and innovative demonstration projects in end-of-life care [1, 6, 7]. And many professional organizations concerned about aging and dying formulated guidelines for such care.

Those efforts have prompted considerable progress in the intervening years. I firmly believe dying patients get better physical, psychological, social, spiritual, and familial care now than they did in 1995. But gaps in those end-of-life care domains and in their quality still exist. This book tries to assess current gaps in the last four domains and to point toward possible future improvements. Of course, any attempts at improvement must have clearly defined goals, reflect up-to-date knowledge of the field, and take account of available resources (such as time; material support; and qualified, available HPs).

Because everyone dies, everyone has a stake in these improvements. But the public relies on HPs to lead the way. To help HPs prepare for that leadership role, the next two sections address the questions key to improving end-of-life care, What is dying well? and What is quality end-of-life care? Those questions are similar but not the same. The first question addresses a broad topic about the patient's experience of dying; the second, a narrower topic about what caregivers do for the patient while he or she is dying. I devote a separate section of text to each question. But the answers I give are not my own: They come instead from recent research studies involving the main participants of end-of-life care—dying patients, their family survivors, doctors, and other HPs. I believe the direct involvement of these people in end-of-life care earns them the greatest say in how such care should evolve.

These next sections rely on three tables—Tables 15.2, 15.3, and 15.4—to summarize people's answers to the two questions. The tables categorize answers under either the four familiar patient care domains (the physical, the psychological, the social, and the spiritual) or two additional domains (the familial and the procedural). While some answers may logically fall under several domains, I have placed each answer under the domain I judged suits the answer best. The tables are admittedly detailed. I, therefore, urge the reader to avoid becoming lost in the details but to try instead to see general patterns. The boldfaced type and shading that highlight the high-importance items should help. The accompanying text, too, describes some patterns I see in the data.

A Look Now: What Is Dying Well?

Tables 15.2 and 15.3 give answers to this question from various end-of-life care stakeholders. The tables focus on five studies, published from 1993 to 2007. One study used national surveys of doctors, other HPs (including nurses, social workers, chaplains, and hospice volunteers), seriously ill patients, and recently bereaved family members [7]. Three other studies used local surveys of terminal patients, most of whom had cancer [8–10]. Some patients in two of those studies were already receiving hospice care [9, 10]. The fifth study used a local survey of family survivors of recently deceased patients [11]. Drawing on these studies, Table 15.2 summarizes answers about items of dying well that mainstream medicine already addresses well, and Table 15.3 summarizes answers about items that mainstream medicine does not yet address well (and might never be able to address) [12].

Table 15.2 Dying well: importance ratings by various respondents for items that mainstream medicine already addresses well

Items of dying well	Doctors	Other health professionals	Patients				Family survivors	
	Steinhauser[a] 2000 [21]	Steinhauser[a] 2000 [21]	Fryback[a] 1993 [8]	McMillan[a] 1994 [9]	Byock[a] 1998 [10]	Steinhauser[a] 2000 [21]	Steinhauser[a] 2000 [21]	Norris[a] 2007 [11]
The patient's physical domain								
Staying clean	**High**	**High**	–[b]	–	–	**High**	**High**	–
Feeling no pain	**High**	**High**	**High**	Mod.[b]	–	Mod.	**High**	Mod.
Feeling no breathlessness	Mod.	Mod.	–	–	–	Mod.	Mod.	–
Having no constipation	–	–	–	Mod.	–	–	–	–
Sleeping well	–	–	–	Mod.	–	–	–	–
Functioning well	–	–	**High**	Mod.	–	–	–	Mod.
Receiving physical touch	Mod.	**High**	**High**	Mod.	–	Mod.	Mod.	–
The patient's psychological domain								
Maintaining dignity	**High**	**High**	**High**	–	–	**High**	**High**	–
Having a doctor who knows one as a person	Mod.	**High**	–	–	–	Mod.	Mod.	–
Being supported by the doctor	–	–	–	**High**	–	–	–	–
Feeling no depression/anger	Low	Mod.	–	Low	–	Mod.	Mod.	–
Feeling no anxiety	Mod.	Mod.	–	Low/Mod.	**High**	Mod.	Mod.	–
The patient's social domain								
Trusting the doctor	**High**	**High**	**High**	–	–	Mod.	**High**	–
Being comfortable with the nurse	Mod.	**High**	–	–	–	**High**	**High**	–
Not dying alone	Low	Mod.	–	Low	–	Low	Mod.	–
The patient's spiritual domain								
Knowing the doctor can talk about death	Mod.	**High**	–	–	–	Mod.	Mod.	–

(continued)

Table 15.2 (continued)

Items of dying well	Doctors	Other health professionals	Patients				Family survivors	
	Steinhauser[a] 2000 [21]	Steinhauser[a] 2000 [21]	Fryback[a] 1993 [8]	McMillan[a] 1994 [9]	Byock[a] 1998 [10]	Steinhauser[a] 2000 [21]	Steinhauser[a] 2000 [21]	Norris[a] 2007 [11]
The family domain								
Knowing the family is prepared for one's death	Low	Mod.	–	–	–	Mod.	Mod.	–
The process-of-care domain								
Being cared for by skilled professionals	–	–	**High**	**High**	–	–	–	–
Being cared for by one's personal doctor	Low	Low	–	–	–	Low	Low	–
Having access to doctor or nurse after hours	–	–	–	–	–	–	–	**High**
Knowing what to expect from the illness	Mod.	Mod.	–	–	–	**High**	Mod.	–
Having named a proxy	**High**	**High**	–	–	–	**High**	**High**	NS[b]
Having written out treatment wishes	Low	Mod.	–	–	–	Low	Mod.	NS
Having some control over care	–	–	**High**	Mod.	–	–	–	NS

[a]Four of the five studies cited in the table asked respondents to rate the importance of items against each other. The other study, the one by Norris et al., compared importance ratings for items among decedents with poor, fair, or good qualities of life during the week before death (as reported by family). The Norris items listed here differed significantly among the three quality-of-life groups except for items labelled "NS." The Norris items labelled highly important had large positive associations with the good quality-of-life group. Other items showing no significant differences among the quality-of-life groups and not included in this table were receiving visits from clergy, praying, listening to music, and enjoying nature. The other studies reported no data on those items.

[b]"Mod." means moderate; a dash means no data; and "NS" means no significant differences in ratings among decedents with poor, fair, or good qualities of life during the week before death (as reported by family informants to Norris et al.)

Table 15.3 Dying well: importance ratings by various respondents for items that mainstream medicine does not yet address well

Items of dying well	Doctors	Other health professionals	Patients				Family survivors	
	Steinhauser[a] 2000 [21]	Steinhauser[a] 2000 [21]	Fryback[a] 1993 [8]	McMillan[a] 1994 [9]	Byock[a] 1998 [10]	Steinhauser[a] 2000 [21]	Steinhauser[a] 2000 [21]	Norris[a] 2007 [11]
The patient's physical domain								
Having no fatigue	–[b]	–	**High**	Low	–	–	–	–
Eating well	–	–	–	Mod.[b]	–	–	–	–
The patient's psychological domain								
Having someone to listen	Mod.	**High**	–	–	–	**High**	**High**	–
Being allowed to discuss fears	Mod.	Mod.	–	–	–	Mod.	Mod.	–
Reviewing accomplishments in life	Low	Mod.	–	–	–	Low	Low	–
Enjoying some activities	–	–	–	Mod.	–	–	–	Mod.
Being mentally prepared to die	Low	Mod.	**High**	–	**High**	Low	Low	–
The patient's social domain								
Enjoying personal relationships	Mod.	**High**	–	–	**High**	Mod.	Mod.	**High**
Being visited by family	**High**	**High**	–	–	–	Low	**High**	**High**
Feeling supported by family and friends	–	–	**High**	**High**	–	–	–	**High**
Saying things that need to be said	**High**	**High**	–	–	–	Mod.	Mod.	**High**

(continued)

Table 15.3 (continued)

Items of dying well	Doctors	Other health professionals	Patients				Family survivors	
	Steinhauser[a] 2000 [21]	Steinhauser[a] 2000 [21]	Fryback[a] 1993 [8]	McMillan[a] 1994 [9]	Byock[a] 1998 [10]	Steinhauser[a] 2000 [21]	Steinhauser[a] 2000 [21]	Norris[a] 2007 [11]
The patient's spiritual domain								
Communicating with God	–	–	**High**	**High**	–	–	–	–
Believing life is still good	–	–	**High**	**High**	–	–	–	Low
Feeling at peace	–	–	–	–	**High**	–	–	–
The familial domain								
Having finished business with others	Mod.	**High**	–	–	–	Mod.	Mod.	–
Having financial affairs in order	Mod.	Mod.	—	–	–	Mod.	Mod.	–

[a]Four of the five studies cited in the table asked respondents to rate the importance of items against each other. The other study, the one by Norris et al., compared importance ratings for items among decedents with poor, fair, or good qualities of life during the week before death (as reported by family). The Norris items listed here differed significantly among the three quality-of-life groups. The Norris items labelled highly important had large positive associations with the good quality-of-life group.

[b]"Mod." means moderate; a dash means no data.

All four kinds of respondents in Table 15.2 recognized important items about dying well that mainstream medicine already addresses well in five of the six end-of-life domains. Not surprisingly, more of those items fall into the patient's physical domain than into any other. They include staying clean, feeling no pain, feeling no breathlessness, and receiving physical touch. The physical domain may include the most items for dying well because that domain fits best the established physical care models of medicine and nursing [13–15] and because the new palliative care specialty further emphasizes those items. Items in the psychological domain for dying well include maintaining the patient's dignity, having a doctor who knows the patient as a person, and feeling no anxiety; such items in the social domain include trusting one's doctor and being comfortable with one's nurse. Items of dying well in the spiritual domain include knowing one's doctor can talk about death; items in the process-of-care domain include knowing what to expect from the illness and having named a proxy. According to these data, only in the familial domain is there not at least one important item for dying well that mainstream medicine addresses well.

The items with unexpectedly few data or low ratings here warrant further research. Those items include having no constipation, sleeping well, functioning well, being supported by one's doctor, having no depression or anger, not dying alone, knowing one's family is prepared for one's death, having confidence about being cared for by skilled HPs, being cared for by one's personal doctor, having access to HPs after hours, and having prepared an instructional advance directive. Some of these items need precise definitions as well as prevalence assessments. Items unexpectedly appearing rarely in the table may also warrant future research. One example is the patient's having some say, or control, in his or her own end-of-life care. That item's infrequent mention surprises me due to the recent political and legal movements promoting patient decision-making. Further research should test the intuition that that item and others are indeed important. The research should also help clarify the relative importance of *all* items for dying well so that HPs can set priorities in care accordingly.

Unlike Table 15.2, Table 15.3 lists items of dying that mainstream medicine does not yet address well [15]. Few of those items, as judged by all respondent groups, occur in the patient's physical domain, and none occur in the process-of-care domain. However, two of these items occur in the patient's psychological domain: having someone to listen and being allowed to discuss fears. Two more occur in the patient's social domain: enjoying personal relationships and saying things that need to be said; a third, being visited by family, occurs there for all respondent groups except, surprisingly, patients. Three others, rated highly only by patients, occur in the spiritual domain: communicating with God, believing life is still good, and feeling at peace. Two other items occur in the familial domain: finishing business matters involving others and having one's financial affairs in order. Some of these items certainly lie beyond mainstream medicine's ability to change. For example, mainstream medicine probably cannot boost patients' ability to enjoy interpersonal relationships. But I believe mainstream medicine *can* help relieve more of other problem areas than HPs might immediately think. HPs can invite patients to discuss their fears of dying, and be ready to listen when patients do so. HPs can also encourage families to visit, patients and families to say things that need to be said, and patients to finish up important business and financial matters before the patients die. Such coaching on how to die well, I believe, should be a basic part of HPs' end-of-life care.

Both Tables 15.2 and 15.3 can serve as informal checklists for HPs as they monitor a dying patient's overall well-being. If some items in these checklists appear neglected or otherwise problematic, HPs might address them with the patient, family members, or HP consultants and implement interventions to try to help [16, 17]. Still, clinicians often demur on nonmedical matters, not appreciating how much good sense they as clinicians have about human nature and how much influence they have with patients and families. A clinician's simply expressing concern about a nonmedical problem may be impetus enough to prompt patient or family to seek help about it. The clinician may sometimes even have sufficient informal expertise to provide all the help the patient or family need. If not, the clinician can arrange for professionals with special expertise to consult. For example, if the patient wants spiritual guidance, the clinician can ask the patient's personal clergyman or the hospital's chaplain to visit.

Table 15.3, like Table 15.2, contains items that deserve further study. Items mentioned as highly important by only one or two groups are prime examples. I wonder whether most studies to date have simply not asked about these items, not asked about them in revealing ways, or have asked and found only negative results that have not been published. Examples drawn from Table 15.3 include being free from fatigue, being mentally prepared to die, feeling supported by family and friends, communicating with God, and feeling at peace. HPs need future research to clarify the importance of these items to dying patients.

What Is Quality End-of-Life Care?

Table 15.4 gives answers of various groups to the other important question for improving end-of-life care: What is quality end-of-life care? Drawing on several prominent articles, the table summarizes people's views about 17 aspects of that care. Those views come from "experts" (writing in consensus guidelines or giving their personal opinions), practicing doctors, other HPs, patients, and family survivors [18–20]. I classify 5 of those aspects of care as general "domains of care," 10 as individual care items under "the process-of-care domain," and 2 as types of satisfaction—the patient's and the family's. The table displays the breadth of recognition and the importance ratings of each aspect among the various respondent groups.

For the five general domains of care at least some of *all* respondent groups said attending to the patient's physical well-being and psychological well-being is moderately or highly important in quality end-of-life care. Several groups, but oddly not the experts or the nondoctor HPs, said attending to the family's

Table 15.4 Quality end-of-life care: importance ratings by various respondents of care domains and of individual care items

Domains or items of care	Experts (Teno 2001) [18][a]	Practicing doctors (Curtis 2001) [19][b]	Other health professionals (Curtis 2001) [19][b]	Patients (Singer 1999) [20][c]	Patients (Curtis 2001) [19][b]	Family survivors (Curtis 2001) [19][b]	Family survivors (Teno 2001) [18][d]
Domains of care							
Attending to the patient's physical well-being	**High**	Mod.[e]	Mod.	Mod.	Low	Mod.	**High**
Attending to the patient's psychological well-being	**High**	**High**	**High**	–[e]	**High**	**High**	–
Attending to the patient's social well-being	Mod.	–	–	Mod.	–	–	–
Attending to the patient's spiritual well-being	Mod.	–	–	–	–	–	–
Attending to the family's overall well-being	Low	**High**	Low	Mod.	Low	**High**	**High**
The process-of-care domain: individual care items							
Arranging for ready access to care	Low	Mod.	Mod.	–	**High**	**High**	–
Demonstrating technical competence	–	Low	Mod.	–	**High**	Mod.	–
Providing coordination and continuity of care	Mod.	**High**	**High**	–	Low	Low	–
Communicating well with the patient[d]	–	**High**	**High**	–	**High**	**High**	**High**
Educating the patient and family about what to expect in dying	–	Low	Mod.	–	Mod.	Mod.	**High**
Respecting the patient's values about dying[b,c]	Low	Mod.	Low	**High**	Mod.	Low	–
Respecting the patient's decisions about dying[c,d]	**High**	Low	Low	Mod.	Low	Low	**High**

(continued)

Table 15.4 (continued)

Domains or items of care	Experts (Teno 2001) [18][a]	Practicing doctors (Curtis 2001) [19][b]	Other health professionals (Curtis 2001) [19][b]	Patients (Singer 1999) [20][c]	Patients (Curtis 2001) [19][b]	Family survivors (Curtis 2001) [19][b]	Family survivors (Teno 2001) [18][d]
Helping the patient prepare to die	Mod.	–	–	–	–	–	–
Attending to family grief	Mod.	–	–	–	–	–	**High**
Addressing the financial impact of the illness	Low	–	–	–	–	–	–
Overall satisfaction with care							
Paying attention to patient satisfaction	**High**	–	–	–	–	–	–
Paying attention to family satisfaction	Mod.	–	–	–	–	–	–

[a]Teno et al. reviewed the published guidelines of professional organizations and the opinions of individual experts about the domains and individual items of quality. The authors identified 14 domains or items mentioned by at least 16 of the 30 sources. I rated these domains or items as having high, moderate, or low importance according to whether they had more than 70% mention (i.e., in 22–30 sources), 66–70% mention (i.e., in 20-21 sources) or 50-65% mention (i.e., in 16–19 sources), respectively.

[b]Curtis et al. asked focus groups of health professionals (including doctors, nurses, and social workers), potentially fatally ill cancer or AIDS patients, and family survivors of recently deceased patients to describe doctor attitudes and behaviors that make for quality care. Transcript analyses revealed 12 frequently mentioned domains or items. The authors ranked these domains or items by the number of spontaneous comments about each one within participant groups. I grouped the domains or items according to their rankings, high importance (rankings, 1–4), moderate importance (rankings, 5–8), and low importance (rankings, 9–12). These labels correspond roughly to greater than 10%, 6–10%, and less than 6%, respectively, of comments within a particular participant group. I combined three of the authors' original items—"respect and humility," "personalization," and "attention to patient values"—into this table's "respecting the patient's values about dying." I then calculated that item's ranking by averaging the rankings of the original three items.

[c]Singer et al. asked seriously ill patients and nursing home residents to name key domains or individual items of quality of care. Only five received more than 5% mention among participants. The table lists those five. I labeled them as having high, moderate, or low importance according to whether they were mentioned by greater than 40%, 20–39%, or 5–19% of participants, respectively. Deciding based on the example quotes in the authors' text, I classified the authors' original item "avoiding inappropriate prolongation of dying" under "respecting the patient's values about dying," and their original item "achieving a sense of control" under "respecting the patient's decisions about dying."

[d]Teno et al. elicited from recently bereaved family members their views on domains or individual items of quality care. The authors identified five. I split one of the original items into two, based on the authors' descriptions in the text: "Achieving control over everyday decisions and shared control over treatment care decisions" became "respecting the patient's decisions about dying" and "communicating well with the patient." Because the authors provided no rankings of care domains or items, I labeled all six as having"high importance."

[e]"Mod." means moderate; a dash means no data.

well-being domain is also moderately or highly important. (I doubt that experts or nondoctor HPs *really* consider attending to family well-being *un*important. These unexpected results, coming from one study, may be only happenstance and deserve further testing.) For the other two general domains of care, the patient's social well-being and spiritual well-being, only the experts rated both of those domains at least moderately important; most respondent groups did not rate those domains at all.

For individual care items under the process-of-care domain, the groups showed no universal agreement about items' importance in end-of-life care. Four items—arranging for ready access to care, communicating well with the patient, educating the patient and family about what to expect in dying, and respecting the patient's decisions about dying—came closest to universal agreement.

The experts were generally the rating outliers over all 10 care items. For example, unlike the other respondent groups the experts rated (moderately) important helping the patient prepare to die (including performing advance care planning) and attending to family grief, but rated relatively unimportant arranging for ready access to care. The experts surprisingly did not rate at all demonstrating technical competence, communicating well with the patient, or educating the patient and family about what to expect in dying. Yet the experts, and they alone, *did rate* as important paying attention to patient and family satisfaction.

Despite the apparent puzzlements of the Table 15.4 results, they may actually show some coherence. For example, among all domains of care, attending to the patient's psychological well-being received consistently the highest importance ratings. And among individual care items, arranging for ready access to care, communicating well with the patient, educating the patient and family about what to expect in dying, and respecting the patient's decisions about dying also had consistently the highest importance ratings. Many respondents, therefore, may consider these care items as key ways to achieve a dying patient's psychological well-being.

Still, I urge readers to interpret all these importance ratings cautiously. Personal perspectives may influence them greatly. For example, experts may rate high the domains of attending to the patient's social well-being and spiritual well-being because those experts enthusiastically support the corresponding hospice principles. In contrast, the experts may *not* rate the care items of demonstrating technical competence, communicating well with the patient, and educating the patient and family about what to expect in dying because they—the experts—think of those items as an inherent part of all end-of-life care. Family survivors, on the other hand, may rate as important attending to the family's overall well-being, and the HPs' communicating well with the patient, educating the patient and the family about what to expect in dying, respecting the patient's decisions about dying, and attending to family grief because those aspects affect families most in an end-of-life experience. Table 15.4, then, may serve mostly as a reminder that the same end-of-life experience can have different important aspects for different people. Care should adapt accordingly.

A Look Forward: Comparing Views on Dying Well and Quality End-of-Life Care

Comparing views about dying well and quality end-of-life care (Tables 15.2, 15.3, and 15.4) gives clues about how end-of-life care already serves patients and families well and how it might serve them even better in the future. Among the domains of end-of-life care, the patient's physical well-being and psychological well-being already have a firm foothold in today's mainstream clinical medicine [14, 21–23]. The family's overall well-being has a somewhat firm foothold. Many stakeholders in the dying process agree on the importance of these domains and surely spend considerable time, effort, and material resources on them. In contrast, the patient's social well-being and spiritual well-being domains rest on shakier grounds. Despite consistent expressions of support in the professional literature, those two domains command relatively little attention in practice [7, 14, 21–26].

Why this difference? I believe it arises largely from the "fit" of the various patient-focused end-of-life care domains within mainstream clinical medicine: The patient's physical well-being and psychological well-being fit relatively neatly into the priorities of that medicine; the patient's social well-being and spiritual well-being do not [15].

As I have described it before, the history of twentieth-century clinical medicine provides some support for this idea. Medical science over the twentieth century made wondrous advances in explaining, diagnosing, and treating physical and psychological diseases. At first that science achieved cures for some physical diseases and significant palliation of symptoms for many others. Then medical science discovered the chemical imbalances that cause many psychiatric diseases. Advances in diagnosing and treating them followed as new treatments proved able to correct those imbalances. All the amazing successes stoked clinicians' confidence in the new, scientifically based diagnostics and therapeutics, and the scientific model came to dominate mainstream clinical medicine.

But, as acute diseases (especially the infections) yielded to treatment, the chronic and terminal diseases emerged in prominence late in the century. Clinicians responded by carrying over the previously successful scientific medical model to those diseases. End-of-life care often benefitted. New treatments, for example, reduced the pain, nausea, shortness of breath, depression, and anxiety that dying patients may experience. I believe those successes have raised the physical and psychological domains of care to dominance in end-of-life care just as those successes did earlier in acute care. But the social and spiritual domains have been simultaneously pushed into the background.

Meanwhile the domain of the family's well-being has also gained some prominence in end-of-life care but for other reasons. Clinicians directly engage the family during a patient's dying process: Clinicians notify family members about the patient's life-threatening illness, explain it to them, often secure from them decisions about treatment, notify family members again at or near the death, and initiate through them the disposition of the body after the patient dies. Furthermore, attending clinicians and other HPs often witness directly the family's distresses during the patient's dying experience. Family grieve the progressive losses of the patient's function and companionship. Family may also have to confront long-dormant relational pathologies, which surface under the stress [27]. HPs may identify with those family distresses. And, of course, patient well-being and family well-being are invariably intertwined. HPs, therefore, often must help the family in order to help the patient.

I conclude that modern mainstream clinical medicine and the social and spiritual domains of end-of-life care differ sharply. In fact, clinical medicine and those domains seem to occupy totally separate worlds. Most clinicians are uncomfortable with those domains and feel inadequately prepared to address problems in them. Clinicians also worry that trying to address such problems only opens a Pandora's box of patient questions clinicians cannot answer and patient emotions they cannot control [6]. Clinicians, therefore, build for themselves a mental barrier between the physical, psychological, and familial domains of care on one side—the turf of traditionally trained, scientifically oriented clinicians—and the social and spiritual domains of care on the other—the turf of social workers, psychotherapists, and clergy. During care of the dying, clinicians tend to keep to the physical, psychological, and familial side; to refer social and spiritual problems to specialists on the other side; and eventually to lose track of those problems altogether.

My literature review, however, weighs strongly against this compartmentalized approach to end-of-life care. The dying patient's social well-being and spiritual well-being should concern *all* attending HPs, even the scientifically trained clinicians. I agree with the experts who rate the patient's social well-being and spiritual well-being as *at least* moderately important (and roughly as important as the family's well-being) in end-of-life care (Table 15.4). I believe the opinions of the general public support that view. Furthermore, some process-of-care items *surely do* affect the social and spiritual domains of dying (Table 15.2): trusting one's doctor, feeling comfortable with one's nurse, knowing one's doctor can talk comfortably about death, and not dying alone. And other process-of-care items *may* affect those domains: enjoying personal relationships (including with HPs), saying important things (such as good-byes or expressions of forgiveness to others) perhaps at the encouragement of clinicians, being assured that life is still good, and feeling at peace (Table 15.3). I also believe that the attitudes and behaviors of HPs can powerfully affect the social and spiritual experiences of dying patients.

Even if recognizing their opportunity to affect the social and spiritual domains of dying, some clinicians may disagree with my point that they have a duty to try to act on that opportunity. Clinicians might argue they have only scientifically based clinical expertise and must practice within its bounds. They might also insist that efforts to improve the dying patient's social and spiritual well-being will prove excessively complicated, time-consuming, and emotionally draining. The efforts might also prove ineffective. Clinicians might argue they may not succeed at bolstering the depressed dying patient's belief in the goodness of life or at convincing the disinterested family to visit the patient. Furthermore, clinicians might claim that many patients and family members want clinicians to confine their caregiving to physical and psychological issues and to leave social and spiritual issues alone. These patients or family members may fear intrusions into their personal privacy. The possibility of unwanted proselytizing, after all, is a common argument against clinicians' discussing religion or praying with their patients.

Other factors within medicine may also create obstacles to convincing clinicians to pay as much attention to the social and the spiritual as to the other domains of end-of-life care. Mainstream clinicians, so used to pursuing cure for physical disease, may have difficulty reigning in those deeply ingrained, though well-intentioned impulses and refocusing on other goals when a patient becomes terminal. Furthermore, medicine's current emphasis on "evidence-based" practices and on quantitative performance measures to justify reimbursement may convince clinicians (and insurers) that only observable, measurable outcomes matter. While physical care offers many discrete goals and measurable outcomes and psychological care offers some, social care and spiritual care offer almost none [2, 28–30]. Today's predominantly scientifically based, quantitatively driven medical model simply does not accommodate well those nonmainstream domains of care [31].

Specialization within medicine also creates an obstacle to social and spiritual care at the end of life. That specialization, a clear trend over the last 80 years, creates compartmentalization of care. Most clinicians, especially the doctors, specialize now. Specialist clinicians get used to focusing only on problems within their specialty and to leaving other problems to other HPs. The cardiologist addresses only problems concerning the heart; the colostomy nurse, only problems concerning the colostomy; the discharge clinical social worker, only problems concerning discharge planning. Compartmentalization of care follows, causing compartmentalization of responsibility. Because no clinician has any special expertise in social or spiritual care, no clinician oversees those aspects of the dying patient's care and accepts responsibility for them. They often get lost as a result.

Even forces outside medicine conspire against attending to the dying patient's social well-being and spiritual well-being. High personal mobility, high divorce rates, and other factors often fray Americans' social support networks. HPs may not be able to locate a dying patient's social support people when that patient needs them most. Furthermore, the emphasis of contemporary medicine and Western medical ethics on rationality, personal privacy, and patient self-determination may inadvertently magnify the dying patient's sense of isolation and frustrate caregivers' efforts to nurture social and spiritual support for the patient. In addition, increasing secularization in America, the simultaneous decline in church membership, and ever more idiosyncratic spiritual beliefs may deprive caregivers of readily identifiable ways to engage dying patients in the spiritual domain. Some commentators claim that even hospice no longer adequately addresses spiritual care [32].

Despite all these counterarguments I agree with one prominent end-of-life care expert who says, "Serious illness, dying, caregiving, grieving, and death cannot be completely understood within (such a highly specialized, highly scientific) framework" [15, 22, 33]. End-of-life care, I conclude, must contain two basic components: a "total care" treatment plan encompassing all five domains of care [34, 35], and a "universal responsibility" ethic holding *every* HP attending a dying patient responsible for seeing that each domain is addressed for that patient. The idea of "total care" treatment is, of course, not new. End-of-life care philosophies, especially hospice, have long espoused it. But the idea of "universal responsibility" is new.

Embedding genuine "total care" and "universal responsibility" into everyday end-of-life care will require a significant change in care practices. But end-of-life care must adapt to dying patients' wishes, not the other way around [20]. The next two sections suggest how to accomplish that lofty goal.

Needed: A Different Guiding Vision for End-of-Life Care

Implementing a genuinely "total, universally responsible" end-of-life care requires a different guiding vision than the typical guiding visions of current mainstream medical care. To help formulate that vision, I contrast three images of patients and their caregivers. The broken-down car represents the patient under current acute medical care, the torn cloth represents the patient under current chronic or end-of-life care, and the sick strawberry represents the patient under a genuinely "total, universally responsible" end-of-life care. Each caregiver—the car mechanic, the weaver, and the strawberry farmer—views his or her duties differently. From my perspective the strawberry farmer provides the best guiding image for quality end-of-life care [36].

The Broken-Down Car

This patient image, sometimes called the "mechanical" image, dominates today's acute care medicine. The broken-down car needs repairs, say, replacement of punctured or worn out tires. The owner brings the car to the mechanic at the tire store. The mechanic focuses exclusively on the defective tires. He replaces them but looks for no other problems with the car. He assumes all other parts are functioning perfectly. He engages the car owner only to describe the necessary repairs, get paid, and hand back the keys.

The mechanical image of patients served early modern medicine well, reflecting the first widely curable problems such as infections and traumatic wounds. This image, however, has never suited end-of-life care well. As I suggested in the last section, the mechanical image focuses end-of-life caregivers on the physical domain of dying (by treating symptoms such as pain, shortness of breath, or nausea) but does not much encourage or equip them to address the other three patient domains of dying [35]. This image simply ignores those domains. Under it, once all treatments for physical problems fail, clinicians may think they have nothing more to offer the dying patient. So they just "disappear" as other authors imply [37]. Patient and family feel abandoned despite having strong wishes for ongoing, meaningful, and supportive relationships with the clinicians.

The Torn Cloth

Some recent articles appear to use a patient image like this. A cloth tears when its cross-woven threads weaken and fray. The weaver hired to make the repairs homes in on the tear site. She examines it, identifies the different threads requiring repair or replacement, and performs the necessary reweaving. She may not notice other worn areas. If she does, she may not examine them or attend to them. They may be serviceable at the moment, may never cause problems, and, if so, can be repaired later.

I think this patient image reflects current chronic or end-of-life care. As for end-of-life care, recent improvements in standards and training have prepared caregivers to recognize and address some "worn areas," or problems, in addition to the physical ones. Caregivers may now notice some psychological or familial problems in dying but perhaps not others such as social or spiritual problems. Even if the scientifically trained clinicians do notice social or spiritual problems, the clinicians may consider those problems outside their expertise or their clinical responsibility. They may either not address the problems at all or refer them on to other professionals without further thought. The clinicians thereby disengage themselves from the problem-solving process in those domains. Just as with the broken-down car image, the torn cloth image does not adequately lead to fulfilling patient and family wishes for ongoing, meaningful, and supportive relationships with caregivers. The image surely does not inspire "total" care in the way I foresee it.

The Sick Strawberry

Believe it or not, I think this patient image fits best the idea of total, universally responsible end-of-life care. (I choose strawberries as my concrete image here because an area of South Texas near my home famously grows them.) The farmer knows that a strawberry must have all its defining characteristics to sell as a strawberry in the market. Those characteristics include size, shape, color, texture, taste, and smell. Drought or fungal infestation sickens the strawberry by affecting all those characteristics at once. The farmer must, therefore, "treat" all of them to make the strawberry healthy again. He or she cannot try to treat color without affecting taste, or taste without affecting smell, or size without affecting texture. And, to treat the whole strawberry, the farmer must water, spray, prune, and fertilize the whole strawberry plant. Once the strawberry regains all its defining characteristics, the farmer can sell it as a strawberry.

I believe a person, like the strawberry, consists of many inextricably intertwined characteristics. The four patient domains of end-of-life care reflect those characteristics. A patient has physical characteristics, but they do not define the whole person. The patient also has psychological, social, and spiritual characteristics, and they are inseparable from each other. Therefore, especially with a chronic or terminal illness, treating the patient requires addressing all the domains of the person.

Many studies assume this view of a patient as a multidimensional, indivisibly whole person. One study, seeking to identify "states worse than death" asked patients to name the elements necessary to make life worth living. Patients mentioned conditions from all four patient care domains. Eighty-two percent mentioned high physical function; 73 % mentioned freedom from pain, sleeplessness, and other physical symptoms. Seventy-three percent also mentioned the spirit or desire to live; 61 %, being needed by others, interacting meaningfully with others, or being useful to society, and 48 %, a spiritually meaningful existence (such as living to God's plan or some other good purpose). Sixty-one percent mentioned the lack of being a physical, emotional, or financial burden on one's family or on society in general; and 41 % mentioned being able to give and receive love [38]. As these study patients suggested, everyone lives in the psychological, social, and spiritual domains that along with the physical domain define our lives.

I believe these domains of a person are indivisible. Each domain affects the others. Thus, the dying patient does not fall ill in just one domain but in all of them simultaneously. Physical pain, for example, affects the patient's psychological, social, and spiritual domains. It may cause depression or despair (the psychological domain), hinder communication with others (the social domain), and raise existential questions about why suffering occurs (the spiritual domain). The patient's physical pain also affects the familial domain by causing family members to suffer vicariously. Though present in every illness, these nonphysical domains gain special prominence near the end of life.

The New Expansive Role for End-of-Life Health Professionals

No HP giving end-of-life care, therefore, can ignore the four patient domains and one family domain involved in it. One cannot give adequate care by addressing only some of the domains; one must address them all even if only to refer and then to monitor from afar the problems associated with them.

But the idea of total, universally responsible end-of-life care may trouble some HPs. It obviously requires expanding each HP's role within the end-of-life care team. But I believe such care entails little extra attentiveness, time, and effort. Each HP participating in that care—whether doctor, nurse, social worker, chaplain, administrator, or someone else—tracks all five care domains for the patient. Such universal responsibility for total care sometimes lands the HP outside his or her area of expertise. A doctor may observe harmful family dynamics, a nurse may perceive a patient's spiritual crisis, a social worker may notice increasing physical symptoms, or a chaplain may discover depression as the cause for a patient's hopelessness. Nonetheless, all HPs attending the patient accept *some* responsibility for all domains of that patient's care. No one abandons any domain automatically to other HPs. Furthermore, total, universally responsible care requires that every HP's contribution in any domain be taken seriously. Naturally, each HP must appreciate the limits of his or her expertise and, when necessary, request help from others in addressing specific problems [2, 7, 9]. For that reason the care team provides each HP an important professional safety net, giving collective support and ready access to others' expertise.

I realize that a patient or family member might occasionally object to a HP's crossing usual professional boundaries to address an end-of-life domain outside the HP's formal training. The HP, therefore, might need to request permission to address sensitive matters, whether inside or outside one's discipline. Chaplains already set a good example in this way: chaplains routinely ask patients for permission to pray with them. Nonetheless, I believe serious objections occur rarely so long as the HP behaves with sensitivity and respect. However, whenever firm objections do arise, the HP should honor the bounds the patient or the family member sets.

Suggestions for Future Practice and Research

What specifically, then, does total, universally responsible end-of-life care suggest that HPs attending dying patients do? Tables 15.2, 15.3, and 15.4 provide ideas in all four patient domains and in the familial domain. These ideas address some of people's most common worries about dying.

First, HPs should, of course, ensure the best physical care possible. Clinicians (and other HPs to the extent they can) should monitor dying patients' physical needs. Unrelieved pain and soiled, incontinent states must be considered emergencies. Any HP who notices such needs must take immediate action to get them met. Some doctors, for example, change soiled patient linens themselves without making patients wait for nurses or nurse's aides to do so. And every end-of-life HP should practice respectful therapeutic touch when appropriate. It need only be holding the patient's hand or resting one's own hand on the patient's shoulder during a conversation. Still, that touch powerfully communicates sincere caring and comforts patients.

Second, HPs should cultivate good end-of-life communication skills [1, 39]. Doctors, in particular, must learn to discuss death comfortably with patients [7, 40, 41]. Partly to that end, communication training is gaining prominence in undergraduate and postgraduate medical education curricula. Such training should be grounded in patient preference [42] and expert experience, should offer opportunities for supervised practice, and should provide constructive feedback on performance.

Third, HPs should address the anxieties people have about dying. HPs—especially doctors—must be willing to address the uncertainties that patients and families have [39]. Many people want to know what lies ahead. Though unable to give *exact* predications in specific cases, HPs can often give general predications based on past experience. For example, they might answer concerns about terminal suffering by saying, "I have seen many people die. They do not appear to suffer at the end. In fact, they seem to die just by going peacefully to sleep." HPs may sometimes need to repeat such comments more than once over time because they may not "sink in" for patients and family members the first time. And HPs should update for patients and family members any clinical predictions that change dramatically as patients' conditions change [10, 12].

Fourth, HPs should conscientiously try to learn dying patients' choices about dying and to implement them [20, 39, 43] even though Table 15.4 suggests doing so does not always rate a high priority for doctors or other HPs. Allowing patients some say helps personalize their care. For example, liberalizing visiting hours for family, allowing pets to visit, or granting a patient's wish to die at home can give patients enormous satisfaction.

Fifth, all HPs should emphasize to dying patients and their families the importance of the nonphysical domains of care [15, 21, 34, 44, 45]. In that way doctors and other HPs can legitimize those domains. HPs might explain that despite physical symptoms and short remaining life spans the end of life can offer patients and families significant opportunities for personal growth. Accordingly, HPs might invite questions, problems, and fears for discussion. HPs might also encourage patients to say important good-byes, to name trustworthy medical proxies, to put their financial affairs in order, and to tap into spiritual resources to the extent they wish to. And HPs might urge families to visit patients and to conduct life reviews with them.

Sixth, HPs should do everything possible to eliminate barriers to access, coordination, and continuity of end-of-life care [1, 12]. Dying patients and their distressed families do not deserve unreasonable institutional or professional barriers to care. Simply providing a reliable 24-h contact number for getting clinical advice can relieve people's anxieties and prevent unnecessary hospitalizations.

Conclusion

End-of-life care has improved greatly over the last 40 years but is still not ideal [6]. The impetus for this book came partly from my sense that the nonphysical domains do not yet get the attention they deserve in end-of-life care. The summary of the professional literature for this chapter supports that view.

Many articles in that literature urge "total" end-of-life care including adequate attention to the psychological, social, spiritual, and familial domains as well as the physical domain. But actual care—especially that delivered by traditionally trained clinicians—often focuses exclusively on the physical domain with its concerns about symptoms, palliative treatment regimens, chemotherapy, intensive care unit stays, and resuscitation decisions [28–30, 45]. Only some of the care addresses patient psychological problems such as depression or anxiety [46]; only a little addresses familial problems such as the emotional burden of proxy decision-making [47], and almost none addresses patient social or spiritual problems such as saying important good-byes or achieving inner peace. Too often, nonphysical problems at the end of life are overlooked, given only cursory attention, or considered adequately addressed strictly as psychological problems. This underemphasis dissipates the sense of responsibility many HPs, especially clinicians, feel for those nonphysical problems. HPs' involvement with them may reduce to the social worker's processing a hospice placement, the doctor's notifying the family when the patient's death is imminent, or the chaplain's saying the occasional brief prayer [30, 48]. Such an approach robs end-of-life care of its comprehensiveness, balance, and fulfillment.

We must do better. Again, approximately 3.5 million Americans die every year. Each death is unique. It presents its own combination of unanswered questions and mysteries, many of which touch the psychological, social, spiritual, and familial domains this book emphasizes. Not all those questions or mysteries are impenetrable. This book suggests areas for further exploration either as formal research studies or as conscious observations by individual practicing Health professionals. Those of us who attend dying patients owe it to them to engage the learning opportunities within both the physical and nonphysical domains of care.

But we must not limit our learning to techniques—even those, say, to treat symptoms, to communicate sensitively, or to recognize social and spiritual factors in dying. Such know-how learning can easily entrance us into forgetting the equally important learning about the meanings of end-of-life care as we give it and will eventually receive it ourselves. We must constantly broaden our perspectives to include techniques *and* meanings alike. The gravity of end-of-life care demands we address both.

The best learning experiences will require us to travel with our dying patients along their paths to death even though we cannot fully understand what is happening. We must pledge full involvement in that journey. That involvement must incorporate *all* domains of end-of-life care. And we must constantly try to improve our care. Past successes must not blind us to the need for future advances [11, 49]. We must continually rededicate ourselves to testing our fundamental ideas about end-of-life care [2, 13], to stoking anew our passions for it, and to increasing our efforts to improve it [24, 50]. We should start now.

Summary Points

1. Current end-of-life literature emphasizes patient physical and psychological care over patient social and spiritual care and family care.
2. Health professionals who attend dying patients should ensure that all five domains of care are addressed to the extent patients and family members wish.
3. Future research and care innovations should give special, increased attention to patients' social and spiritual care while continuing to address the other domains, too.

To Read More …

1. Curtis JR, Wenrich MD, Carline JD, et al. Understanding physicians' skills at providing end-of-life care: perspectives of patients, families, and health care workers. J Gen Intern Med. 2001;16(1):41–9
2. Steinhauser KE, Christakis NA, Clipp EC, et al. Factors considered important at the end of life by patients, family, physicians, and other care providers. JAMA. 2000;284(19):2476–82 (3).
3. Sulmasy DP. A biopsychosocial-spiritual model for the care of patients at the end of life. Gerontologist. 2002;42(special issue III):24–33.

References

1. Lynn J. Learning to care for people with chronic illness facing the end of life. JAMA. 2000;284(19):2508–11.
2. Wallston KA, Burger C, Smith RA, Baugher RJ. Comparing the quality of death for hospice and non-hospice cancer patients. Med Care. 1988;26(2):177–82.
3. The SUPPORT Investigators. A trial to improve outcomes for seriously ill patients: the study to understand prognoses and preferences for outcomes and risks of treatment. JAMA. 1995;274(20):1591–8.
4. Moskowitz EH, Nelson JL. The best laid plans. Hast Cen Rep. 25(6):S3–6.
5. Guyatt GH, Mitchell A, Molloy DW, et al. Measuring patient and relative satisfaction with level or aggressiveness of care and involvement in care decisions in the context of life-threatening illness. J Clin Epidemiol. 1995;48(1):1215–24.
6. McPhee SJ, Rabow MW, Pantilat SZ, Markowitz AJ, Winker MA. Finding our way—perspectives at the close of life. JAMA. 2000;284(19):2512–3.
7. Steinhauser KE, Christakis NA, Clipp EC, et al. Factors considered important at the end of life by patients, family, physicians, and other care providers. JAMA. 2000;284(19):2476–82.
8. Fryback PB. Health for people with a terminal diagnosis. Nurs Sci Q. 1993;6(3):147–59.
9. McMillan SC, Mahon M. Measuring quality of life in hospice patients using a newly developed hospice quality of life index. Qual Life Res. 1994;3(6):437–47.
10. Byock IR, Merriman MP. Measuring quality of life for patients with terminal illness: the Missoula-VITAS quality of life index. Palliat Med. 1998;12(4):231–44.
11. Norris K, Merriman MP, Curtis JR, et al. Next of kin perspectives on the experience of end-of-life care in a community setting. J Palliat Med. 2007;10(5):1101–15.
12. Stewart AL, Teno J, Patrick DL, Lynn J. The concept of quality of dying persons in the context of health care. J Pain Symptom Manage. 1999;17(2):93–108.
13. Payne SA, Langley-Evans A, Hillier R. Perceptions of a "good" death: a comparative study of the views of hospice staff and patients. Palliat Med. 1996;10(4):307–12.
14. Bailey FA, Williams BR, Woodby LL, et al. Intervention to improve care at life's end in inpatient settings: the BEACON trail. J Gen Intern Med. 2014;29(6):836–43.
15. Cohen LM, McCue JD, Germain M, Kjellstrand CM. Dialysis discontinuation: a "good" death? Arch Intern Med. 1995;155(1):42–7.
16. Perkins HS, Cortez JD, Hazuda HP. Diversity of patients' beliefs about the soul after death and their importance in end-of-life care. South Med J. 2012;105(5):266–72.
17. Perkins HS. The varied understandings of "terminal cancer" and their implications for clinical research and patient care. Med Decis Making. 2014;34(6):696–8.

18. Teno JM, Casey VA, Welch LC, Edgman-Levitan S. Patient-focused, family-centered end-of life medical care: views of the guidelines and bereaved family members. J Pain Symptom Manage. 2001;22(3):738–51.
19. Curtis JR, Wenrich MD, Carline JD, et al. Understanding physicians' skills at providing end-of-life care: perspectives of patients, families, and health care workers. J Gen Intern Med. 2001;16(1):41–9.
20. Singer PA, Martin DK, Kelner M. Quality of end-of-life care: patients' perspectives. JAMA. 1999;281(2):163–8.
21. Steinhauser KE, Clipp EC, McNeilly M, et al. In search of a good death: observations of patients, families, and providers. Ann Intern Med. 2000;132(10):825–32.
22. Donovan K, Sanson-Fisher RW, Redman S. Measuring quality of life in cancer patients. J Clin Oncol. 1989;7(7):959–68.
23. Cohen SR, Mount BM, Tomas JJN, Mount LF. Existential well-being is an important determinant of quality of life: evidence from the McGill Quality of Life Questionnaire. Cancer. 1996;77(3):576–86.
24. Smith AK, Buss MK, Giansiracusa DF, Block SD. On being fired: experiences of patient-initiated termination of the patient-physician relationship in palliative medicine. J Palliat Med. 2007;10(4):938–47.
25. Sachs GA. Sometimes dying still stings. JAMA. 2000;284(19):2423.
26. Perkins HS, Jonsen AR. Dying right in theory and practice: what do we really know of terminal care? Arch Intern Med. 1985;145(8):1460–3.
27. Perkins HS. Controlling death: the false promise of advance directives. Ann Intern Med. 2007;147(1):51–7.
28. Wright AA, Zhang B, Ray A, et al. Associations between end-of-life discussions, patient mental health, medical care near death, and caregiver bereavement adjustment. JAMA. 2008;300(14):1665–73.
29. Wright AA, Zhang B, Keating NL, Weeks JC, Prigerson HG. Associations between palliative chemotherapy and adult cancer patients' end of life care and place of death: prospective cohort study. BMJ. 2014;348(g1219, 04 March 2014):1–10.
30. Bailey FA, Burgio KL, Woodby LL, et al. Improving processes of hospital care during the last hours of life. Arch Intern Med. 2005;165(15):1722–7.
31. Kaasa S, Loge JH. Quality of life in palliative care: principles and practice. Palliat Med. 2003;17(1):11–20.
32. Daaleman TP, VandeCreek L. Placing religion and spirituality in end-of-life care. JAMA. 2000;284(19):2514–7.
33. Byock I, Norris K, Curtis JR, Patrick DL. Improving end-of-life experience and care in the community: a conceptual framework. J Pain Symptom Manage. 2001;22(3):759–72.
34. Wenrich MD, Curtis JR, Ambrozy DA, et al. Dying patients' need for emotional support and personalized care from physicians: perspectives of patients with terminal illness, families, and health care providers. J Pain Symptom Manage. 2003;25(3):236–46.
35. Sulmasy DP. A biopsychosocial-spiritual model for the care of patients at the end of life. Gerontologist. 2002;42(special issue III):24–33.
36. Berlinger N. Cueing "the conversation". Hast Cen Rep. 2013;43(4):29–30.
37. Prigerson HG, Jacobs SC. Caring for bereaved patients: "All the doctors suddenly go." JAMA. 2001;286(11):1369–76.
38. Pearlman RA, Cain KC, Patrick DL, et al. Insights pertaining to patient assessments of states worse than death. J Clin Ethics. 1993;4(1):33–40.
39. Hanson LC, Danis M, Garrett J. What is wrong with end-of-life care? Opinions of bereaved family members. J Am Geriatr Soc. 1997;45(11):1339–44.
40. Gilligan T, Raffin T. Whose death is it, anyway? Ann Intern Med. 1996;125(2):137–41.
41. Jackson VA, Mack J, Matsuyama R, et al. A qualitative study of oncologists' approaches to end-of-life care. J Palliat Med. 2008;11(6):893–903.
42. Emanuel EJ, Fairclough DL, Wolfe P, Emanuel LL. Talking with terminally ill patients and their caregivers about death, dying, and bereavement: Is it stressful? Is it helpful? Arch Intern Med. 2004;164(18):1999–2004.

43. Pfifferling JH. A cultural prescription for medicocentrism. In: Eisenberg L, Kleinman A, editors. The relevance of social science for medicine. Dordrecht: D. Reidel Publishing Company; 1981. p. 197–222.
44. Bryce CL, Loewenstein G, Arnold RM, et al. Quality of death: assessing the importance placed on end-of-life treatment in the intensive-care unit. Med Care. 2004;42(5):423–31.
45. Curtis JR, Patrick DL, Engelberg RA, et al. A measure of the quality of dying and death: initial validation using after-death interviews with family members. J Pain Symptom Manage. 2002;24(1):17–31.
46. Chibnall JT, Videen SD, Duckro PN, Miller DK. Psychosocial-spiritual correlates of death distress in patients with life-threatening medical conditions. Palliat Med. 2002;16(4):331–8.
47. Braun UK, Naik AD, McCullough LB. Reconceptualizing the experience of surrogate decision making: reports vs. genuine decisions. Ann Fam Pract. 2009;7(3):249–53.
48. Abbott KH, Sago JG, Breen CM, Abernathy AP, Tulsky JA. Families looking back: one year after discussion of withdrawal or withholding of life-sustaining support. Crit Care Med. 2001;29(1):197–201.
49. Teno JM, Field MJ, Byock I. Preface: the road taken and to be traveled in improving end-of-life care. J Pain Symptom Manage. 2001;22(3):713–5.
50. Weissman DE, Block SD. ACGME requirements for end-of-life training in selected residency and fellowship programs: a status report. Acad Med. 2002;77(4):299–304.

Index

Note: Page numbers followed by f and t indicate figures and tables, respectively.

H.S. Perkins, *A Guide to Psychosocial and Spiritual Care at the End of Life*,
DOI 10.1007/978-1-4939-6804-6

N

O

P

CPSIA information can be obtained
at www.ICGtesting.com
Printed in the USA
LVHW080011201119
637974LV00001B/1/P

9 781493 968022